Plants and the Human Brain

Plants and the Human Brain

David O. Kennedy

OXFORD
UNIVERSITY PRESS

Oxford University Press is a department of the University of Oxford.
It furthers the University's objective of excellence in research, scholarship,
and education by publishing worldwide.

Oxford New York
Auckland Cape Town Dar es Salaam Hong Kong Karachi
Kuala Lumpur Madrid Melbourne Mexico City Nairobi
New Delhi Shanghai Taipei Toronto

With offices in
Argentina Austria Brazil Chile Czech Republic France Greece
Guatemala Hungary Italy Japan Poland Portugal Singapore
South Korea Switzerland Thailand Turkey Ukraine Vietnam

Published in the United States of America by
Oxford University Press
198 Madison Avenue, New York, NY 10016

Library of Congress Cataloging-in-Publication Data
Kennedy, David O.
Plants and the human brain / David O. Kennedy.
p. cm.
ISBN 978–0–19–991401–2 (alk. paper) 1. Phytochemicals. 2. Brain chemistry. I. Title.
QK861.K46 2014
572'.2—dc23
2013031617

9780199914012

9 8 7 6 5 4 3 2

Printed in the United States of America on acid-free paper

To Oliver and Tabitha, with lots of love

CONTENTS

PART ONE ■ Why Do Plant Secondary Metabolites Affect Human Brain Function?

PART TWO ■ The Alkaloids

■ Plants and the Human Brain

PART ONE

Why Do Plant Secondary Metabolites Affect Human Brain Function?

1 From Shamans to Starbucks

The settlers arrived from the East. They had crossed the Mediterranean in large dugout boats and probed the coastline for the ideal landfall before paddling their flotilla inland for days on a river that wound lazily through a temptingly fertile flood plain. As the river widened suddenly into the large lake that had filled the bowl of a long-extinguished volcanic caldera, they must have scanned the distant shores and picked out a low, tree-crowded island, separated from the mainland by 300 meters of protective water. Their sturdy, seaworthy boats were crammed with everything they needed to make this place their home: domesticated animals, two breeds of dogs, pottery, implements fashioned from wood and stone, and the seeds of the crops that grew in the settlements of their homeland. The time was 7,700 years ago, and these travelers had journeyed a long way. They had traversed hundreds of miles of sea from the far-distant lands bordering the Eastern Mediterranean, a region that has been described as the crucible of the Neolithic Age.

Their new home was on Lake Bracciano, 20 miles outside of modern-day Rome. This sophisticated new settlement, bringing with it a foretaste of the agricultural lifestyle that would sweep across Europe over the next 2,000 years, was a near-instant success. The tree rings of the timbers used to build the structures tell us that the settlement grew steadily, with timber houses eventually spreading across the whole of the island over the course of the next four hundred years. Then disaster: an increasingly wet climate, maybe a silted river outlet, possibly a flood, and the lake filled higher with water. The settlers' island home sank beneath the surface.[1,2]

For the past three decades divers have been slowly sucking away the stubborn but protective mantel of mud that settled over La Marmotta, as the archaeological site is known today.[2] Among the many thousands of artifacts, boats, timbers, bones, and tools that have been retrieved to date from the lake bed there are two finds that are of particular interest. The first comprises several fragments of the polypore fungus *Daedaleopsis tricolor* found inside dwelling structures; these fungi were probably harvested for their pharmacological properties and used in ritual or medicine.[3] The second comprises the contents of a single room that contained both the seeds of the opium poppy *Papaver somniferum*, and, most importantly, a religious "mother" idol. Together these artifacts suggest that these early seafaring immigrants were already growing opium for its psychotropic properties and consuming it in a ritualistic setting.[4]

If, during the period that La Marmotta was occupied, we could have turned 400 miles east to a rock shelter in the Spanish Sierra de las Cuerdas Mountains, or indeed 500 miles south to a deep cave overlooking the verdant savannah that would become the parched Sahara Desert of today, we would without doubt have found contemporaries of the immigrants to La Marmotta coming together for social gatherings or rituals. In both places they would have been watched over by shamanic rock art depicting local species of the hallucinogenic *Psilocybe* mushroom genus.[5,6] Alternatively, if we had traversed the Atlantic and crossed the widest part of the vast continent to the coast of Peru, we would have found the Nanchoc Valley folded into the tropical western slopes of the Andes. There we might have come across the inhabitants of the region's mixed foraging and agricultural villages, who would chew coca leaves and carelessly discard them on the earth floors of their homes, where they would remain, stamped into the dirt, for nearly 8,000 years before being unearthed. This, however, was not the casual or opportunistic use of the coca plant (*Erythroxylum coca/novogranatense*), but rather an organized facet of community life; the coca was purposefully cultivated and the villagers had created communal facilities for the burning of calcium-rich rocks, which could be crushed and ground to produce powdered lime, which they chewed along with the leaves to release the active alkaloid chemical, cocaine.[7]

These first signs of man's relationship with psychoactive plants were captured by chance in the archaeological record of three continents. They suggest both that wherever humans lived they sought out plants and fungi with mind-altering properties, and that this search for plant-derived chemicals was already an enduring one. Exactly how enduring and how this relationship was first forged might be suggested by contemporary research that shows that, among the plants consumed by wild chimpanzees, there are several dozen that have no nutritional value but do have specific medicinal properties. A number of these plants are used by local indigenous peoples for their antiparasite, antibacterial, and antimalarial bioactivity.[8,9] Sporadic, intentional self-medication with specific medicinal plants has also been observed in ill chimpanzees suffering from, for instance, parasites, wounds, and bacterial infections. Further study of the range of plants consumed medicinally by primates has also led to the discovery of a number of novel compounds, including triterpenes with antiparasitic,[10] antitumor,[8] and antimalarial properties[11] that might be useful in human medicine. How our primate relatives came to be consuming these plants is suggested by simple feeding experiments, whereby novel medicinal plants with low nutritional value are introduced to captive groups of orangutans. Once they've sniffed the plants and rejected a small proportion, they typically sample the remaining plants and then share the favored ones, usually by stealing them but often by offering the less palatable portions to group-mates.[12] It is readily apparent that similar plant-sampling behavior in our distant ancestors would eventually have uncovered medicinal or psychotropic plants. However, it should also be noted that there is only anecdotal evidence of non-human mammals using plants on account of their psychoactivity per se,—for instance, the use of ibogaine roots by gorillas, porcupines, and wild boar purportedly for their mind-altering properties.[13]

So pervasive is the relationship between humans and psychotropic plants and fungi that they have played a major part in shaping mankind's history. The most obvious and all-encompassing role has been in the development of religions. As we'll see below, historical and archaeological records are scattered with rich evidence of the roles played by euphoriants, deliriants, and hallucinogens in early religion. For instance, plants have been associated with their own gods in many cultures; the "Poppy Goddess" of Crete, a figure with a trance-like expression and a crown of moveable poppy-shaped pins; the ancient Greek goddess of fertility and the harvest, Demeter, whose emblem was an opium poppy[4]; the Celtic god Bel, the Norse/Germanic god Thor, and the Roman god Jupiter, all associated with henbane (*Hyoscamus niger*); the Germanic goddess of love Freya, inextricably linked to cannabis (*Cannabis sativa*); and the Germanic god Odin, associated with opium, deadly nightshade (*Atropa belladonna*), and fly agaric (*Amanita muscaria*). The drugs themselves have also often been attributed with being a direct personification of a god or goddess and have been deified accordingly,—for instance, the *kykeón*, a mysterious drink deified by the cult of Demeter and

Persephone and used in their "Eleusinian Mysteries" initiation rite[14]; soma, or haoma, a psychotropic drink deified in Hinduism and Zoroastrianism, respectively[4]; and the Egyptian god of spiritual rebirth, Osiris, who was putatively the personification of the hallucinogenic *Psilocybe* mushroom.[15]

This use of psychotropic plants/fungi is echoed in the shamanic traditions distributed throughout every corner of the globe. The most vibrant illustration is Central and South America, which is home to a rich palette of traditionally consumed psychoactive plants/fungi, including hallucinogens such as the jungle vine *Banisteriopsis caapi* and the shrub *Psychotria viridis*, which were combined in the drink *ayahuasca*; members of the *Anadenanthera* genus, which were ground up and snuffed; the peyote cactus (*Lophophora williamsii*) and *Psilocybe* mushrooms; and trance inducers and deliriants such as *Salvia divinorum*, tobacco (*Nicotiana tabacum/rustica*), the seeds of *Turbina corymbosa*, and members of the *Brugmansia* and *Datura* genera of the *Solanacea* family, to name but a few of many. There are few inhabited places on the continent that would not provide ready access to several of these plants.

The common feature tying the relationship between these natural psychotropics and religions throughout the world is that the plant or fungus is always used to engender an altered state of consciousness, either a trance or a hallucinatory state that serves as a gateway to a spiritual or religious experience and/or communication with the spirit worlds. This theme of "entheogenic"[a] psychotropic plants and mushrooms allowing the consumer to experience "the god within" and communicate with spirits, other worlds or ancestors, has echoed throughout the world. Consequently, deeply embedded traditions of shamanism can be seen in the pre-monotheistic traditions and histories across the continents.[16]

Contemporary scientific research shows that entheogens such as the hallucinogen psilocybin (the principal active chemical in *Psilocybe* mushrooms) consistently engender altered states of consciousness that are described by consumers as mystical or spiritual and that often include experiences of a "higher reality." They have this effect irrespective of the nature of the consumer's religious or spiritual beliefs, or, indeed, lack of belief.[17–20] The simple fact is that, courtesy of self-consciousness, mankind is inherently spiritual.[21] Therefore, it isn't difficult to conceive that entheogenic mystical experiences must have been interpreted by our distant ancestors as unveiling or confirming the existence of spirit worlds, gods, or an afterlife. In light of this it seems reasonable to conclude that plant and fungal chemicals may well underlie the origin of many spiritual concepts. It also doesn't take a great leap of imagination to see the roots of our contemporary religions, including the dominant monotheistic religions, firmly embedded in the soil of the polytheistic traditions that originated in plant-derived experiences.[22–25]

So, where have all the shamans gone? Mostly they've been swept away in developed parts of the world by the tides of monotheistic religions. Certainly the use of entheogenic plants was directly suppressed by Christianity, with inquisitions and witch hunts raging through Europe in the 13th to 17th centuries, while the gradual proscription of most psychoactive plants followed a similar time course in the Islamic world.[26] Inevitably, as Europeans colonized large sections of the world they took with them their intolerance of entheogenic drugs, at least as and when it suited their purposes.

a. An entheogen, from the Greek "entheos," meaning "god within," is any psychoactive substance used to engender a spiritual experience.

Of course, religion is not the only sphere in which psychotropic plants have helped to shape mankind's history. Within recent centuries they have, on one hand, also driven the redrawing of the world map through colonization and economic dominance, and they've fostered wars, mass persecution, murder, slavery, and self-inflicted disease. On the other hand they have also been a key factor in the development of modern medicine and the pharmaceutical sciences. The best way to appreciate how profoundly these natural compounds have obstructed or channeled the tides of history is to consider a brief linear history.

A BRIEF HISTORY OF PSYCHOACTIVE PLANTS AND FUNGI

Antiquity

Much of the earliest archaeological evidence of psychotropic plant use is open to interpretation. Deposits of opium poppy seeds found at La Marmotta (5700 BC), or in Neolithic and Bronze Age settlements between the Jura Mountains and the French/Swiss Alps (4000–3000 BC), or interred alongside poppy heads in a grass bag found in a burial cave (2500 BC) in Granada, Spain, may well simply reflect the alternative uses of the opium poppy as a source of oil and food. Similarly, Taiwanese and Chinese pottery bearing the imprint of hemp cloth and rope, dating as far back as 8000 BC, might simply reflect the practical use of tough cannabis stem fibers.[4] However, the first unequivocal written evidence of mankind's enduring relationship with psychotropic plants is provided by clay tablets bearing the "cuneiform" script, indentations of abstract patterns made by pressing a wedge-tipped stylus into damp clay. These clay tablets originated in the Sumerian civilization that flourished from the 4th to the 1st millennia BC in Mesopotamia, the region between the Euphrates and Tigris rivers in modern-day Iraq. The tablets, dated to the middle part of the 3rd millennium BC, record the use of some 250 plants, including the opium poppy, mandrake (*Mandragora officinarum*), and deadly nightshade.[27] The use of the opium poppy is seen most clearly in tablets from the settlement of Nippur, an important seat of worship, in which the plant is denoted by the ideogram "Hul Gil," translated as "joy plant." The text includes reference to the cultivation and harvesting of opium, and given that the "Hul Gil" ideogram had cropped up in texts dating to the 4th millennium BC, it seems likely that opium use was well embedded in Sumerian society, at least in terms of ritual or religious use.[28,29] Similarly, several tablets among a vast horde found among the ruins of the Royal Library of Ashurbanipal in the city in Nineveh attest to the popularity of cannabis.[30] While the debris was a result of the sacking of Nineveh by the Scythians in 612 BC, the tablets are thought to contain the collected knowledge that the Sumerian and Akkadian civilizations had accumulated over the preceding 2,000 years.

If we shifted some 700 miles east and a touch north of Mesopotamia to the land surrounding the Oxus River (the intersection of modern-day Turkmenistan, Iran, and Afghanistan), plant residues in pottery found in temple buildings of Gonur Depe dating from the early 3rd millennium BC show that the contemporaneous civilization in that region was also using plants of the psychoactive *Ephedra* genus mixed with either cannabis or opium and possibly fly agaric, in a combination that may well represent the ritualistic and religious drink deified by the two religious splinter groups that would radiate from the Oxus River area. The drink, called "soma" or "haoma," appears in the *Rig Veda* and *Avesta* texts that laid the foundations for Hinduism in India and Zoroastrianism in Iran, respectively.[4]

If, on the other hand, we shifted 700 miles in the opposite direction, we would find the ancient Egyptian civilization approaching the end of its Old Kingdom period. A plethora of indirect evidence drawn from sculptures, reliefs, hieroglyphs, jewelry, clothing, and burial practices suggests that the entheogenic properties of *Psilocybe cubensis* and *Amanita muscaria* mushrooms may well have underpinned some of the key tenets of early Egyptian religious beliefs, including both the "Eye of Horus" and the god Osiris. If this is the case, the hallucinogenic experience of consuming these mushrooms was believed to confer immortality and divinity to the consumer and allow direct communication with the gods.[15] By the time the Sumerian tablets were being tooled, the first representations of other potentially psychoactive plants, such as the blue water lily (*Nymphaea caerulea*) and cannabis, were beginning to appear within tombs.[4,30] By the beginning of the New Kingdom epoch (1550 BC), such plants would be commonly depicted in tomb engravings and offerings.[4] So, for instance, the tomb of Tutankhamun included a wall depiction of the pharaoh with water lily and mandrake in hand, sculpted in relief on a gold-plated shrine,[31] and carved ceiling coffers show his wife administering the sick pharaoh with water lily, mandrake, and opium.[4] These same plants also featured in the magico-religious *Book of the Dead*. Indeed, the Egyptian priesthood served a dual purpose as priests and physicians, and would have been well versed in the use of many medicinal and psychoactive plants. Much of their medical knowledge survives in the form of papyri deposited in the tombs of priests/physicians in order for them to accompany the deceased into the afterlife. The most detailed, the *Ebers Papyrus*, written circa 1600 BC, and reputedly found at the feet of a mummified corpse in the Theban necropolis, incorporated knowledge accrued throughout the history of the Egyptian civilization, including some 700 medical and magical remedies. Many of these involved plants, including henbane, the opium poppy, and cannabis.[30]

The Egyptians owed much of their wealth to regional trade, including with South Asia, and, at approximately the time that the *Ebers Papyrus* was being written, the late Minoan civilization centered on Crete was adding to its own extensive export roster by growing and exporting opium. Findings from a subterranean temple chamber in Gazi show that the Minoan islanders worshipped the "Poppy Goddess." Indeed, as the Mycenaean civilization became predominant in the region over the next few hundred years, an opium trade network, centered on Cyprus, with connections to many of the civilizations in the Mediterranean and Middle East, flourished. The trademark Cypriot base-ring opium jugs fashioned to look like incised poppy heads have been found scattered throughout the region, including in Syria, Palestine, and Egypt.[4]

The Mycenaean period also provided the setting for the myths and legends, many of them incorporated from other regional cultures, that underpinned the religious beliefs of the later ancient Greek civilization. Psychoactive plants feature prominently. For instance, Hecate was the underworld goddess both of witchcraft and poisonous plants and was associated with specific plants such as wolf's bane (genus *Aconitum*), mandrake, opium, and deadly nightshade,[27] whereas Circe, a minor goddess and witch, administered a poison that was probably deadly nightshade to the crew of Odysseus' ship. The antidote, described as moly, that then saved Odysseus was most likely to have been a member of the snowdrop family such as *Galanthus nivalis*.[b,32] Deadly nightshade was

b. Deadly nightshade and *Galanthus nivalis* have opposite effects on the acetylcholine neurotransmitter system: the former antagonizes muscarinic receptors and the latter increases acetylcholine levels, thus compensating.

also associated with one of the three mythical "fates," Atropa (after whom the plant is named), who severs the thread of life at the point of death, and it was used by the Greek cult of Dionysius (Bacchus to the Romans). Likewise, opium was associated with the mythological twins Hypnos and Thanatos, representing sleep and death, and had its own deity, Demeter, the goddess of fertility and the harvest, whose emblem was an opium poppy[33]; and henbane garlanded the wraith-like spirits of the dead that roamed the banks of the river Styx at the entrance to Hades.[34] One further notable psychotropic was the secret ingredient of *kykeón*, a mysterious potion to which was attributed godlike properties; it was reputed to engender an ecstatic experience of death and resurrection. The drink formed the cornerstone of the Eleusinian Mysteries, the initiation ceremony of the cults of Demeter and Persephone, which survived right through from the Mycenaean Age to the Roman Empire and counted among its members many of the notable members of Greek and Roman society. The identity of the *kykeón* is shrouded in mystery, but it is known that the ingredients included flour and mint, and it seems likely that the active ingredients were hallucinogenic, water-soluble lysergic acid amides from the fungus ergot, which grows symbiotically on cereal in the region around Athens.[14,26,35,36] With the exception of the *kykeón*, these plants, along with other plants such as hemp and mandrake, also featured extensively in the burgeoning medicine of the Greeks, along with their many ceremonial and social roles. Indeed, opium was generally regarded as the drug of choice for the populace.[26]

Paradoxically, despite the intertwining of psychotropic and medicinal plants with religion, myth, and magic, one major contribution of the Classical Greek era was to begin the process of dissociating medicine from the supernatural. The key figures in this regard were Hippocrates (460–377 BC), the "father" of Western medicine, who, along with his own tutor Herodicus, believed that disease and illness had natural, rather than supernatural, causes. They believed that these included diet, lack of exercise, and lifestyle. The *Corpus Hippocraticum*, a body of work drawn together by the followers of Hippocrates, is notable in that it is devoid of any mention of supernatural causes for disease. However, it does include a critical review of the effectiveness of over 200 plant-based treatments. It therefore provided the first step toward rational pharmacognosy. Hippocrates' work also furnished the starting point for several other pivotal figures. The first of these was the philosopher and botanist Theophrastus (372–287 BC), a polymath student of Plato and Aristotle, whose many works included the *Historia Plantarum*, a series of 10 books that established the science of botany and remained a core reference work well into the Middle Ages. This work introduced the first taxonomy of plants and included a description of their medicinal properties. Naturally, the Ancient Greeks' medical knowledge and partiality to opium and other drugs was absorbed by the Romans, so the next pivotal figure in the development of medicine was the Greek physician Pedanius Dioscorides (circa 40–90 AD), who practiced in Rome, where he laid the foundations of pharmacology in the West with his treatise *De Materia Medica*.[37] This tome described the dosage and efficacy of about 600 plant-derived medicines, including a number with central nervous system effects, including members of several genera from the *Lamiaceae* family, such as species of *Salvia, Melissa,* and *Rosmarinus* (to which he attributed beneficial brain and memory effects), valerian (*Valeriana officinalis*), St. John's wort (*Hypericum perforatum*), and cannabis (to which he didn't attribute psychotropic effects). He also noted the medicinal effects of a variety of spices, such as saffron (the stigma of *Crocus sativus*), turmeric (*Curcuma longa*), and cinnamon (the inner bark of *Cinnamomum verum*), all of which would have been imported from India via Alexandria. Naturally he also included the usual suspects: opium, deadly nightshade, and henbane, the last of which he prescribed in the form of a salve to be rubbed on the skin, a much safer mode of administration than oral consumption. His recommendation to use mandrake as an anesthetic, combined with hemlock (*Conium maculatum*) as a muscle relaxant, was also the first example of a balanced, combined surgical anesthetic.[38] Dioscorides was followed by a series of pharmacists and clinicians of note, but Galen (129–200 AD), a Roman philosopher and physician of Greek descent, was the most celebrated. He was the first to suggest that the brain was the seat of the intellect and made an unrivalled contribution to the understanding of brain anatomy at that time. He also described over 500 plant-derived medicines, including "galenic" formulae composed of multiple plant extracts and parts, and demonstrated that herbal/plant extracts could contain both beneficial and harmful components.[39] Despite the many advances that he made Galen also supported the misplaced belief that the "humors" (blood, phlegm, black bile, and yellow bile) corresponding to the four types of human (phlegmatic, sanguine, melancholic, and choleric) underlay health and disease and could be coordinated to rectify illness. This was accomplished by bloodletting, purging, vomiting, and poly-treatments with plant extracts. These beliefs informed medical practice well into the early modern era.[40]

Of course, the Western tradition was not the only well-developed medical system. In the East there were two alternative traditions of medicinal plant use that survive into the present day. In India the Hindu tradition, Ayurveda, originated in hymns contained within the *Arthavaveda*, one of four Hindu texts that were compiled possibly as early as 1500 BC. Both the *Rig Veda*, another of the four texts, and the *Atharvaveda* make reference to the mysterious "soma" that lay at the heart of the Hindu religious beliefs. The first formal, written description of medicinal plant use within Ayurveda came with the medical treatises *Charaka Samhita* (circa 900 BC) and *Sushruta Samhita* (circa 600 BC). Both texts reflected knowledge accumulated over the preceding millennium, and they described 341 and 395 medicinal plants, respectively. Examples included the use during surgery of cannabis and "soma," as well as the use of sammohin and sanjivan as anesthetics and reversal plant extracts, respectively. The identities of all of these plants, with the obvious exception of cannabis, have been lost.[41,42] While there was some interchange of knowledge between the Ayurvedic and Western medicinal systems via the Greek/Roman import/export system centered in Alexandria, further to the east the Chinese medical system developed largely independently. The oldest Chinese medicinal book, *Wushi-er Bingfang* (Recipes for Fifty-Two Ailments), which included numerous natural treatments, was compiled around 350 BC. The more comprehensive *Shen Nong Ben Cao Jing* was a treatise on agriculture and medicinal plants written during the Eastern Han dynasty (25–220 AD) that putatively described the verbal traditions passed down from first emperor and herbalist Shen Nong, who reigned 3,000 years previously. This text included descriptions of 252 medicinal plants and their therapeutic effects.[39]

The use of medicinal and psychotropic plants across the African, Australian, and American continents is unrecorded during antiquity, but archaeological evidence suggests that psychoactive plant use was alive and well. For instance, Central/South America is lightly scattered with archaeological evidence of the indigenous population's use of a variety of psychoactive plants in the far past. Beyond the chewed and compressed coca leaves in the Nanchoc Valley, there are many examples of smoking pipes and *Anadenanthera* seeds (circa 2130 BC), snuff trays and tubes (circa 1200 BC), stone carvings of mythological beings with cactus slices in hand (circa 1300 BC), and representations of cacti and other psychoactive plants in ceramics (circa 1000 BC), to name but a few.[16,43]

Middle Ages (5th to 15th Centuries)

Galen's life coincided with the zenith of the Roman civilization. During the two centuries following his death the Roman empire went into slow decline, and eventually fractured into the Eastern and Western empires. The subsequent collapse of the Western Empire led to an extended period that is often described as the "Dark Ages" in Western Europe, an extended epoch of cultural and economic deterioration that was typified by an obscured historical record, plagues, barbarism, invasions, and a decline into feudalism. During this period many of the cultural advances of the past faltered and receded, including the study and understanding of psychoactive plants. Indeed, as the renamed Christian Roman Empire disintegrated, the edicts of the emperors included proscription, on penalty of death, of "nocturnal ceremonies" and the use of diabolic psychoactive plants.[26] The Eastern Empire, centered on Constantinople (now Istanbul) and often referred to as the Byzantine Empire, survived for a further 1,000 years, although this was without any further dramatic expansion in knowledge. However, the understanding of medicinal plants contained in the *Corpus Hippocraticum* and texts by Dioscorides, Galen, and Theophrastus, along with knowledge from the Ayurvedic tradition, were integrated and augmented within the emerging Islamic medicine that burgeoned during the Islamic Golden Age (mid-7th to mid-13th centuries AD).[44] Many of the treatises of the Roman and Greek scholars were lost from the Western record and only survived in Arabic until they were retranslated back into Latin or Greek many hundreds of years later.[45] The Islamic Golden Age saw the establishment of pharmacology, and the work of the Greek/Roman physicians was developed and expanded by scholars such as Rhazes (Muhammad ibn Zakariyā Rāzī) (865–925 AD), possibly the most prolific and accomplished physician of the Islamic Golden Age, who introduced several major improvements to the methods of medicinal chemistry. Similarly, Avicenna (Ibn Sina) (980–1037) compiled a methodical treatise containing information on 800 simple drugs and many compounds, the majority of which originated from plants. He also introduced the medicinal use of many minerals. His work was particularly important in that it loosened the reliance on the Galenic "humors" approach and formally introduced the concepts of reasoning and experimentation.[46]

The flow of knowledge from West to East began to be reversed in the 12th and 13th centuries as works, such as those of Avicenna, were translated into Latin and began to erode the stalemate in Western medicine.[46] Crusaders also returned from their extended sojourn in the Near East with medical knowledge acquired on their travels.[44] Less positively, Islamic strictures on the use of many psychoactive drugs, which had been viewed in a neutral light up to this point, were introduced.[26]

While this period didn't see any dramatic advances in knowledge, several developments were under way that would be inextricably linked to the remapping of the globe by European colonists. First, the psychoactive properties of the coffee berry (from the *Coffea* genus, e.g., *Coffea arabica*) were discovered in Ethiopia, with cultivation of the plant spreading to Yemen as a consequence of an invasion by the Ethiopians in the 6th century AD. Coffee subsequently arrived in the Arabic literature, with a first mention of the berries of the plant in the works of Rhazes circa 900 AD. However, it wouldn't be until the 15th century that the true potential of the coffee bean would be unlocked by the revelation that roasting the seeds from the center of the berry produced the drink we know today.[47] Second, an infusion of the leaves of the bush *Camellia sinensis*, which had originated in Southeast Asia at the intersection of Burma, India, China, and Tibet, received its first unambiguous mentions in Chinese writings around 600 AD. It would be used medicinally until the Ming Dynasty (1368–1644), when it would become popular as a social drink[48] and would spread rapidly to the West in the form of tea at the end of that period. The third event, which would play a more minor role as a supporting factor in the story of tea, was the arrival of opium in China by overland caravan during the Tang Dynasty (618–917 AD). The use of opium would remain largely medicinal in China until the introduction of tobacco in the late 16th century.[49]

Renaissance and the Age of Discovery

To a certain extent, during the "Dark Ages," alchemists, traditional healers, and herbalists had been the keepers of the accumulated Western European knowledge of medicinal plants. Unfortunately this group was also the target of a widespread persecution that bridged the period from the late Middle Ages and well into the Renaissance.

Until the late 12th century the Christian church took a relatively ambivalent view with regard to witchcraft and specifically prohibited the torture or execution of suspected witches. However, the Inquisitions, the activity of Catholic bodies charged with the rooting out and brutal suppression of heresy, formally commenced with the Episcopal Inquisition in 1184. Initially, within Europe, the Inquisitions concerned themselves with stamping out nonconformist Christian movements such as the Cathars and Waldensians. However, when Pope Gregory IX extended the remit to witchcraft and sanctioned the confiscation of the belongings of those that confessed, the persecution rapidly spread and grew to encompass the use of plant-derived drugs. This made any practitioners of folk healing, which represented the principal medicine for much of the population, particularly vulnerable.[26] Indeed, even pharmacists/chemists were prone to inquisitorial attention. For instance, Pietro d'Abano, a professor of medicine who, among many achievements, had translated several Islamic medical texts into Latin, died in jail as a consequence of rigorous questioning just prior to his being found guilty of heresy in 1316.[50] While the Catholic inquisitors did not concern themselves primarily with witchcraft, a series of witch-hunt "crazes" swept across regions of Europe throughout the 15th to 17th centuries. Knowledge of notable psychoactive plants had been passed down from the Roman/Greek tradition and the eventual recipient, the European herbalist, healer, or soothsayer (often a wise female "Hag," meaning "hedge," denoting the place where she found her medicinal plants), would have commanded the properties of hemlock, henbane, deadly nightshade, members of the *Scopolia* genus, and the exotic mandrake where it was available (more often local replacements such as black and white briony were used).[33,34,51] The stock in trade of the herbalist may well have included the provision of "love potions" and "flying potions." Certainly testimony from the various witch trials suggests that the archetypal notion of the witch flying on her broom originated in the practice of applying a psychoactive plant salve to the skin, with a broom handle or similar phallus then used to insert the drugs into the vagina or anus, where it would be rapidly absorbed by the mucous membranes. The combination of deliriant plant extracts and physical stimulation would then elicit the experience of flying and sexual hallucinations.[34] Naturally, this use of potent drugs may have created a whole class of drug-abusers whose strange lifestyle and behavior would have marked them out from the normal,[27] and herbalism and the use of deliriant drugs would sometimes be combined with pagan practices, including group fertility/orgiastic celebrations. The witch hunts peaked in the 16th and 17th centuries and involved the execution of an estimated 60,000 people in Europe over the course of three centuries.[52] By the late 16th century experts such as Jean Bodin, the French jurist and political philosopher, whose works were influential in other countries, including Britain, advocated torture for anyone simply found with ointments on their body, and of course the same fate might have awaited the user of any of the psychoactive plants and fungi discovered and used since antiquity.[26]

The witch hunts were typically triggered by inexplicable events, deaths, or ailments and as such reflected to a certain extent the poor medical knowledge of the time. One pivotal figure in the history of medicine who rejected the general tendency to see witchcraft at every turn was the flawed genius, chemist, and physician Theophrastus Bombastus von Hohenheim (1493–1541), better known as Paracelsus. He single-handedly ushered in the start of a medical revolution as he publicly and vehemently refuted the Galenic notion of the "humors" that had held sway for more than 1,000 years, in favor of a new system of medical and scientific knowledge based on sound empirical foundations. His copious writings laid the foundation for toxicology, antisepsis, modern anesthesia, chemotherapy, military trauma surgery, and psychiatry, and he rationalized the pharmacopeia.[40,53–55] Paracelsus, for his part, saw many manifestations of "witchcraft" as psychiatric conditions,[55] and he was well acquainted with the various plants employed by the typical "witch." Through careful experimentation Paracelsus also discovered that the active chemicals in opium were more soluble in alcohol than water and developed his own opium tincture preparation, which he named laudanum. Given the limited alternatives, any treatment that could suppress coughs and diarrhea (in an age where these two symptoms often presaged death), and was furthermore a painkiller and soporific, was bound to be popular. Laudanum, and numerous variants, would become one of the most widely taken medicines for several centuries, and a modern variant continues to be prescribed occasionally to this day.[c] Paracelsus certainly featured as a key figure in the rejection of superstition, tradition, and unquestioned medical authority, and he instigated the move toward experimentation that marked the medical step-change that would be continued by notable Renaissance physicians such as Andreas Vesalius and William Harvey.[56] Having said this, the worst excesses of the witch hunts still took place after Paracelsus and his successors had started the process of modernizing medicine. It may well be the case that the knocking away of the foundations of Galenic medicine without a fully developed alternative explanation for many diseases revealed dark gaps in medical knowledge that would be filled by suppositions of witchcraft and diabolic forces. In this way the start of the modernization process may well have exacerbated the problem.[57]

Paracelsus was born just one year after Christopher Columbus discovered the Americas (1492). Unfortunately, when he died at the age of 47 he was probably totally unaware of the veritable treasure chest of psychoactive plants that the continent represented. The earliest Spanish conquistadors had been largely ignorant of the local

c. At least in its simplified form—that is, without Paracelsus's additions to the simple tincture, which included crushed pearls, musk, and amber.

flora, but they were followed within a few years by members of the Franciscan religious order, who were intent on the Catholic evangelization of the indigenous population. Many of these friars carried out important work describing and chronicling the societies that they found in Middle and South America. The most prominent among them was Bernardino de Sahagún, who has been described as the first anthropologist.[58] Sahagún believed that the best way to approach the conversion of the indigenous peoples to Christianity was to first understand them, their worldview, and their idolatrous ways. His works, including the *Historia General (de las cosas de la Nueva España*—often referred to as the Florentine Codex), provided the first written descriptions of the ritual use of a number of entheogenic and hallucinatory plants unknown in Europe, including ololiuhqui (seeds of *Turbina corymbosa*), péyotl (*Lophophora williamsii* cactus), tlápatl (*Datura stramonium*), tzintzintlápatl (*Salvia divinorum*), and míxitl (*Datura anoxia*). He also described the use of the local species of *Psilocybe* mushrooms, called teonanácatl in the local Nahuatl language.[59] In 1536 Sahagún also helped found the first college in the Americas in Mexico City, a metropolis built on the ruins of the Aztec city of Tenochtitlan. The college was highly unusual in that it integrated both Spanish and Aztec knowledge. Notable works created in the college included the herbal pharmacopoeia *Libellus de Medicinalibus Indorum Herbis* (Little Book of the Medicinal Herbs of the Indians), completed in 1552 by an Aztec staff member and subsequently translated into Latin.[60] This book was a condensation of Aztec knowledge of the local flora, describing plants and their medicinal or ritual uses. Back in Spain, Nicolas Monardes, a physician from Seville who had never journeyed to the New World, published the widely read second part of his history of the medicinal plants of the New World, *Historia medicinal de las cosas que se traen de nuestras Indias Occidentales,* in 1571. News of the bountiful flora of the New World, and the opportunity to capitalize on new medicinal discoveries and replace the expensive medicinal plants that were being imported into Spain, spurred King Philip II of Spain to persuade one of his own physicians, the naturalist Francisco Hernandez, to organize the first scientific expedition to the Americas, tasked with obtaining "an account of all medical herbs, trees, plants and seeds that exist in a given place".[61] The place was Mexico City and its environs, but in reality Hernandez, with a coterie of botanists, native doctors, painters, and a geographer, toured Mexico and Mesoamerica for three years, collecting and describing some 3,000 species of the local flora along with details of their medicinal or ritual use by the indigenous peoples. Contrast this with the few hundred plants described in Dioscorides' *Materia Medica*. Hernandez's work even extended to clinical trials of plants conducted with Indian patients in Mexico City's only hospital.[61] Hernandez eventually returned to Spain with many novel seeds and plants that would go on to be introduced to the rest of the known world.

Naturally the Inquisitions were also exported to the New World. In Central and South America Inquisitions were set up by the new Spanish and Portuguese overlords to seek out heretics within their own communities, to mold the existing beliefs of the indigenous population toward Christianity, and to suppress and punish the use of the "diabolical" plants. The entheogenic plants described by the early Franciscan friar Bernardino de Sahagún would become one focus of a crusade against the infernal practices of the natives, and widespread persecution and the destruction of crops followed for the next 200 years.[26,59] Even within his own lifetime Sahagún became disillusioned as the Spanish continued to brutalize the populace and the largely pro-native Franciscans were replaced by a hardline secular clergy. Sahagún's masterwork, *Historia*

General, and the comprehensive herbal pharmacopoeia produced within his college were subsequently lost for 200 years as they languished on the shelves of the King's and the Pope's libraries.[58] This loss of knowledge and the vicious suppression of the use of ceremonial plants and fungi meant that many psychoactive plants and fungi simply disappeared from the record, with some rediscovered only in the mid-20th century. For example, the use of both the *Psilocybe* genus of mushrooms and *Salvia divinorum* survived only in remote mountainous regions in Mexico, well away from the interference of the Spanish. They were rediscovered by Western travelers only some 60 years ago.[62,63] Knowledge of many plants may still be lost. Unfortunately, even Philip II's attempt, via Francisco Hernandez, to translocate the rich tradition of medicinal plant use from the New World to the Old World was met with suspicion and questionable success. In many cases the novel treatments simply didn't fit within the Galenic medical system.[61] The exceptions were plants with specific commercial potential, including cocoa (seeds of *Theobroma cacao*) and quinine (bark of the *Cinchona* genus of trees), and a range of plants from the nightshade (*Solanaceae)* family, including foods such as the potato, tomato, and peppers (the *Capsicum* genus) and the scourge of human health, tobacco. The discovery of tobacco heralded a period in history in which the human craving for psychoactive plants would literally provide the motive for the remapping of the globe, and colonization, subjugation, warfare, and mass slavery.

By the time the conquistadors arrived, tobacco use, in the form of the leaves of *Nicotiana tabacum/rustica*, had pervaded the entire American continent and the Caribbean for many hundreds of years. The plant was cultivated ubiquitously, even among otherwise hunter-gatherer societies, and was one of the staple psychotropic plants of shamanic rituals due to its "entheogenic" properties at high doses. Methods of consumption differed and included snuffing dried leaves, drinking juices, and inserting suppositories. However, the most common mode of administration was smoking; by pipe in North America and rolled into cigars in South America.[61]

Although tobacco had featured in many of the earliest interactions of Christopher Columbus's crew with the natives of the Americas, it was close to 50 years before it made its first recorded appearance in Europe, when it arrived at the Portuguese court. Despite this slow start, by 1570 tobacco was being grown in comparatively small quantities in Belgium, England, Italy, Spain, and Switzerland, and by 1600 it had been transplanted via trade and colonization to China, India, Java, Japan, the Philippines, and West Africa, and via Chinese traders on into Mongolia, Siberia, and Tibet. Within little more than a century of Columbus's discovery of the Americas, tobacco had circumnavigated the globe and was being grown and used all over the known world.[61] The impetus for this rapid expansion in tobacco use was not simply the highly addictive properties of nicotine, but rather the claimed medicinal properties of the plant. It could be used as a topical poultice, chewed with lime, or less often smoked in the form of dried leaves. Just as the availability of tobacco began to rise, the notable physicians Nicolas Monardes, Matthias de L'Obel, King James I's physician, and the French Ambassador in Lisbon, Jean Nicot (after whom the plant is named) became the first of many to eulogize the medicinal virtues of tobacco. They claimed that it was a panacea that cured ailments ranging from toothache, through coughs and asthma, to malignant tumors.[49,61,64] Monardes also assigned tobacco its humoral essence—hot and dry in the second degree—and this allowed it to fit into the Galenic medicinal system. Following its introduction to the French court by Jean Nicot, tobacco, taken in the form of snuff, also began to rapidly circulate within the royal courts of Europe and farther afield,[61] with pipe smoking initially lagging behind until it was popularized in the late 16th century by Thomas Harriot (prior to his dying of smoking-related cancer) and his friend Walter Raleigh. As tobacco was introduced to Asia and the Far East by Spanish and Portuguese traders operating from colonies in the Philippines and China (Macao), respectively, it took with it the label of panacea, quickly establishing itself in the Chinese pharmacopeia and penetrating all levels of society.[49,61] As elsewhere, smoking tobacco would become an important social pastime in China, mixing as it did an addictive drug with smoke, which played a pronounced part in the rituals and magic of the region. It would also eventually provide an entrée to the smoking of opium, at first in combination with tobacco and later, following a temporary ban on tobacco, by itself.[49]

The colonization of large swathes of the world by the Spanish, Portuguese, Dutch, French, and British in the 17th and 18th centuries was driven simply by commercial motives. Often the exploration and establishment of new colonies was funded by private investors, with governments taking over only when the resultant colony became financially viable. The widespread adoption of tobacco, driven by nicotine addiction, made it a major impetus for the annexation of land and the establishment of tobacco-producing colonies throughout the Caribbean and much of South America. However, as growing tobacco is a labor-intensive business that requires plenty of land for crop rotation (due to autotoxicity), within half a century many of the smaller producers switched to sugar production. The major exception was the English colony in Virginia, which was struggling to survive until John Rolfe imported *Nicotiana tabacum* to replace the smaller, shrubbier local variety of *Nicotiana rustica*. The colony expanded exportation from a meager 170 pounds of tobacco in 1614 to 370,000 pounds in 1628. The Maryland colony saw an equally rapid growth of tobacco cultivation and export back to the Old World. By the end of the 17th century some 3 million pounds of tobacco were being imported into England from the American colonies and then being re-exported worldwide, generating by far the largest proportion of English customs revenues. As Charles I memorably noted in the mid-17th century, the English colonies in the Americas were "Wholly built upon smoke." By 1719 France alone imported 6 million pounds of tobacco from London, and the labor-intensive industry was becoming increasingly reliant on slave labor.[61] At this point, the other major tobacco-producing colony was Brazil, which had started the process of exporting tobacco direct to the African slave coast to fund the purchase and shipment of slaves. Between 1700 and 1800 alone the Portuguese rulers of Brazil imported some 2 million slaves from the African continent. Some of these were destined to work in goldmines, but the majority ended up toiling on tobacco and, increasingly, coffee plantations. By the end of the 18th century Brazil and North America accounted for the bulk of global tobacco production; in both cases the industry was driven by slave labor.[61]

Although tobacco was to prove a scourge, it did prove useful in one respect. In Peru, the leaves of the Coca plant (*Erythroxylum novogranatense/coca*) occupied the same "panacea" medicinal ground as did tobacco, but it failed to catch on with the Europeans. This was partly because, unlike tobacco, it failed to shake off its connotations of ritual use. Indeed, coca use would be cited as evidence in every witchcraft trial in Peru throughout the 17th century, and missionaries would encourage the wholesale destruction of the coca crop. However, the main reason was that it was "discovered" some three decades after tobacco. It had simply missed the boat; Europe and the world already had its panacea.[61]

The two other psychoactive plant products that helped shape the world as we know it had several things in common; they were usually taken in the form of a hot drink, often with that other novel, staple product of slavery, sugar; and they exerted their hold over humans on account of the presence of caffeine, a mildly addictive and reinforcing plant chemical. The products were, of course, coffee and tea, made from the beans of *Coffea arabica* and the leaves of *Camellia sinensis*, respectively.

It wasn't until the early 15th century that the true potential of the coffee bean was unlocked by roasting and grinding, which allowed the grinds to be steeped or boiled in hot water. The resultant drink was first mentioned in Arabic writings in the middle of the 15th century, and it managed to survive the religious intolerance of psychoactive plants and alcohol that had swept through the Islamic world in previous centuries.[26,65] In 1536 the Turks occupied Yemen and began exporting coffee back to Turkey.[47] When Suleiman the Magnificent granted legal permits to coffee houses within the Ottoman Empire in 1551, the Yemeni coffee bean quickly became a major import item.[26,47] By the mid-17th century coffee had reached Europe. In Britain the first coffee shop was opened at Oxford University in 1650. Fifty years later there were more than 2,000 coffee houses in London, and both here and in France these egalitarian meeting places provided the fertile environment for the seeds of social change to grow. In France this mixing and discourse would lead to revolution, while in England coffee houses would spawn Lloyds of London, the stock market, and newspapers, including *The Tatler* and *The Spectator*.[47] As coffee consumption was spreading, the coffee-growing monopoly had been jealously guarded within Yemen. However, the embargo on exporting coffee plants was broken in the early 17th century, both by the revered Sufi Baba Budan, who smuggled a few seeds to India, and by Dutch merchants, who transported a bush to the Botanic Garden of Amsterdam. The majority of coffee produced in the world today originated from these few plants as the Dutch first transported plants from the stock introduced by Baba Budan in India to their colonial plantations in Java and other islands in the Dutch East Indies in the mid-1600s.[47] The progeny of the Amsterdam plant was subsequently transported to Ceylon in 1658 and to Dutch Surinam in South America, with a first crop of beans being harvested in 1718.[65] A single plant was also nursed on the voyage to French Martinique in 1714. This plant provided the stock for colonial plantations spread across Middle and South America and the Caribbean. In the same year a single plant was exported to the Portuguese colony of Brazil, and 20 years later the cultivation of coffee commenced in Saint-Domingue, the French colony that made up the Western part of the island of Hispaniola. These two last colonies would dominate world coffee production in succession, starting with Saint-Domingue. In both cases their coffee industries were fueled by the import of millions of slaves from the African continent. In the case of Brazil, the human traffic was paid for with tobacco. By 1788 Saint-Domingue was growing 50% of the world's coffee, but with a population that was composed of 90% slaves or their freed descendants. The huge slave population was living in appalling conditions, literally being worked to death. In 1791, inspired by the French Revolution of two years previously, the slaves revolted. The subsequent 12-year armed struggle was to become the only successful slave uprising in history. However, the country, now renamed Haiti, never regained its preeminent position as a coffee producer. Brazil's turn to dominate the coffee markets came somewhat later, after the country gained its independence from Portugal in 1822. Dominance in coffee production came at a cost. The vast Atlantic Forest, which covered up to 1.5 million square kilometers when the Portuguese had arrived, was slowly eaten away as a continuous cycle of tree clearance, coffee planting, and resultant soil toxicity rolled deep into the previously verdant forest, leaving pastures and croplands in its wake. Today only 4,000 square kilometers or so of the forest survive. The cost was also seen in human traffic. When the ban on slave transportation was enforced by the Brazilian government in 1851, largely thanks to a blockade of slaving ships by the British, 2 million slaves were already working in Brazil, and it was the last "Western" country to abolish slavery. When it did so in 1871, the government attempted to temporarily protect the coffee industry. The resultant Rio Branco law or "law of the free womb" meant that only the children of existing slaves would be freed, with the millions of existing slaves intended to die in harness.[47] Similar but smaller-scale scenarios would play out in post-independence Mexico and other Mesoamerican countries in the late 19th century. For instance, in Guatemala the coffee industry required the unspoiled lands of the indigenous Mayans, who were dispossessed and subjected to forced migration to the coffee fields and enforced labor by armed militia. This was slavery in all but name, and influenza and cholera were common killers in a population crowded onto plantations and lacking immunity to Western diseases. Similar processes in countries such as El Salvador and Nicaragua would lead to police states and authoritarian military regimes, economic and social inequality, and the slavery of the indigenous populations, setting the pattern for the dysfunctional Central American governments of more recent times.[47]

The modern history of tea was similarly rooted in the exploitation of other races, but the key events took part largely in the Eastern Hemisphere. Tea was first imported to Europe by the Dutch East India Company, reaching mainland Europe by 1638 and Britain by 1650.[48] Tea drinking became particularly popular in Britain due to a number of factors. The first was that the British East India Company began to import huge amounts of tea almost by accident; they used tea simply as a potentially profitable ballast for their return journeys from the Orient. Faced with an excess of tea, they launched a concerted marketing campaign extolling its virtues in order to stimulate demand.[66] Second, tea was easier to make, cheaper, and a more profitable drink than coffee. By the 1730s most of the several thousand London coffee houses had gone, replaced by chophouses, pubs, and tea shops.[47] By the mid-1700s tea had become the principal cargo of the British East India Company, with some 3 million pounds shipped per year to London. From there it was exported to any areas of the world that the British had colonized, settled in, or traded with. Tea became, and remains, the second most widely consumed drink globally, after water.

The key to understanding the story of tea is to understand the British East India Company. The Company was privately owned by wealthy shareholders and, in 1600, was granted a Royal Charter by Elizabeth I giving it monopoly rights on the exploitation of commercial opportunities in Asia and the Far East. The British government, in theory, had no direct control over its activities. However, the Company became synonymous with colonialism, exploitation, and the abuse of monopoly power and acted as a *de facto* arm of government. This included operating a huge private army.[66,67] The Company had, in the words of one senior employee, a "voracious appetite for lands and fortresses",[67] and it was instrumental in the conquest and annexation of the vast swathes of territory that would transform the British Empire into the largest in history and would make Britain the dominant global power for over a century. The insatiable appetite for tea, predicated of course on the psychoactive properties of caffeine, was a key driver for the expansion of the British Empire across Asia, and it funded both the government, via prodigious import duties, and the acquisitive Company via vast profits.[66,67]

For the first century or more of their relationship with the Company, the Chinese had enjoyed an absolute monopoly on tea production. However, the Company had acquired Bengal as a consequence of victory by its private army at the Battle of Plassey in 1757, and Chinese tea bushes were transported to India and the newly occupied coastal areas of Sri Lanka (known as Ceylon) in the last years of the 18th century. Tea production was subsequently a primary motive for the expansion of the Company's dominions to the entire Indian subcontinent, which it accomplished by the 1820s. Similarly, tea motivated the subjugation of the remainder of Sri Lanka in 1815. When slavery was abolished in the British Empire by order of Parliament in 1833, the vast territories under the direct control of the East India Company were specifically excluded. Business could carry on as previously. The depredations of the Company didn't stop there. The Bengal famine of 1770, in which millions died, was attributable indirectly to the tea trade, as the Company had enforced an agricultural switch from food production to growing opium (see below) and other trading goods in order to pay for tea. As the famine developed the Company compounded the situation by hoarding what food resources there were in order to profit from rising prices.[66,67]

During the mid- to late 1700s, the tea trade had another unintended consequence. By the mid-18th century the 1.75 million colonists of America drank more tea than the entire 7 million population of Britain.[66] The tea came from China, and, as part of their deal to share the revenues from Bengal, the Company persuaded the government to waive the duty it paid on tea that passed through London on its way to North America. However, to compensate, the government introduced an import tax on tea arriving in America, prompting an increase in smuggling. When the Company sought to flood America with cut-price tea the colonists were enraged, about both "taxation without representation" and the monopoly power of the Company. The resultant dispute escalated, via the Boston Tea Party—which saw 90,000 pounds of tea dumped in Boston Harbor—into the American War of Independence (1775–1783). At one stroke Britain lost both a valuable colony and the tax revenues from the booming tobacco trade.[66,67] The subsequent switch in North American drinking habits away from tea subsequently inflated demand for coffee, driving on Brazil's slave-fueled coffee economy.[47]

Beyond the subjugation of entire nations, the trade in tea also had another unfortunate, but not unintended, consequence. The Chinese initially enjoyed an absolute monopoly on tea production and export via Canton, the only port open to foreigners. Even after the annexation of India and Sri Lanka, the booming market for tea meant that Chinese tea was essential to satisfy soaring demand. The East India Company had few Western commodities that interested the Chinese, so they were running a huge trade deficit in goods and were being forced, by decree of the Chinese emperor, to pay the balance in silver bullion. This situation was commercially untenable, so the Company resorted to criminal means to balance the books. The acquisition of Bengal had ceded a Dutch monopoly on opium production to the Company, who began to pay huge bribes to Chinese customs officials in order to smuggle the drug, still in chests bearing the Company crest, into China via willing British merchants based in Hong Kong. Initially the Company restricted production in order to keep the price high, but under competitive pressure from "illicit" opium producers they eventually increased opium imports. In 1831 alone company records show 19,000 chests of opium, each weighing 63 kg, being smuggled into China, with the overall figure for British opium imports rising to 105,000 chests by 1879. The Company had effectively balanced its trade deficit with China by 1828, with income from illegal opium outweighing the cost of tea.[67] Prior to the East India Company's intervention a small minority of the Chinese population already consumed opium. Where other societies tended to take their many opium formulations orally, with few reported problems, the Chinese smoked the drug. Comparatively low-quality Chinese opium was initially simply one of many plant products added to tobacco, but over time, partly as a consequence of health concerns and an official ban on tobacco, opium switched to being smoked alone.[49] The opium being imported into China by the British East India Company was of much better quality than the Chinese variety and was seen initially as a luxury good, imbued with a European exoticness, and was consumed almost exclusively by the upper classes. However, over time, the vast amounts of opium being shipped into the country penetrated all strata of Chinese society. By 1839, with an estimated 12.5 million people regularly smoking opium, the Chinese government finally took action to reverse the flow.[67] It seems unlikely that this was a direct consequence of the damage that widespread addiction may have wrought on Chinese society, but rather that opium, the supply of which was illegal, was associated with a malignant penetration of the closely guarded Chinese society by the Europeans.[49] Plus, of course, the smuggling entailed a huge loss of silver bullion and was untaxed due to its illicit nature. In any event, Chinese forces surrounded the British trading enclave of Canton and destroyed some 20,000 chests of opium, triggering an escalating series of events that resulted in the first "Opium War." Within 3 years the British Navy had bombarded and harried the Chinese into submission, and the Treaty of Nanking saw reparations paid to the opium smugglers. At the same time a number of Chinese ports were forcibly opened to British traders (who had been confined to trading through Canton), while Hong Kong was transferred to Britain as a colony. Opium smuggling quickly resumed and increased.[67] Subsequently, a second "Opium War," fought over similar issues, started in 1856 and ended in Chinese capitulation after Beijing was captured by land forces in 1860. This led to a further opening of China's borders. Opium would continue to be a societal problem in China until the Cultural Revolution. However, the country exacted a small revenge with the exportation of the "opium den" to the Western world with the radiating Chinese diaspora. Interestingly, not long after the Opium Wars the British government awoke to the monster it had spawned and absorbed the dominions and army of the Company into the state, finally dissolving the British East India Company as an entity in 1874. On its demise, the English politician Horace Walpole voiced a commonly held view, noting that Britain, in the guise of the Company, had "murdered, deposed, plundered, usurped".[67] Much of this abhorrent behavior was driven by the enormous profits of the tea business.

The Modern Era

The modern era for psychoactive plants, and indeed medicine, dawned in 1806 when the German pharmacist's apprentice Friedrich Wilhelm Sertürner reported the isolation of the "principium opii" from opium[68] and created the first pure, naturally derived medicinal compound. His subsequent paper, describing the isolation, crystallization, crystal structure, and pharmacological effects of morphine, was published in 1817 and triggered an international torrent of research. Organic and medicinal chemistry as disciplines had arrived.[69] Within a decade and a half a wide range of compounds from medicinal plants had been isolated and identified, and in some cases their structure elucidated. These included compounds that affected brain function such as the painkiller salicylic acid from the bark of the willow *Salix alba*, caffeine, nicotine, atropine from *Atropa belladonna*, and noscapine, also from the opium poppy, to name but a few. Cocaine followed in 1855, although, as with morphine and quinine, the ability to synthesize this compound would lag behind the first isolation by many decades.[39,70] Less than a decade after Sertürner's paper reported the isolation of opium, Emanuel Merck began to manufacture and wholesale morphine. The modern pharmaceutical industry had been born.

The nascent discipline of pharmacology, the study of drug action, which had originated in experiments assessing the effects of whole plant extracts such as *Strychnos nux-vomica* and *Strychnos toxifera* on animals and physiological tissue, was revolutionized by the availability of the pure active components (brucine/strychnine and tubocurarine, respectively). Similarly, in the early years of the 20th century the plant chemicals nicotine and muscarine would play a key role in the identification of neurotransmission as a fundamental process in the brain and would enable the identification of the first neurotransmitter, acetylcholine.[71] Many other plant chemicals have been, and continue to be, employed in neuroscience on the basis of their very specific effects on neurotransmitter receptors or enzymes (e.g., N, N-dimethyltryptamine, muscimol, bicuculline, scopolamine, pilocarpine, atropine, etc.). Even in the present era of synthetic drug development it is interesting to note that more than a quarter of all new drugs registered with governmental authorities worldwide for all indications within the last 25 years were either natural products or directly derived from natural products produced by organisms such as plants, fungi, and bacteria.[72]

While great strides were being made on the scientific front, the general population of the Western world was becoming ever more familiar with potent psychotropic plant extracts. As we've seen, opium had been the cornerstone of European medicine for 400 years, first in the form of Paracelsus's original "laudanum" alcohol tincture of opium, and later in popular formulae created by notable 17th-/18th-century physicians and chemists such as Thomas Sydenham, Thomas Dover,[d] and Jakob Le Mort, whose "paregoric" was a camphorated tincture of opium.[29] The regular medicinal uses of opium ranged from sleep, to pain control, anesthesia, and the treatment of endemic diseases such as coughs and tuberculosis, diarrhea, and other gastrointestinal conditions. The many established medicines were joined by legions of over-the-counter patent medicines, many of which were dangerously opaque with regard to the narcotic nature of their contents. For instance, opium featured as an ingredient in "quieteners" for children, such as the quaintly named "Mrs Winslow's Soothing Syrup" and the "Pennyworth of Peace." Opium was also used as an ingredient in confectionery, and to it was generally attributed a vast range of beneficial psychological effects.[73]

The 19th century represented the opiates' boom years, with per capita consumption increasing 10-fold from start to finish, boosted firstly by the nascent pharmaceutical giants such as Merck KGaA, who began to market morphine, and secondly by the invention and introduction of the hypodermic needle in 1853.[73] It was also a century of insidious drug dependence. In 1888 it was recorded that 15% of the 10,000

d. Dover was originally a privateer, and his exploits before retiring to become a doctor included rescuing Alexander Selkirk, the model for Robinson Crusoe. His Dover's Powder combined opium with syrup of ipecac (*Carapichea ipecacuanha*), a South American plant with emetic and expectorant properties.

prescriptions dispensed by 35 pharmacies in Boston contained opium or derivatives, with this figure rising to 78% of the prescriptions that were repeated more than twice.[74]

Nor was it just opiates that were proving popular by the end of the 19th century. Somehow, since its discovery in the early 16th century, coca had been largely overlooked. In the late 18th century Spanish officials in the New World had even pressed, unsuccessfully, for coca to be cultivated and marketed as an answer to the exceedingly popular and hugely profitable tea and coffee.[61] However, it was not until the mid-19th century before the first coca-based commercial products appeared.[75] The most popular of these products, introduced in 1863, was a drink called Vin Mariani, manufactured by steeping coca leaves in wine, with the alcohol content working as a solvent for the cocaine. The resultant beverage became popular throughout Europe and the United States. However, by 1886 the competition was heating up with the introduction of "Coca-Cola," a temperance drink that had no alcohol but included a refreshing combination of cocaine from coca leaf and caffeine derived from the Kola nut. Coca-Cola was a roaring success, but it wasn't until 1903 that the cocaine was removed, although to this day the drink contains a non-narcotic coca leaf extract for flavoring. Products containing semi-refined cocaine continued to be sold over the counter in the United States until 1916 in forms ranging from tonics, through cigarettes, confectionery, and patent medicines to purely medicinal products. This last category included a number sold for the treatment of opium addiction. As the marketing blurb from the American cocaine manufacturer Parke-Davis warned rather temptingly, cocaine "could make the coward brave, the silent eloquent, and render the sufferer insensitive to pain".[75] The popularity of cocaine was bolstered further by a paper extolling the virtues of the wonder drug published in 1884 by Sigmund Freud,[76] who, as a well-known user of cocaine, was paid to endorse the products of both Parke-Davis and Merck, their European equivalent.[75,77] In addition to Freud's paper, new techniques to easily prepare semi-refined cocaine, and the publication of papers describing the anesthetic effects of cocaine, caused demand to skyrocket. Merck alone saw production soar from 3,000 to 158,000 pounds of cocaine between 1884 and 1886.[77] In the United States, nonmedical cocaine found increasing uses as an aid to productivity for manual laborers, who were often supplied with the drug by their employers, and its use drifted down the social scale. As its use swelled to include the lower strata of society and it became associated with urban criminals, youths, and African Americans, it also began to be seen as a social threat.[78] By the turn of the century cocaine's addictive nature was also becoming readily apparent, and the scene was set for the first moves toward prohibition, although nonmedical cocaine would remain freely available until 1916.[75]

A further development at the end of the 19th century also increased the eventual likelihood of prohibition. The acetylation of morphine, first carried out by the English chemist Alder Wright in 1874, allowed the resultant compound to cross the blood–brain barrier more rapidly than morphine, creating a characteristic "rush" of euphoria and increasing its addictive properties. Alder Wright was entirely unaware of this particular property of his new compound. It was only when it was synthesized independently by Friedrich Bayer and Co., just two weeks after they had first acetylated salicylic acid to create aspirin, that it became available as a drug.[69,73] It reached the market in 1898, under the trade name "Heroin," and was marketed initially as a cough remedy and nonaddictive substitute for treating morphine addiction.

Even as its medicinal worth began to be questioned, heroin began to be manufactured on an industrial scale in many countries. However, it was British companies such as Whiffens, T & H Smith, and Macfarlans that initially capitalized most on the new discovery, exporting vast amounts of heroin to feed the huge demand in China, which had been created in the first place by the opium smuggling of the British East India Company and the Hong Kong trading houses during the preceding 150 years.[73] Britain's behavior, and the economic jealousy of other nations at the lucrative opium trade, would eventually lead to the first international drug prohibition agreement, the International Opium Convention signed in the Hague in 1912[e] and brokered by the League of Nations.[74] There had been previous attempts to control drugs, including the Pure Food and Drug Act of 1906 in the United States, which controlled the labeling of patent and over-the-counter products, and which, combined with a public information campaign, led to a sharp contraction in the narcotics market. The subsequent Harrison Narcotics Act (1914) added further controls and introduced taxation of narcotics, although at a level that, in effect, introduced prohibition.[79] The Paris Convention of 1931[f] (and the later Single Convention on Narcotic Drugs in 1961) widened the participating countries, tightened controls, and broadened the definitions of drugs. Prohibition had arrived, and over time the list of proscribed drugs has been, and continues to be, widened. The rest, as they say, is history.

Judge James Gray, speaking about the United States, but with words that apply just as readily to the rest of the world, described the subsequent and ongoing drug prohibition as "the biggest failed policy in the history of our country, second only to slavery." It certainly has parallels with the United States' dalliance with the prohibition of alcohol between 1920 and 1933, which saw a boom in criminality, the emergence of organized crime, a tripling of the prison population, and an increase in the per capita consumption of liquor.[79] In the case of the "war on drugs" the annual worldwide turnover in illicit drugs had risen by 2011 to an estimated $320 billion.[79] If "illicit drugs" were a nation this figure would have made it the 30th largest economy in the world, according to United Nations figures. Globally, about 5% of adults use illicit drugs, and the levels of consumption generally increase along with the stringency of a nation's antidrug laws. So, for instance, in the United States, which has arguably one of the most punitive approaches to drug sales and possession, 8.9% of the population over 12 years old, which works out to an estimated 22.6 million citizens, admitted taking illicit drugs within the month prior to the 2010 *National Survey on Drug Use and Health.* This represented a significant increase over two years previously, but this was largely attributable to increased cannabis use. In the same survey an estimated 7.1 million of these drug-taking Americans also fitted the diagnostic criteria for substance use disorders as defined in the psychiatric manual DSM-IV.[80]

While having no effect on levels of drug taking, prohibition has had a number of intractable and self-perpetuating negative side effects: the mass criminalizing of otherwise law-abiding citizens (particularly in the United States[g]); the creation of huge

e. The supply of large amounts of heroin continued for several decades as loopholes in the laws of member nations were closed down and production shifted around the globe, finally settling in Japan at a point where they were intent on aggressive domination of their region and, indeed, the world. There was no need for illicit production until the end of the war and the subjugation of Japan.[73]

f. Convention for Limiting the Manufacture and Regulating the Distribution of Narcotic Drugs.

g. In 2009, out of every 100,000 adults in the United States a staggering 756 were imprisoned, the majority for drug-related offenses.

criminal enterprises; the corruption of people, institutions, and governments; murder on an industrial scale in some countries; the exit of vast amounts of currency; and the expenditure of phenomenal sums of money on drug enforcement (the United States alone spent $48 billion on drug enforcement in 2008).[79] However, as James Gray notes, throughout it all the estimated proportion of the population (of the United States) that are "addicted" per se to drugs has remained largely static since before any attempt at prohibition, when an estimated 1.3% of the population had accidentally addicted themselves to the many medicinal, patent, and over-the-counter opium and cocaine preparations.[79]

WHAT DOES IT ALL MEAN?

This history has been, by necessity, brief and selective. It barely touches on the many medicinal plants that modulate brain function less dramatically than our social and illicit drugs, simply because they tend not to figure so dramatically in the affairs of the human race. It also says nothing of the food plants and their component chemicals that promote brain function as a part of our diet. It does, however, illustrate several things. The relationship between humans and psychoactive plants (and less frequently fungi) is an enduring one, extending well into our prehistory, and it is a relationship that has literally shaped mankind's history. Even if we ignore the role of "entheogenic" deliriant and hallucinogenic plants in the very conception of religions, we can see clearly that plants containing psychotropic compounds, such as the opium poppy, tobacco, tea, and coffee, have driven many, often unsavory, aspects of global history. They have fostered the global predominance of a small subset of nations; fueled their economic power; and triggered forced colonization, the displacement of peoples, mass slavery, murder, and war.

Psychotropic plants continue to hold sway over us. While high-income Western nations have started the slow process of weaning themselves off tobacco, its consumption continues to increase steadily in developing nations: the World Health Organization estimates that over 1.2 billion people smoke tobacco worldwide, that tobacco accounts for 10% of all deaths globally, and that the habit will inexorably kill 50% of smokers prematurely. Tea and coffee represent the second and third most consumed drinks in the world (the most consumed being water), with gross production in 2010 in the region of 4.5 and 8 million tons of leaves or beans, respectively. The coffee house is resurgent and ubiquitous on the Western High Street. Starbucks alone has over 20,000 outlets and had a turnover in 2012 ($13.29 billion) that, had it been a country, would have placed it as the 121st largest economy out of the 184 countries monitored by the International Monetary Fund. And, of course, the "gross domestic product" of the "nation" of illicit drugs continues to rise relentlessly, creating a havoc of murder and lawlessness in whole regions of the world and fostering criminality in otherwise generally law-abiding societies.

We know a great deal about "how" plant chemicals have such a hold over us and about "how" they modify human brain function in terms of the physiological mechanisms involved. But the question of "why" plants have these effects on human brain function has been almost entirely ignored. What is in it for the plant? What drives the plant to synthesize chemicals that interact with the human brain? As well as describing "how" plants affect brain function, the rest of this book will also try to provide an answer, or several answers, to the "why" question.

2 Secondary Metabolites and the Life of Plants

Plants are the most successful form of life on earth. The plant kingdom comprises more than 300,000 species, 85% of which are flowering plants (angiosperms).[81] Courtesy of rich levels of specialization this kingdom has spread its members over the vast majority of the land, with plant leaves collecting the light falling on 75% of the terrestrial surface of the earth. Plants have colonized all but the most inhospitable of climes, surviving in a range of temperatures from a freezing Siberian –56° C to over 40° C in deserts.[82] As a family plants make an exponentially larger contribution to terrestrial biomass, by volume and weight, than all other forms of life combined,[83] and they have made an incalculable contribution to the hospitality of the gaseous environment that supports life on earth as we know it.[82]

But plants have a problem: they are stationary, rooted in place, and they are autotrophs, feeding themselves by synthesizing the complex organic compounds, the carbohydrates, fats, and proteins they require for life, from simple inorganic molecules that they find in their immediate vicinity. As part of this process they harvest light to generate their energy requirements via photosynthesis. The resultant plant material forms the energy source, precursors, and essential nutrients for the vast majority of the other forms of life on terrestrial earth. Their lack of movement brings plants a number of specific challenges, including the engineering of their own pollination and seed dispersal, local fluctuations in the supply of the simple nutrients they require to synthesize their food, variations in ultraviolet light and hydration, and the coexistence of herbivores and pathogens, intent on their destruction, in their immediate environment. Plants have therefore evolved a number of characteristics that facilitate survival. Many of these adaptations are physical: root systems that collect nutrients, leaves that collect light, flowers that attract pollinators, and physical barriers such as tough coats, spines, and bristles that dissuade herbivores. However, they have also evolved biochemical pathways that allow them to synthesize a raft of chemicals. These phytochemicals (i.e., plant chemicals) have been called "secondary metabolites" because they don't play any direct role in the "primary" metabolic processes that underpin the plant's minute-by-minute, day-by-day survival, growth, development, and reproduction. However, they do increase the plant's overall ability to survive and overcome local challenges by allowing it to interact with its environments in a multitude of ways.[84] Some of the roles of secondary metabolites are relatively straightforward; for instance, they play a host of general, protective roles, for example as antioxidant, free-radical scavenging, ultraviolet light-absorbing and anti-proliferative agents; and they defend the plant against microorganisms such as bacteria, fungi, and viruses. They also manage inter-plant relationships, acting as allelopathic defenders of the plant's growing space against competitor plants. More complex roles include dictating or modifying the plant's relationships with other complex organisms.[84–86] Their primary role here is often viewed as being one of feeding deterrence, and to this end many phytochemicals are bitter and/or toxic to potential herbivores, with this toxicity often extending to direct interactions with the herbivore's central and peripheral nervous systems.[87] However, equally importantly, plants also have to foster a number of symbiotic relationships in order to survive. The most obvious role here is attraction of pollinators and other symbionts via colors, scents, and chemical signals, or the provision of indirect defenses for the plant by attracting the natural enemies of their herbivorous attackers.[84–86]

Most of these more complex chemical interactions involve the second most dominant life form on earth; insects. This subgroup of the invertebrate arthropod clan numbers a prodigious 1 million species, representing well over half of all of the species of life identified on earth to date.[88,89] Insects typically live within the comparatively warm and humid microclimates, rich in chemical emissions, that surround plants, and nearly half of all insect species are herbivorous, with the remainder living courtesy of either a direct symbiotic relationship with plants or by predation on other herbivorous insects and animals.[90] On the other side of the coin 85% of angiosperm species, and therefore more than 70% of all species of plants, are entirely reliant on symbiotic insect interactions for pollination.[91] Not surprisingly, plants and insects have co-evolved in terms of physical and chemical diversity over their 400-million-year common history.[90,92]

In contrast to the pivotal role of insects in the life of plants, and therefore the Earth, it is notable that vertebrates make up a mere 4% of species on the planet and are physically outweighed by insects by a factor of 10 to 1 across temperate areas of the earth.[83] Whereas insects are essential to life on earth, vertebrates are not. As Grimaldi and Engel[91] note when discussing the essential pivotal role of insects in the Earth's biosphere, "Remove all vertebrates from earth, by contrast, and ecosystems would function flawlessly (particularly if humans were among them)." In these terms humans have been inconsequential to the plant kingdom until the very recent past, and certainly until the advent of agriculture some 12,000 years ago, with the ensuing deforestation and transformation of the earth's surface.

And yet, a wide range of plant secondary metabolites modulate the functioning of the human brain. Most of these chemicals fall into the three most abundant groups of secondary metabolite structures: the phenolics, the terpenes, and the alkaloids. Indeed, on the basis of the previous chapter, humans as a species could be described as being in thrall to a number of these chemicals. To start to unravel why plant chemicals have diverse effects on the human nervous system, it is probably best to start at the beginning.

■ THE EVOLUTION OF PLANTS AND INSECTS

If we wind back the clock 465 million years we would find an Earth of contrasts. On one hand the land was barren, sterile, devoid of life save an occasional crust of microbes, and, on the other hand, the seas teemed with the complex life forms that had unfolded in the preceding 100 million years. This "Cambrian explosion" of marine diversity had seen the basic architecture of the major animal phyla being constructed, including the supremely successful arthropod body plan that would eventually provide us with insects.[93] At this point green photosynthesizing algae had already existed in the sea for half a billion years and floating algal mats clung to the shorelines. However, now they were making their first tentative incursions onto the land, colonizing the shorelines and edges of the oceans, in the newly evolved form of the bryophytes, small, nonvascular plants. It would be another 40 million years before plants began to spread further inland, but then, in a dazzling 65-million-year burst of evolutionary activity (from 425 to 360 million years ago), plants diversified.[82] The vascular plants appeared courtesy of the ability to synthesize lignin, a complex polymer that toughens cell walls, providing both the structural strength required for increasingly complex physical structures and a protective layer against desiccation. The plants could now start to move away from the water and colonize drier areas.[94] For the first 30 million years of this ensuing period the shrubs and small trees were leafless, with photosynthesis being carried out in their bare branches and stems. Gradually, over the next 10 million years at least four plant groups, the extinct genus *Archaeopteris* and the ancestors of modern horsetails, ferns, and seed plants, independently developed leaves. These unfolding solar panels, developed courtesy of the homeobox gene networks that had driven evolution across phyla and that were now activated in these plants by the stress of falling global carbon dioxide levels, dramatically increased the ability of plants to capture sunlight and their photosynthetic capability.[82] The next major development was the divergence of the gymnosperms, seed-bearing plants that are still common today in the guise of conifers, cycads, gnetales, and the ginkgo tree. This taxon subsequently became the dominant plant family, making up some 60% of global flora by 260 million years ago.[95] It was only comparatively recently, around 145 million years ago, that the world's currently dominant group of plants, the flowering angiosperms, emerged from within the gnetales. It was only after the drawn-out evolution of first the carpel, then double fertilization, and finally the flower itself, that the angiosperms were poised for an explosive diversification and radiation. This epoch (described by Darwin as an "abominable mystery") saw all of the major lineages of angiosperm plants established within the 10 million or so years after the first entry of the angiosperms into the fossil record.[96] This period was then followed by the most productive period of speciation in the history of plant life. Between 115 and 90 million years ago structural innovations originated independently in separate clades, and the early small, weedy shrubs and herbs gave way to the full diversity of angiosperm forms seen today.[97] As part of this process the small, structurally simple flowers of the early angiosperms, which must have been pollinated opportunistically by a wide variety of insects, gave way to larger, more complex and individual flowers whose features were molded by the plant's relationship with specialized insect pollinators.[96] As color was added to the green background of the global flora, the angiosperms colonized and attained ecological domination across the full spectrum of environments, habitats, and regions throughout the world.[91,95]

Naturally the plants were not alone in their colonization of the Earth. A number of arthropod species, such as the arachnids and myriapods, migrated from the sea into the nascent, miniature plant life at the water's edge. The first six-legged hexapods are preserved in the fossil record from some 410 million years ago.[91] Their appearance heralds the emergence of the insects, a taxon of animals that would coexist and co-evolve with plants almost from the outset, sharing the plants' journey as they colonized the Earth. These two taxa would remain largely alone on land for tens of millions of years. It would be another 50 million years before the first amphibious vertebrates began to inhabit the margins of the water, and a further 50 million years again before the reptiles emerged.[98]

From the outset the fossil record captures evidence of the many complex and specialized relationships between plants and insects, with evidence of insects first feeding on plants by piercing and sucking (circa 400 million years), with a subsequent progression through chewing to mining and galling over a period of more than 100 million years.[90] This evidence shows that the pattern of plant damage also changed over time, with continuous marginal feeding on plant material giving way to smaller areas of more selective feeding damage that may reflect the evolution of specialized mouth parts by the insects, or alternatively the development of chemical defense strategies by the plant.[99] While the causal relationship underlying the diversification of plants and insects is difficult to resolve, it is notable that the diversification of insects coincided with the increase in the size, complexity, and variety of plant architecture. For instance, flight may well have developed in response to the evolution of upright shrub-like plants.[100] Certainly insects, once airborne, were alone in the skies. It would be another 170 million years before the first vertebrates would fly, and as atmospheric oxygen levels increased, so did insect size, with wingspans reaching 70 cm during an 70-million-year (320 to 250 million years ago) spike in oxygen levels, before they shrank again as oxygen levels fell and predatory birds and bats eventually appeared.[101]

The greatest period of speciation of insect families was still to come. It followed the explosive radiation of the angiosperms, with their physical variety, architectural complexity, and multiplicity of new ecological roles. Half of our current insect families, including the speciose *Lepidoptera, Coleoptera, Diptera,* and *Hymenoptera* orders, either appeared or increased dramatically in terms of constituent families during this period. However, this process of insect radiation had also started for several other orders of insects some 100 million years before the emergence of the angiosperms,[90] and it may be that the increased herbivory associated with insect diversification provided the selective pressures for the evolution of the angiosperms.

Having said this, the co-evolution of plants and insects is actually seen most clearly in terms of biotic pollination. Some 85% of angiosperm species, and therefore more than 70% of all species of plants, are pollinated by insects. Prior to the angiosperms, plants were largely wind-pollinated, with the exception of the later gymnosperm gnetales, which exuded sticky pollination droplets that attracted unspecialized insects, including flies, butterflies, beetles, and wasps.[91] The earliest specialied clade of pollinating insects identified to date were the Eurasian scorpionflies (order *Mecoptera*), who were equipped with a fluid-syphoning proboscis and probably fed on nectar-like ovary fluids from gymnosperms some 30 million years before the emergence of the angiosperms.[102] These symbiotic, specialized, and unspecialized pollination relationships may also have provided the selective pressure for the first angiosperms.[91] Certainly, during a 30-million-year period starting 120 million years ago, the physical features that facilitated insect pollination evolved. These ranged from simple sepals and petals to the complex floral characteristics, designed to deliver nectar and pollen most economically to specialist insects, that are typical of the pollination systems seen in contemporary plants. Along with this, insects evolved characteristics concordant with the symbiotic relationships of pollination, including appropriately modified mouthparts (e.g., a proboscis of the correct length), pollen-collecting mechanisms, and wings suitable for hovering. Nectar robbers would hopefully be disbarred by the lack of the same structural features.[103,104] Attraction and pollination strategies were therefore a major factor in the diversification of plant and insect genera during the angiosperm radiation.[104] For instance, evidence from the fossil record and from molecular phylogenetic studies suggests that the 130 million-year-relationship between flowering plants and bees (which today number 20,000-plus species and are the single most important genera for pollination) has fueled their respective diversification as they have undergone a series of parallel "explosive" co-radiations.[105,106]

Of course, pollination is not just about the physical structure of flowers; its evolution, and that of the angiosperms, also required the elaboration of the synthesis of a host of secondary metabolites, including the pigments that provide color to flowers, volatile chemical emissions that signal the presence of the flower to insects, a raft of nectar components, and an arsenal of defense chemicals to ward off the rapidly radiating herbivorous clades of insects.

■ THE EVOLUTION OF THE SYNTHESIS OF SECONDARY METABOLITES

Land plants as a collective group of 300,000-plus species synthesize and employ less than 10,000 separate "primary metabolite" chemicals. These compounds carry out the many biosynthetic and metabolic processes necessary for the plant to survive on a minute-by-minute or day-by-day basis and complete its life cycle. In comparison, a conservatively estimated 200,000 or more secondary metabolite compounds are employed across plant species.[107] These chemicals are not involved in primary metabolism, but rather they carry out ecological roles for the plant, providing protection from biotic and abiotic stressors and facilitating the plant's multifarious interactions with its environment. The disparity in numbers between primary and secondary metabolites is so wide because the synthesis and function of individual primary metabolites originated in the common unicellular ancestors of all plants, and they are therefore common across all plants. In contrast, the diversification of secondary metabolites has proceeded alongside the speciation of plants, with synthesis of the majority of secondary metabolites being of relatively recent origin. So, a single angiosperm plant may be able to synthesize in the region of only 1,750 to 3,500 secondary metabolites, but different plant lineages have evolved the capacity to synthesize qualitatively and quantitatively different sets of compounds.[107] As examples, the nightshade (*Solanaceae*) family of plants employ alkaloids (tropane and steroidal) as their defense weapons of choice (see Chapter 7), whereas the mint (*Lamiaceae*) family synthesize a wide range of terpenes with diverse ecological functions, including defense (see Chapter 11). Generally, the closer the relationship between plant species, the more closely they rely on a similar selection of chemicals.[86]

Naturally, the process underlying the expansion of the pool of secondary metabolites, and the establishment of the synthesis of chemicals with specific functions, is natural selection. This fundamental evolutionary process is fueled by random genetic mutations that give rise to novel variations within the population in terms of their fitness to survive. The central mechanism underpinning the creation of evolutionary novelty is the process of gene duplication, whereby new copies of existing genes are created by a number of processes, including unequal crossing over during meiosis and the duplication of entire chromosomes and genomes. Once created, the duplicate genes can remain functionally unchanged, mirroring the function of the original gene, or they can fall silent, becoming a pseudogene. Alternatively, they can either begin to share the functions of the original gene, providing an opportunity for refinement of the function (sub-functionalization), or they can diverge in terms of function, free from functional constraints, as the original gene continues in its ancestral role (neo-functionalization).[108]

These gene duplication events, and the subsequent refinement or diversification of function, have been particularly common in plants. As an example, evidence suggests that a whole-genome duplication event, dated approximately 320 million years ago, left the ancestor of all gymnosperms with spare genetic material. This in turn may have allowed the development of seeds as a mode of reproduction. A subsequent whole-genome duplication event in a member of the gymnosperm division of gnetales, dated approximately 200 million years ago, provided the spare genetic capacity for the increased biological complexity and the diversification of regulatory genes underlying the eventual development of flowers, ushering in the globally dominant angiosperms.[109] We can see these duplications in single plant species. For instance, the genome of the angiosperm *Arabidopsis thaliana*[a] shows evidence of both the whole-genome duplication event that preceded the diversification of the angiosperms[109] and two further major genome duplication events, followed by subsequent loss of many of the extra genes.[108]

In the case of secondary metabolites, any direct mutation of the original genes underpinning primary metabolite synthetic pathways would probably have drastically reduced the plant's chance of survival. However, the various duplication events provided spare genetic material that could mutate freely, without endangering the plant. In very rare cases a mutation in the duplicates of the genes underpinning a synthetic pathway could potentially trigger the synthesis of a range of novel chemicals that were slightly different from, but related to, the original primary metabolite(s). If one of these novel chemicals increased the fitness of the plant to survive, then natural selection would see that the new chemicals were retained.

It is notable that only a small proportion of the several thousand individual secondary metabolites found in the average plant have been shown to have bioactive properties that would seem to readily increase the synthesizing plant's fitness. Even groups of related secondary chemicals that share common synthetic pathways often feature one dominant bioactive compound, or a few, along with a greater number of apparently inactive compounds. For instance, *Cannabis sativa* synthesizes more than 60 separate cannabinoids via a common pathway, of which the principal bioactive compounds, Δ9-tetrahydrocannabinol (THC), cannabinol, and cannabidiol, are by far the most abundant. Many of the remaining cannabinoids have no apparent bioactivity.[110] Correspondingly, salvinorin A is the most abundant of a range of neoclerodane diterpenes from *Salvia divinorum*,[111] and morphine is the most abundant among 40 similar alkaloids synthesized by *Papaver somniferum*.[29] In both of the latter cases, as with the cannabinoids, many of the related chemicals have little or no known bioactivity. This certainly lends support to the notion that selection operates at the level of synthetic pathways, rather than individual chemicals, and that the evolution of an enzymatic pathway capable of economically generating large numbers of related, novel molecules is more likely to increase the potential fitness of the plant than synthesis of a single novel chemical.[112] By this account a single genetic mutation that changes the function of a single enzyme may provide one or more novel, slightly modified chemicals, which will then become an additional substrate for existing, promiscuous enzymes, which will then catalyze a whole range of additional novel chemicals.[113] Only one of these chemicals has to provide an adaptive advantage, provided this isn't outweighed by any disadvantages inherent in the synthesis of the entire group of novel chemicals. If it survives the early stages of natural selection, the novel molecule(s) may be sculpted further, in a process described as "evolutionary molecular modeling," which will maximize its adaptive bioactive properties.[86]

a. Arabidopsis was the first plant to have its genome sequenced and remains the "model" plant for much biomolecular research in plants.

We can see the process of gene duplication followed by the elaboration of novel biosynthetic pathways at work in examples drawn from each of the major groups of secondary metabolites. So, for instance, with regard to alkaloids, in the nightshade (*Solanaceae*) family (Chapter 7), the gene for the essential primary metabolism enzyme spermidine synthase, which converts putrescine to spermidine, was duplicated. The spare copy of the gene, free from selective pressure, evolved a slightly modified function coding for a novel enzyme that catalyzed the conversion of putrescine to the new compound, N-methylputrescine, shortly after the diversification of the angiosperms. This novel compound then provided the starting substrate for the synthesis of the large group of secondary metabolite tropane alkaloids.[114] Similarly, in at least three separate lineages of plants, the duplicate of the gene coding for the enzyme deoxyhypusine synthase developed a new function, coding for homospermidine synthase, and this became the first unique enzyme required for the synthesis of the entire pyrrolizidine group of alkaloids.[113]

In the case of terpenes, many of this group of chemicals function as primary metabolites, and the underlying synthetic pathways are common to all eukaryotes. The entire roster of plant terpene synthases can be divided into seven genetically distinct subfamilies that were created by successive instances of gene duplication, followed by neo-functionalization and sub-functionalization, during the speciation of plants. So, the two most ancient of these subfamilies synthesize the terpenes involved exclusively in primary metabolism across all plants: the remaining five subfamilies synthesize terpenes that have purely secondary metabolite properties. One of these subfamilies is found in the primitive lycophytes, one is restricted to the gymnosperms, and the remaining three are found solely in the angiosperms.[115] This suggests that the gymnosperm and angiosperm terpene synthase subfamilies developed comparatively recently—after the divergence of the angiosperms—and subsequently both groups of plants independently evolved monoterpene and sesquiterpene synthases from duplicates of the more ancient diterpene synthases. It is notable that any one given synthase has been shown to be capable of synthesizing multiple different terpenes from a single substrate,[115] so the particularly rich palette of angiosperm terpene synthases provided them with the potential to synthesize an extensive arsenal of novel secondary metabolite terpenes.[116]

Likewise, in the case of phenolics, the enzymes that underpin the synthesis of stilbenes, chalcones, and flavonoids evolved a single time in an ancestor of the land plants, from duplicates of the "primary metabolite" enzymes that control polyketide and fatty acid synthesis across eukaryotes. The irregular distribution of genes for these secondary metabolites within plants was then dictated by genus-specific redundancy and diversification.[117]

Naturally, the evolution and diversification of secondary metabolite synthesis has been driven by the need to overcome ecological challenges. Indeed, the very existence of plants as the dominant form of life on land may be attributed to the evolution of two separate secondary metabolite biosynthetic pathways that enabled plants to adapt to the terrestrial environment. If we return to the sea maybe 500 or 600 million years ago we would find a rich soup of multicellular algae living safely in an environment of buoyant water that filtered out damaging ultraviolet light from the sun. At the edge of the sea some of these algae were being rhythmically washed up onto the barren shoreline. Any algae, or their haploid spores, stranded on the shoreline would have fallen victim to the increased ultraviolet B (UV-B) radiation. This provided an ideal population for Darwinian evolution. At some point an ancestor of the nonvascular bryophyte plants evolved the ability to synthesize the phenolic chalcones and flavonoids.[94] These compounds are integral to plants' responses to many stressors, but for the pioneering plants their primary function may have been in absorbing light in the UV-B spectra and, in the case of the subgroup of flavonols, regulating cellular levels of oxidative stress, allowing survival on land.[94,118] Armed with these new protective chemicals the bryophytes began to colonize the shorelines. The second set of pivotal, novel biosynthetic pathways created lignin, a tough polymer made of phenolic compounds that is incorporated into the plant's cell walls. This process involved the evolution of multiple enzymes, all of which were derived from closely related primary metabolism enzymes.[94] Lignin provided structural strength and therefore facilitated bigger and stronger plants and trees. It also allowed the development of a vasculature, and it provided physical protection against desiccation and pathogens. With these modifications the resultant vascular plants (tracheophytes) became free to propagate away from the wet haunts of the bryophytes near water. They went on to colonize the majority of the planet.[94] The evolution of lignin also had another fortunate consequence. Lignin makes up approximately 30% of the planet's biomass, much of it in the form of wood, but it also degrades very slowly; as such, it is a major sink for carbon. The explosion in lignin synthesis and its subsequent interment underground, where it would eventually form fossil fuels, served to reduce CO_2 levels while increased levels of photosynthesis augmented oxygen levels. The atmospheric conditions of the earth became more favorable for the development of the leaf and the huge diversification of life.[82,94]

■ THE CO-EVOLUTION OF SECONDARY METABOLITES AND INSECTS

To the plant the insect represents both the major threat, in terms of potential herbivory, and also the answer to a number of key ecological problems, most notably the need of the dominant angiosperms to pollinate themselves efficiently and economically. Much of the huge diversity in secondary metabolite synthesis can therefore be directly attributed to the co-evolution of plants and insects. This relationship is seen most clearly in the evolution of the vast array of defense chemicals employed by plants to deter herbivores, and the corresponding speciation of herbivorous insects. Ehrlich and Raven[92] first noted the ongoing relationships between groups of phylogenetically related insects with groups of phylogenetically related plants and proposed that insect/plant co-evolution was as a consequence of what they termed a reciprocal "arms race," typified by alternating adaptive radiations of the two taxa. In this scenario a genetic mutation initially confers protection on the plant species from its insect herbivores by allowing the synthesis of a novel chemical that deters feeding, giving the plant an adaptive advantage over other plants and allowing it to thrive and colonize territory. Subsequently a genetic event allows one insect species to re-escalate herbivory on the plant, for instance via the detoxification of the chemical. Without competition from other herbivores, the insect is then free to feed and prosper throughout the range of the plant. This process then repeats itself, with the plant's palette of defensive secondary chemicals becoming more complex or potent over time as insects adapt by developing novel ways to detoxify or tolerate plant material. As an example, a wide range of specialist and semi-specialist insects from a number of orders, including the *Coleoptera*, *Hymenoptera*, and *Lepidoptera*, feed on plants that have evolved the synthesis of defensive pyrrolizidine alkaloids. These insects, in turn, have evolved the ability to actively sequester the toxic alkaloids, accumulating them in their external cuticles and wings as a defense against predation by other insects, birds, and mammals. A number of male moths and butterflies have even evolved the equally useful trick of slightly modifying the alkaloid for use as an airborne pheromone, which is wafted at females during courtship, indicating the protective load of alkaloids that the male will endow to the eggs during the mating process.[119]

Overall the "arms race" or "phytochemical co-evolution" theory of plant/insect co-evolution finds support in the observation of a high proportion of specialized insect feeding and the comparative toxicity of specific defense chemicals for generalist, as opposed to specialist, insect herbivores.[120] There are also many examples of a close correspondence in speciation between insect genera and their host plants as assessed through cladogram and molecular clock analyses.[90] The theory also leads to the prediction that secondary chemicals with more recent origins should be less widespread and more effective than those with longer provenance, which indeed can be the case.[121] Similarly, the "enemy release hypothesis" suggests that invasive plant species from a different geographical location will benefit from a lack of evolved tolerance in insects to their arsenal of secondary metabolites, and subsequently a comparative lack of herbivores. This can indeed be the case with, for instance, *Hypericum perforatum* suffering less herbivory in introduced areas than in its home range.[122] However, the exact nature of the plant/insect chemical relationship in evolutionary terms has been debated extensively, and the key questions revolve around whether the co-evolution is symmetrical, as envisaged by Ehrlich and Raven, or asymmetrical, whereby plants evolve and exert unidirectional evolutionary pressure on insects, but not vice versa.[90,123] While a number of competing theories have been proposed, there is little disagreement over the central tenet that co-evolution, in one form or another, took place. The theories also share one other common feature, the notion that the key relationship is that between insects and plants,[121] with little weight given to the possible pressures exerted by other herbivores.

Much of the discussion around insect/plant co-evolution has centered on defense chemicals and counter-defenses. However, pollination by insects, rather than an "arms race," drove the rampant diversification of the angiosperms,[124] and this, understandably, coincided with a similar diversification of insects.[90] The evolution of the insect/plant relationship can be readily observed in the development of physically complex flowers and matched pollinators. However, this evolutionary process also included the elaboration of a number of non-defensive secondary metabolites, including those contributing to color, nectar, and, most importantly, chemical attraction. It may be the case that the first chemical attractants emitted by plants were the volatile components of cocktails of defensive chemicals that had evolved to deter feeding on reproductive organs, ovules, and pollen. Emitted at levels too low to defend the plant, these components may have been inadvertently used as a signal indicating the presence of nutritious food by passing insects. The insects may have subsequently transferred pollen between flowers, providing a service for the flower that increased its fitness.[116] Certainly many of the volatile attractants emitted by flowers, for instance the low-molecular-weight monoterpenes and sesquiterpenes, have dual purposes as attractants and defense chemicals, with the specific function dictated by the tissue, mode of delivery, and quantity delivered (see Chapter 10).

The volatile terpenes are ubiquitous in complex floral scents,[125] and curiously, the majority of the most commonly encountered plant monoterpenes and sesquiterpenes are also synthesized endogenously by insects, who use them as interspecies pheromones and intraspecies allomones; chemical communication signals with functions in mating, aggregation and recruitment, trail making, signaling alarm or danger, and marking sources of food. This is no coincidence: the cocktail of volatile organic compounds emitted by insect-pollinated angiosperms has been shown to be closely matched to those synthesized by pollinating insects in general, and their own clades of specialist pollinators in particular. However, the volatiles emitted by wind-pollinated gymnosperms do not match those of pollinators but do match those of herbivorous insects.[126] Less direct evolved relationships also exist, with, for instance, orchids synthesizing insect alarm pheromones that function as homing signals for predatory hornets who visit the flower looking for prey and unwittingly take part in pollination.[127] Similarly, sexually deceptive orchids mimic the sex pheromones of female insects in order to lure the male of their specialized species into the flower for pollination purposes.[105] As the synthesis of terpenes was an ancestrally endowed ability in both plants and arthropods, but the need for attractant chemicals in plants evolved only with the angiosperms, it seems likely that the synthesis of volatile terpenes by plants originally arose as a defensive mechanism. High levels of terpenes would certainly be able to interact deleteriously with existing insect sensory equipment, and, when attractants became necessary, the same chemicals, so readily detected by insects in a social context, could take on the dual functions they enjoy today.[126]

Naturally, the close correspondence of terpenes between clades of plants with their partner clades of insects,[126] and the frequently observed specificity of bouquets of attractant chemicals to one clade of insects,[128] again suggests close chemical co-evolution between insects and plants. One marker that illustrates the level of chemical co-evolution of the two clades is their respective complements of cytochrome P450 (CYP) enzymes. This superfamily of hemoprotein enzymes is ubiquitous across all life forms, where its member enzymes function as catalysts in the synthesis and metabolism of biochemicals. Given that the common ancestors of plants and animals had a very limited number of CYP enzymes, the number and diversity of these enzymes in contemporary taxa of organisms give an indication of the evolutionary history of the organism in terms of chemical complexity. Beyond roles in primary metabolism, plants' CYPs also govern the synthesis of secondary metabolites and the signaling molecules (such as hormones) that trigger their production. In insects they manage the many chemical relationships with plants, including detoxification of defense chemicals.[129] So, in the case of flowering plants an original complement of 11 ancestral genes common to all plants has been expanded, via genetic duplication and diversification of function in the resultant "spares," to a complement, for a typical single plant, of approximately 300 cytochrome P450 genes, clustered into approximately 50 families of genetically related enzymes. Across all plants, over 5,100 cytochrome P450 enzymes have been identified, clustered in a total of 127 families. Similarly, insects as a group have an extensive complement of some 2,130 CYPs clustered in 69 families. In comparison to these chemically sophisticated taxa, vertebrates possess in the region of only 1,460 CYPs in total, clustered in a modest 19 families.[130] These comparative figures tell us several things. First, plants, as sessile organisms, rely on a highly evolved synthesis of diverse chemicals for their success, and insects (but not vertebrates) similarly live in a world dominated by chemicals, both in terms of their own pheromone communications and their chemical interactions with plants. Given that CYPs are integral to the synthesis of secondary metabolites, the complexity of the CYP profiles in plants and insects hints at the factors that have driven the evolution of secondary metabolite synthesis in plants.[129]

■ THE ROLE OF SECONDARY METABOLITES IN THE LIFE OF PLANTS

Defending the plant against herbivores and attracting pollinators are just two of many ecological roles that secondary metabolites play for the plant; in reality these chemicals are intrinsically involved in every interaction that the plant has with its environment, in many ways compensating for its lack of movement. The specific and differing palettes of roles played by the major groups of secondary chemicals—the alkaloids, phenolics, and terpenes—will be discussed in greater detail in later chapters, but for now a brief synopsis of the broad ecological roles played by secondary metabolites will be useful. In the following section these are broken down into ecological functions related to defense against herbivores, attraction, interactions with other plants, interactions with microbes, and responses to microbial infection and abiotic stressors. Figure 2.1 summarizes the range of ecological roles played by secondary metabolites.

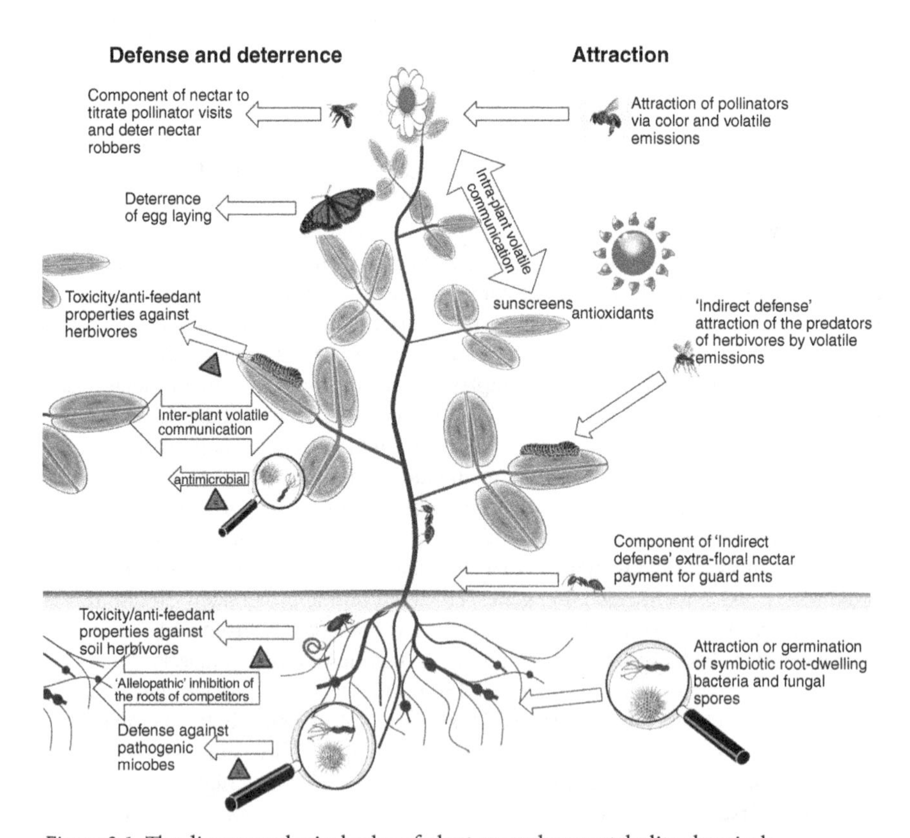

Figure 2.1 The diverse ecological roles of plant secondary metabolite chemicals.

Defense Against Herbivory

Toxicity

All three major groups of secondary metabolites play roles in the defense against herbivory. For the alkaloids this is their primary role, and they provide the primary form of defense for approximately 20% to 30% of higher plants.[131] In contrast, terpenes are typically multifunctional, with their defensive properties dependent on the tissue, mode of delivery, and concentration, whereas phenolics, as a group, are comparatively benign; where they do provide defense against herbivory it tends to be a secondary role. However, there are also plenty of exceptions proving this rule. In general, as noted above, a series of related compounds from one chemical group tends to predominate in the defensive capability of any given plant, with the nature and concentration of the compounds depending on individual plants, species, and populations.[86,132]

The synthesis and accumulation of defense chemicals can be dependent both on the age of tissue and its importance in terms of reproduction; young, growing tissue and fruits, seeds, and flowers, and the most accessible outer layers of the plant, are typically rich in defense chemicals.[132] In the case of many alkaloids, synthesis takes place in distant parts of the plant, with the compounds then actively transported to target tissue. For instance, tropane alkaloids are often synthesized in the root but then transported to herbivory-prone aerial tissue for storage.[114] Synthesis of defensive chemicals can also be constitutive, with the plant retaining a constant background level of the chemical, or it can be induced by a number of triggers,—for instance, the herbivore physically damaging or secreting saliva into plant tissue, laying its eggs, or simply moving about on the plant's surface, thereby disturbing sensory outgrowths such as hairs or scales, which trigger the transcription of defense-related genes and the synthesis of defense compounds.[132,133] In some cases wounding leads to the rupturing of separate compartments containing the defense chemical, safely stored in an inert form, and the enzymes required to activate it.[132] An alternative strategy is to either store or directly synthesize the defense chemical in specialized structures that, in effect, externalize the chemical, preventing autotoxicity. The most common examples are glandular trichomes, which might comprise a basal cell(s) that acts as a "chemical factory",[134] a stalk cell, and one or more bladder-like head cells that store the potentially toxic chemicals, safely sealed off from the vasculature of the plant.[135] Physical contact by a visiting insect then triggers the release of a sticky cocktail of chemicals.[134] These structures are particularly common in plants that use terpenes in a defense capacity, for instance *Cannabis sativa* and the mint (*Mentheae*) family. Similarly, laticifers, cells or groups of cells that exude latex when physically compromised, are found in approximately 10% of angiosperms and a variety of gymnosperms. As with glandular trichomes these structures deliver a "double jeopardy" response whereby the insect is physically compromised by the sticky exudate, which can immobilize/drown the insect or gum up its mouth parts, while at the same time the latex delivers a variety of toxic secondary metabolites. The most obvious examples are the opiate alkaloids such as morphine in the latex of the opium poppy. Latex also serves to seal the damaged area and delivers a variety of antimicrobial compounds.[136] In practice many defense chemicals, and indeed the concentration of storage structures such as glandular trichomes on the plant surface, are retained at constitutive levels and then increased by herbivory-related signals, generally via the actions of the jasmonate family of plant hormones (see below).[133]

Once encountered by the herbivore, defense chemicals exert their toxic effects via a number of interactions with molecular targets, including biomembranes, nucleic acids (DNA, RNA), and proteins.[132,137] Given the central role of proteins in every cellular process, this last category of targets includes many components that contribute both to a plethora of peripheral physiological functions and to neuronal function, including interactions with receptors for neurotransmitters/chemicals, enzymes that catalyze metabolism and the synthesis of neurochemicals, ion channels that regulate the electrical signals in cells and neurons, transcription factors that activate specific genes, components of the cellular cytoskeleton, and transporters that move ions and molecules across membranes.[132,137] DNA is a particularly vulnerable molecular target, and a range of secondary metabolites can interfere with the replication, transcription, and translation of DNA/RNA in microbes and animals. As an example, pyrrolizidine alkaloids can bind to functional groups of DNA bases, and a range of other alkaloids (berberine, sanguinarine, harmine, emetine, quinine, etc.) intercalate themselves between DNA base pairs, inhibiting replication and protein biosynthesis.[137]

Naturally, given the co-evolution of plants and their herbivorous neighbors, many insects have adapted to either tolerate or actively benefit from specific plant defense chemicals. For instance, more than 250 species of insects sequester defense chemicals from over 40 families of plants, with many of these either storing the chemicals for later secretion or incorporating them into their own tissue and bodily fluids in order to decrease their own palatability or increase their toxicity to predators.[138]

Hormonal Mechanisms

Insects synthesize a number of triterpene steroidal hormones, or ecdysteroids, that modulate their life course, metamorphoses, and social interactions.[139] However, they can't synthesize the steroidal nucleus of these hormones in sufficient quantities and therefore have to sequester the precursors cholesterol and sitoserol from their diet. As well as providing these precursors, a variety of plants also synthesize compounds that are identical or structurally similar to these insect hormones. In some cases these "phytoecdysteroids" (or phytoecdysones) seem to have the sole function of mimicking the effects of these insect hormones, disrupting the functioning of the insect's endocrine system,[123,140] whereas for other compounds this may be a secondary function. Either way, interference with the insect ecdysteroid system leads to developmental abnormalities and increased mortality.[141] As an example, phytoecdysteroids that mimic the insect molting hormone 20-hydroxyecdysone prevent the pupa from casting off its old cuticle, leading to death.[142] A number of isoflavones (a subset of the phenolic flavonoids) can also compete with estrogen for binding to vertebrate estrogen receptors, modulating hormonal reproductive parameters, such as the onset of estrus, uterine function, and genital formation.[139] However, it is unclear whether vertebrates are the intended target, as insects possess recently discovered orthologs of these same estrogen receptors.[143]

Attraction

Pollination

Approximately 85% of flowering plants are pollinated by insects. For instance, the ubiquitous, monophyletic bee family pollinates the majority of human food crops, and they have had an evolutionary relationship with flowering plants spanning 130 million years.[144] As with most pollinators the relationship is facilitated by the plant providing multimodal signals[145]; of these, color, scent, and the chemical components of nectar food rewards are directly provided by secondary metabolites.

The colors of flowers allow insects to identify appropriate target flowers, with the color of the flower typically matched to the color spectrum detected by the visual systems of the potential pollinator. Color itself is provided by a number of secondary metabolites concentrated in the epidermis of leaves or cuticle of flowers and fruits, including a wide range of phenolic flavonoids, in particular anthocyanins and anthocyanidins, which contribute rich colors that range from oranges through reds to blues and shades between, and flavones and flavonols, which contribute whites and yellows. The terpene carotenoids provide yellow and some red and orange colors. Insects also perceive ultraviolet light, and nectar guides advertising the direction to the food reward, invisible to the mammalian eye, are drawn in the petal in flavonols.[84] All of these pigments, synthesized as they are in the outer layers of plant tissue, are also multifunctional, acting as sunscreens, antioxidants,[146] anti-feedants, digestibility reducers, and toxins and antimicrobial chemicals.[147] Fruit color can also serve to attract larger animals involved in seed dispersal, which is often accomplished by initial ingestion, followed by expulsion of the undigested seed in a handy quantity of manure.

Most angiosperm orders release a complex bouquet of floral volatiles that are usually synthesized within floral organs themselves, with all of the requisite biosynthetic genes and enzymes expressed locally.[125] The scent usually includes a range of terpenes, phenolics, benzoic acid derivatives, and aliphatics, along with other minor components. In total, more than 1,700 volatile chemical compounds have been identified in floral scents to date.[125] However, monoterpenes feature in all plant emissions, and most of the individual chemicals that are commonly released by the majority of plants are monoterpenes or benzoic acid derivatives.[125] As discussed above, the most commonly encountered plant volatile chemicals are also synthesized endogenously by insects for use as pheromones and allomones. As an example, the monoterpene linalool is found in 70% of plant emissions.[125] At low concentrations linalool also serves as an endogenously synthesized aggregation hormone, which is released by bees to attract nest-mates, while at high concentrations it is used as a signal of distress.[126]

As a general rule floral scents are complex mixtures, and the specific functionality of the scent in terms of interactions with specific or general pollinators can be dictated by a single unusual compound, or mixtures and ratios of more common constituents.[128] The resultant scent can have both attractive and deterrent properties depending on the relationship between the insect (obligate vs. facultative) and plant.[148] Scents can also vary according to the fertility of the flower in terms of pollen ripeness, and by time of day. As an example, flowers that are pollinated by moths have a strong scent, but only during the hours of darkness.[84] The highly adaptive nature of floral scents can most readily be seen in deceitful plants, such as the many species of sexually deceptive orchids that trick pollinating wasps and bees into copulating with a female decoy flower by releasing a floral bouquet that includes the corresponding female insect's sex pheromone chemicals. These flowers are highly specialized and often include mechanisms for the temporary entrapment of the pollinator.[105,149] Similarly, many carnivorous plants, such as flytraps and pitcher plants, mimic floral attractants to lure their victims into their traps.[150]

Nectar is a sweet secretion produced by a diverse range of nectaries, either within flowers or "extra-florally," by any other plant surface, with the exception of the roots. It is made up of water and sugars, with additional amino acids and other diverse secondary metabolites. The exact chemistry has to fulfill two opposite functions: the attraction of mutualist insects (and sometimes hummingbirds and bats) and protection against microbial infections and the incursions of non-mutualist nectar robbers. In terms of attraction the sugars and amino acids are the key elements as they provide a food reward that can be modified according to the nutritive preference or requirement of the specific target pollinator.[151] However, nectar also contains volatile compounds such as alcohols, aldehydes, and terpenes, which provide it with its own independent attraction "bouquet." The exact constituents can be the same as the plant's floral bouquet, or unique to the nectar.[151,152] In terms of attraction, nectar plays two separate roles; floral nectar attracts and rewards pollinators, and extra-floral nectar attracts and rewards mutualist "guard" insects, such as ants and parasitoid wasps that live on or congregate near the plant, deterring herbivores and nectar robbers. In this role nectar secretion can be "constitutive," with secretion rates often diurnally matched to the pattern of activity of the mutualist, or it can be stimulated as a defense mechanism by herbivory or mechanical damage (via a jasmonic acid pathway – see below). Nectar can also contain potentially toxic levels of diverse secondary metabolites that might be tolerated by specialist pollinators but make the nectar unpalatable to nectar robbers.[151] This toxic load can also have a further benefit in that it can regulate the amount of nectar distributed to each pollinator. For instance, an increased nicotine level in nectar from the tobacco plant leads to more visits by pollinators but the removal of less nectar per visit, thereby increasing the efficiency of pollination.[152] Finally, many of the secondary metabolites in nectar, including terpenes and phenolics and "nectar proteins" or nectarines, have antimicrobial properties that protect the nectar from infection.[151]

Indirect Defense

The secretion of extra-floral nectar and the provision of structural features such as food bodies and domatia, chambers that serve as shelter or nesting sites for insects, encourage "guard" ants and other symbiotic insects to live on a plant. These physical traits are widespread and can be found on a third of woody plants.[153] Many plants also synthesize and emit a cocktail of volatile chemicals in response to leaf damage and egg-laying by a wide range of herbivores. This chemical information, in turn, attracts a variety of predatory insects, mites, nematodes (in the case of roots), and even birds, which then consume or parasitize the invading herbivores.[153] As an example, direct damage to the leaves of maize plants lead to the synthesis and emission of a number of green leaf volatiles (the chemical components that give green leaves their typical smell), while chemicals within the invading caterpillars' saliva provoke the synthesis and release of further volatile chemicals, predominantly terpenes and phenolics. The net effect of this bouquet of volatiles is the attraction of parasitic wasps that lay their eggs in the caterpillars.[154] Similarly, the act of depositing eggs on a leaf by insects can lead to the emission of volatiles that attract egg parasitoids.[155] In general the complex blend of volatiles depends on the genotype of the plant and the species and developmental stage of the herbivore. Carnivores and parasites can then use this information to discriminate plants being damaged by a specific herbivore.[156]

Experimental evidence certainly confirms the fitness benefits of resource-mediated indirect defense, with, for instance, the exclusion of specialist "guard" insects from a plant leading to greatly increased herbivory, and increased extra-floral nectar rewards leading to greater defense.[153,156] The many obligate ant–plant interactions also strongly suggest reciprocal natural selection and the possibility of co-evolution between specific plant and insect clades. However, the case for information-mediated indirect defense via the emission of volatiles is less strong at present, as these signals, by their nature, play multiple functions for the plant. Herbivory-related volatile emissions certainly attract predatory/parasitic insects, but this may simply reflect a learned response.[153]

Plant–Microbe Symbiotic Chemical Interactions

The unseen root system of the plant tends to be overlooked, but in reality the immediate vicinity of the spreading net of coarse and fine roots, termed the rhizosphere, is rich in secondary metabolites and chemical interactions.[157] As well as providing defense against soil pathogens and herbivores, these interactions include both the chemical management of relationships with mutualistic fungi (mycorrhiza) and bacteria that have symbiotic relationships with the plant predicated on mutualistic exchange of nutrients. For instance, in the case of the legumes (Fabaceae), flavonoids released by the plant into the rhizosphere induce nitrogen-fixing bacteria to synthesize and excrete "nodulation factors" into the soil and these, in turn, are detected by the plant, leading to the creation of root nodules for the bacteria to colonize.[158,159] Other flavonoids manage similar nitrogen-fixing relationships (but without nodulation) for other families of plants[147] and direct other aspects of symbiotic behavior by mimicking the bacteria's "quorum-sensing" chemical signals that govern local bacterial concentrations.[159] Flavonoids also contribute to the essential symbiotic relationship with the arbuscular mycorrhizal fungi that contribute to plant phosphorus uptake by colonizing the surface of roots in the vast majority of plant species.[147,159]

Plant–Plant Chemical Interactions

Naturally, one key factor dictating the success of any plant is the nature and extent of the community of plants in its immediate vicinity. Plants can gauge competition from a number of environmental signals, such as nutrient levels and the background ratio of red to far-red light, which indicates whether the leaves of neighboring plants are absorbing light for photosynthesis. They can then adapt their below- and aboveground growth correspondingly.[160] They can also engage in chemical interactions with neighbors. A rapid way for a plant to communicate the presence of a stressor to its own distributed aerial parts is to release volatile signaling compounds into the air, which can then be absorbed by its own distant leaves, triggering a physiological response throughout the plant.[161] Other plants in the immediate vicinity can also absorb the volatiles, triggering or priming the appropriate response to the same stressor. As an example, when undamaged lima beans are exposed to volatiles released by neighboring plants in response to beetle damage they increase their own secretions of extra-floral nectar, attracting the beetles' natural predators and priming a more vigorous response if they are subsequently also attacked.[162] This process can extend to a plant absorbing volatiles specific to a neighboring plant through its leaves, followed by their re-release, spreading increased resistance across the neighborhood.[163] While the adaptiveness of this mode of communication is clear for the recipient plant, the benefits for the emitting plant are less so, and the phenomenon may simply represent "eavesdropping," whereby only the recipient benefits, rather than this representing mutually beneficial communication.[164]

Allelopathy—Chemical Warfare

The other form of inter-plant chemical interaction has a rather more obvious benefit to the emitting plant, as it involves the poisoning of potential competitors.

The most straightforward way for a plant to capture adequate nutrients, water and light is to dissuade local competition from growing in its immediate vicinity. Many plants accomplish this by poisoning the surrounding soil with "allelopathic" secondary metabolites that interfere with the germination, growth, or development of local competitors. These allelopaths can either be emitted into the rhizosphere or leached into the soil from leaves and fallen plant tissue. Typically, these chemicals are relatively simple multifunctional terpenes and phenolics,[84] with concentration, as always, dictating the eventual effect. However, they can also belong to other chemical classes (for instance, this is one role of the alkaloid caffeine).

The classic examples of allelopathy are plants and trees that sit in patches of bare earth. These include the walnut tree, which was reputed by Pliny the Elder to have a poisonous shadow, but which actually secretes the phenolic chemical juglone (5-hydroxy-l,4-naphthoquinone) into the soil, preventing root elongation in competitors by interfering with gene expression in the relevant signaling pathways.[165] Similarly, sagebrush (*Salvia leucophylla*) roots exude a range of monoterpenes that have been shown to inhibit germination and constrain the growth of other competing plant species.[166] The characteristic pattern of vegetation in the Californian chaparral, featuring island sagebrushes in a sea of barren earth, is periodically reset by natural fires that sweep across the landscape, fed in part by the highly volatile terpenes in the soil. A more varied flora then establishes itself until the sagebrush reasserts its chemical dominance over a period of years.[84] Invasive plant species also provide some striking examples of allelopathy in action. For instance, Japanese knotweed (*Polygonum cuspidatum*), an aggressive invader species spreading across Europe and the United States, owes its domination of its environs partly to its ability to suppress the germination and growth of seeds of competitors by emitting a cocktail of chemicals, including volatile anthraquinones and a number of phenolic stilbenes and catechins, into the surrounding soil.[167–169] Similarly, the invasive (to North America) spotted knapweed (*Centaurea maculosa*) secretes catechins into the surrounding soil, and these inhibit the growth of competitor species by inducing oxidative stress and cell death in any encroaching roots.[170]

Of course, allelopathy isn't restricted to these extreme examples; many mundane plants, such as members of the wheat family (*Triticum* genus), use simple phenolic acids to suppress competition. These ubiquitous chemicals can influence nutrient uptake, plant morphology, enzyme function, and protein synthesis in competitor plants, potentially via a modulation of hormone levels and subsequent gene expression.[157,171] Allelopathy also doesn't necessarily involve direct interactions with competitors. As an example, garlic mustard (*Alliaria petiolata*) emits glucosinolates, which inhibit the growth of mycorrhiza fungi in the rhizosphere of neighboring plants, disrupting the mutualistic exchange of nutrients. A lack of defense by the native soil fungi against this novel threat contributes to the plant species' aggressive invasiveness when introduced into new environments.[172]

Defense Against Microbes and Other Stressors

Plants synthesize a wide variety of chemicals to protect themselves from microbial infection. It has been suggested that these compounds fall into two distinct categories: phytoanticipins, which are synthesized constitutively in anticipation of attack and perform as a chemical barrier, and phytoalexins, which are synthesized only in direct response to the detection of a microbial attack.[173] However, the same chemicals are often synthesized as part of a general defense and repair toolkit in the face of a wide range of abiotic stressors such as lack of water, increased salinity, leaf damage, heat, or ultraviolet light. Many of these secondary chemicals, particularly those expressed near the surface of the plant, are also multifunctional, with antioxidant and antimicrobial functions along with their primary function. For instance, anthocyanins primarily provide color but also function as antioxidants and antifungal agents.[84] This, of course, makes it generally difficult to disentangle the induced/constitutive and biotic/abiotic stressor drivers of chemical synthesis and ensures that both phytoanticipins and phytoalexins are usually present, at least at background concentrations, in all plants.[84] The multipurpose nature of many of these chemicals also means that plants exposed to one stressor, for instance the abiotic stress of ultraviolet light, are subsequently better protected against biotic stressors such as microbes and insect herbivores.[174]

The exact nature, in terms of chemical group, of a phytoalexin response is specific to the plant division or family and dictated largely by the nature of the secondary metabolites that the plant utilizes as a constitutive first defense. As this role most often falls to phenolics and terpenes, they are also the most commonly employed response to infection. The deployment of alkaloids is the exception rather than the rule, with phytoalexins from this group seen in only two families, the *Papaveraceae* and the *Cruciferae*.[175] In the case of the crucifers the many indole alkaloids are unusual in that they contain a sulfur atom derived from cysteine and are therefore related to the family-specific glucosinolates.[176]

■ A WORD ABOUT SECONDARY METABOLITE SYNTHETIC PATHWAYS

The synthesis of most secondary metabolites, even those synthesized "constitutively" as part of the normal metabolism of the unthreatened plant, is induced in response to signals from the environment. As we can see from the above, these environmental signals can include those related to a wide range of biotic and abiotic stressors, which, in turn, activate a number of interacting hormonal stress-signaling pathways in the plant. The most prevalent and best understood of these pathways is the jasmonate pathway. This hormonal system underpins the induced synthesis of diverse secondary metabolites, and floral and extra-floral nectar, in the face of abiotic stressors, physical wounding, microbial attack, herbivory, and ovipositioning.[114,151,156,177–179]

The jasmonate group of hormones comprise jasmonic acid, a metabolite of the fatty acid linolenic acid, and a variety of its conjugated and hydroxylated derivatives, such as methyl-jasmonate and jasmonoyl—isoleucine. The jasmonates contribute to a number of aspects of normal physiological functioning in plants, including many aspects of developmental plasticity, including flowering, organ development, and the growth and elongation of shoots and roots.[180,181] However, the most pertinent function of the jasmonates here is in the regulation of stress responses, and, in particular, the induced synthesis of secondary metabolites, including many terpenes, alkaloids, and phenolics. The jasmonates are synthesized when linolenic acid is released by chloroplast membranes in response to external stressors and is metabolized to jasmonic acid and derivatives via the octadecanoid pathway.[178,181] The jasmonate derivatives then function by binding to intracellular receptors (the F-box protein CORONATINE-INSENSITIVE 1). One component of this receptor complex is a protein (Jasmonate ZIM-domain [JAZ]) that has bound to and suppressed the transcription factors involved in stress responses. This protein is degraded, and the net result of this removal of inhibition is the activation of the transcription factors and subsequent changes in gene expression,[182–184] leading, among other effects, to increased production of defensive or protective secondary metabolites.

Naturally, the jasmonate signaling pathway does not act alone in this; indeed, it is merely one component of a complex signaling system that incorporates the plant hormones salicylic acid, abscisic acid, and ethylene, and the ubiquitous gaseous signaling molecule nitric oxide, among other factors. These various components interact via a complex "cross-talk" between pathways[185] that modulate, among other things, levels of JAZ proteins.[182] As an example, abscisic acid, ethylene, and salicylic acid do not generally directly induce secondary metabolite synthesis (with some exceptions) themselves, but they are all expressed as a consequence of stressors. Ethylene and abscisic acid, by and large, work in synergy with the jasmonates and tend to be particularly active in the face of herbivory, physical tissue damage, and necrotrophic pathogens. However, salicylic acid, which is a key mediator of immune reactions to biotrophic and viral microbial pathogens, tends to be antagonistic to the jasmonate and abscisic acid pathways—although paradoxically the opposite can also be true. Either way, all three hormones are intrinsic to the final makeup of the palette of secondary chemicals produced following activation of the jasmonate pathway, and their synergistic and antagonistic cross-talk fine-tunes the plant's response to match the stressor.[178,186]

There are several intriguing examples of an evolved tactic by microbes that take advantage of this cross-talk to unbalance the defensive response of the plant. For instance, the necrotrophic plant fungus *Botrytis cinerea* excretes into the plant a polysaccharide that specifically activates the salicylic acid pathway, thereby suppressing the jasmonate pathway.[186] In contrast, the bacteria *Pseudomonas syringae* introduces the chemical coronatine, a structural and functional analog of jasmonic acid, and in effect activates the plant's jasmonate pathways, which in turn downregulates the salicylic acid pathways and therefore the immune response to infection by bacteria such as itself. This does offer one compensation to the plant: the increased jasmonate activity, and presumably augmented synthesis of defense chemicals, increases resistance to herbivorous insects.[187] In an even more complex example, herbivory by the Western corn rootworm beetle (*Diabrotica virgifera virgifera*) was found to downregulate genes involved in synthesizing both defensive polyphenols and jasmonic acid. However, it transpired that these effects were actually attributable to chemicals excreted by *Wolbachia*, a bacteria introduced into the plant by the beetle. *Wolbachia*, it turns out, lives in a species-wide symbiotic relationship with the beetle.[188]

■ SUMMARY

Plants and insects are the dominant forms of life on terrestrial earth. They have co-evolved alongside each other since plants began to colonize the earth, and they share an intimate relationship that is partly governed by the synthesis by plants of a range of secondary metabolite chemicals. As well as both deterring insect herbivores and attracting insect pollinators, these secondary metabolites also help manage the other interactions between the plant and its environment. In this regard the phytochemicals function in stress responses to a range of biotic and abiotic stressors; they act as intra- and inter-plant chemical communication agents, including a role as allelopathic toxins that prevent encroachment by competitor plants; and they both manage symbiotic relationships with bacteria and fungi and function as antimicrobials. In comparison to the pivotal relationship that plants enjoy with insects, the vertebrates, including humans and other mammals, are somewhat irrelevant to plants. As we'll see later, the ecological functions of secondary metabolites, and in particular the role that these chemicals play in the close relationship between insects and plants, underlie many of the psychotropic effects of these phytochemicals within the human central nervous system.

3 More Alike Than We Are Unalike—Why Do Plant Chemicals Affect the Human Brain?

If we were to wind back the clock again to a little over 1 billion years ago,[189] we would find that the single-celled ancestor of all of the eukaryotes was a "surprisingly complex" organism. It already possessed all of the hallmark cellular features and metabolic processes of contemporary plants and animals.[190] This very complexity may be attributable to a turbulent phase in the environment, and therefore evolutionary history,[190] that saw a series of large-scale genetic duplication events.[191] These left the ancestral organism with abundant spare genes. As these were free from selective pressure they could mutate freely and take on new functions without endangering the survival of the organism. This could then drive forward rapid evolution.

Prior to and during the Cambrian explosion (~580 to 500 million years ago) this complex organism, with its genetic potential for evolution, was going to differentiate and eventually give rise to the branches of the eukaryote kingdom that include plants and the animals. How much this primordial ancestral single-celled organism endowed to these descendants can be seen by comparing the genomes of diverse eukaryotes and identifying the common genes that must have been handed down through the generations. As an example, the single-celled protist *Naegleria gruberi* (which can surprisingly transform itself from amoeba to flagellate), has a small, compact 41 million base-pair genome that nevertheless packs in some 15,000 genes, of which more than 4,000 are directly traceable to the last common ancestor of all eukaryotes.[192] The human, on the other hand, has a bulky genome of approximately 2580 million base-pairs, mainly composed of lengthy introns, that contains only some 23,000 genes. Of these, more than 3,400 genes are traceable to the same common ancestor. A similar percentage of the genome of plants also comes from the same origin. So, for instance, the model plant *Arabidopsis thalania* has some 26,000 genes in total, of which 3,200 are from the common eukaryote ancestor.[190,192] This shared genetic inheritance comprises a large core of common "housekeeping" genes that are essential for survival and that dictate a raft of processes, including genome replication and expression, and central metabolism.[193] This genetic conservation underlies the many commonalities that we see in the molecular and physiological properties of plants, animals, and insects, and it has dictated the direction of development for all of the clades. To give one of the earliest examples, even before life took shape on Earth adenosine triphosphate (ATP) was almost certainly synthesized within the sterile prebiotic environment. This ubiquitous molecule was then selected very early by the first organisms to become their main source of energy, and it now occupies a pivotal position as the predominant energy currency of all life forms.[194] The adoption of this role for ATP also constrained the potential routes of chemical development going forward, dictating the direction of the evolution of "primary metabolism." This collective term encompasses the pathways that modify and synthesize proteins, fats, carbohydrates, and nucleic acids to provide the specific compounds required for the organism's physical construction and the production of energy. These processes are essentially the same across all life forms, with only a few minor variations. As an example, processes such as glycolysis and the citric acid ("tricarboxylic acid" or "Krebs") cycle comprise complex series of enzymatic reactions that underpin the metabolic pathways that generate useable energy from carbohydrates, fats, and proteins and that also produce a host of substrates for other enzymatic processes. These pathways, which take place in the mitochondria of eukaryotes, are common to all living cells.[195]

We can also directly trace the origins of the animal nervous system back 600 million years[189] to the single-celled common ancestor of the human and the chronoflagellate. This organism already possessed the "genetic repertoire" and key molecular components that form the basis of neuronal function.[196] The many ancestral components common to contemporary vertebrates and chronoflagellates include the same calcium and sodium channels[196] that underpin neuronal electrical signals, and the proteins that control the release of neuronal neurotransmitters in animals.[197] Of course the common ancestor of animals and chronoflagellates didn't arrive fully formed, and if we step farther back in time we can likewise discern many of the genetic origins of the animal nervous system in the last common ancestor of humans and plants. By a process of both divergent and convergent evolution these commonalities have, in turn, given rise to some surprising similarities between plants and humans in terms of the parameters that we tend to think of, rather egotistically, as being solely related to the functioning of the mammalian brain.

It is also possible to start to answer the question "Why do plant chemicals affect the human brain?" by considering some of these similarities; firstly those between plants and humans, and secondly those between humans and invertebrates, and, most particularly, the plants' most intimate neighbors, the insects.

■ THE SIMILARITIES BETWEEN PLANTS AND ANIMALS

In very recent years there has been a growing appreciation of the complexity of the mechanisms and processes that dictate a plant's perception of its surroundings, and how it responds to environmental stimuli. Indeed a whole new field of research, "plant neurobiology," which considers the molecular, chemical, electrical, and hydraulic properties underlying plant perception and behavior, has been gradually emerging as a discipline.[198] An increased focus on these processes has thrown up some fascinating similarities between the "neurobiology" of the plant and the human. It seems apt to summarize some of the many similarities between plants and animals that may underpin, in part, the central nervous system effects of phytochemicals.

Receptors and Signaling Cascades

All cells, whether they exist in the form of a single-celled organism or as a single component of a more complex multicellular organism, have to be able to receive information from their environment. For a single-celled organism this process might involve detecting a change in a chemical concentration gradient followed by an appropriate movement; whereasfor the most complex organisms this process might involve signaling between many millions of cells, with a resultant change in physiological parameters or behavior. For instance, the average human brain contains something in the region of 23 billion neurons, each of which is interconnected to other neurons by some 5,000 to 10,000 synapses. On top of this there are also a slightly larger number of the glial cells that support neuronal functioning, and these again are interconnected with other glial cells and neurons (although this is poorly understood to date).[199] All of these cells communicate via a bewildering array of neurotransmitters, neuromodulators, hormones, and signaling molecules, all of which function as part of multiple signal transduction systems. Typically, irrespective of the complexity of the organism, at a cellular level the signal transduction process comprises some form of extracellular signaling molecule binding to and activating one or more proteinaceous cellular receptors on the external surface of the cell's membrane (in the case of signaling molecules that can penetrate the cell membrane, the receptors may be in the cell's cytoplasm or nucleus) with enough specificity to differentiate the signal from other background signals. This is followed by transmission of the signal across the membrane and into the cell, where it is amplified and disseminated within the cell, resulting in a physical response, often dictated by changed gene expression in the cell nucleus.[200]

Having arisen in a common single-celled ancestor, much of the basic machinery and many of the processes of cellular signal transduction are conserved across phylogenetic kingdoms. The animal nervous system functions courtesy of two main types of membrane receptors for neurochemicals. The most abundant are "metabotropic" receptors, such as the "G-protein-coupled" receptor that winds in and out of the membrane. The external portion of this receptor is bound to by the signaling molecule, and the signal is then transmitted across the membrane to the internal portion of the receptor. It is then propagated within the cell via "second messenger" signal transduction pathways. In contrast, "ionotropic" receptors, such as the "ligand-gated ion channel," open or close a membrane channel in response to binding by the signaling molecule, allowing ions (sodium, calcium, potassium, or chloride) to flow directly into the cell. These ions either change the electrical properties of the cell or trigger second-messenger signal transduction pathways. As we'll see below, both of these receptor types are conserved in humans and plants[201–204]; in some cases the plant receptors are activated by the same chemicals that we typically think of as "neurotransmitters." Other receptors commonly expressed in both plants and animals include the leucine-rich repeat receptors[200] and a range of protein kinase receptors. This latter, ubiquitous group of receptors include the "receptor tyrosine kinases," which are found in cells throughout the animal body, including the brain. In plants the conserved group of receptors are termed "receptor-like kinases." In both cases these receptors are involved in transducing stress-related signals, which are delivered to the cell by hormones and other stress-related ligands.[205]

The similarity in signal transduction pathways extends to the inside of the cell, with the conserved receptors propagating their signal via the same signaling cascades.[200] For instance, in the case of protein kinase receptors, the ubiquitous mitogen-activated protein kinase (MAPK) cascade, in which a series of protein kinases activate (phosphorylate) each other in a chain reaction that amplifies and transmits the signal within the cell, is conserved in plants and humans.[206,207] Indeed, plants, due to duplication of the original common ancestral genes, have a particularly rich complement of more than 1,000 protein kinase genes; this is approximately double the number of kinase genes seen in the human genome.[208] Intriguingly, while the downstream products of

MAPK activity in animal cells include the synthesis of a number of inflammatory factors, including prostaglandins,[209,210] plants also enjoy an orthologous, genetically conserved response, but in their case the modified pathway leads to the synthesis of the jasmonate hormones, rather than prostaglandins, and ultimately the synthesis of a raft of defensive secondary chemicals[211–213] (see "Immune and Stress Responses" below, and Chapter 8).

Mammalian Neurotransmitters, Neurochemicals, and Receptors?

We immediately associate a number of chemicals solely with the functioning of the human or mammalian brain. However, many of the neurotransmitters, neuromodulators, and hormones that are integral to the functioning of our own central nervous system are in fact the chemical products of metabolic processes that existed before the differentiation of plants and animals over a billion years ago. Many of these "neurochemicals" were synthesized and served cellular functions in our common unicellular ancestors,[203,214,215] and they have gone on to occupy key roles in the lives of plants, often sharing functional similarities with the roles they play as signaling molecules in mammals. The following describes the comparative plant and mammalian roles of a number of chemicals that are typically thought of as "neurochemicals."

Acetylcholine

As a key mammalian neurotransmitter acetylcholine plays a part in the functioning of muscles and the parasympathetic nervous system and contributes to the majority of brain processes, including long-term potentiation and neuronal plasticity, sensory perception, arousal, and attention. It functions in animals by binding to ionotropic 'nicotinic' receptors and G-protein-coupled 'muscarinic' receptors. However, more recently, the role of acetylcholine both in non-neuronal physiological processes in animals and in its ubiquitous role as a genetically conserved signaling molecule across all forms of life, including plants, has attracted increasing attention.[216] In plants acetylcholine contributes to the regulation of growth, germination, flowering, water homeostasis, and photosynthesis.[203,216,217] It has been shown to carry out some of these functions by binding to receptors analogous to the nicotinic[218] and muscarinic receptors[219,220] found in animals. The latter receptors have been shown to exert their effects on, for instance, stomatal opening via the same signal transduction mechanisms that contribute to the effects of the analogous receptor in animal cells (i.e., an increase in cytosolic calcium).

Glutamate and Gamma-Aminobutyric Acid (GABA)

These amino acids (non-proteinogenic in the case of GABA) are the most abundant mammalian neurotransmitters and play approximately opposite roles in the vertebrate nervous systems. Glutamate is the most prevalent excitatory neurotransmitter, with profound roles in cognitive function, and in particular memory, due to its role in synaptic plasticity. GABA, on the other hand, is the major inhibitory neurotransmitter[221] with receptors on most, if not all, neurons in the central nervous system.[222] Both glutamate and GABA bind directly to their own ionotropic receptors and modulate the flow of ions (chloride in the case of GABA, and sodium and calcium in the case of glutamate) across neuronal membranes, thereby modulating neuronal excitability. They both also influence internal cellular processes by binding to G-protein-coupled membrane receptors.[215]

In plants both glutamate and GABA also serve signaling functions. Glutamate plays a central role in many aspects of amino acid and nitrogen metabolism and acts as a precursor to chlorophyll in developing leaves. It also acts as a signaling molecule and plays multifarious roles in germination, growth, light-mediated development, carbon/nitrogen sensing, and responses to stressors. These roles include a contribution to the jasmonic acid hormonal pathways.[223] Many of these functions are as a consequence of direct binding by glutamate to phylogenetically conserved receptors in plant membranes. Evidence suggests that these receptors exert the same ionotropic effects as the animal homologues[215,223] and result in action potentials broadly similar to those generated in mammalian cells.[224] GABA is synthesized from glutamate by the same enzymatic pathway in plants and animals.[225] In plants GABA plays a central role in general metabolism and is rapidly produced in response to a variety of stressors.[226,227] Its secondary roles include the attenuation of oxidative stress and defense against insects and nematodes.[225,228] It also functions as an intracellular signaling molecule, particularly during plant development. While unidentified as yet, the possibility of specific plant GABA receptors remains open, although it is also possible that GABA may also bind to a domain of the plant ionotropic glutamate receptor and modulate its function allosterically.[227] GABA also features in emissions designed to communicate with other plants and organisms[229] and may function as an induced defense compound in its own right.[230]

The Indoleamines—Serotonin and Melatonin

The ubiquitous amino acid tryptophan played a key role in the evolution of life due to its ability to absorb light to generate biological energy. This property, which is courtesy of its indole ring structure, made tryptophan and its derivatives key components of all light-capturing proteins, such as chlorophyll and rhodopsin. The same property underlies many of its functions in plant and animal physiology.[214]

Across vertebrates the L-tryptophan–derived neurotransmitter serotonin (5-hydroxytryptamine [5-HT]) plays a role in most basic biological processes, including movement, breathing, sexual reproduction, and temperature regulation. In higher animals it also modulates more complex behaviors such as sleeping, eating, memory and learning, attention, sexual activity, and, in humans, mood.[214] The closely related neurohormone melatonin (N-acetyl-5-methoxytryptamine) is synthesized by the pineal gland in the mammalian brain during darkness and plays diverse roles related to its key function in regulating circadian rhythms throughout the brain and body, including the modulation of sleep, mood, sexual behavior, and aspects of seasonal reproduction.[231,232]

Plants also synthesize serotonin, melatonin, and a number of structurally related indoleamines, including the ubiquitous plant hormone auxin (indole-3-acetic acid).[233] Although the role of these indoleamines in plants has received comparatively little attention, serotonin, often in interaction with auxin,[234] has been shown to help regulate root system growth and architecture, the maturation and ripening of fruit, and senescence. It is also induced in response to microbial attack and as a defense against herbivores.[235–238]

Melatonin has been isolated in a diverse range of angiosperms, with quantities varying as a consequence of light levels as in animals. In plants its functions include an analogous role in regulating circadian rhythms (see below) and other photoperiod responses, and it has related effects on the flowering, germination, and growth of the plant above and below ground.[232,239]

The Catecholamines—Dopamine, Adrenaline, and Noradrenaline

In both animals and plants dopamine and its derivatives adrenaline and noradrenaline (also known as epinephrine and norepinephrine) are synthesized from the amino acid L-phenylalanine via a conserved synthetic pathway. These three catecholamines are most closely associated with their role as animal neurochemicals. Dopamine has wide-ranging mammalian functions as a key neurotransmitter, with roles including the modulation of diverse aspects of cognitive function and behavior, motivation and reward, sexual gratification, and sleep. Adrenaline and noradrenaline are both neurotransmitters and neurohormones, with roles encompassing modulation of aspects of brain function, heart rate, gluco-regulation, the "fight-or-flight" response dictated by the sympathetic nervous system, and stress-related responses in most tissues of the body.

In plants dopamine is a key intermediary in the synthesis of a wide range of alkaloid secondary metabolites, including morphine and mescaline. The other catecholamines have also been isolated at varying concentrations in a wide range of plants, including many fruits and green food plants.[202] To date, their endogenous roles in the life of plants are poorly delineated. However, it is notable that dopamine is a potent antioxidant and that the catecholamines as a group may function as intermediaries in photosynthesis, and in the responses to stress and infection. They may also interact with auxin to modulate growth and influence flowering. Adrenaline and noradrenaline have also been shown to modulate sugar metabolism in plant tissue via similar mechanisms as seen in mammalian cells (i.e., inactivation of glycogen synthase and activation of phosphorylation). A number of indirect lines of evidence also suggest that the catecholamines may function by binding directly to as-yet-unidentified receptors in plant tissue.[202]

Purines

The purinergic signaling system employs purines such as adenosine, ATP, and pyrimidines. These molecules are released from the cell into the extracellular matrix to function as the most basic extracellular signaling molecules in both animals and plants. In animals, purines modulate the activity of neurons by acting as co-transmitters within all of the neurotransmitter systems, throughout both the central and peripheral nervous systems. They therefore play a plethora of roles in diverse tissues and organs. They do this via three discrete receptor subtypes. The most familiar of these, and the earliest to be characterized, are the G-protein-coupled "P1" adenosine receptors, but these have now been joined by two subtypes of "P2" purinoreceptors, the ionotropic P2X and G-protein-coupled P2Y receptors that are bound by ATP and other purinergic signaling nucleotides.[240] In plants, purines and pyrimidines function in a conserved capacity, triggering similar cellular transduction pathways involving nitric oxide, calcium, and reactive oxygen species. Common roles across the taxa include modulation of growth, cell death, responses within the respective immune systems, and responses to biotic and abiotic stressors.[241] However, only indirect evidence points to the existence of membrane-bound purine receptors in plant cells to date.[242]

Steroid Hormones

Mammals express five types of steroidal hormones: mineralocorticoids and glucocorticoids, and the three groups of female and male sex hormones, the progestogens, estrogens and androgens. These hormones contribute to a raft of physiological processes, including physical homeostasis, metabolism, development of sexual characteristics, brain function, and stress responses to illness and injury, including inflammatory and immune function responses. They do this via two distinct signal transduction mechanisms: they diffuse into cells and bind to cytoplasm or nucleus receptors, which generate slow cellular responses, and they also bind directly to membrane receptors, which generate a faster response.[243,244] The most abundant progestogen is progesterone, which when synthesized in the ovaries, placenta, and adrenal glands is primarily associated with multiple aspects of female reproduction and lactation. However, progesterone is also synthesized throughout the central and peripheral nervous system in both males and females; acting as a "neurosteroid," playing multiple roles, including in myelination and neuroprotection.[245,246] Progesterone receptors are also expressed in diverse brain regions, including the hypothalamus, pituitary, and cortex, and progesterone modulates sexual behavior, parental behavior and aggression, and mood.[246,247]

Among the many steroids that plants synthesize, only the brassinosteroids were thought to exert hormonal effects until recently.[244] However, a number of "mammalian" hormones, including progesterone, androgens, and estrogens, have been identified in a wide range of tissues taken from a variety of plants.[243,248–250] These steroids may function as precursors for other steroidal secondary metabolites.[248] However, it is also notable that progesterone, in particular, exhibits many of the characteristics of an endogenous plant hormone.[244] In this respect it is notable that in plants both cytoplasm/nucleus and membrane receptors have been identified for brassinosteroids, and these include several receptors that bind endogenously circulating progesterone (and other "mammalian" sex steroids) in plants.[244] These include two membrane receptors, "membrane steroid binding protein 1" (MSBP1) and "steroid binding protein" (SBP), that are widely distributed throughout plant tissues. These receptors represent partially conserved orthologues of the mammalian "progesterone receptor membrane component 1" (PGRMC1) progesterone receptor.[243] Similarly, circulating estrogens and specific estrogen-binding sites similar to the mammalian nuclear "estrogen receptor α" have been identified in various tissues from two members of the Solanaceae.[251] The enzymatic pathways that result in the sex steroids in plants are also partially conserved.[249] Unsurprisingly, the "mammalian" sex hormones exert a wide range of effects on plant growth and development. For instance, progesterone influences root elongation, germination and pollen tube growth, flowering, seedling and plant growth, and antioxidative stress responses,[244] in some cases showing strongly biphasic effects (e.g., increasing growth parameters at low doses and impairing them at higher doses).[243]

Nitric Oxide

Nitric oxide (NO) is a gaseous molecule whose small size and neutral charge make it ideal for rapidly diffusing across membranes, while its possession of an unpaired electron allows it to interact with other ubiquitous biological molecules. It therefore plays a key role as a local signaling molecule across all forms of life, with particularly concordant functions in terms of the orchestration of physiological processes across the lifespan of both animals and plants. In both phyla NO has key signaling roles in mitochondrial and immune function and the physiological response to biotic and abiotic stressors.[252–254] It also contributes to the regulation of a plethora of other processes related to the lifestyle and life course of the clade. So, in animals NO plays a modulatory role in platelet function, smooth muscle contraction and relaxation, defense against pathogenic microorganisms, egg fertilization, peripheral and cerebral vasodilation, neurotransmission, and the neurovascular coupling of neural activity to local cerebral blood flow.[255–259] In plants it contributes to metabolism, germination, stomatal function, development of roots and pollen tubes, flowering, leaf senescence, and disease resistance.[255,260]

The understanding of the NO synthetic pathways in both plants and animals is incomplete.[261] In plants a number of synthetic pathways have been partially investigated, and, while synthesis from L-arginine is present, these primarily involve the reduction of nitrate, via nitrite, to NO by a variety of enzymes, including nitrate reductase and xanthine reductase.[261] In animals it had been assumed that the primary pathway involves synthesis from L-arginine by several isoforms of nitric oxide synthase (NOS). However, recent evidence in animals shows that the pathway described above for plants is also present, with dietary nitrate being reduced to nitrite and eventually NO, which in turn is also rapidly oxidized back to nitrite and nitrate after release. Nitrite can then be reduced back to NO by a wide range of proteins and enzymes in blood and tissue, including NOS.[262] These latter observations have to be taken in context with several lines of evidence suggesting the presence of mammalian-like NOS in plants,[261] which have been bolstered by the recent discovery of a genetically conserved NOS in green algae.[263] As more information is gathered, the clearer become the similarities between plant and animal NO synthesis.

Plant Secondary Metabolites?

Turning the tables somewhat, a number of compounds typically thought of solely as plant secondary metabolites have also recently been identified as endogenous mammalian neurochemicals.

Opiates

Morphine and a range of related opiates are usually thought of as alkaloid plant secondary metabolites, and their profound central nervous system effects in humans have almost exclusively been attributed to their opportunistic binding to the receptors for endogenous mammalian opioid peptides, primarily the enkephalins, endorphins, and dynorphins (see Chapter 5). However, recent evidence has shown that morphine and a number of other opiates and their precursors are synthesized endogenously in a range of tissues by both invertebrates and vertebrates via synthetic pathways similar to those seen in plants. In mammals morphine has been isolated from the heart, adrenal glands, white blood cells, and neural tissue, including the human brain. It is also present in plasma, suggesting a hormonal mode of action.[264,265] Along with this, recent research has also identified the novel μ^3 G-protein-coupled receptor, identified to date in the heart, vascular tissue, white blood cells, stem cells, and brain, which has high affinity for morphine and related opiates but negligible or low affinity for the 'traditional' endogenous opioid peptides.[264] Morphine's endogenous role, via the μ^3 receptor, appears to be to modulate constitutive nitric oxide release, resulting in downregulation of excitability in the relevant tissues, leading to modulation of vascular and immune function and stress responses.[264,265] Within the brain this function may include a modulatory effect on the activity of the catecholamine neurotransmitters, which are structural relatives of morphine.[266]

The Tryptamines

The hallucinogenic tryptamines—N,N-dimethyltryptamine (DMT), 5-methoxy-DMT (5-MeO-DMT), and 5-hydroxy-dimethyltryptamine (5-HO-DMT or bufotenin)—are indole alkaloid secondary metabolites synthesized by plants from a wide range of families. They represent the active components of a number of traditional Meso/South American hallucinogenic preparations, including ayahuasca and yopo (see Chapter 6), and are controlled drugs in their own right. All three tryptamines are also endogenously synthesized in small quantities by vertebrates, including humans, where they function as neurotransmitters,[267] binding to recently discovered G-protein-coupled trace amine-associated receptors (TAARs).[268] These receptors are implicated in a host of mood and neurological disorders,[269] including playing a role in the hallucinogenic and delusional phenomena experienced by schizophrenics.[270] They are also expressed throughout areas of the brain that underpin aspects of conscious awareness and sensory perception, such as the prefrontal cortex, hippocampus, substantia nigra, amygdala, and basal ganglia.[267] Interactions by the tryptamine plant chemicals, and closely related chemicals, such as psilocin (4-hydroxy-N,N-DMT or 4-HO-DMT) from the *Psilocybe* mushroom, with their own endogenous mammalian TAARs, must therefore contribute to their hallucinogenic and perception-distorting effects.[267]

Immune and Stress Responses

The immune response to infection by pathogens originated in early single-celled organisms, and many aspects are therefore conserved across the phylogenetic kingdoms.[271] In both plants and animals the innate response to infection is activated by partially conserved "pattern recognition receptors"[207] that identify the invading pathogen's molecular fingerprint, also known as its "pathogen associated molecular pattern" (PAMP). Detection of the presence of the pathogen activates immune responses that are similar in a number of ways.[272] For instance, the responses across the taxa include a highly conserved mitogen-activated protein kinase (MAPK) signaling cascade; autophagy; RNA interference, or "gene silencing"; the production of defensins (antimicrobial peptides that disrupt the membrane integrity of the microbe); the generation of pathogen-killing reactive oxygen species that are produced via similar mechanisms during the inflammatory response in animals and by individual plant cells; programmed cell death, whereby an infected cell sacrifices itself[207,271,273]; and the production of the oxylipin groups of chemicals that play a role in the response to both biotic and physical stressors in their respective taxa.[177,274]

The Oxylipins and Other Plant Hormones

The oxylipin pathways are of particular interest. In animals the metabolites from these pathways, termed eicosanoids, are produced by the oxidation of 20 carbon polyunsaturated fatty acids by cyclooxygenase and lipoxygenase enzymes. The eicosanoids comprise the leukotrienes, prostacyclins, prostaglandins, and thromboxanes, all of which are local autocrine/paracrine signaling molecules. As a group they play a regulatory role in cellular stress responses, including the modulation of mammalian immune responses, inflammatory processes, and cardiovascular and brain function.[177,272,275] To take an individual but pertinent example, prostaglandins are synthesized in animals from arachidonic acid and are central to inflammatory and immune responses, they modulate blood flow by dilating or constricting blood vessels and by determining the aggregation, or stickiness, of blood platelets, and they contract or relax bronchial and smooth muscle and modulate gastric acid secretion.[275] The specific effect of the prostaglandin is dictated by the tissue in question and in turn the many types or subtypes of G-protein-coupled receptor to which it binds.[276]

Of potentially pivotal relevance, the jasmonate family of plant signaling oxylipins are synthesized via the same genetically conserved pathways as the mammalian prostaglandins, and as a consequence the two groups of signaling molecules are structurally and functionally related.[277] In plants lipoxygenases convert the 18 carbon polyunsaturated fatty acid "α-linolenic acid" to the ubiquitous jasmonates. These hormones include jasmonic acid and its methyl ester, methyl jasmonate, and a number of bioactive jasmonic acid/amino acid conjugates, for instance jasmonoyl–isoleucine.[181] As noted in the previous chapter, the jasmonate signaling pathways play key functions in a variety of plant developmental processes, including growth, fertility, fruit ripening, senescence, and root elongation. They are also absolutely central to the plant's defensive signaling system against both abiotic stressors such as physical damage, ultraviolet light, dehydration, and salinity, and biotic stressors such as microbial pathogens and herbivores. In the case of biotic stressors some jasmonates can function in their own right as direct and indirect defense agents,[181,278] but their principal role is in the induction of the synthesis of secondary metabolites from all of the chemical groups.[212,213,279,280] They accomplish this by binding directly to intracellular jasmonate receptors, triggering the activation of transcription factors for the plant-specific synthetic pathways.[184,213] Jasmonate signaling can also take place outside of the plant, with airborne methyl jasmonate emissions being absorbed by the surface of leaves and other plant tissues, triggering gene transcription and defensive responses either in distant parts of the emitting plant or in neighboring plants.[181]

The jasmonates also work synergistically with another conserved signaling molecule, the growth and stress hormone abscisic acid. In combination these two compounds trigger many secondary metabolite responses elicited by a number of stressors.[178] Abscisic acid itself is a universal signaling molecule with a shared signaling pathway across phyla,[281] underpinned, in plants, animals, and other taxa, by a wide range of orthologous genes.[282] It was first isolated from mammalian brain tissue in the mid-1980s[283] and has more recently been shown to be synthesized endogenously within a variety of mammalian tissues and cells.[281] In plants the independent (from the jasmonates) influence of abscisic acid is also seen in the processes of growth and development and in the response to pathogens and abiotic stressors, such as variations in temperature, hydration, and salinity. In animals abscisic acid not only stimulates similar cellular responses to abiotic stressors as those seen in plants, but also has wide-ranging functions within the immune system, including modulation of insulin release from the pancreas and the proliferation of stem cells. A number of abscisic acid's effects are related to its modulation of prostaglandin synthesis and interactions with peroxisome proliferator-activated receptor (PPAR) nuclear receptors.[281,284]

Before leaving the topic of plant hormones, it is also worth noting that salicylic acid, the principal hormonal antagonist of the jasmonate pathways in plants, is also present at measurable levels across animal phyla. In part this may simply reflect sequestration from the diet, with vegetarian humans and animal species exhibiting higher circulating levels of salicylic acid than non-vegetarians as a consequence of fruit and vegetable consumption. In humans the range of "normal" levels of salicylic acid actually overlaps with the concentrations seen in the sizeable subgroup of humans regularly taking salicylic acid in its acetylated form, better known as aspirin. However, carnivorous animals and humans on restricted diets that have no natural sources of salicylic acid still have measurable levels of this phenolic compound, and preliminary evidence from a study that fed volunteers with a radiolabeled precursor, benzoic acid, suggests that salicylic acid may be synthesized *de novo* by humans.[285] This in turn hints at an endogenous regulatory role. We'll return to the topic of the conserved "plant" hormones in Chapter 8.

Molecular Clocks

Before leaving this topic it is also worth looking at similarities in another major biochemical and biomolecular mechanism as an illustration of how deep the concordance between the taxa runs. Almost all living organisms live their lives to the ticking of biological clocks, and the vast majority live in an environment that is subject to profound rhythmic changes. The most obvious examples are the daily changes in light and dark or the seasonal changes in day length and temperature. To survive and prosper any organism needs not only to perceive these changes, but also to anticipate them and modulate its behavior accordingly. Consequently, the majority of terrestrial life forms exhibit circadian rhythms, genetically based 24-hour rhythms that dictate aspects of their biochemistry, physiological processes, and behavior. These rhythms are typically reset by environmental cues such as changes in the dark/light cycle and temperature. In humans the most familiar example of the many circadian rhythms is regulation of the sleep/wake cycle, but body temperature, hormonal and immune function, digestion, physical activity, and alertness are all dictated in part by biological clocks. Longer "supradian" biological rhythms include the menstrual cycle and, in many nonhuman mammals, seasonal reproductive cycles and hibernation. In plants rhythmicity is seen in most aspects of their life cycle and daily life. Examples of circadian processes include regulation of daily patterns of growth and flowering, photosynthesis, stomatal opening and gas exchange, and the emission of chemical attractants.[286] Longer supradian rhythms include seasonal germination, patterns of flowering, senescence, and leaf fall.

In humans the majority of circadian effects are dictated by the activity of the suprachiasmatic nucleus, an area of the brain composed of approximately 20,000 cells that exhibit summed activity in a pattern of approximately 24 hours that is coordinated by signals from light-sensitive cells in the eye. In mammals the key circadian neurochemical is the hormone melatonin, which is released by the pineal gland in the brain during darkness, irrespective of whether the animal is nocturnal or diurnal.[231] Recent research has also demonstrated the existence of ubiquitous peripheral oscillators in cells throughout the body.[287] In plants, melatonin plays an analogous role in directing circadian rhythms,[239] and, while plants have no central control of rhythmicity, peripheral oscillators are distributed throughout the cells of the entire plant.[286]

Irrespective of the life form, the cellular molecular oscillators that drive these rhythms are broadly the same.[231,286] They comprise an autoregulatory negative feedback loop, consisting of proteins generated by several genes in response to an external stimuli, that interlock and interact in a temporally consistent pattern of synthesis and degradation, or phosphorylation. This ultimately generates a self-sustaining cellular rhythm that approximates to 24 hours and that is reset by environmental cues such as the onset of light in the morning. While it has been proposed that the genetic feedback loops that drive circadian clocks have arisen independently in different phylogenetic groups of organisms during eukaryotic evolution[288], it is also notable that many functional molecular elements of circadian clocks appear to be conserved from common ancestors that precede the divergence of plants and animals.[289,290] As an example recent research has identified orthologous circadian clock-related genes (JMJD5) in plants and humans that play a central role in regulating cellular rhythmicity via histone methylation.[289] Plant and human cell lines lacking the gene both show significantly speeded circadian rhythms. When the plant gene is subsequently transfected into a human cell line lacking the gene, the normal rhythm is re-established; this demonstrates that the plant protein can function within the human circadian system.

Similarly, both plant and animal circadian systems are reliant on cryptochromes, expressed by the conserved genes Cry1 and Cry2. These blue-light–sensitive proteins play numerous roles related to light sensitivity in plants, including resetting the cellular circadian clocks in response to light. In animals they play the same pivotal role in the molecular oscillations driving cellular circadian clocks as they do in plants and may also function as photoreceptors within the retinal ganglion cells that project to the suprachiasmatic nucleus.[291]

It is clear from the abover, that all told, with regard to the chemicals and physiological systems that underpin mammalian brain function, plants and humans are more similar than we might think.

■ SIMILARITIES IN THE INSECT AND HUMAN CENTRAL NERVOUS SYSTEMS

Whereas plants and animals diverged on the evolutionary tree from a common ancestor more than 1 billion years ago, the branching in the bilateria that would lead to deuterostomes, and eventually modern humans, on one hand, and protostomes such as the insects, on the other, occurred a relatively recent 620 million or so years ago.[189] Naturally, given the much closer phylogenetic relationship between insects and humans, it is unsurprising to find that the majority of molecular and cellular processes within the taxa are either identical or very similar. You will recall from the start of the chapter that plants and humans have a common inheritance of some 3,000 or more essential “housekeeping” genes that dictate cellular and molecular processes from their last common ancestor. This figure rises to over 5,000 “housekeeping” genes shared by man and the fruit fly, *Drosophila melanogaster*.[193] In terms of the whole genome, insects and humans bear striking similarities. As an example, the fruit fly has a genome that encodes about 14,000 genes, and among these are identifiable homologues for more than 75% of human disease-related genes.[292,293]

The common ancestry and the similar genetic endowment underlie substantial similarities in the nervous systems of insects and humans. These similarities start with the entire development and structure of the nervous system. To take one example, the eyes of insects and mammals are outwardly different; on the one hand insects have compound eyes made up of thousands of directional photoreceptor units, while mammals have spherical lensed eyes. However, the visual systems of both taxa are derived from their common ancestor, and they develop courtesy of the same cascade of molecular events and retain conserved genetic underpinnings and functional mechanisms, including modes of neural transmission. Likewise, the ventral cord of insects and the spinal cord of humans undergo the same developmental processes and serve the same functions, and the differentiation of brain areas and axon connectivity is guided by conserved genes and mediated by the same cell-to-cell communication and molecular events.[294] Even the cognitive architecture of the insect brain shares similarities with that of the mammals, with common principles of modularity within the central nervous system in terms of specific sensory domains, and higher-order structures integrating information; in the case of insects this function is served by the mushroom bodies.[295,296] Indeed, recent research using novel techniques to map cellular genetic expression has shown that both the invertebrate mushroom body and the vertebrate pallium (the cerebral cortex in mammals) evolved from the same structure in a common ancestor. Their developmental elaboration within different taxa involves the same conserved genetic patterning mechanisms and resultant neuron types.[297]

In terms of the physiology of brain function, the molecular architecture of the two systems is in many parts identical, with common neuronal functional apparatus governing the synaptic release and recycling of neurotransmitters, and receptor interactions and signal transduction mechanisms.[298] Naturally the latter includes the use of NO as an important secondary messenger in the nervous system.[299,300] The same specific molecular processes also underlie higher-order brain function. As a single example, insect long-term memory is dependent on the same process of long-term potentiation as seen in mammals,[301] and this is underpinned by the same molecular processes, such as NO and cyclic adenosine monophosphate (cAMP) signaling[302,303] and the involvement of glutamate α-amino-3-hydroxy-5-methyl-4-isoxazole propionic acid (AMPA) and N-methyl-D-aspartate (NMDA) receptors.[304] In much the same way, the architecture and functionality of the insect visceral nervous system is analogous with the vertebrate autonomic nervous system.[299]

Neurochemicals and Receptors

Overall the insect and mammalian nervous systems function courtesy of the same neurochemicals, including many neuropeptides,[305] hormones,[299] and neurotransmitters and their associated receptors. The relevant signal transduction systems also function in the same manner and exert similar physiological or behavioral effects. As the neurotransmitter systems, in particular, are the putative target of many plant chemicals, it is worth describing the congruity in some detail.

Acetylcholine

Acetylcholine is the principal excitatory neurotransmitter in the insect central nervous system (as opposed to glutamate in mammals). It is also the principal neurotransmitter in sensory neurons,[306] and it contributes to olfactory and tactile learning.[302,307] Acetylcholinesterase, the enzyme that breaks down acetylcholine following its synaptic release, is therefore present throughout all insect brain structures.[308] Insects also express both of the isoforms of "mammalian" acetylcholine receptors, the ionotropic nicotinic receptors and the G-protein-coupled muscarinic receptors.[294] Nicotinic receptors are seen in greater concentration, and they play a specific role in the acquisition and retrieval of long-term memories.[309] Muscarinic receptors have a more restricted distribution but are expressed throughout cognition-relevant areas of the insect brain, including the mushroom body. They also play a role in memory processes, although agonist/antagonist studies suggest that their role is more prevalent in the recall of previously learned information.[310,311]

Dopamine

This neurotransmitter plays a wide range of common roles in mammals and insects, including controlling movement, cognition, and development.[294] Dopaminergic activity in the insect mushroom bodies also modulates learning and memory, and it works, in concert with GABA, to contribute to the modulation of both arousal and sleep.[312,313] Insects express members of the same two families of dopaminergic G-protein-coupled receptors as mammals; in insects these are termed D_1-like and D_2-like receptors,[294] although the D_2 receptors have greater retained homology between the taxa.[313] These receptors function by either activating (D_1) or inhibiting (D_2) adenylyl cyclase, leading to increased or decreased intracellular cAMP levels and modulation of calcium and potassium ion channels. Insects also have a further class of "invertebrate" dopamine receptors that are derived from octopamine receptors and that function in the same manner as D_1 receptors.[313]

Octopamine/Tyramine

It was originally thought that octopamine and tyramine were exclusively invertebrate neurochemicals. They are structurally related to adrenaline and noradrenaline respectively, and in many ways they functionally replace these neurotransmitters in the invertebrate central and peripheral nervous systems.[314,315] In this regard they act as stress hormones that adjust the body's energy demands to the situation and play a role in the control of behavior, learning and memory, immune responses, appetite, olfaction, and metabolism.[316] However, emerging evidence has also identified octopamine and tyramine as endogenous mammalian neurotransmitters that belong to the "trace amines" group (as with the tryptamines above). Despite being present in small quantities, these neurochemicals have far-reaching peripheral and neuromodulatory functions as a consequence of binding to the recently identified[317] TAAR family, which themselves are homologues of the insect octopamine/tyramine receptors.[267,269]

Serotonin

In humans the effects of serotonin are mediated by seven families of receptors, designated 5-HT_1 to 5-HT_7. All of these are G-protein-coupled receptors with the exception of the ionotropic 5-HT_3 receptors. The 5-HT_1 and 5-HT_2 families have five and three members respectively (denoted by letters; e.g., 5-HT_{1A}). Insects (*Drosophila*) have three orthologous groups of receptors that correspond to 5-HT_2, 5-HT_7, and 5-HT_{1A} receptors. These receptors modulate a number of insect behaviors that are analogous to those modified by this system in mammals, including sensory processing, circadian behaviors, learning and memory, aggression, and courtship and mating.[294,318–320]

Glutamate

The predominant excitatory neurotransmitter in the vertebrate central nervous system plays a similar, if less pronounced, excitatory role in invertebrates. Ionotropic glutamate receptors are found across most forms of life, and in plants glutamate plays a host of roles (see above). Insects express homologous members of each of the three mammalian ionotropic glutamate receptor families: AMPA, NMDA, and kainite.[321] One of the less common G-protein-coupled glutamate receptors is also widespread throughout the insect nervous system.[322] Glutamate plays a specific excitatory role at the insect neuromuscular junction, and its receptors are expressed throughout the nervous system, where its functions include modulation of olfactory, gustatory, and visual perception.[323] One key function that is shared between mammals and insects is glutamate's role in learning and memory: AMPA and NMDA receptors play an identical role in the process of long-term potentiation that leads to the consolidation of memory.[304]

GABA

GABA is the principal inhibitory neurotransmitter in both vertebrate and invertebrate nervous systems, with both ionotropic $GABA_A$ receptors and G-protein-coupled $GABA_B$ receptors expressed throughout both taxa. In the case of $GABA_A$ receptors, these function largely as they do in mammals and are expressed in insect brain areas associated with sensory perception and higher-order behaviors[294]. As a key inhibitory molecule, GABA, working via $GABA_A$ receptors, plays a role in both arousal and sleep in conjunction with dopamine.[312] $GABA_B$ receptors, on the other hand, while orthologous and triggering similar cellular signaling cascades, have diverged somewhat from vertebrate receptors in terms of ligands[294] but play similar roles in olfaction and circadian rhythms.[324]

Opioids

Both the classic μ and κ opioid receptors and the endogenous opioid receptor ligand enkephalin are widely expressed in the insect brain and nervous system and play analogous roles.[325–328] Both endogenously synthesized morphine and its recently identified μ_3 receptors have also been identified across invertebrate phyla,[264] including insects.[329]

Steroid Receptors

The "nuclear receptor" superfamily, members of which typically respond to steroid hormones, are conserved across the animal kingdom from a common ancestor. Insects express nuclear receptors from each of the six identified subfamilies, and these function in the same manner as the vertebrate homologues. For insects the majority of nuclear receptor ligands identified to date are the juvenile hormones and ecdysteroids that are analogous to the mammalian steroid hormones. However, many nuclear receptors from both taxa, but particularly those of insects, currently remain "orphan receptors" in that their endogenous ligand has not yet been identified. These include for both taxa the "estrogen-related receptors" that are close relatives of the estrogen receptor.

Naturally, given that the evolutionary divergence of insects and animals took place some 600 million years ago, the same nervous system systems can vary in function between phyla. As an example, in mammals acetylcholine is the principal neurotransmitter at peripheral muscular junction synapses and glutamate is the principal excitatory neurotransmitter throughout the central nervous system. In insects these roles are largely reversed.[298,330] Similarly, changes in serotonergic activity can have opposite effects on aggression in vertebrates and invertebrates,[299] and dopaminergic neurons are implicated in aversive learning in insects but reward in mammals.[331,332]

Insect Models of Human Systems and Behavior

Courtesy of broad genetic similarities, and the possession of a relatively simple nervous system that bears striking similarities to the mammalian system in terms of molecular processes, neurochemistry, functionality, and architecture, invertebrates such as the fruit fly (*Drosophila melanogaster*) and the honey bee (*Apis mellifera*) have been deployed for more than half a century as model organisms for unraveling many of the fundamental processes in the central nervous system.[306] The findings from these studies are often extrapolated directly to mammalian systems. *Drosophila* actually started its laboratory life in the 1930s in research that established the field of developmental genetics, work that eventually culminated in a Nobel Prize for Edward Lewis. Since that time genetic mutants of this small fly have been used in research into the etiology and behavioral changes associated with a range of neurological diseases, including Huntington's disease, Parkinson's disease, Alzheimer's disease, and other dementias.[333] They have also served as genetic models for fundamental processes shared with mammals, such as the workings of circadian and supradian rhythms[334] and the molecular underpinnings of sleep and sleep disorders.[335]

Given the similarity between the nervous systems of insects and mammals, it is not surprising to find that they can exhibit similar behaviors to mammals. Insects have therefore also been employed as standard models for the investigation of cognitive processes,[294,295] with the honey bee serving as a model for the study of "an intermediate level of cognitive complexity" both in terms of behavior and neural mechanisms.[300] In this capacity insects have been used as models to study short- and long-term behavioral consequences of, among others, addictive drugs,[298] alcohol,[336] diet,[337] sleep,[338] and age-associated cognitive decline.[339] Insect behavior can also be modified in surprisingly similar ways to those seen in mammals by a wide range of pharmacological agents. For instance, drugs that upregulate activity in the acetylcholine neurotransmitter system improve memory processes in both mammals and insects, while downregulation of the same system also has the opposite effects in both.[310] Insects have therefore been used to investigate the behavioral and mechanistic effects of pharmacological agents that modify the functioning of neurotransmitter systems such as serotonin,[340] dopamine,[313] glutamate, GABA,[341,342] and acetylcholine.[309,310,343] Unintentionally, these pharmacological investigations have included the administration of a wide range of plant secondary metabolites to insects—not through a curiosity as to the effects of plant chemicals in insects, but rather on the basis that most of our drugs of abuse and many of our psychotropic medicines are plant secondary metabolites. A number of plant secondary metabolites also have agonist or antagonist effects within an individual neurotransmitter system or even at a single receptor, and they are therefore widely used as probes in neuropharmacological studies; again, many of these studies have used insect models. The results across these studies show surprising parallels between both the mechanisms of action and behavioral effects of plant secondary metabolites in insects and mammals. This topic will be returned to in more detail in the next chapter.

WHY DO PLANT SECONDARY METABOLITES AFFECT HUMAN BRAIN FUNCTION?

This chapter and the preceding chapter suggest two broad hypotheses as to why plant secondary metabolites modify human brain function:

> Hypothesis 1: They modify human brain function because plants and humans share multifarious cellular, biochemical, and molecular similarities.

It is evident from the above that, courtesy of a mutual endowment from their last common ancestor, plants and humans exhibit a wide range of similarities at a molecular and cellular level. The similarities extend to shared signal transduction pathways, conserved cellular stress-response systems, and the synthesis and utilization by plants of a plethora of functional chemicals that we tend to think of as mammalian neurochemicals. These often function in the plant and mammal via similar mechanisms.

This all suggests that some of the effects of plant chemicals on human brain function may simply be a consequence of some of these similarities. In this regard it's possible to identify a number of separate, but potentially contiguous, avenues in which straightforward similarity might drive chemical interactions between plants and humans. These include the following:

- The *similarity in signal transduction pathways:* Phytochemicals, such as secondary metabolites, that fulfill endogenous functions in plants may interact with their intended, conserved, molecular targets in the consuming humans.
- The *similarity between plant stress-signaling hormonal pathways and the mammalian stress-signaling pathways:* Interactions and cross-talk within plant hormonal pathways dictate the synthesis of secondary metabolites. Some aspects of these pathways are conserved in humans, allowing the possibility that functional plant chemicals, including some classes of secondary metabolite, may interact with their intended molecular targets within the analogous human system.
- *Simple chemical similarity:* Plants and humans employ many of the same chemicals in functional roles, and secondary metabolites are often structurally related to these chemicals.

- *A transfer of bioactivity:* The bioactive properties of plant chemicals that are not necessarily similar to those found in humans may be transferred wholesale to the consumer.

Hypothesis 2: They modify human brain function because insects and humans share strikingly similar nervous systems.

In evolutionary terms humans have been almost entirely irrelevant to plants. Even the impact of vertebrates as a family pales into insignificance when set against the enduring and intimate relationship enjoyed by the insect and plant clades. As discussed in the previous chapter, insects and plants have co-evolved side by side for over 400 million years, and it is this relationship that has driven the evolution of many secondary metabolite synthetic pathways. Secondary metabolites play a number of crucial roles for the plant, including, but not limited to, deterrence of the 50% of the 1 million species of insect that survive by eating plant tissue, and attraction of the insects that pollinate more than 70% of all plants. In both roles the imperative driving the plant's evolution has often been interaction with the insect's nervous system, both in terms of interfering with its functioning and co-opting chemicals that have functional meaning in the insect's life (e.g., the many insect pheromones that plants synthesize as attractants). The simple reason, therefore, that many secondary metabolites affect human brain function is that the central nervous system of humans is biochemically, architecturally, and functionally very similar to that of the insect (and indeed other invertebrates). Indeed, given that insects are in many ways the dominant taxa of terrestrial animal, it could be said that humans have an elaborated version of the insect nervous system. Certainly, with only a few minor exceptions the two taxa share the same neurotransmitters and a range of conserved receptors for them to bind to, and more often than not they express similar behavioral modification when these systems are modulated. The effects of plant secondary metabolites on human brain function are therefore simply the unintended echo of the chemicals' intended interactions with the insect nervous system.

Of course, the two hypotheses—"plant/human" and "insect/human" similarities—are not mutually exclusive. In many cases biological concordance between the plant and target organism would be a necessary condition for a novel molecule to interact with the target in the first place. Only if it did so, and did so to the advantage of the plant, would selective pressure for its retention come into play. It is also worth observing that the extent to which the properties of a given phytochemical conform to these two hypotheses depends very much on the chemical class to which it belongs. For instance, the alkaloids (Chapters 4–7) could be described as toxic defense chemicals, and the vast majority of their effects on the human brain function simply reflect their evolved function as toxins and deterrents, primarily targeting insects and other invertebrates. The phenolics (Chapter 8 and 9), on the other hand, play multiple roles in plants, including as signaling molecules, antioxidants, and antimicrobials, and many of their effects would appear to be predicated either on cross-kingdom signaling or a wholesale transfer of bioactivity as a consequence of simple similarities between the taxa. The terpenes (Chapters 10–13) then occupy the middle ground, combining both defense against, and attraction of, insects, along with a number of properties shared with the phenolics. The two hypotheses will therefore be expanded on and illustrated throughout the remaining chapters of the book.

PART TWO
The Alkaloids

4 Alkaloids and the Lives of Plants and Humans

No group of chemicals is more closely interwoven with the history of the human race than the psychotropic alkaloid secondary metabolites. As we saw in Chapter 1, the plant chemicals that have taken humans on their entheogenic journeys to discover their gods within and without, and which have potentially laid the very foundations of the world's monotheistic religions, are almost exclusively alkaloids. On the one hand, caffeine, nicotine, and morphine have bankrolled empires and driven colonization, the subjugation of entire continents, wars, mass slavery, and murder, while at the same time the craving for a single alkaloid, nicotine, continues to kill more humans than any other cause. On the other hand, alkaloids have also formed the bedrock of modern pharmacology and medicine, and they continue to play multifarious important medicinal roles in our everyday lives, while still providing a rich seam for the discovery of new drugs.

Ever since Friedrich Sertürner published his paper describing the isolation, crystallization, crystal structure, and in vivo pharmacological effects of morphine in 1817—in effect creating the first pure, naturally derived medicinal compound—alkaloids have been the subject of intense research. Within a few decades of Sertürner's discovery dozens of important alkaloids had been isolated from plants and characterized, including caffeine, nicotine, atropine, quinine, strychnine, and cocaine.[70] The pharmaceutical industry had been born, and its stock in trade for many decades was largely the extraction of pure alkaloids from plant material. Indeed, it wasn't until 1886 that the first alkaloid, coniine, the poisonous compound from hemlock (*Conium maculatum*), was directly synthesized. When it was, a wave of optimism swept through the research community. The primary targets of the research effort became the synthesis of morphine and quinine. The latter, extracted from the bark of trees in the *Cinchona* genus, had been one of the few medicinal discoveries successfully introduced to Europe following the Spanish conquest of South America. Used as a muscle relaxant and treatment for fever by the indigenous Quecha people, *Cinchona* bark was already in use by Europeans by the beginning of the 17th century. It was isolated in 1820 and proved pivotal to the history of the 19th and 20th century, as it represented the only treatment or preventive measure for malaria. It therefore facilitated the colonization of huge swathes of the tropics and played an important role in World War II. Being able to synthesise and manufacture quinine, rather than relying on the natural *Cinchona* source, would have represented a golden commercial opportunity.[70] However, over time the optimism of the phytochemists dissipated as the magnitude of the task became clear. After 150 years of research the first synthesis of morphine, and definite confirmation of its structure, finally came in the 1950s,[344] although even today two aspects of its synthesis in the plant remain unclear.[70] The first paper describing an unambiguous, stereoselective synthesis of quinine was published in 2001.[345] To date, both morphine and quinine have eluded commercially viable synthesis.[70]

Following the enormous strides taken in the 19th century, interest in the alkaloids waned somewhat during the interwar period. However, isolated alkaloids continued to be identified from traditional herbal treatments. For instance, *Rauwolfia serpentina* (Indian snakeroot), a traditional treatment for insanity, yielded the antipsychotic and antihypertensive indole alkaloid reserpine.[70] The introduction of reserpine, and the synthetic chlorpromazine, both in the early 1950s, revolutionized the treatment of schizophrenia, as they represented the first mainstream pharmacological treatments of any kind for psychosis.[346] Similarly, physostigmine, from the Calabar bean, provided the first treatment for the cognitive deficits of Alzheimer's disease in the 1970s, and its semisynthetic derivatives, such as rivastigmine, and several other plant-derived alkaloid cholinesterase inhibitors, such as galanthamine and huperzine, continue to be the only viable treatments for the disease's symptoms to this day.[195] Psychotropic alkaloids also continued to provide the underlying structure for the development of semisynthetic and synthetic drugs. So, for instance, cocaine's structure gave us a wide range of local anesthetics, including procaine and lidocaine; curare gave us a selection of muscle relaxants; hyoscyamine and atropine formed the basis for an assortment of muscarinic receptor antagonists, including the Parkinson's disease treatment benztropine; and morphine's structure provided an extensive range of analgesics, including pethidine and tramadol.[195] Alkaloid drug discovery naturally encompasses many non-psychotropics as well. For instance, in the early 1960s the traditional West Indian treatment for diabetes, *Catharanthus roseus*, yielded the antidiabetic vinblastine and its close relative, vincristine. It transpired that both of these compounds, plus several other *Catharanthus* alkaloids, had potent anticancer effects, inhibiting and reversing tumor growth. They continue in widespread use in oncology to the present day.[70]

From the 1950s onwards the ethnopharmacognosy approach to drug discovery gave way to programs that involved the systematic mass screening of plants for components that were active in in vitro assays. Typically these screening programs looked for bioactivity relevant to cancer and compounds that could be either used in their natural state or structurally modified for increased potency. This novel approach led to the identification of camptothecin (from *Camptotheca acuminata*), a quinoline alkaloid with cytotoxic properties as a consequence of its ability to inhibit topoisomerase I, an enzyme required for DNA replication.[70] On closer inspection this tree, whose name translates from Chinese as "happy tree," happened to have had a long history of traditional Chinese medicinal use as a cancer treatment.[347]

The drug discovery process rolls on, and the alkaloids synthesized by plants and fungi continue to provide a particularly rich seam of compounds. It is interesting to note that to date, in terms of both ethnopharmacognosy and mass screening, the bioactivities of the phytochemicals present in only about 15% of plants have been investigated.[348]

■ STRUCTURES AND SYNTHESIS

Alkaloids are a structurally diverse group of low-molecular-weight compounds that contain one or more nitrogen atoms, typically as part of an amine group. While no single classification exists, alkaloids are often grouped on the basis of the identity of their nitrogen-containing structural component. Most of the familiar psychotropic alkaloids belong to a handful of these alkaloid groups: tropane (e.g., atropine, scopolamine, hyoscyamine, cocaine); pyridine (e.g., nicotine, arecoline); indole (e.g., ibogaine, psilocybin, dimethyltryptamine, bufotenin, physostigmine, ergotamine); isoquinoline (e.g., morphine); phenethylamine (mescaline); imidazole (e.g., pilocarpine); and purine alkaloids (e.g., caffeine, theobromine). The underlying alkaloid structure for these groups is typically derived from an amino acid, which provides the nitrogen atoms and often leaves its own carbon skeleton largely intact within the structure of the alkaloid. The principal amino acid precursors here are lysine, phenylalanine/tyrosine, tryptophan, and histidine and derivatives such as nicotinic acid and anthranilic acid.[195] A number of alkaloids are also synthesized from non-amino acid precursors, such as acetate and purine.[195]

Of the 27,000-plus alkaloids identified to date in nature, more than 75% are synthesized by plants.[195] They feature as secondary metabolites in approximately 30% of higher plants and are most often found in angiosperms, although other taxa, including ferns, mosses, and gymnosperms, also express alkaloids. Naturally, plants do not have a monopoly on alkaloid synthesis, and a number of alkaloids with specific effects on brain function are synthesized by fungi, including psilocin and lysergic acid. Plants, vertebrates, and invertebrates also share the synthesis of numerous chemicals that have an "alkaloid" structure but tend to be classified as "amines." These include several compounds that we tend to think of primarily as human neurotransmitters, including serotonin, dopamine, histamine, and noradrenaline.[131] Conversely, a number of alkaloids that we think of predominantly as plant secondary metabolites are also synthesized endogenously by animals in small quantities and play modulatory roles in brain function. These include several β-carboline alkaloids,[349] morphine,[266] and several of the tryptamine "hallucinogens," including N,N-dimethyltryptamine (DMT), 5-methoxy-DMT (5-MeO-DMT), and 5-hydroxy-dimethyltryptamine (5-HO-DMT or bufotenin).[267] These chemicals typically play "primary" roles in animal nervous system function. However, there are also many examples of alkaloids being used by animals in secondary or ecological roles. For instance, a wide variety of insects and frogs sequester alkaloids from their diet either to protect themselves from predators or alternatively to use in their chemical communications with their own and other species (see below). Occasionally these plant-derived chemicals are slightly modified by the animal prior to use, but there are also a number of examples of animals synthesizing alkaloids de novo for use in ecological roles such as defense and chemical communication. For instance, the ladybird beetle family (*Coccinellidae*) synthesizes up to 50 distinct alkaloids, including adaline, coccinelline, and harmonine. These compounds are used as defensive compounds that leak from joints in a process termed "reflex bleeding" when the insect is threatened.[350] While there are few mammalian examples, it is interesting to note that the musk deer synthesizes the alkaloid muscopyridine as a component of its musk pheromone excretion,[351] whereas beavers sequester the alkaloid castoramine from the water lily and excrete it as a component of a pheromone secretion that they use to communicate with other beavers.[352]

Some of the alkaloids with particular relevance to brain function are shown in Figures 4.2, 4.3, 5.1, and 7.1.

EVOLUTION OF ALKALOID SYNTHESIS

As described in the previous chapter, the evolution of the plant secondary metabolite pathways took place courtesy of the expansion of plant genomes due to a number of large-scale genetic duplication events. These events generated spare genetic material that could mutate freely without endangering the survival of the plant.[109] The alkaloid biosynthetic pathways then typically arose from mutations in duplicates of the genes coding for enzymes involved in primary synthetic pathways. The novel enzyme would then produce slightly modified chemical products that in turn would be modified by existing, promiscuous enzymes. This process would produce novel chemicals that might, in rare cases, increase the fitness of the plant by filling an ecological need.[113] The distribution of the alkaloid group would then be dictated by the point on the evolutionary tree at which it arose, as well as whether it arose once (monophyletically) or multiple times (polyphyletically), and whether the clades further up the evolutionary tree retained the synthetic capacity.

The benzylisoquinoline family of alkaloids, which includes morphine, codeine, and thebaine, provides a monophyletic example. This group comprises some 2,500 known structures that are synthesized by a diverse minority of clades of eudicot plants. The first committed step in the synthesis of this group involves the condensation of two L-tyrosine derivatives, dopamine and 4-hydroxyphenylacetaldehyde, by (S)-norcoclaurine synthase.[353] The ability to synthesize this enzyme was therefore the first key unique development that gave plants the potential to produce benzylisoquinoline alkaloids. Molecular phylogenetic research shows that the genetic event that potentiated the synthesis of the enzyme, the mutation of one of several primary metabolite proteins, occurred just prior to the diversification of the eudicot and monocot clades, with a minority of groups of eudicot plants further up the ancestral tree from the group subsequently evolving elaborations of the benzylisoquinoline structure to lesser or greater extents.[353] But why has synthesis of benzylisoquinoline alkaloids been restricted to only a small proportion of eudicot species? One possibility is that following the evolution of the biosynthetic pathway the capacity for synthesis remains in many plants but is switched off,[86] or alternatively, mutations in the biosynthetic genes, or other genes playing regulatory or transport protein roles, lead to inactivation of the pathway.[353] In line with this a "latent molecular fingerprint" for benzylisoquinoline alkaloids can be detected in eudicot plants that do not synthesize this group of chemicals.[353] Naturally, the factors that lead to the switching on of synthesis, or its perseverance, and the elaboration of specific alkaloid structures must be rooted in the selective pressure of environmental factors.

In contrast, the synthesis of tropane alkaloids developed polyphyletically following independent genetic events in a number of families of plants. Polyamines, such as putrescine, spermine, and spermidine, are primary metabolites synthesized by virtually all living species via conserved pathways that include the conversion of putrescine to spermidine by spermidine synthase. The resulting compounds contribute to RNA transcription and the regulation of cell growth, differentiation, and cell death.[354] However, in plants of the *Solanaceae* (nightshade) family, which includes genera containing plants such as tobacco, mandrake, henbane, deadly nightshade, and *Datura*, a genetic event following the divergence of the *Solanaceae* from other taxa led to the creation of a duplicate of the gene for spermidine synthase. Free from selective pressure, this gene mutated and evolved a slightly modified function as putrescine N-methyltransferase, an enzyme that catalyzes the conversion of putrescine to N-methylputrescine. This compound represents the first substrate for the synthesis of nicotine and the other tropane alkaloids that give the *Solanaceae* their distinctive psychotropic properties.[114] In general, the *Solanaceae* synthesize tropane alkaloids in their roots and transport them to aerial parts. However, there are some 200 tropane alkaloids, distributed within seven distantly related families. Recent evidence suggests that plants from the *Erythroxylum* genus, whose last common ancestor with the *Solanaceae* existed over 100 million years ago, also independently developed the synthetic pathways that lead to the synthesis of tropane alkaloids, most notably cocaine. In this genus tropane alkaloids are synthesized in young leaves rather than the roots.[355,356]

Similarly, the pyrrolizidine alkaloids are scattered among diverse lineages, with 400 different structures identified in 600 angiosperm species. In this case, the duplication and mutation of another gene involved in polyamine biosynthesis, the gene encoding deoxyhypusine synthase, led to the novel enzyme homospermidine synthase, which participates as the first unique step in pyrrolizidine alkaloid synthesis.[113] This same genetic mutation has occurred at least four times in different angiosperm families.[357,358] Across the species expressing this group of alkaloids there are wide variations within and between species in the specific compounds produced by the pathway, with individual subgroups of compounds typically restricted to one clade. The sites of alkaloid synthesis also vary between clades, with the homospermidine synthase gene expressed in diverse cell types and tissues, ranging from root tips to reproductive organs, suggesting individual evolution of the regulatory elements controlling the enzyme's gene expression.[113] Overall the diversity in structures and sites of synthesis suggests that the first and evolutionarily most ancient steps in pyrrolizidine alkaloid synthesis are highly conserved, whereas the subsequent diversification of their synthesis across species is exceedingly plastic.[358]

The retention of any novel chemical would require that it increases the fitness of the plant to survive. In the case of alkaloids, defense is clearly the driver underlying the evolution of their synthesis. Ehrlich and Raven[92] were the first to suggest that plant chemical defenses arose as a consequence of a complex insect/plant co-evolution driven primarily by a reciprocal "arms race." By this account evolution proceeded by a series of alternating adaptive radiations, with plants developing novel defenses, thereby taking the advantage, and insects countering with their own adaptations, taking back the advantage. The exact nature of the plant/insect relationship in evolutionary terms has since been hotly contested, with the key point of contention being whether the co-evolutionary relationship was symmetrical, as envisaged by Ehrlich and Raven, or asymmetrical. In this latter scenario plants would have evolved and exerted unidirectional evolutionary pressure on insects, who were then engaged in a perpetual game of catch-up, but not vice versa.[90,123] While a number of competing theories have been proposed to explain the evolution of defensive secondary metabolite synthesis, they tend to share one common feature, the notion that the key relationship is that between insects and plants, with little weight given to the possible pressures exerted by other herbivores.

■ ECOLOGICAL ROLES OF ALKALOIDS

Defense Against Herbivory

Evidence suggests that the primary *raison d'être* of the alkaloids as a class of plant chemicals is to function as toxic defense agents against herbivorous animals.[137,359] In particular, given the close relationship between the taxa, most alkaloids act specifically as toxins or feeding deterrents to most insects.[90] Typically, alkaloid biosynthetic pathways will involve a variety of cell types with intermediates in the pathway trafficked from cell to cell. These processes can take place in a variety of cell types and locations within the plant,[360] with many alkaloids being synthesized at a distance from the site of storage, for instance in the roots, and then being transported via the phloem and xylem throughout the plant, accumulating in the tissue that is most valuable to the plant in terms of survival and reproduction.[119,361] The following three chapters will describe some of the evidence confirming a defensive role against insect herbivores for a number of the key alkaloids that modulate brain function.

At this point it is worth visiting some of the phenomena thrown up by the close relationship between insects and plants that support the notion of their co-evolution. Many of these involve alkaloid secondary metabolites. The simplest examples are the many instances of evolved tolerance to specific alkaloids exhibited by the numerous specialist insects that live via herbivory on a single species or clade of plants. A classic example is the larvae of the tobacco hornworm moth (*Manduca sexta*), which consume tobacco leaf but excrete more than 90% of the nicotine they have consumed within 2 hours. In comparison, when exposed to food containing nicotine, the housefly (*Musca domestica*) excretes just 10% of the nicotine it consumes within 18 hours.[90] This evolved tolerance by the specialist can make increasing the synthesis of alkaloids in the face of specialist attack futile, so plants have developed the capacity to differentiate the nature of the attacker, for instance through an "immune system"-like reaction to the individual species' saliva. They can then reduce defense chemical production when faced with specialist attack and instead increase growth to compensate for lost tissue.[362] However, the plant must also strike a balance. So, for instance, the response to the tobacco hornworm saliva and regurgitant can still be to increase the synthesis of nicotine and reduce root growth, as the nicotine does suppress the growth of the hornworm larvae despite its evolved tolerance. Plus, any reduction in nicotine content will make the plant's foliage more attractive to a wide range of specialist herbivores, offsetting any advantage.[363] For the hornworm, despite the detrimental effect of nicotine on its development, it still makes sense to consume tobacco leaf, as any unexcreted nicotine is sequestered and stored in its cuticle as a defense against its own predators.[364] These include a number of generalist parasitoid wasps that lay their eggs inside the hornworm's body. However, in a further twist of the various evolved relationships, specialist parasitoids such as *Cotesia congregata* have themselves evolved tolerance to the nicotine sequestered by the hornworm.[364] For the plant, the many evolved relationships to its alkaloid load at higher trophic levels mean that whereas specific alkaloids act as deterrents to generalist insect herbivores, they can have an opposite, attractant, effect on their own specialist herbivores.[359,365]

Sequestration of alkaloids is also the driver underlying many more complex evolutionary and ecological relationships among diverse organisms. As an example, the pyrrolizidine alkaloids either can be synthesized *de novo* by the plant, or alternatively, in the case of many species of grass, can be sequestered from endophytic fungi that live symbiotically with the grass in a relationship that includes the beneficial transfer of the fungal alkaloids to the grass. These alkaloids then act as feeding deterrents, growth inhibitors, and toxins to the vast majority of generalist herbivore insects. However, a variety of specialist herbivore insects, including species of moth, butterfly, and beetle, feed on the pyrrolizidine-expressing plants and actively sequester the alkaloids, accumulating them in turn in their cuticles and wings as a defense against predation by other insects, birds, and mammals. Of course, some predators have, in turn, evolved the capacity to tolerate the pyrrolizidines. For instance, the harvestman spider (*Mitopus morio*) can eat the larvae of the specialist herbivore leaf beetle courtesy of its own ability to metabolize or excrete the beetle's deadly load of alkaloids.[119] The protective effect of the alkaloids can also ascend another level of the food chain. So, for instance, the "poison dart" frogs from the South American genus *Dendrobates* possess a dietary uptake system for a wide range of alkaloids, including the pyrrolizidines, which they accumulate from their diet of specialist herbivore ants that in turn had consumed the alkaloids from plants and accumulated them in their cuticles. The frogs then express the alkaloids on their skin as a defense against predation.[366] In a final twist, indigenous human tribespeople then use the toxic alkaloid secretions from the frogs' skins to poison the tips of blow darts that are then used to incapacitate their own hunted prey.

To add one final level of complexity, pyrrolizidines from the diet can also function as either pheromones or their precursors for numerous specialist butterfly and moth species. The males excrete the chemicals from hair-like structures on their wings and waft them at the female during courtship. This signals to the female how well endowed the male is with pyrrolizidines, indicating its own fitness and the level of protective alkaloids that will be transferred to any eggs that result from the relationship, offering them, in turn, protection from predation.[119]

Antimicrobial Properties

Alkaloids from a wide range of structural groups have been demonstrated to possess antibacterial and antifungal properties, although it is notable that both the phenolic and terpene groups of phytochemicals have markedly more bioactive members in this respect.[367] Most of the evidence with regard to bacteria is garnered from studies assessing the in vitro effects of plant chemicals on microbes that have a particular relevance as human pathogens, such as *Escherichia coli* and *Streptococcus pneumoniae*. However, evidence does suggest that the effects in these pathogens are related to the ecological role of alkaloids as antimicrobial agents in the plant.[368] The case for an evolved role in this respect is bolstered by evidence of fungi and bacteria evolving the ability to detoxify alkaloids. As an example, endophytic fungi, whose partially symbiotic relationship with the plant resembles that of a specialist herbivore, can convert toxic alkaloids to sources of energy and nitrogen.[368]

Allelopathic Properties

Representatives of all alkaloid classes possess allelopathic properties, and when excreted into the rhizosphere or leached into the soil surrounding the plant at high enough concentrations, they either kill or retard the growth of nearby competitor plants.[137,359] The alkaloids with allelopathic properties include a number that exert specific central nervous system effects in herbivores. A classic example is the purine alkaloid caffeine, which exerts a wealth of behavioral effects in both vertebrates and invertebrates. In its allelopathic role, caffeine-containing berries fall to the ground and "caffeinate" the soil surrounding the plant, reducing and, at higher doses, completely preventing the root formation of encroaching plants by interfering with protein metabolism.[369] Similarly, the tropane alkaloids scopolamine and hyoscyamine, the purine alkaloid theophylline, and the benzylisoquinoline alkaloid berberine all inhibit the germination and growth of other plants.[369,370]

MODES OF ACTION

The toxic effects of alkaloids typically relate to interactions with molecular targets, including proteins and nucleic acids (DNA, RNA), in the consuming organism.[132,137] Given the central role of proteins in every cellular process, this category of targets includes many components that contribute both to a plethora of peripheral physiological functions and also to neuronal function. Interactions in this respect include those with receptors for neurotransmitters/neurochemicals; enzymes that catalyze metabolism and the synthesis of neurochemicals; ion channels that regulate the electrical signals in cells and neurons; transcription factors that activate specific genes; components of the cellular cytoskeleton; and transporters that move ions and molecules across membranes.[132,137] In general terms DNA is a particularly vulnerable molecular target, and a range of alkaloids can interfere with replication, transcription, and translation of DNA/RNA in microbes and animals. As an example, pyrrolizidine alkaloids can bind to functional groups of DNA bases, and a range of other alkaloids (including berberine, sanguinarine, harmine, emetine, quinine, etc.) intercalate themselves between DNA base-pairs, inhibiting replication and protein biosynthesis and instigating apoptosis (programmed cell death).[137] Naturally, as with DNA, many of the target cellular components and biochemical pathways are common to animals and microbes. The psychoactive alkaloids, by definition, are not usually deadly at the doses that humans consume, and their bioactivity is generally attributable to a restricted subset of mechanisms of action. These will be described for the most prevalent psychotropic alkaloids in general terms below, and in more detail in the following chapters.

Modes of Action Related to Brain Function

As we saw in Chapter 3, most of the chemicals that we almost exclusively associate with animal neurotransmission are also present and serve endogenous functions in plants. Many secondary metabolite alkaloids are synthesized from the same amino acid precursors and via the same enzymatic pathways as these "neurochemicals." The resultant compounds often share structural features with the neurochemical. This allows them to directly mimic or disrupt the activity of the neurochemical or otherwise interact at a number of stages in its functional life cycle. They often do this by binding to proteins within the neurotransmission system that would normally be occupied by the endogenous ligand. The clearest example of this confluence of synthetic pathways, structure, and function is manifested in the direct binding of the secondary metabolite to the target neurotransmitter's membrane receptors, leading to either increased or blocked neuronal activity. However, they can also interfere with the synthesis, storage, release, or reuptake of the neurotransmitter—for instance, by interfering with the functioning of the vesicular transporters that package the neurotransmitter, ready for release at the synapse. This effectively either reduces or increases the amount of neurotransmitter released into the synapse. Similarly, they can competitively occupy and either block or reverse the action of the transporter that normally takes the neurotransmitter back into the cytoplasm of the presynaptic neuron. This usually leads to a net increase in the amount of neurotransmitter in the synapse. The principal neuronal modes of action are illustrated in Figure 4.1.

The descriptions of the mechanisms of action of some common psychoactive alkaloids are organized below with respect to their common precursor.

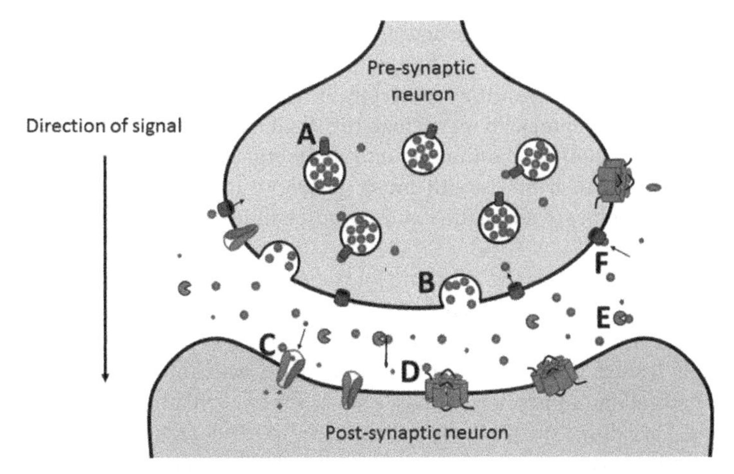

Figure 4.1 Neurotransmission and some mechanisms of action of psychotropic alkaloids. In normal functioning the neurotransmitter is synthesized in the cytoplasm of the presynaptic neuron and packaged into balloon like vesicles by vesicular transporters (A). These are then trafficked to the synapse at a rate dictated by the neuron's activity, where they spill their contents into the synaptic cleft (B). The neurotransmitter that is floating in the synapse can then bind to and activate receptors embedded in the membrane of the postsynaptic neuron (or the presynaptic neuron it has just come from). In the case of an ionotropic receptor (C) this will open a channel to allow negatively or positively charged ions into the neuron, changing its electrical properties, whereas in the case of a G-protein-coupled receptor (D) the signal will be transduced across the membrane, triggering a signaling cascade inside the neuron. The neurotransmitter released into the synaptic cleft is then either metabolized by enzymes (E) and its component parts are reabsorbed, or it is actively pumped back into the cytoplasm of the presynaptic neuron that originally released it by reuptake transporters (F). It is then ready for repacking into a vesicle. The mechanisms by which alkaloids (and other secondary metabolites) interfere with neurotransmission typically include by blocking or reversing the functioning of vesicular transporters (A); by increasing or decreasing neurotransmitter release at the synapse (B); by binding to the receptors (C or D) and mimicking or blocking the activity of the endogenous neurotransmitter; by binding to the recycling enzyme (E), preventing it from metabolizing the neurotransmitter, thereby increasing levels available in the synapse; and by blocking the reuptake transporters (F), again increasing the amounts of the neurotransmitter remaining in the synapse.

Alkaloids Derived from L-phenylalanine/L-tyrosine

In animals the amino acid L-phenylalanine can be metabolized to L-tyrosine by the addition of a single hydroxyl group. L-tyrosine is then the precursor for the monoamine neurotransmitters dopamine and tyramine and their derivatives octopamine, adrenaline, and noradrenaline. Octopamine and tyramine are particularly prevalent in invertebrates, where they substitute for adrenaline and noradrenaline as major neurochemicals, although they also act as "trace amine" neurotransmitters in mammals. In plants the synthesis of these catecholamine "neurotransmitters" proceeds directly from L-tyrosine and they then perform roles in sugar metabolism, photosynthesis, and hormone function. Dopamine also goes on to provide the substrate for a large group of alkaloid secondary metabolites. These include the phenylethylamine alkaloid mescaline, the principal hallucinogenic compounds from the peyote cactus *(Lophophora williamsii)*, and related tetrahydroisoquinoline cactus alkaloids.[195] The contribution of dopamine is readily apparent in the structure of mescaline (Fig. 4.2). While the hallucinogenic effects of mescaline are typically attributed to its agonist effects at serotonin (5-HT) receptors, in particular the "hallucinogenic" 5-HT_{2A} receptor, mescaline also has both dopaminergic and adrenergic receptor binding properties.[371,372] In particular, it binds with more specificity to α_2 adrenergic receptors, whose natural ligands are adrenaline and noradrenaline, than it does to serotonin receptors.[372]

Dopamine also provides the substrate for the more complex benzylisoquinoline alkaloids, which include morphine and related opiates.[202] Once again, the structural residue of L-tyrosine can be detected in morphine's structure (see Fig. 4.2), and it is this feature that morphine has in common with the μ, κ, and δ-opioid receptors' natural opioid peptide ligands, such as the enkephalins, endorphins, and dynorphins. This allows morphine to mimic their activity at the same receptors.[195] Similarly the L-tyrosine/dopamine structure can also be seen in the structure of the protoberberines, a group of benzylisoquinoline alkaloids found in members of the *Berberis* genus. This group includes a number of compounds that interfere with biogenic amine reuptake transporters and exhibit antagonistic dopaminergic and adrenergic receptor binding properties.[373] As an example, berberine, a potential antidepressant and anxiolytic, inhibits the synaptic reuptake of both dopamine and noradrenaline and binds antagonistically to α_2-adrenergic and D_2 dopamine receptors.[374]

L-phenylalanine itself is also the direct precursor of a smaller number of other plant alkaloids, once again endowing them with a structural similarity to the catecholamine neurotransmitters. These sympathomimetics, so called because they mimic the effects of sympathetic nervous system neurochemicals adrenaline and noradrenaline, include cathinone, the principal psychoactive component of khat (leaves of *Catha edulis*), and ephedrine and related alkaloids that are found in several *Ephedra* species. Both cathinone and ephedrine directly mimic the activity of noradrenaline at α- and β-adrenergic receptors and increase the level of the neurotransmitter in the synaptic cleft by displacing it from neuronal storage vesicles.[195,375,376] Cathinone has the additional ability of being able to release dopamine into the synaptic cleft.[375] Cathinone, norephedrine, and ephedrine also provide the starting point and basic chemical structure for the semisynthetic drugs mephedrone, amphetamine, and methamphetamine respectively. All of these drugs also share functional similarities with their precursors, in particular exerting many of their effects in mammals via interactions with the adrenergic system.[195] Amphetamine in particular shares the pharmacological profile of cathinone.[375,376] Both the naturally occurring and semisynthetic sympathomimetic drugs exert similar physiological effects on the sympathetic nervous system, for instance increasing blood pressure in a similar manner.[377]

It is also worth noting that, although it is a tropane alkaloid and technically derived from the amino acid ornithine, cocaine has a benzoyl moiety that is derived from L-phenylalanine, lending it structural similarity to the catecholamines. The principal pharmacological property underlying cocaine's stimulant and euphoriant effects is its ability to bind to and prevent the functioning of monoamine reuptake transporters, leading to increased concentrations of dopamine, noradrenaline, and serotonin at the synapse.[378] The structural similarities between the amino acid precursors, neurotransmitters, and secondary metabolites (and semisynthetic derivatives) derived from L-tyrosine and L-phenylalanine are illustrated in Figure 4.2.

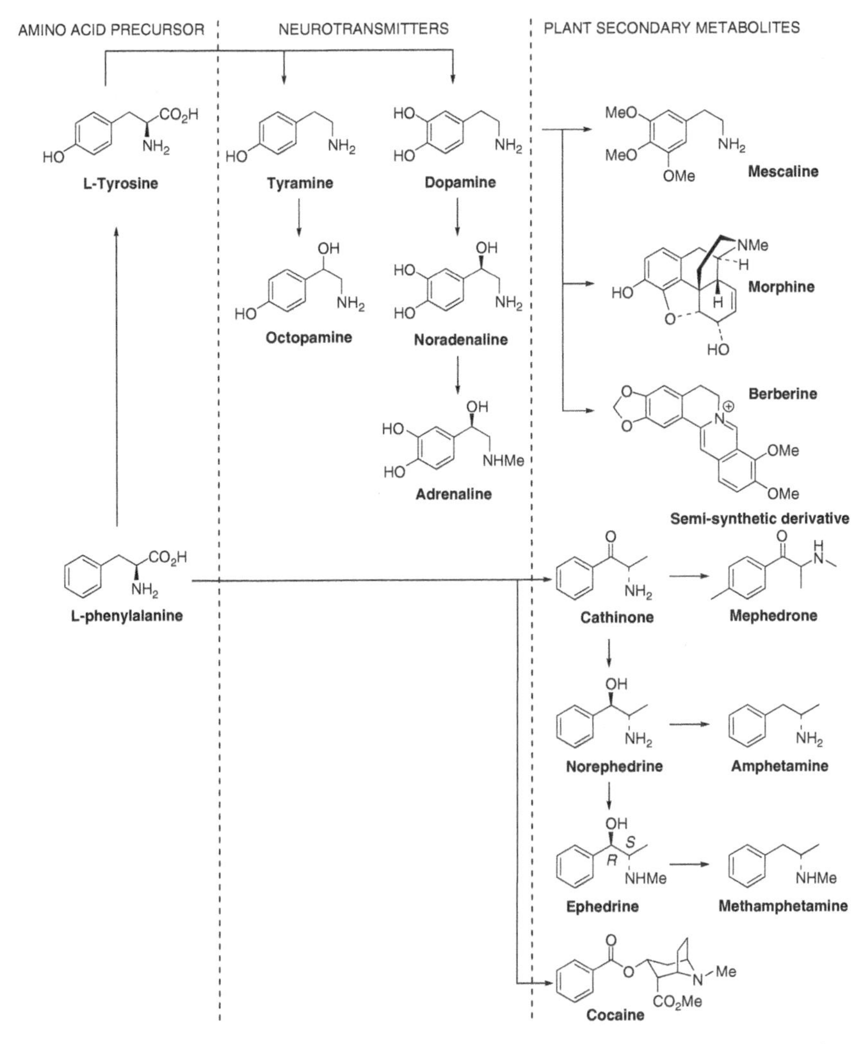

Figure 4.2 Structural similarity between the neurotransmitters and secondary metabolite alkaloids derived from the amino acids L-phenylalanine and/or L-tyrosine. The synthetic pathways are represented by the arrows. In the case of morphine it is the residue of tyrosine which is the structural element that it has in common with the endogenous opioid ligands for the opioid receptors it binds to. The psychostimulants adrenaline, noradrenaline, and cathinone are derived directly from L-phenylalanine, and they then form the substrates for the semisynthetic compounds mephedrone, amphetamine, and methamphetamine, respectively. Cocaine is included here as half of its structure is derived from L-phenylalanine.

Alkaloids Derived from L-tryptophan

The amino acid L-tryptophan is also the precursor to a number of neurotransmitters with an alkaloid structure, including serotonin (5-hydroxytryptamine [5-HT]), melatonin, and the trace amine tryptamine.[268] It also serves as the precursor for a wide range of alkaloids in plants.[195] A number of the psychoactive indole alkaloids are derived from L-tryptophan via tryptamine. These include DMT, 5-MeO-DMT, and bufotenin, all of which are both endogenous mammalian "trace amine" neurotransmitters and hallucinogenic plant secondary metabolites.[267] In addition, psilocin (4-HO-DMT), the active compound from the *Psilocybe* genus of "magic" mushrooms, is also derived from tryptamine, via DMT, and is a close structural analogue of this group. All of these "tryptamine" hallucinogens bear a close structural relationship to serotonin (Fig. 4.3) and have their most pronounced effects through their direct agonistic binding to serotonin receptors, including the "hallucinogenic" 5-HT_{2A} receptor. However, they also bind to a variety of catecholamine receptors, including α-adrenergic and dopamine $D_{1,2,3}$ receptors.[372,379] Naturally, given the endogenous role of trace amines, all of these hallucinogenic tryptamines may owe some of their effects to interactions with trace amine receptors.[267]

Tryptamine also provides one portion of the many terpene-indole alkaloids, with the remaining portion a C9 or c10 fragment of terpene origin. Examples include ibogaine, from the root bark of *Tabernanthe iboga*, a stimulant with hallucinogenic properties at high doses,[73] which has a particularly complex pharmacological profile. It noncompetitively binds to and inhibits both serotonin and dopamine reuptake transporters and directly interacts with a complex array of receptors, including serotonin, dopamine, acetylcholine, noradrenaline, and κ/μ-opioid receptors.[372,380,381] Similarly, the stimulant and aphrodisiac compound yohimbine and the antipsychotic reserpine, both of which are found in the Indian/Chinese traditional medicinal plant snakeroot (*Rauwolfia serpentina*), have complex pharmacological profiles related to their tryptamine structure. Yohimbine has binding affinity for subtypes of α-adrenergic, serotonin, and dopamine receptors, where it typically acts as an antagonist.[382] Reserpine, on the other hand, binds to the monoamine recognition site of vesicular monoamine transporters in peripheral and nervous system tissue, irreversibly preventing serotonin, dopamine, and noradrenaline from being stored in neuronal vesicles prior to release into the synapse, and requiring the regeneration of vesicles before neurotransmitter release can recommence.[383]

L-tryptophan also serves as the direct precursor for a number of other fungal secondary metabolites, including lysergic acid and its derivatives, which are collectively known as the "ergot" alkaloids. This group includes a number of compounds, such as the vasoconstrictor ergotamine, that predominantly interact with adrenergic and serotonin receptors.[195] Psychoactive members of this group include lysergic acid amide, which owes its consciousness-distorting effects to serotonergic activity, including at the 5-HT_{2A} receptor, although it may also interact with dopamine D_2 receptors.[384,385] The most celebrated member of this group, however, is the archetypal semisynthetic hallucinogen lysergic acid diethylamide (LSD), which has a potency some 30 or 40 times that of lysergic acid amide. It is also the most promiscuous of the hallucinogens, exerting excitatory and inhibitory effects at all subtypes of 5-HT_1 and 5-HT_2 receptors, several noradrenaline receptors, and the five subtypes of dopamine receptor.[372] All of the ergot alkaloids owe their major, multifarious activities within the brain to common structural components shared between lysergic acid, the parent compound of the ergot alkaloids, and both serotonin and the catecholamine neurotransmitters. The structural similarities between the amino acid precursors, neurotransmitters, and secondary metabolites derived from L-tryptophan are illustrated in Figure 4.3.

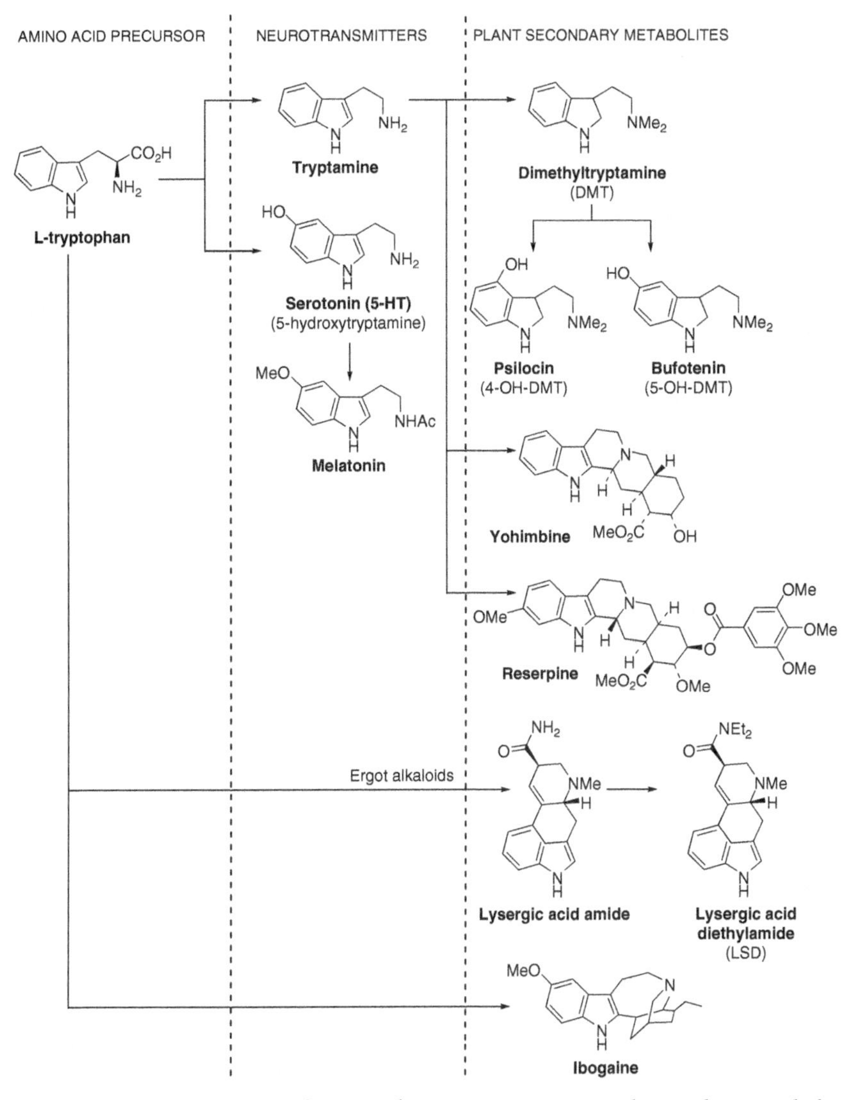

Figure 4.3 Structural similarity between the neurotransmitters and secondary metabolite alkaloids derived from the amino acid L-tryptophan. The synthetic pathways are represented by the arrows. The "tryptamine" hallucinogens DMT, psilocin, and bufotenin, as well as yohimbine and reserpine are synthesized via tryptamine. The ergot alkaloids are synthesized directly from tryptophan and incorporate both the basic structure of dopamine and serotonin.

Alkaloids Derived from Purine

The purine alkaloids have a limited distribution but include caffeine, the most widely consumed psychoactive phytochemical, which is synthesized via several routes of *de novo* synthesis or recycling involving the ubiquitous purine bases adenine and guanine.[195,386] The inhibitory neurotransmitter adenosine is constructed from a molecule of adenine attached to a ribose sugar molecule and therefore shares structural elements with caffeine and the other methylxanthines, such as theobromine and theophylline. Indeed, recycled adenosine can be one potential substrate for the synthesis of caffeine's xanthine skeleton.[386] As a consequence caffeine has its psychoactive effects primarily by binding to adenosine A_1 and A_{2A} receptors.[387] Blockade of the effects of the inhibitory neurotransmitter adenosine at the A_1 receptor leads to increased levels of dopamine, acetylcholine, noradrenaline, serotonin, glutamate, and GABA,[388] potentially leading to improved alertness and cognitive function. On the other hand blockade of the vasodilating effects of adenosine at the A_{2A} receptor leads to decreased cerebral blood flow.[389] The structural similarity between caffeine and adenosine is shown in Figure 5.1 in Chapter 5.

Alkaloids with Activity Unrelated to Common Precursors

Other psychoactive alkaloids simply bear structural motifs similar to those seen in neurotransmitters despite having different precursors. For example, several compounds owe their receptor binding effects to structural elements they have in common with the neurotransmitter acetylcholine, which itself is formed from choline and acetyl-CoA. Key examples here are tropane alkaloids such as nicotine, which is derived from ornithine and nicotinic acid, and hyoscyamine, scopolamine, and related compounds.[195] In the case of nicotine, its psychoactive properties are partially predicated on structural similarities with acetylcholine. Agonistic binding to nicotinic receptors has the knock-on effect of increasing the release of a number of neurotransmitters, including acetylcholine, glutamate, and serotonin. Increased dopaminergic activity in the ventral tegmental area of the brain underlies nicotine's potent addictive properties.[390] Scopolamine and hyoscyamine, on the other hand, compete with acetylcholine to bind to muscarinic receptors due to a similar spatial relationship between an ester linkage and nitrogen atom in all three molecules.[195] Scopolamine then blocks the action of acetylcholine, disturbing brain function and modulating functioning of the parasympathetic nervous system.[391] A similar blockade of nicotinic receptors, with resultant paralysis of voluntary muscles and therefore respiration, underlies the deadly, toxic effects of tubocurarine, the principal alkaloid in curare dart poisons.[195] The structural similarities to acetylcholine of nicotine, hyoscyamine, scopolamine, atropine, and the muscarinic receptor agonist pilocarpine acetylcholine are shown in Figure 7.1 in Chapter 7.

Alkaloids can also modulate neuronal function by occupying and inhibiting the enzymes that degrade neurotransmitters, thereby effectively increasing the amount of the neurochemical and therefore its activity. The most obvious example here is the inhibition of the acetylcholine degrading enzyme acetylcholinesterase (AChE), leading to a net increase in synaptic levels of acetylcholine. This is a property associated with disparate alkaloids, including the isiquinoline alkaloid galanthamine, the sesquiterpene alkaloid huperzine, and the indole alkaloid physostigmine. All of these compounds have been used as treatments for the cholinergic deficits associated with Alzheimer's disease in humans. Similarly, a variety of alkaloids inhibit the monoamine neurotransmitter degrading enzyme monoamine oxidase (MAO), leading to a net increase in the availability of monoamine neurotransmitters such as noradrenaline, dopamine, and serotonin. These include the indole alkaloids harmaline and harmine, which inhibit both AChE and MAO and cause hallucinations at high doses.[137,392,393] Harmine, derived from the vine *Banisteriopsis caapi*, is also used to potentiate the hallucinogenic effects of DMT in decoctions such as ayahuasca by preventing DMT's breakdown by MAO.

A number of psychoactive alkaloids also modulate the functioning of the ion channels that foster the electrical impulses within and between neurons. In common with many animal venoms, the activation of voltage-gated channels by phytochemicals, resulting in the inhibition of subsequent neuronal and neuromuscular signals, can be an effective and deadly defense, and these properties are exhibited by a number of highly toxic alkaloids. For instance, tetrodotoxin, which is best known for its defensive role in the puffer fish and its potentially deadly contribution to the Japanese delicacy *fugu*, is synthesized by bacteria in the gut of the fish as well as other marine microorganisms, and it owes its high toxicity to the blockade of sodium channels.[195] However, there are also many examples of nondeadly ion channel effects, including those exerted by phytochemicals such as cocaine and caffeine as part of their plethora of molecular effects. Similarly ibogaine, which exhibits a complex pattern of receptor and transporter effects in the central nervous system, also dose-dependently blocks potassium channels in the heart.[394]

■ ALKALOIDS AS NEUROPHARMACOLOGICAL PROBES

To investigate the functioning of the nervous system it is necessary to have a set of chemical tools that have very focused effects—compounds that can bind to or interact with a specific neuronal component and modify specific aspects of neuronal functioning in a predictable manner. A number of alkaloids have a long history of use as exactly this type of tool in mammalian neuropharmacological studies. This is predicated on their ability to bind to and modulate the function of specific receptors, neuronal mechanisms, or enzymes. In terms of direct receptor interactions, the classic examples are the alkaloids nicotine and muscarine, both of which were pivotal in establishing the existence, distribution, and function of nicotinic and muscarinic acetylcholine receptors. These compounds were isolated from tobacco (*Nicotiana tabacum*) and the fly agaric mushroom (*Amanita muscaria*) in 1828 and 1869 respectively. In the opening years of the 20th century researchers found that they could trigger physiological responses in nerves. They then went on to play a pivotal role in the elucidation of neurotransmission as a process and the identification of the first neurotransmitter, acetylcholine.[71] The alkaloids subsequently gave their names to the receptor to which they bind and have their agonistic effects, and they continue to be used as neuropharmacological probes. Muscimol, another alkaloid from *Amanita muscaria*, was also used from the late 1960s onwards to identify and characterize the functioning of the GABA neurotransmitter system, and it continues to be widely used in vitro as a potent and selective vertebrate $GABA_A$ receptor agonist.[395,396] Similarly, the benzylisoquinoline alkaloid bicuculline, from several species of the *Corydalis* genus, is routinely used as an in vitro probe that blocks vertebrate ligand-gated ionotropic $GABA_A$ receptors.[397]

The ability of reserpine, one of the first antipsychotics, to deplete monoamine levels in the brain by binding to and blocking vesicular monoamine transporters also made it a useful tool not only as a photo-label to probe the distribution and function of vesicular monoamine transporters[398] but also for assessing the role of differing levels of biogenic amines, for instance on mood and in the response of humans to cocaine.[399] Indeed, reserpine's negative effect on mood, its tendency to create depression in humans, combined with the knowledge of its vesicular transporter effects was a key factor in the development of the influential "monoamine hypothesis" of depression, which attributes mood disorders to abnormal reductions in the levels of monoamine neurotransmitters such as serotonin.[346]

In terms of enzymatic effects, a number of alkaloids, for instance physostigmine, an indole alkaloid from the Calabar bean (*Physostigma venenosum*), have been employed on the basis of their cholinesterase-inhibiting properties to elucidate the functioning of cholinergic neurotransmission. Similarly, harmaline, the β-carboline alkaloid from *Banisteriopsis caapi*, is used on the basis of its MAO-A inhibitory properties to increase monoamine levels in in vivo animal models.[400]

■ ALKALOIDS AND THE INSECT NERVOUS SYSTEM

The preceding description of some of the key mechanisms by which alkaloids have their effects on brain function was drawn from research almost exclusively involving mammalian neuronal systems. As noted previously, it is unlikely that vertebrates are the intended ecological target of many of these secondary metabolites. However, insects and mammals share nervous systems that are identical or very similar in almost all respects, with the sole exception of their respective levels of complexity.

The only notable exceptions to this are the neurotransmitters adrenaline and noradrenaline, which are absent in insects but are functionally replaced by octopamine and tyramine. Even this does not represent a major difference, as the four neurotransmitters are closely related in terms of structure (see Fig. 4.2). Recent evidence has also demonstrated that both of the "insect" neurotransmitters also function as mammalian "trace amine" neurotransmitters, and that mammalian adrenergic receptor ligands often have comparable effects within the insect tyramine/octopamine system.[269] It is also notable that the secondary metabolites that exert effects on synaptic/vesicular release or storage of monoamine neurotransmitters, such as reserpine, cocaine, and ephedrine, also have the same effects with regard to octopamine/tyramine (Table 4.1). Additionally, while there is little information about the binding properties of secondary metabolites at insect octopamine/tyramine receptors, many psychoactive secondary metabolites interact with mammalian trace amine receptors,[317,401] so it's a fair assumption that they also interact with insect octopamine/tyramine receptors.

The mechanisms outlined in the preceding sections above, including the many interactions of alkaloids with neurotransmitters, vesicular and membrane neurotransmitter transporters, and enzymes, are equally prevalent within insect nervous systems. However, overall, the neuropharmacological effects of secondary metabolites in insects, their intended targets, have received very little direct research attention. The evidence that we do have regarding the comparative effects of secondary metabolites in insects and mammals has not generally been accumulated because of a desire to understand the underlying functional *raison d'être* of secondary metabolites; rather, it has accrued as a consequence of two coincidental and tangential lines of research.

In the first place the plant-derived alkaloids used as neuropharmacological probes in mammalian studies have also been employed to characterize the existence and function of receptors, neurotransmitters, enzymes, and other functional apparatus in the insect nervous system. Second, given the many similarities between insect and

TABLE 4.1. *Comparison of the mechanisms of action and behavioural effects of psychotropic alkaloids in insects and humans*

Alkaloid (Group and *example plant*)	Insect models		Human/mammalian models	
	Mechanism	Effect on behavior	Mechanism	Effect on behaviour
Atropine (Tropane—*Atropa belladonna*)	Muscarinic receptor antagonist[414,415]	Impaired memory retrieval.[416] Reduced aggression and brain electrical activity in ants.[417] Knock down and death in *Drosophila*.[415]	Muscarinic receptor antagonist[418]	Impairs memory in primates and humans.[419] Anticholinergic symptoms and death with ascending dose.
Caffeine (Purine—*Coffea arabica*)	Competitive antagonist at inhibitory adenosine receptor[420,421] leading to increased dopaminergic activity,[422] specifically involving dopamine D_1 receptor signaling[420]	Decreased visual learning performance,[423] increased arousal, decreased sleep,[421,424,425] decreased tonic immobility,[426] increased locomotor activity and decreased feeding.[427] Death with ascending dose.[421]	Competitive antagonist of inhibitory adenosine A_1 and A_2 receptors leading to increased dopaminergic and glutamatergic activity[428]	Increased wakefulness and alertness[429] and improved attention in humans.[430] Increased anxiety, restlessness, insomnia, psychomotor disturbance and death with ascending dose.[431]
Cocaine (Tropane—*Erythroxylum coca*)	Inhibits presynaptic reuptake of octopamine,[432] dopamine,[433,434] and serotonin[435] at membrane reuptake transporters. Modulates dopamine synthesis and binding—plus interaction with tyramine.[436]	Increased dancing in bees[437] and grooming in *Drosophila* at low doses. Larval tremors, rearing, and "walk-off"; increased and erratic activity, paralysis, and death with increasing dose.[438,439,440]	Inhibits presynaptic reuptake of dopamine, noradrenaline, and serotonin at membrane reuptake transporters[441]	Intense euphoria, heightened alertness, increased energy, self-confidence, and fearlessness.[442] Paranoid ideation and psychosis with extended use.[441] Hyperthermia, elevated blood pressure, and death with ascending dose.

Continued

TABLE 4.1. *(Continued)*

Alkaloid	Insect mechanisms	Insect behavior	Human mechanisms	Human behavior
Ephedrine (alkaloid—*Ephedra sinica*)	Adrenergic mechanism as physiological effects abolished by β-adrenergic blockade[443]	Disturbs bee thermoregulation[443]	Disrupts vesicular storage, increasing the release of noradrenaline, while also disrupting its reuptake (weak effect on dopamine).[441,444] Some direct effects at α-adrenergic receptors.[441,445]	Mild euphoriant and stimulant. Insomnia/anxiety and possibly psychosis with higher dose or extended use.[446]
Amphetamine and Methamph-etamine (semisynthetic derivates of ephedrine)	Amphetamine: increased synaptic release of octopamine, serotonin, and dopamine[447] Methamphetamine: increased dopaminergic signaling[448] via dopamine D_1 receptors,[420] reduced dopamine reuptake[433]	Increased intraspecific aggression in ants,[449] decreased response to food stimuli.[447] Increased arousal/decreased sleep.[420,448]	Disrupts vesicular storage, increasing release, and inhibits presynaptic reuptake of noradrenaline and dopamine.[441]	Increased alertness, euphoria, increased movement and self-confidence. As dose increases: exaggerated euphoria or dysphoria, paranoia, compulsive behavior. Psychosis with sustained administration.[450]
LSD (semisynthetic derivates of lysergic acid)	LSD: binds agonistically[451] to 5-HT_1 and 5-HT_2 receptors[405,406,451,452,453] and dopamine receptors[453,454] in bee and/or locust. Effects abolished by $5\text{-HT}_1/5\text{-HT}_2$ receptor antagonists.[455]	Ants: fumbling foreleg and antenna movements.[456] Decreased aggression and modulated brain electrical activity.[457] After an initial increase, decreased food sharing, foraging, and social activity.[456,458] Dose-dependent impaired locomotion and death at higher doses.[449] *Drosophila*: reduced locomotor activity and distortion of visual processing abilities.[445]	LSD: partial agonist at the $5\text{-HT}_2/5\text{-HT}_1$ receptors, including 5-HT_{2A} receptors, plus modulation of activity at adrenergic receptors, and dopamine D_{1-5} receptors[371,459]	Impaired cognitive function. Distortions in perception, including time and physical perception of the body, visual hallucinations, changed affect, increased imagination and biographical memory, mystical experiences.[459]
Morphine (Isoquinoline—*Papaver somniferum*)	μ- and κ-opioid receptor agonist.[460] Effects abolished by μ- and κ-opioid receptor antagonists/agonists.[327,461,462,463,464]	Decreased protective reaction to noxious stimuli,[460] including in praying mantis,[462] cricket,[461,463] and honey bee.[464] Increased feeding behavior.[465]	μ- and κ-opioid receptor agonist[466]	Analgesia, euphoria, sedation, reduced gastrointestinal activity, and respiratory depression and death following overdose[69]

Alkaloid	Insect mechanisms	Insect behavior	Human mechanisms	Human behavior
Nicotine (Pyridine—*Nicotiana tabacum*)	Agonist nicotinic receptor binding[467] and increased dopaminergic activity[297]	Improved bee olfactory memory[468] and synaptic plasticity and memory in *Drosophila*.[469] Hyperactivity at low doses—hypokinesis and akinesis at higher doses,[470] loss of "geotaxic" response to gravity and death with ascending dose.[439]	Agonist nicotinic receptor binding leading to increased acetylcholine, glutamate, 5-HT, and dopamine activity[389]	Stimulant effects, increased alertness and improved attention and memory at low doses.[471] Delirium, dizziness, nausea, headache, perspiration, salivation, and fluctuations in blood pressure or heart rate, and eventually death, with higher doses.[472]
Physostigmine (Indole—*Physostigma venenosum*)	Cholinesterase inhibitor and nicotinic receptor agonist[473]	Increased trials to habituation of proboscis extension reflex[408]	Cholinesterase inhibitor and allosteric nicotinic receptor agonist, but at higher doses blocks open channels[474]	Improves memory in aged primates and humans[419]
Pilocarpine (Imidazole—*Pilocarpus* genus)	Muscarinic receptor agonist[414]	Improved memory (nestmate recognition).[342] Knock down and death.[415]	Muscarinic receptor agonist[418]	Improved cognitive function including memory in old rats[475] and young primates (but high level of cholinergic side effects)[476]
Reserpine (Indole—*Rauwolfia serpentina*)	Depletion of monoamines (dopamine, serotonin, tyramine, octopamine) via inhibition of vesicular monoamine transporter in presynaptic terminal[477,478]	Lethargy and decreased locomotor activity[478,479] and decreased response to food stimuli[408,447]	Depletion of monoamines (dopamine, serotonin, noradrenaline) via inhibition of vesicular monoamine transporter in presynaptic terminal[382,397]	Antipsychotic, but tardive dyskinesia with extended use[480]
Scopolamine (Tropane—*Atropa belladonna*)	Muscarinic receptor antagonist[414,415]	Memory decrements in bees.[310,342,414] Reduced aggression and brain electrical activity in ants.[417] Knock down and death.[415]	Muscarinic receptor antagonist[390]	Reduced alertness; psychomotor, cognitive, and memory decrements.[481] Anticholinergic side effects and death with ascending dose.

mammalian nervous systems (and the many advantages of using cheap, rapidly reproducing and aging organisms in less ethically contentious research), insect models have been used in the elucidation of the mechanisms of action and behavioral effects of a wide range of medicinal and social drugs derived from plants, with the results often extrapolated back to vertebrates, including humans.

Examples of alkaloids used as probes include the use of nicotine and muscarine, from the 1960s onwards, to identify the existence and function of both muscarinic and nicotinic receptor types in the fly, and to characterize the properties of the insect cholinergic system.[402] Muscimol was more recently used as a ligand in the identification of the existence of insect GABA receptors,[403] at which it produces the same agonistic effects as at the analogous mammalian receptor.[404] Similarly, both LSD and the mammalian serotonin receptor ligand DMT, which is both an endogenous mammalian trace amine and a "tryptamine" alkaloid synthesized by *Psychotria viridis* and several other plants, have been used as ligands in the characterization of serotonin receptors in the insect brain.[405,406] Reserpine has also been used to assess the contribution of monoamine neurotransmitters to insect behavior, for instance in the aggressive response of crickets to interlopers[407] and the proboscis extension reflex in bees.[408] While the above examples all demonstrated comparable effects across the taxa, the use of these alkaloid probes has also, on occasion, illustrated the divergence in morphology and function of neurotransmitter systems in vertebrates and invertebrates. For instance the benzylisoquinoline alkaloid bicuculline is routinely used as an in vitro probe that blocks vertebrate ligand-gated ionotropic $GABA_A$ receptors, thereby isolating their effects. However, the vast majority of insect GABA receptors (with the only recorded exception to date being found in the moth Manduca sexta[409]) are insensitive to bicuculline due to morphological differences in the receptor.[410] However, bicuculline does modulate the function of insect glutamate receptors[411] and inhibits nicotinic receptor function in both vertebrates[412] and insects.[413]

If we turn to the use of insect models in investigations of the nervous system and behavioral effects of a range of medicinal and social drugs, we find a clear correspondence in insects and mammals in the functional effects of activating or antagonizing a variety of receptors and neurotransmitters with alkaloid secondary metabolites (see Table 4.1). For instance, administering nicotine at low doses to insects leads to similar improvements in cognitive function and the same increases in physical activity as those seen in mammals (including humans), with decrements and paralysis creeping in with ascending dose. Similarly, the muscarinic receptor agonist pilocarpine and antagonists atropine and scopolamine improve or disrupt memory and other aspects of cognitive function in insects in the same manner and via the same mechanisms as those seen in mammals.

Our other social and illicit drugs also have comparable effects in insects. For instance, in both mammals and insects caffeine increases arousal and disturbs sleep; cocaine and ephedrine and semisynthetic derivatives such as amphetamine and methamphetamine increase activity and arousal; and morphine increases tolerance to noxious stimuli. Similarly, the lysergic acid derivative LSD also has profound effects on insect behavior, disturbing movement and social behavior and distorting visual perception. The evidence showing common mechanisms of action and behavioral correlates in insects and mammals following the administration of alkaloids is summarized in Table 4.1. A more detailed description of the effects of these compounds on insect brain function is provided in the relevant sections of the next three chapters.

■ SOME CONCLUSIONS

For now it is probably adequate to note that the evidence summarized in Table 4.1. demonstrates quite clearly that both the mechanisms of action and the physiological or behavioral effects of many psychotropic alkaloids can be strikingly similar in humans and insects. Given that the evolutionary pressure driving the diversification of secondary metabolite synthesis in plants would appear to be overwhelmingly predicated on the relationship between plants and insects, it is possible to conclude that where alkaloids affect brain function in humans this is simply because we share strikingly similar nervous systems to their intended insect targets. This theme will be returned to in more detail in the next three chapters.

5 The Rewarding or Addictive Drugs

Only a very small subset of plant secondary metabolites have "rewarding" effects in mammals. This somewhat exclusive group comprises primarily a disparate group of alkaloids including morphine, nicotine, cocaine, ephedrine, cathinone, and caffeine, plus the terpenophenolic tetrahydrocannabinol from cannabis. To a lesser or greater extent each of these compounds actively reinforces its own consumption, making the consumer want to consume them again, in turn making them habit-forming or addictive. Because of this property this small group of compounds, and several of their semisynthetic derivatives, have exerted a disproportionate influence over the modern history of mankind (see Chapter 1). This rewarding property is not shared by the many other psychoactive secondary metabolites reviewed in later chapters (cannabis will be reviewed in Chapter 12).

The reason that these compounds reinforce their own consumption in humans is rooted in the survival of species. To prosper, the members of any species of animal must carry out a wide range of behaviors that increase their ability to survive and reproduce. In the case of humans these behaviors include adaptive activities such as eating, drinking liquids, social and sexual activity, and avoiding danger. Whether or not a member of a species carries out these activities is not left to chance. To make sure that they behave appropriately, animals have evolved specific brain mechanisms that provide the motivation to carry out the adaptive activities. These can best be described as the "reward pathways," a collection of interconnected brain structures and regions that identify the behavior or stimuli with salience for the preservation of the species as being rewarding, or pleasurable, and stamps its memory with this information. The attachment of the reward information then reinforces the behavior, motivating approach and consumption behaviors, and allowing a subconscious prediction of the future reward associated with the behavior through associative conditioning.[482,483]

In essence this pathway makes us want to carry out the rewarding behavior, whatever it might be. In mammals, the brain regions involved in the reward process are described as the 'mesolimbic dopamine pathway' and include the following: the amygdala and frontal cortex, which contribute the reward/pleasure and emotional salience to the stimuli; the hippocampus, which encodes the subsequent memory; and, most importantly, the ventral tegmental area and the nucleus accumbens, which stamp the memory of the stimulus with the rewarding information. Activity in the nucleus accumbens, in particular, is the central defining property of the reward pathways,[484] and activity in this brain structure is linked, for instance, to aberrant overconsumption of rewarding behaviors such as eating, sexual behavior, and gambling.[485,486]

The reason that the "rewarding drugs" reinforce their own consumption is that their many effects within neurotransmitter systems includes "hijacking" the reward pathways by activating the key reward pathway structures (the ventral tegmental area and the nucleus accumbens). The activity in these specific brain areas lends these drugs their reinforcing properties, consistently stamping their use with the erroneous information that they are rewarding, often to a greater extent than natural stimuli, thereby making the consumer want them.[484] The establishment of the relationship between the drug and the reward is learned by a process of operant conditioning. Therefore, the strength, frequency, and speed of onset of the effects of the compound in question (i.e., how strongly and quickly the reward follows the behavior) are all key factors in establishing the association between taking the drug and its rewarding properties and therefore how much it is wanted or, alternatively, its addictiveness. Typically, drugs that have a faster onset or are taken by a route associated with faster bioavailability in the brain are more addictive.[487,488] For the more potent drugs the subsequent descent into addiction can then be seen as a chain of events. This starts with the activation of the mesolimbic dopamine reward pathways the first time the drug is taken and then progresses with continued use via a wide-ranging and complex series of neuro-adaptations, including habituation and sensitization to the effects of the drug within the relevant neurotransmitter systems. These culminate in the negative physical and affective symptoms associated with withdrawal. Eventually, disruption of the functioning of the prefrontal cortex and activation of the brain's stress system can also lead to a powerful drive to seek the drug, which is coupled with a lack of overriding, prefrontal control over the inadvisable behavior.[484]

Each of the rewarding drugs exerts its own individual palette of neuropharmacological effects, and the mode by which they activate the ventral tegmental area and the nucleus accumbens is therefore different in each case.[484] These mechanisms of action will be described in the individual sections for each alkaloid secondary metabolite below. Nicotine has been included alongside the other deliriants from the *Solanaceae* family in Chapter 6, and tetrahydrocannabinol, which belongs to the terpene group of phytochemicals, will be reviewed separately in Chapter 12.

■ MORPHINE

The home range of the opium poppy (*Papaver somniferum*) is on the Anatolian coast of the Black Sea.[489] The exudate that oozes from the cut skin of the poppy's immature seed pod has been collected, air-dried, and used as a euphoric, analgesic, and medicinal panacea since at least the epoch of the Sumerian culture that flourished from ~4000 to 3000 BC in Mesopotamia.[29] The subsequent use of opium followed the tides of migration and the ebb and flow of trade, civilizations, and empires as it radiated across the Eurasian continent (see Chapter 1).

The milky latex of the opium poppy contains approximately 40 alkaloids, but the bulk of the alkaloid content is attributable to just five: morphine, the principal constituent, which makes up 8% to 17% of the dried latex, codeine (0.7% to 5%), narcotine (1% to 10%), papaverine (0.5% to 1.5%), and thebaine (0.1% to 2.5%).[29]

The opium alkaloids themselves belong to a wider group of over 2,500 benzylisoquinoline alkaloids that are derived from the amino acid tyrosine via dopamine. The structure of morphine and its precursors are shown in Figure 4.2 in Chapter 4. This group of alkaloids comprise a wide range of typically species-specific compounds, synthesized by a diverse minority of species within a number of orders of angiosperm. The ability to synthesize benzylisoquinoline alkaloids arose monophyletically prior to the differentiation of the eudicot and monocot clades, and the "latent molecular fingerprint" that would have allowed their synthesis can still be detected in many non-synthesizing eudicot plants.[353] Naturally, the factors that lead to the switching on of synthesis in the minority of eudicot species that express benzylisoquinoline alkaloids must be rooted in the selective pressure of environmental factors or the emergence of other facilitating physiological mechanisms. In the case of *Papaver somniferum* the selective pressure for the synthesis of the opiate alkaloids can be assumed to be defense against insect herbivores. However, it may well have been the case that the emergence of specialized cells, known as laticifers, which are external to the vasculature of the plant and which produce toxic alkaloids and store them safely in a sticky latex (see below), ultimately tilted the cost/benefit balance in favor of the synthetic pathway being activated.[353,490]

The opiates and their many semisynthetic derivatives have provided us with a treasure trove of contemporary medicinal drugs. However, the opium poppy is probably most notorious for providing us with drugs of abuse. While the air-dried latex, known as opium, has no doubt been used in a social or recreational context throughout its history, three separate events propelled its principal alkaloid, morphine, to the front rank of the addictive drugs. The first was the isolation of morphine as a single compound by Friedrich Sertürner in 1806. This was followed by the invention of the hypodermic needle and syringe in 1853 by the Scottish physician Alexander Wood, who modeled his creation on the bee sting and originally intended it solely for the administration of morphine. The much faster mode of intravenous administration increased morphine's rewarding properties and led to it becoming widely preferred over previous oral opium preparations.[491] The final event was the synthesis of diacetylmorphine by the English chemist C. R. Alder, who failed to appreciate the importance of his discovery. It was subsequent rediscovered two decades later by the Bayer AG pharmaceutical company. Diacetylmorphine was originally marketed in Germany as a cough remedy and later as a supposedly nonaddictive treatment for morphine dependence. Its original brand name, Heroin, remains in common usage when describing the drug when used illicitly. The key property of heroin that makes it so potent is the addition of the two acetyl groups. These increase the molecule's lipophilicity and allow it to cross the blood–brain barrier far more quickly than morphine. Once in the brain it is rapidly metabolized back to morphine and mono-acetylmorphine.[466] The introduction of heroin at the turn of the 20th century created a huge illicit demand for the drug, which persists today.[69] Indeed, the global market value of the illicit opium and heroin trade is estimated by the United Nations at $68 billion, of which heroin represents approximately 90%. More than 16 million people around the world use illegal opiate products each year.[492]

In terms of human pharmacology, morphine is primarily a selective agonist at the G-protein-coupled μ-opioid receptors that are the natural binding sites for the body's endogenous painkillers, the endorphins and enkephalins. This binding affinity is due to the structural residue of tyrosine that can be detected in the structure of both morphine and the endogenous opioids (see Fig. 4.2).[195] The site of morphine's activity was actually discovered before the natural ligands for the receptor were first identified, so μ, the Greek letter "mu," refers to morphine. When the first endogenous ligands for these receptors were identified in 1974 they were then named endorphins, a contraction of "endogenous morphine."

The μ-opioid receptors themselves are distributed throughout the key areas of the nervous system involved in the perception and management of pain, including the dorsal horn of the spinal cord, the amygdala, and brainstem regions such as the periaqueductal gray area. They are also expressed in several layers of the cerebral cortex, the mesolimbic reward pathways, the nucleus accumbens, and peripherally in the intestinal tract. This widespread distribution means that, as well as analgesic properties, μ-opioid receptor activation also leads to euphoria, sedation, reduced gastrointestinal activity, and, courtesy of the brainstem, the respiratory depression that kills in the event of a morphine overdose.[69] Within the mesolimbic dopamine reward pathways the μ-opioid receptors themselves function by inhibiting the release of the inhibitory neurotransmitter GABA, leading to a net increase in the release of other neurotransmitters, including dopamine, and therefore neuronal excitability. The rewarding properties of morphine are then dependent on its direct binding to μ-opioid receptors within the ventral tegmental area and nucleus accumbens, leading to increased activation via dopamine release.[466,484] As noted above the comparative rewarding properties, and therefore addictiveness, of differing modes of morphine delivery depend on the amount of morphine delivered and the tightness of the relationship between administration of the drug and activation of the reward pathways. So, pure morphine is more potent than opium; smoking, which bypasses the digestive system, provides a quicker reward than oral consumption; injection directly into the bloodstream provides the reward faster again than smoking; and diacetylmorphine (heroin) crosses the blood–brain barrier more rapidly than morphine but is then rapidly metabolized back to morphine in the brain. The multifarious permutations of drug and delivery dictate the potential for abuse of morphine, with injected heroin the most potent combination in terms of potential for dependence and abuse.[487,488]

The extended use of morphine (and the "psychostimulants," see below) leads to long-term regulatory changes in neurotransmitter systems that far outlive cessation of drug taking and that lead to a multitude of unpleasant withdrawal symptoms. These changes include a particularly long-term upregulation of the κ-opioid receptor system at which the endogenous painkiller dynorphin acts. Although both systems produce analgesia, the κ-opioid/dynorphin system naturally antagonizes the effects of the μ-opioid/endorphin system, and therefore it naturally downregulates the effects of morphine, both in terms of its euphoric effects and the dopaminergic activation of the nucleus accumbens. Unfortunately, continued use of morphine leads to a compensatory upregulation of the κ-opioid system that persists long after any μ-opioid–related effects have dissipated. The net effect is the long-term dysphoria, depression, and anxiety that outlives physical withdrawal symptoms and provides ongoing negative reinforcement for the reinstatement of morphine consumption.[466]

Naturally we tend to think of morphine and related opiates as chemicals synthesized by a plant that just happen to usurp the functioning of opioid receptors. However, endogenous synthesis of morphine and several related alkaloids, by similar enzymatic pathways to those seen in plants, has recently been confirmed in both invertebrate and vertebrate animal tissue, including the brain.[264,266] Alongside this, specialized "μ_3"-opioid receptors have been identified that are insensitive to the classic endogenous opioids (endorphins/enkephalins) but that strongly bind morphine (and the synthetic μ-opioid receptor antagonist naloxone).[264,329] It transpires that endogenously synthesized morphine, acting via μ_3-opioid receptors, plays a variety of signaling roles across phyla, including modulation of vascular and immune function and the modulation of stress responses. All of these effects can be attributed to interaction with constitutive nitric oxide synthesis.[264] In this endogenous role morphine has been shown to extend the lifespan of rodents and insects and to have a wide range of protective metabolic, cellular, and physiological properties, including enhanced synaptic plasticity and neuroprotection in mammals and a modulatory effect on the activity of the structurally related catecholamine neurotransmitters (see Fig. 4.2).[266,493]

Ecological Roles of Opiates and Brain Function

The first steps of opiate synthesis take place within the vascular bundles throughout the opium poppy, with the final enzymatic stages and alkaloid accumulation taking place in the laticifer network itself, where the alkaloids are stored in vesicles.[494] Laticifers in the aerial parts of plant, and their associated latex, are specifically mechanisms for the defense against insect herbivory. They deliver the double defense of physically compromising the insect with their sticky exudate, which can immobilize/drown the insect or gum up its mouth parts, and at the same time they deliver a variety of toxic secondary metabolites. In the case of the opium poppy, morphine represents up to 17% of the dry weight of latex.[29] It can therefore be assumed that morphine and the other opiates primarily function as chemical defense agents targeting insects, although as latex also provides a physical barrier that hardens over any wound they may also function as antimicrobial agents.[136,490]

Research using invertebrate models confirms that morphine affects behavior in insects and other invertebrates in a similar manner to mammals. For instance, administration of morphine increases pain thresholds and decreases protective behaviors and "fearful" responses to noxious stimuli.[465,495,496] As an example, it suppresses the escape behavior of crickets from a heated box to the extent that they suffer irreparable damage.[496] It also modulates insect feeding behavior and motor activity, with demonstrations of both increased activity and long periods of immobility, although these effects depend on the species and administration regime.[328,465] The similarities with humans extend to the development of both tolerance and behavioral sensitization to the drug's effects with extended administration.[496] Morphine also plays some distinct ecological roles specifically relevant to insects, in particular disrupting their life course by producing a dose-dependent delay in pupation and in the eclosion of adults, and significantly reducing survival to adulthood.[327] Similar effects are also seen in the offspring of adult flies administered morphine prior to egg laying.[497]

In terms of mechanisms, both μ- and κ-opioid receptors and the endogenous opioid receptor ligand enkephalin are widely expressed in the insect brain and nervous system,[325–328] and this system has been confirmed as the site of morphine's activity by studies involving the co-administration of the opioid receptor antagonist naloxone, which blocks its analgesic and behavioral effects.[326,328,496,498] As noted above, endogenously synthesized morphine and its recently identified μ_3-receptors have also been identified across invertebrate phyla,[264] including insects.[329]

The potential deleterious effects of compromising the insect opioid system is nicely illustrated by the parasitoid wasp *Ampulex compressa*, which injects an opioid receptor agonist directly into the brain of its cockroach victim, reducing its volitional control of its own movement, and allowing the wasp to lead it by an antenna to its fate as an edible, living egg depository within its burrow.[325]

■ THE PSYCHOSTIMULANTS (COCAINE, EPHEDRINE, CATHINONE, AMPHETAMINE, METHAMPHETAMINE)

The psychostimulant plant alkaloids—cocaine, ephedrine, and cathinone, and their semisynthetic derivatives, amphetamine and methamphetamine—exert a common pattern of psychological effects in humans. To a greater or lesser extent they all produce elation, euphoria, mood elevation, increased alertness and concentration, and reductions in fatigue.[499] They also exert common "sympathomimetic" physiological effects related to their noradrenergic activity, for instance engendering similar increases in blood pressure and heart rate.[377] Finally they are all also "rewarding," with a potential to become addictive and engender negative symptoms of withdrawal. They owe these effects to either one or both of two mechanisms. The first is the binding to and blocking of one or more of the reuptake transporters that actively transport molecules of the monoamine neurotransmitters dopamine, noradrenaline, or serotonin back into the cytoplasm of the presynaptic terminal from which it was originally released. The second mechanism is by increasing neuronal release of the same neurotransmitters by disrupting vesicular storage, thereby increasing cytoplasmic levels of the neurotransmitter and promoting efflux into the synapse by reversing transporter-mediated exchange. In both cases the net effect is increased levels of the neurotransmitter in the synapse.[441,444]

The profile of these uptake and storage effects varies between the compounds. Cocaine is solely a reuptake inhibitor, blocking dopamine, noradrenaline, and serotonin transporters. Ephedrine, on the other hand, primarily increases noradrenaline release and inhibits its reuptake and has modest similar effects with regard to dopamine and no effect on serotonin levels. Amphetamine and methamphetamine release and inhibit the uptake of both noradrenaline and dopamine, with markedly more modest effects on serotonin.[441] Cathinone similarly has noradrenergic and dopaminergic properties. While these compounds do not typically bind directly to the classic monoamine receptors, emerging evidence does suggest that the amphetamine-like psychostimulants do bind directly to the recently identified "trace amine associated receptors" (TAAR1 to TAAR9; see also Chapter 3). These receptors co-localize with monoamine transporters and play a modulatory role by binding endogenous monoamine neurotransmitters and other trace amines such as tyramine and octopamine. Direct agonist properties at these receptors may well modulate the psychostimulants' effects on neurotransmitter uptake and release.[500]

The above stimulants also have one other related common property: potent activation of the dopaminergic mesolimbic reward pathways, in particular the nucleus accumbens, via their direct actions on extracellular dopamine levels at nerve terminals within these areas.[484,499,501] The reward pathways are also particularly rich in TAAR1 receptors, which appear to have a natural function as mediators of the modulatory effects of trace amines on dopamine neuron firing activity. This may go some way toward explaining the relative specificity of these compounds with regard to these brain regions and may provide a possible mechanism underlying the behavioral sensitization produced by extended use of psychostimulants.[502] Overall, the euphoriant and reinforcing or addictive potential of each psychostimulant is predicated on its comparative dopaminergic potency. So, for instance, ephedrine, which has only modest effects on the release and reuptake of dopamine, is a comparatively mild euphoriant and has low potential for addiction in humans.[503] The differences in mechanisms of action and the resultant behavioral/psychological effects of each alkaloid will be reviewed below.

Of course, the ability of the psychostimulants to bind to and modulate the activity of vesicular and membrane monoamine transporters and TAARs is directly related to their structural similarity to the monoamine or trace amine neurotransmitter in question. In turn, this similarity is predicated on their common precursors. In animals the trace amines/neurotransmitters tyramine, octopamine, dopamine, adrenaline, and noradrenaline are synthesized from the amino acid L-phenylalanine via the decarboxylation of L-tyrosine. In plants L-phenylalanine serves as the precursor for the synthesis of the phenethylamine alkaloids cathinone and ephedrine and a suite of related alkaloids found in the plants from the *Ephedra* genus and *Celastraceae* family.[195,377] The archetypal members of these groups of plants in terms of human consumption are *Ephedra sinica* and *Catha edulis* (khat). The alkaloids from these plants, in turn, serve as precursors in the illicit manufacture of semisynthetic drugs of abuse such as amphetamine and methamphetamine. While cocaine belongs to the tropane alkaloids, a separate group partially derived from the amino acid ornithine, its structure incorporates a benzoyl moiety that is derived from L-phenylalanine, lending it structural similarity to the phenethylamines and the catecholamine neurotransmitters.[195] The amino acid precursors, neurotransmitters, and psychostimulant alkaloids are shown in Figure 4.2.

Cocaine

Erythroxylum, the genus of South American tropical flowering shrubs, includes more than 200 species of coca plant. However, the leaves associated with chewing coca and cocaine production are obtained from either of two geographically separated species, *E. coca* and *E. novogranatense*, that grow naturally on the Eastern slopes and Western slopes and highlands of the Andes respectively.[195] Coca leaves contain a cocktail of alkaloids that represent between 0.7% and 2.5% of their dry weight. Cocaine is the most abundant alkaloid, but other alkaloids include cynnamoylcocaine, α- and β-truxilline, tropococaine, hygrine, hygroline, and cuscohygrine. In keeping with a protective role the alkaloids are preferentially synthesized in young leaves at the ends of branches and reproductive tissues.[355,504] Cocaine itself is a tropane alkaloid, derived from the amino acid L-orthinine. Most of mankind's 8,000-plus–year relationship with cocaine was played out in the Northern reaches of the Andes (see Chapter 1). It only really came to international attention in the mid- to late 19th century, when cocaine was isolated and a number of patent medicines and consumer products containing cocaine or coca extracts, including Coca-Cola, were introduced into the European and American markets. The boom years for cocaine began in the 1980s with the emergence of the drug cartels, ruthless criminal gangs based within the coca-growing regions and the transit countries of Middle/South America, who flooded the North American market with what had been a relatively obscure drug. In the wake of the vast profits being drained from the United States came escalating, industrial-scale violence, with an estimated 12,000 drug industry-related murders in Mexico alone in 2011.

By the mid-1990s cocaine use had actually been in decline in the United States, but this trend was reversed with the introduction of the much more addictive "crack" cocaine.[77] To give an idea of the scale of the market, the value of illicit cocaine globally was estimated at $85 billion in 2009, with the two biggest markets for cocaine, the United States and Europe, contributing $38 billion and $34 billion of this figure respectively.[492] This meant that cocaine represented a substantial proportion of the estimated $320 billion global drug market[79] and was the second most valuable drug commodity (after cannabis). The global number of regular users was estimated in the same year at 17 million people. Data from 2006 show that in the 12 months leading up to their being surveyed, some 6 million Americans over 12 years of age reported using cocaine, and 1.5 million of these had used crack cocaine, with more than 5% of students in the last year of high school using cocaine products.[505]

Cocaine is generally consumed in two forms: cocaine hydrochloride salt (following treatment with hydrochloric acid), which is water-soluble and therefore readily absorbed through the nasal mucosa, and the "crack" cocaine alluded to above. This is manufactured by dissolving cocaine hydrochloride in water with sodium bicarbonate and then heating. The eventual crystalized cocaine has a melting point at 98°C, making it ideal for smoking, with the name derived from the popping and cracking sound of the crystals as they melt.[442] Both cocaine hydrochloride and crack cocaine activate the dopaminergic mesolimbic reward pathways and nucleus accumbens via their direct actions on dopamine release in these areas.[484,499] However, epidemiological evidence shows that smoked crack cocaine is three times more likely to result in clinical levels of dependence than cocaine hydrochloride.[506] This, again, is due to the rapid onset (6 to 8 seconds) of the euphoria induced by crack cocaine, which, in effect, strengthens the learned association between the drug stimulus (the act of smoking) and the reward (euphoria related to activation of the reward pathways). In contrast, the subjective effects of cocaine hydrochloride, which are the same as crack cocaine, are only felt several minutes after snorting, promoting a weaker, but still potent, association.[442,487]

As well as its primary effect of blocking monoamine neurotransmitter reuptake transporters, leading to increased dopaminergic, noradrenergic, and serotonergic activity, cocaine also blocks fast sodium channels in neuronal and heart cells. This mechanism underlies the anesthetic effects of cocaine and synthetic derivatives but also leads to potentially fatal disturbance in cardiac function.[442]

The acute effects of cocaine include intense euphoria, heightened alertness, increased energy, self-confidence, and fearlessness.[442] However, extended acute intoxication with cocaine can lead to paranoid ideation and psychosis. Withdrawal is then often associated with symptoms of "dopaminergic" deficits, such as anhedonia and psychomotor disturbances, and "serotonergic" symptoms such as depression, obsessive thoughts, and lack of impulse control.[441] Other potential side effects of acute or chronic use include ischemic or hemorrhagic strokes, seizures, and cardiac, pulmonary, and gastrointestinal disorders.[442]

Ephedrine

The *Ephedra* genus of gymnosperms is indigenous to temperate and subtropical regions of Asia, Europe, and North and Central America and comprises approximately 45 species of shrubs with slender aerial stems and tiny leaves. Plant material typically contains 0.5% to 2% of alkaloids, with ephedrine, the principal alkaloid, typically representing 30% to 90% of the total alkaloid content.[195]

Extracts of the stems or whole plant of *Ephedra sinica*, and up to 11 other species, collectively known as Ma huang, have reportedly been used in Chinese medicine for more than 5,000 years. Extracts from *Ephedra* were described as stimulants and anti-asthmatic treatments in pharmacopoeias during the Chinese Han dynasty (ca. 207 BC—220 AD). Similarly, *Ephedra gerardiana* has a long history of use in Indian folk medicine, and European species were employed throughout the period of the Roman empire.[446]

While the overall pharmacology of *Ephedra* is complex, its effects would seem to be related to ephedrine and its diastereomeric relative pseudoephedrine, which are lipophilic molecules that readily cross the blood–brain barrier. In the West *Ephedra* extracts have been used as stimulants and as aids to weight loss and in the enhancement of sports performance, often in combination with caffeine.[377] However, the evidence of efficacy in this regard is very weak. In general, the significant risks of psychiatric symptoms such as euphoria and mood disturbances, and cardiovascular or cerebrovascular side effects related to adrenergic activity, far outweigh any benefits.[507,508] The most common symptoms associated with *Ephedra* abuse reported to the U.S. Food and Drug Administration (FDA) were psychosis, depression, and agitation.[509] This led the FDA to ban the use of weight-loss products containing *Ephedra* alkaloids in 2004. They are, however, still available in a plethora of over-the-counter products intended for the relief of nasal and chest congestion, flu, fever, and headache, and in the Chinese traditional medicine Ma huang. Pseudoephedrine, in particular, is often a component of nasal decongestants due to its vasoconstricting effect.[446]

As above, ephedrine primarily fosters the release of noradrenaline and inhibits its reuptake, with only modest effect on dopamine levels.[444] Evidence also suggest that some effects of ephedrine, including vasoconstriction and resultant increased blood pressure, are attributable to a direct agonist effect at α-adrenergic receptors.[441,445] Some of the resultant noradrenergic effects, for instance raised blood pressure, can be attenuated by α-adrenergic receptor antagonists.[377]

Research involving animals suggest that ephedrine has reinforcing properties, that it can substitute for amphetamine and methamphetamine, and that it increases locomotion in rats.[510] These properties are related to increases in extracellular dopamine in the nucleus accumbens.[501] However, on the whole ephedrine has comparatively mild subjective effects in humans in comparison with amphetamine and methamphetamine. These typically include increased ratings of being "high," and mild euphoria and anxiety, along with comparatively modest reinforcing properties.[503,511]

Amphetamine/Methamphetamine

Ephedra sinica actually expresses three pairs of related, optically active, diastereomeric alkaloids: ephedrine and pseudoephedrine; norephedrine and norpseudoephedrine: and methylephedrine and methylpseudoephedrine. These three pairs of alkaloids can serve as the starting point for the manufacture of the semisynthetic compounds methamphetamine, amphetamine, and N,N-dimethylamphetamine respectively.[512]

Methamphetamine has become a particular problem as its use has reached epidemic proportions in the United States. This is because whereas amphetamine is typically synthesized by an alternative route using the controlled chemical phenylacetone, methamphetamine can be readily synthesized from ephedrine and pseudoephedrine precursors. These are commonly available in over-the-counter medications such as decongestants, cough medicines, and the Chinese traditional medicine Ma huang.[513] Typically, methamphetamine is made illicitly by the reduction, by protonation of the hydroxyl group, of ephedrine or pseudoephedrine using red phosphorus and iodine.

In common with ephedrine, amphetamine and methamphetamine both increase the release and inhibit the uptake of noradrenaline, but both also have much more pronounced effects on extracellular levels of dopamine.[441] Methamphetamine also has the downstream effect of increasing extracellular glutamate.[513] In rats the potent rewarding and locomotor-stimulating effects of amphetamine are due to increased dopamine release in the shell and core of the nucleus accumbens respectively.[514] Brain imaging studies using techniques such as positron emission tomography (PET) and functional magnetic resonance imaging (fMRI) also show that both amphetamine and methamphetamine activate the classic mesolimbic dopaminergic reward pathways in humans and that this activation is directly related to the euphoric effects of the drugs.[515,516]

The psychoactive effects of both of these psychostimulants are similar. As an example, amphetamine at low doses produces increased alertness, mild euphoria, and increased movement and self-confidence, but with increasing dose this can become either exaggerated euphoria or the opposite (dysphoria), paranoia, suspiciousness, and compulsive behavior. Sustained administration over several days can then result in symptoms of psychosis.[450] Chronic use of amphetamine and methamphetamine is also associated with cognitive deficits, neuronal damage, potentially due to increased oxidative stress and leakage of the blood–brain barrier, and a downregulation of dopamine receptor populations. This last neuro-adaptation underlies the compulsive pattern of subsequent abuse.[513]

Cathinone

Catha edulis is an evergreen tree indigenous to the Horn of Africa and the Arabian Peninsula. It is a member of the large *Celastraceae* family that comprises about 1,210 species in 98 genera spread across the tropics and subtropics. However, the species is most closely related to eight African genera, and it has been suggested that this small phylogenetic group may share *Catha edulis*' potential for the synthesis of its principal alkaloid, cathinone.[517] The fresh leaves of the tree, known as khat, have been chewed for their psychoactive properties for thousands of years. Its use was documented by the ancient Egyptians and it appeared as a medicine in written Arabic pharmacopoeias by the 11th century.[518] The chewing of fresh khat leaves is currently widespread in the Arabian Peninsula and East African countries, both as a recreational drug and in traditional and religious ceremonies. The habit often starts at a young age and continues lifelong. Improvements in transport networks and migration have spread the use of khat around the globe,[519] although it is now illegal in many countries, including the United States, Canada, and a number of European countries.[518]

Khat leaves contain a cocktail of potentially bioactive secondary metabolites, including monoterpenes and a variety of sequiterpenes, tannins, and flavonoids.[518,520] However, the potent psychoactive properties of freshly picked khat leaves are attributable to cathinone, which closely resembles ephedrine and its derivative amphetamine. However, cathinone is structurally unstable, decomposing to norpseudoephedrine and norephedrine, via reduction of the carbonyl group to an alcohol, as the leaf dries after picking.[195,375] Traditional use therefore requires that leaves are picked shortly before the time of consumption, as their psychoactive effects reduce over time.

Cathinone itself has similar pharmacological effects as those seen following amphetamine,[375] and khat users report a wide range of positive amphetamine-like effects. These include euphoria, excitability, improved concentration, alertness, and increased confidence and sociability; however, these effects are replaced after several hours by increased anxiety and irritability, poorer concentration, and lower mood.[519,521] Chronic khat chewing is associated with a plethora of negative central nervous system effects, such as migraine, impaired concentration and motor coordination, insomnia, and stereotypical behavior; it can be a precipitating factor in the development of psychosis and may exacerbate the severity of other preexisting psychiatric disorders.[518] Withdrawal occasions tremors, mild depression, lethargy, and nightmares.[522] Khat chewing is also associated with a range of potential negative, cardiac, cardiovascular, cerebrovascular, and gastrointestinal side effects.[518] While cathinone's cardiovascular effects are in keeping with increased noradrenergic activity, its vasoconstricting effects, including in the heart, do not seem to be related to adrenergic mechanisms[518] but may rather be a consequence of activity at TAARs.[523]

A group of semisynthetic social drugs can also be synthesized from cathinone. These "cathinones," as they are collectively called, include the newly identified social drug 4-methylmethcathinone (mephedrone or "meow-meow"). The effects of mephedrone in terms of pharmacology, behavior, and brain activation fall between amphetamine and the serotonergic empathogen MDMA (Ecstasy or 3,4-methylenedioxy-N-methylamphetamine). Mephedrone's stimulant and empathogenic properties are underpinned by both increased dopaminergic and serotonergic activity and a similar pattern of activation of the brain reward system as amphetamine.[524,525]

Ecological Roles of Psychostimulant Secondary Metabolites and Brain Function

The naturally occurring psychostimulants can all be assumed to have a specific function in deterring insect herbivores. Coca leaves from both of the species grown for their cocaine content are relatively free from herbivores in the wild, with the exception of some specialist herbivores such as the larvae of the moth *Eloria noyesi*.[440] This has been attributed to the insecticidal effects of cocaine[367] and its co-occurring alkaloids. Exposing generalist herbivore larvae to pure cocaine on leaves leads to their developing behavioral abnormalities such as tremors, rearing, and "walk-off," whereby they sample tissue and walk away, disappointed or intoxicated. These effects intensify with escalating doses and eventually result in death. This effect may be attributed to cocaine's ability to inhibit the reuptake of the ubiquitous insect neurotransmitter octopamine as well as the other monoamine neurochemicals.[440] Ephedrine also has marked insecticidal effects,[367] for instance killing seed-eating beetle larvae at naturally occurring concentrations.[526] Khat and other members of the *Celastraceae* family have, on the other hand, been used historically as insect repellents and traditional agricultural insecticides in China and South America.[520]

In trying to draw comparisons between the ecological roles of these plant chemicals in insects and some of their psychotropic effects in mammals, we're fortunate in that a number of pharmacological investigations have used insects as models, with the intention of extrapolating their results back to the mammalian nervous system. This research has demonstrated striking similarities between insects and mammals in terms of both the behavioral consequences and the mechanisms of action of cocaine. In the first behavioral study McLung and Hirsh[438] heated crack cocaine on an electrical element in a fly vial, allowing flies to "breathe" the vapor for 1 minute before transferring them to a glass viewing chamber. The behavior of the cocaine and drug-free control flies was filmed for 5 minutes and scored by researchers who were blind to the treatment. The drugged flies behaved very strangely; at low doses of cocaine they groomed themselves continuously, and as the dose escalated they first began to fly in tight circles and walk erratically, and then their movements became faster and more uncontrolled, eventually leading to akinesia and death. Following non-fatal doses flies also exhibited a subsequent sensitization to the effects of cocaine that also mirrored that seen in rodents.[438] Subsequent studies have used more sophisticated methodology, including a "crackometer" for measuring geotaxis and phototaxis (the fly's innate movement responses to gravity and light), and various automated flight and movement systems, and different drug-delivery methods to confirm these initial observations.[439,527] These studies have demonstrated that the behavioral effects of cocaine were predicated, at least in part, on dopaminergic function.[439] For instance, electrochemical measurements taken in the dopamine-neuron–rich protocerebral anterior-medial brain region of live *Drosophila* flies demonstrated that cocaine, along with amphetamine and methamphetamine, inhibited the reuptake of dopamine, with this effect abolished in mutant flies lacking dopamine transporters.[434,528] A potential role for both inhibition of serotonin uptake[435] and the involvement of the trace amine tyramine in both the acute and sensitization effects of cocaine in insects has also been identified.[529,530] Interestingly, several genes initially identified as being involved in the behavioral response of flies to cocaine have been found to be conserved and serve similar functions in mammals.[529]

Given their more limited relevance as drugs of abuse, ephedrine and cathinone have not received the same attention as cocaine with regard to insect models. A single study assessed the effects of ephedrine in insects, and this showed that it disturbed the thermoregulation of honey bees, with this effect abolished by coadministration of a β-adrenergic antagonist.[443] However, amphetamine and methamphetamine are widely taken drugs of abuse and so have benefited from research that we can partially extrapolate back to their naturally occurring ephedrine/pseudoephedrine precursors. In a series of experiments using *Drosophila*, Andretic et al.[448] administered methamphetamine orally and demonstrated a dose-related increase in sleep latency and a reduction in the frequency and duration of sleep during darkness. These effects were attenuated by coadministration of 3-iodo-tyrosine, which inhibits the activity of the enzyme tyrosine hydroxylase and reduces dopamine synthesis. The flies also exhibited increased sexual arousal, as indexed by premating courtship behavior, but a reduction in successful copulation. Alongside this, electrophysiological recording in the medial proto-cerebrum of wild-type flies showed that methamphetamine impaired visual perception responses that were assumed to be dopaminergic in nature.[448] The involvement of the dopamine system was confirmed by the in vivo electrochemical probe studies mentioned above (with regard to cocaine), which confirmed that both amphetamine and methamphetamine inhibited the reuptake of dopamine.[434,528] While these latter studies have specifically explored dopaminergic function because it has particular relevance to humans, it is possible that any anti-herbivore activity is predicated on interactions with other neurotransmitter systems. An open question is whether the mammalian adrenergic effects of ephedrine and derivatives carry over into activation of the insect's analogous octopamine/tyramine system. Given that recent research has identified octopamine as one of the trace amine neuromodulators in the mammalian brain, and that both amphetamines and octopamine have an affinity for the mammalian TAARs[523] that are homologues of the insect octopamine/tyramine receptors,[266,268] it seems highly likely that ephedrine/amphetamine will have pronounced effects within the insect's key octopamine system. While this possibility has not yet been explored in insects, a close structural relative of ephedrine, p-synephrine, an alkaloid from bitter orange (*Citrus aurantium*), has a similar binding affinity for insect octopamine receptors as octopamine itself.[531]

■ CAFFEINE

Caffeine and the related alkaloids theophylline and theobromine are "purine" alkaloids. They are also often referred to as methylxanthines, as their synthetic pathway incorporates the purine xanthine, which itself is synthesized via several routes of *de novo* synthesis or recycling involving the ubiquitous purine bases adenine and guanine.[195,386] Caffeine is synthesized by approximately 100 plant species[532] from 13 separate orders of dicotyledonous angiosperms.[533] A number of species within these orders are consumed by humans around the globe for their psychoactive properties, including the coffee plant (*Coffea* genus, including, for instance, *Coffea arabica*) and kola nut (*Cola acuminata*), both of which originate in Africa; tea *(Camelia sinensis)*, from Southeast Asia; and yerba maté (*Ilex paraguariensis*), guarana (*Paullinia cupana*), and cocoa (*Theobroma cacao*) from South America. It is interesting to note that these caffeine-synthesizing plants are not all closely related, and indeed the ability to synthesize caffeine has developed independently a number of times. For instance, the pathways for the synthesis of caffeine in *Coffea arabica* and *Camelia sinensis* evolved independently long after the phylogenetic branching of their clades from a common ancestor, in both cases via genetic mutations to different members of a diverse family of inherited primary metabolism (carboxyl methyltransferase) enzymes. This repeated evolution is a reflection both of the ubiquity of caffeine's purine building blocks and the simplicity of its synthetic pathways[107]—and, of course, it also reflects the utility of this compound for the plant (see "Ecological Roles of Caffeine and Brain Function" below). The comparatively short histories of coffee and tea use, which date from the 14th or 15th centuries as social drinks, are described in Chapter 1. The South American caffeine-bearing plants may have been used for longer in a social context. Indeed, the early ritual use of the leaves of the *Ilex* genus as a snuff has been confirmed by its inclusion in the grave goods of a medicine man, dated circa 500 AD in Bolivia,[16] and cocoa has a particularly long and illustrious history.

In terms of pharmacology, caffeine is rapidly absorbed and distributed, peaking in plasma around 30 minutes after ingestion.[534] The action of caffeine is largely the result of its competitively binding to adenosine A_1 and A_{2A} receptors and preventing the activity associated with their endogenous ligand, the animal neurotransmitter and neuromodulator adenosine.[387] Adenosine consists of a molecule of adenine attached to a ribose sugar molecule, and it therefore shares structural elements with caffeine and the other methylxanthines. Indeed, recycled adenosine can be one potential substrate for the synthesis of caffeine's xanthine skeleton in the plant.[386] The purine substrate of both adenosine and caffeine and their structural similarities are illustrated in Figure 5.1.

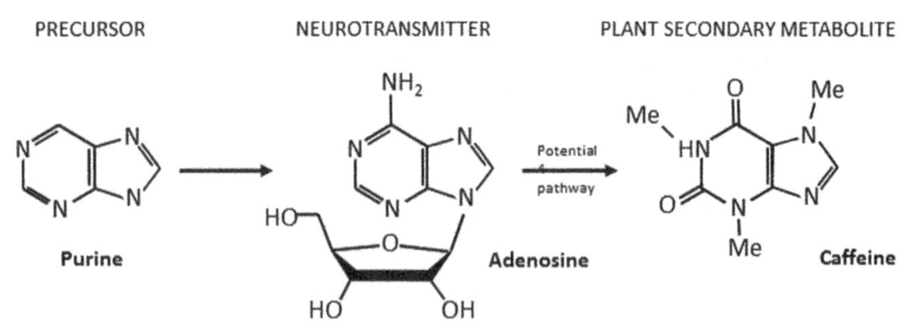

Figure 5.1 Structural similarity between purine, adenosine, and caffeine. The potential plant synthetic pathways are represented by the arrows.

Adenosine itself is an inhibitory neuromodulator that serves to decrease overall neuronal activity. In this role it can be released by neurons in the manner similar to that of classical neurotransmitters,[535] but it is most prevalent as a simple breakdown product of cellular adenine nucleotides. Adenosine levels therefore vary with overall brain activity, building up in the cortex and basal forebrain during wakefulness and then dissipating during sleep.[429] This lends adenosine a number of properties related to the regulation of tiredness and the sleep–wake cycle, and the homeostatic and neuroprotective downregulation of brain activity.[429,535] Caffeine's inhibition of adenosine's inhibitory activity leads to a net increase in brain activity associated with dopamine, acetylcholine, noradrenaline, serotonin, glutamate, and GABA,[536] and in particular it promotes wakefulness and alertness.[429] Caffeine also has a number of competing vascular effects: it exerts endothelium-dependent vasodilation via modulation of nitric oxide synthesis, but its net effect is reduced peripheral and cerebral blood flow mediated by inhibition of the A_{2A} adenosine receptors throughout the vasculature.[537]

As you might expect, the most widely accepted behavioral effects of caffeine are an increase in alertness[538–540] and improved cognitive function, particularly in terms of reaction times[539,541] and attention/vigilance.[542,543] It has been suggested that these effects merely represent alleviation of withdrawal in habitual caffeine consumers.[544] Chronic caffeine consumption may lead to increased adenosine receptor populations and/or sensitivity which may underlie any subsequent withdrawal effects[545],which include increases in basal cerebral blood flow[546] and headaches, fatigue, and lack of energy and concentration.[547] However, caffeine has also been shown to improve cognitive function in individuals who are not habituated to caffeine[430] and in situations where there is no evidence of withdrawal in habitual consumers.[430,545,548] At higher doses caffeine leads to increased anxiety, restlessness, insomnia, tachycardia, and psychomotor agitation, and ultimately to caffeine intoxication, and in extreme cases, death.[549]

Caffeine is the most widely taken psychoactive substance in the world and is consumed regularly by the majority of adults. The typical pattern of daily use suggests that its popularity is due in some way to potentially rewarding or reinforcing effects. However, caffeine has been described as an "atypical drug of dependence"[550] in that it has comparatively mild reinforcing properties in humans and animals, which are seen only at lower doses and which are due to both the positive reinforcement associated with its mild stimulatory effects and negative reinforcement due to the adverse physiological effects of withdrawal following chronic consumption.[550,551] Interestingly, while the evidence of caffeine's ability to reinforce its own consumption is comparatively weak (in relation to the other rewarding drugs described here), when coadministered it has been shown to increase the psychomotor stimulant effects and reinforcing effects of other drugs, including nicotine, cocaine, and amphetamine.[552] It also strengthens the reinforcing effects of non-drug stimuli unrelated to caffeine in rodents[552] and humans that regularly consume caffeine.[553] These results might suggest that caffeine is widely consumed because it enhances the reinforcement associated with the perceptual properties, such as taste or aroma, of the drink and food vehicles that it is consumed in, rather than reinforcing its own consumption per se.

There is also conflicting evidence as to whether caffeine promotes dopamine release in the nucleus accumbens in rats,[554,555] and evidence from an autoradiography study in humans suggests that this brain region is activated only by doses of caffeine higher than those typically consumed on a day-to-day basis.[556] However, it is notable that adenosine receptors are co-localized with dopamine D_2 receptors on neurons in the shell of the nucleus accumbens, and that the removal of inhibition in this brain structure underlies the arousing/wakefulness effects of caffeine in rodents.[557] Interactions between D_2-bearing neurons and D_1-bearing neurons more prevalent to reward may underpin the reinforcement strengthening properties that caffeine has for other drugs and non-drug stimuli.[557] Caffeine certainly appears to owe some of its properties, and the ubiquity of its consumption by humans, to the coopting of elements of the reward pathways.

Naturally caffeine co-occurs in plants with a wide range of other phytochemicals that may also have an impact on brain function. Indeed, caffeinated beverages constitute the single richest source of phenolic compounds for the populations of most developed countries, largely in the form of phenolic acids from coffee and flavanols from tea. The potential beneficial independent effects of these compounds on brain function, including epidemiological evidence surrounding tea and coffee consumption, are discussed in Chapters 8 and 9. In several cases the psychoactive effects of caffeine have also been shown to be modulated by non-phenolic components. For instance, single doses of guarana that contain comparatively high levels of terpenes, but levels of caffeine lower than the minimum established psychoactive dose, have been shown to improve cognitive function,[558] and the amino acid L-theanine, which is found in tea, attenuates the negative effects of caffeine on blood pressure[559] and cerebral blood flow[389] and potentiates its cognitive effects.[560]

Ecological Roles of Caffeine and Brain Function

In plants, given its toxicity, caffeine is synthesized in the cellular cytoplasm and stored safely in vacuoles,[386] with synthesis taking place preferentially in the tissue that has the most value to the plant in terms of survival and reproduction. This includes the young shoots, floral buds, the pericarp of the berry, and the immature beans.[386,532] Caffeine synthesis can also be induced in the vegetative plant parts such as shoots, leaves, and roots by a range of biotic and abiotic stressors.[561]

Caffeine is both a bitter-tasting feeding deterrent [432,562] and an insecticide at naturally occurring concentrations.[432,526] As an example, chronic oral caffeine administration reduces both the lifespan and fecundity of *Drosophila*.[563] Caffeine's synthesis is also directly induced by insect feeding on plant tissue, and its effects on herbivores include an increase in locomotor activity and corresponding reduction in feeding.[427] As with all the best defensive secondary metabolites, individual species of the *Coffea* genus have specialist herbivores that have evolved a tolerance to caffeine. These include the coffee borer beetle (*Hypothenemus hampei*), which lays its eggs within the coffee berries,[564] and the coffee leaf miner moth (*Leucoptera coffeella*), which lays its eggs on leaves in response to their caffeine content.[565] Caffeine released into the soil from roots and decaying beans and foliage also creates a defensive perimeter around the plant,[566] within which it acts as a deadly toxin to mollusks[567] and may inhibit the germination of seeds and the growth of roots and shoots of competitor plants.[369] Unfortunately, caffeine also builds up in the soil, leading to autotoxicity whereby the caffeine-bearing plant itself will also be prevented from growing.[533]

In terms of the mechanisms underlying the anti-herbivore effects of caffeine, it is notable that adenosine plays similar neuromodulatory roles in mammalian and insect central nervous systems, including modulating the sleep/wake cycle. Insects, however, express only a single adenosine receptor (as opposed to four in mammals), which activates the cAMP/PKA (cyclic adenosine monophosphate/protein kinase A) signaling pathway.[568,569] A number of studies in insects have used caffeine on the basis of its adenosine receptor-antagonizing properties to probe the functioning of the insect adenosinergic system, thereby inadvertently elucidating the effects of caffeine at the same time. As an example, Hendricks[421] administered caffeine to *Drosophila* and demonstrated dose-dependent reductions in sleep-like behavior, with death occurring as the dose escalated. An opposite effect on sleep parameters was engendered by the adenosine receptor agonist cyclohexyladenosine. Andretic et al.[420] also demonstrated that the administration of caffeine and two synthetic adenosine receptor antagonists increased locomotor activity during the period corresponding to sleep in *Drosophila*, while the employment of mutant flies suggested that this effect was predicated on adenosinergic modulation of dopamine receptor function. In line with this observation of the role of dopaminergic mechanisms, administration of both dopamine and caffeine have been shown to have an arousing effect in terms of decreasing the duration of tonic immobility (death-feigning) in threatened beetles.[426] It should also be noted that one study demonstrated that mutant flies lacking adenosine receptors still responded to caffeine in a similar manner, and subsequent investigations demonstrated that one potential non-adenosinergic mechanism was the inhibition of phosphodiesterase activity with resultant increases in cAMP and PKA.[570] This mechanism has also been identified in mammals, but only at higher doses.[387]

■ THE PARADOX OF DRUG REWARD

One of the very few attempts to integrate our knowledge of the ecological roles of plant chemicals with our extensive understanding of the mechanisms by which some psychoactive drugs exert their effects on the human brain and behavior revolves around the so-called "paradox of drug reward." In several recent papers[571,572] Roger Sullivan, Edward Hagen, and colleagues discuss the seeming incompatibility between plant chemicals that function specifically to deter herbivores from eating plants, such as nicotine, cocaine, and morphine, and the rewarding properties they exert in the mammalian nervous system through activation of the mesolimbic dopamine reward system. For the plant this would seem to be an outcome entirely opposite to the desired effect, in that they increase the likelihood of the plant chemical's consumption, rather than deterring it. For the herbivores, on the other hand, it seems equally paradoxical that they possess a brain reward system that can so easily be subverted by plant chemicals. In short, one suggested resolution to the paradox is that humans have in some way "evolved to counter-exploit plant neurotoxins".[571,572]

Naturally, this short summary fails to do any justice to the complexity of the arguments and issues covered by these papers. However, there are a number of lines of evidence that suggest a far simpler resolution to the "paradox" and question the validity of the conclusion. In terms of the evolution of humans it is absolutely correct that we, along with our ancestors and vertebrate relatives, have evolved an extensive arsenal of mechanisms for detoxifying plant chemicals, primarily the cytochrome P450 (CYP) enzymes that metabolize a wide range of endogenous and extraneous organic compounds. While it's also true that differing geographical populations of humans exhibit very small differences in their profiles of CYPs, presumably reflecting local historical differences in diets, it seems unlikely that any of the handful of "drugs of reward" have a history of human consumption long enough for them to have had any significant impact in evolutionary terms. While opinions differ, the migration of modern humans out of Africa, a continent with only sparse examples of naturally occurring "rewarding" plant chemicals, can be placed between 125,000 and 60,000 years ago. The first migrants reached Australia only maybe 50,000 years ago[573] and crossed onto the American continent from Siberia via the Bering land bridge some 15,000 years ago, subsequently spreading south into South America, with its incomparably rich flora, over the following 2,000 years.[574] As they traveled, our ancestors would have been entering landscapes replete with an indigenous flora that is largely identical in secondary chemical terms to that which we see today. They certainly would not have exerted any evolutionary pressure on plant chemistry (certainly prior to agriculture). Few would have even directly encountered one of the specific drugs that trigger the mesolimbic reward system: several of the rewarding drugs, including cocaine, caffeine from tea and coffee, and cathinone have enjoyed any use outside of the comparatively small region in which they grow naturally for only a few hundred years at most. Even nicotine and opium have identifiable points of origin on their respective continents, and an archaeological trace that shows that their use and cultivation spread very slowly across their respective continents from their home range within the last few thousand years. In reality, it seems likely that the point where the majority of the human population had encountered one of this small subset of plant chemicals can be measured in terms of a few hundred years, probably with the widespread adoption of tea and coffee and the unfortunate explosive spread of tobacco use around the world. If human evolution plays any part in the "paradox of drug reward" it is merely in terms of the cultural evolution of the human.

There is also scant evidence, beyond a few anecdotal examples, of non-human vertebrates finding the plants that provide us with the drugs in question rewarding. Indeed, with the exception of khat, which can be consumed straight from the bush, the rewarding drugs aren't generally simply picked and consumed. Even coca leaves are typically chewed with lime to release the cocaine, and the rest require human intervention in the form of harvesting and drying, and human methods of consumption such as smoking, infusing, snorting, or injecting to reveal their rewarding properties.

Of course the "paradox of drug reward" is also only a paradox if the intended herbivore targets find the specific plant chemical rewarding enough to increase their consumption of plant tissue. If we accept that humans are not the intended target, then there is an absence of any evidence that suggests that these plants or compounds are reinforcing to their natural consumers. If we return once more to the insects, there is some very slight evidence that bees prefer nectar with naturally occurring levels of nicotine and caffeine, but that they find the concentrations that would be present in plant tissue toxic.[575] Bees administered cocaine topically while feeding on sucrose also show patterns of increased recruitment dancing for hive-mates that is indicative of a more attractive food source. However, this holds true for only very low doses of cocaine, and the effect is identical to that seen following the topical application of octopamine, a neurotransmitter that reliably increases bee dancing in itself.[437] This suggests that cocaine merely has its effects on the dancing of bees via inhibition of the reuptake of octopamine, rather than because the bee is recruiting its nest-mates because of the rewarding properties of the cocaine. While this example tells us little about reward *per se*, it does illustrate some particularly relevant differences between the reward pathways in humans and insects. While the mammalian and insect reward systems have many elements in common, they also differ markedly in terms of brain morphology and the key neurochemicals underpinning reward. Whereas the mammal has a generalized reward system with a single midbrain dopaminergic pathway underwriting the reward associated with diverse stimuli, the insect has more than one reward system. Within this distributed insect reward system octopamine is the neurotransmitter with the clearest role as a mediator of reward, whereas dopaminergic activity can signal reward but is much more likely to signal punishment, resulting in aversion. For the insect it is the interaction between the activity related to both neurotransmitters that dictates the allocation of reward/aversion salience to a stimulus.[576] This does mean that it is theoretically possible that insects may encounter phytochemicals that have the unintended effect of activating their reward pathways, although the increased attractiveness of the plant tissue would reduce the plant's fitness to survive, and presumably would ensure a hasty exit from the evolutionary scene for the plant species. However, there is no more likelihood that the few phytochemicals that humans find so rewarding, due to a specific interaction with a brain system that insects simply do not possess, would be any more rewarding to insects than any other phytochemical that modulated dopaminergic/octopaminergic function.

In the absence of any evidence suggesting that insects find specific secondary metabolites rewarding or reinforcing, it seems more parsimonious to merely conclude that there is no paradox and accept that a handful of plant chemicals (out of hundreds of thousands of secondary metabolites) just happen to unintentionally subvert the functioning of the human dopaminergic reward pathways when prepared and consumed in specific ways. These rewarding drugs have come to prominence only in the very recent past as human culture has adopted them and spread their use around the globe.

6 The Hallucinogens

> "If the doors of perception were cleansed every thing would appear to man as it is, infinite. For man has closed himself up, till he sees all things through narrow chinks of his cavern."
>
> –WILLIAM BLAKE, *The Marriage of Heaven and Hell*

The first difficulty when considering the plant- and fungus-derived compounds described as "hallucinogens" is the definition of the word *hallucinogen* itself. The venerable verb "to hallucinate" is derived from the Latin verb *hallucinari* (to wander in the mind) and was coined in the mid-17th century to describe the experiencing of something that is not present as if it was real. More recent definitions have taken on a certain circularity with drug use, so that according to the *Oxford English Dictionary* a hallucinogen is defined as "a drug that causes hallucinations, such as LSD" and to hallucinate is to "experience a seemingly real perception of something not actually present, typically as a result of a mental disorder or of taking drugs." However, it is notable that compounds of this class do not typically induce perceptions of something that is not present, but rather distort aspects of consciousness or perceptions of things that are present. Where concrete visual images are included in the experience these are typically recognized as being illusory and are therefore best described as "pseudo" or "non-psychotic" hallucinations.[577]

Because of the difficulty of defining an hallucinogen, a number of descriptive names that might better capture the subjective effects of the hallucinogen class of drugs have been suggested or adopted over time. These include "phantastica," a name suggested by the German toxicologist Louis Lewin to encompass drugs that "influence particularly the visual and auditory sphere as well as the general sensibility"[578]; "psychotomimetic," on the basis that these drugs mimic some elements of psychosis; "psychedelic," from the Greek *psyche* and *dēlos* meaning "mind manifesting"; and entheogen, from the Greek word *entheos*, meaning "god within".[371] This last name tries to capture both the spiritual or mind-expanding experiences felt by many users and the traditional use of these drugs across cultures as a means of contacting and communicating with the gods or ancestors (see Chapter 1). However, it becomes obvious that each of these descriptors can encompass a wide range of drugs. For instance, many psychoactive drugs, including Ecstasy, cannabinoids, and anticholinergic drugs, taken in sufficient quantities, lead to an altered state of consciousness, distorted perceptions, and, occasionally, true hallucinations. Similarly, a number of drugs have their historical roots in an "entheogenic" wish to communicate with the gods or ancestors. The obvious candidates here are members of the *Solanaceae* family of plants that contain anticholinergic tropane alkaloids such as scopolamine, which might best be described as "deliriants" (see Chapter 7). The entheogens also include several "dissociatives" such as morphine and salvinorin A (from *Salvia divinorum*), which cause euphoria and feelings of unreality and depersonalization. The key attribute of most of these drugs is that their potential perception-distorting effects are very much secondary to their many other central nervous system effects. These other drugs will be described in greater detail elsewhere in this book. For the purposes of this section it seems appropriate therefore to apply the definition of hallucinogens provided by Hollister,[579] who argued that a true hallucinogen should predominantly affect changes in "thought, perception, and mood" but shouldn't overly impair intellectual or memory function, induce stupor, narcosis, excessive stimulation, or major autonomic nervous system side effects, or generate addictive cravings. Put even more simply, hallucinogens are "physiologically safe molecules whose principal effects are on consciousness".[371] A further key attribute that differentiates these drugs is the "mind-expanding" nature of the subjective experience that they induce at higher doses. So they should "induce states of altered perception, thought, and feeling that are not experienced otherwise except in dreams or at times of religious exaltation".[371,580] These attributes narrow the field of potential drugs to a handful of the "classic" hallucinogens, all of which exert some of their effects via the same mechanism.

Where these chemicals occur in nature they are alkaloid plant and fungal secondary metabolites, or their semisynthetic derivatives. They fall into two broad structural groups, often described as the phenethylamines and the tryptamines.[371] In terms of naturally occurring hallucinogens, the phenethylamines really contains just one structure, mescaline, which is ultimately derived from the amino acid L-tyrosine, via dopamine (see Fig. 4.2 in Chapter 4). The tryptamines, on the other hand, are a broad group of indole alkaloids that are derived from the amino acid L-tryptophan and can be further subdivided. The simple tryptamines are synthesized via tryptamine and share a close structural relationship with the neurotransmitter serotonin (5-hydroxytryptamine), which is synthesized from the same precursor and also incorporates the structure of tryptamine.[195] This group includes N,N-dimethyltryptamine (DMT) from ayahuasca (a South American beverage made from members of the *Psychotria* genus), 5-methoxy-DMT (5-MeO-DMT) and 5-hydroxy-DMT (5-HO-DMT) from Yopo (a South American snuff made from *Anadenanthera peregrina* tree seeds), and psilocybin/psilocin (4-hydroxy-N,N-DMT) from the *Psilocybe* mushroom genus. The second group of tryptamines, the ergot alkaloids, are more complex derivatives of L-tryptophan via lysergic acid, which possess structural motifs that they share in common with both serotonin and the catecholamine neurotransmitters noradrenaline and dopamine, giving this general class of chemicals a diverse range of pharmacological effects.[195] The hallucinogenic lysergic acid derivatives include the semisynthetic lysergic acid diethylamide (LSD) and lysergic acid amide (ergine), an alkaloid that occurs in the seeds of a number of plants that coexist with symbiotic fungi, including *Ipomoea asarifolia/violacea* and *Turbinia corymbosa*.[195] These seeds are collectively categorized as ololiuqui, another traditional South American hallucinogenic preparation. One final member of the "tryptamines" is ibogaine, from the African shrub *Tabernanthe iboga*. This terpene-indole alkaloid is composed of a tryptamine structure with a terpene addition.[195] The structures of the above hallucinogens, their precursors, and related neurotransmitters are shown in Figure 4.3 in Chapter 4.

■ A COMMON MECHANISM OF ACTION?

The general consensus up to this time has been that the defining common pharmacological characteristic of all of these hallucinogens is their ability to bind to

G-protein-coupled serotonin receptors, and in particular to bind agonistically at the 5-HT_{2A} subtype of receptors. This receptor is the main excitatory serotonin receptor in the brain and is distributed throughout the cortex, with the highest concentrations on richly networked frontal cortex pyramidal cells. It is also expressed in sensory processing areas of the brain, including the thalamus. Stimulation of 5-HT_{2A} receptors also leads to an increase in synaptic activity mediated by glutamate,[371,581] the most abundant excitatory neurotransmitter in the vertebrate nervous system, which in effect "turns up the volume" of synaptic activity and modulates the activity of a raft of other neurotransmitter systems. While unidentified as yet, the precise mechanism underlying the glutamatergic effect may be one of the following: the occurrence of 5-HT_{2A} receptors on glutamatergic neurons exiting from the thalamus[371]; the occurrence of receptor complexes incorporating 5-HT_{2A} and glutamate receptors together[582]; or the presence of an unidentified retrograde messenger that triggers the release of glutamate.[581] Activity within the thalamus may itself play a key role in the action of hallucinogens. This brain structure is highly interconnected with the cortex and could be visualized as acting as a switchboard for perceptual information, receiving, processing, and relaying information from each sensory modality backwards and forwards with the associated primary areas of the cortex. The experience of waking consciousness may ultimately represent the oscillation within this network of interconnected thalamocortical neurons, with sensory inputs from the "real world" and attentional mechanisms merely constraining the intrinsic underlying activity of these networks.[583] This intrinsic activity can be most readily appreciated when we consider that the thalamus plays a key role during REM sleep, when the lack of inputs from the sensory modalities, and the absence of control via attentional mechanisms, leads to the rich imagery and thought patterns of dreaming. Within this model the hallucinations inherent in schizophrenia represent an inability of sensory input from the "real" world to modulate intrinsic thalamocortical activity.[583] Similarly, drug-induced hallucinations would be predicated on the direct and indirect modulation of thalamus-driven activity and outputs by direct 5-HT_{2A} receptor binding in the cortex and thalamus, and the downstream consequences of this receptor binding.[371] It is notable that the hallucinations caused by hallucinogens and acute psychosis bear similarities; that effective antipsychotics may owe their activity to interactions with 5-HT_{2A} receptors; and that downregulated frontal cortex 5-HT_{2A} receptor populations are implicated in the etiology of the visual hallucinations associated with neurological diseases such as Parkinson's and schizophrenia.[371,584–586]

Interestingly, a recent functional magnetic resonance imaging (fMRI) study showed that ingestion of the hallucinogenic fungal alkaloid psilocybin led to reduced blood flow within a number of regions of the brain that act as communication hubs for activity, including the thalamus. The reduced activity in the thalamus also correlated negatively with the intensity of the experience for the participant. Overall the study demonstrated that psilocybin enabled "a state of unconstrained cognition" by reducing the coupling of activity between areas that function as connector hubs.[587] Intriguingly, these findings map quite nicely onto Aldous Huxley's metaphor that hallucinogens function by removing the constraints imposed on reality by the "reducing valve" of the brain, ultimately flinging open William Blake's "doors of perception".[587]

The common receptor binding mechanism of action across the hallucinogens is evident in the ability of differing hallucinogens to inculcate cross-tolerance to each other, putatively as a consequence of drug-induced downregulation of receptor populations.[581] However, alongside this single common mechanism the various naturally occurring hallucinogens also exhibit wide differences in their receptor-binding properties. They all also bind to 5-HT_{2C} receptors, and the relevance of this to their net potency is not yet clear. Beyond this their differing profiles of receptor-binding properties and their resulting downstream effects must underlie the differences observed in their potency and psychological effects.[371,588] As an example, variations in dopaminergic effects mediated by dopamine D_1 receptors, either via direct binding in the case of LSD, or via a downstream increase in dopamine release in the case of the other hallucinogens, may also differentially contribute to the primary effects of these drugs.[589] One key emerging hypothesis is also that these drugs may partly owe their effects to their binding affinity for the trace amine associated receptors (TAARs) that were identified as recently as 2001.[268] Indeed, the hallucinogenic plant-derived tryptamines DMT, 5-MeO-DMT, and bufotenin have all been identified as endogenously synthesized mammalian ligands for these receptors[267] (see below).

One key property of the classic hallucinogens, due to the specificity of their pharmacology, is that they are physically safe for humans, engendering an altered state of consciousness at doses that have no short- or long-term negative effects on the physical functioning of the user. They generally have little or no affinity for neurotransmitter targets within the vital physiological systems that sustain life.[371] As an example, there is a huge variation in the toxicity of the most potent and widely taken hallucinogen, LSD, in studies involving mammals. While half of rabbits injected with 0.3 mg/kg have been shown to expire, mice can tolerate more than 100 times this dose, and there have been no documented human deaths at all from an LSD overdose, even following the accidental ingestion of huge doses of the drug.[459] It has been suggested that the only likely negative side effect of the hallucinogens is the re-experiencing of perceptual distortions or "flashbacks" to the drug state at some point in the future, which may be experienced to a lesser or greater extent by a very small minority of users.[590] A recent review conducted in Holland, where *Pscilocybe* mushroom use was legal until 2008, mirrored these observations.[591] On the other hand, while considered generally safe in physical terms, the hallucinogens can be associated with accidents due to inadvisable behavior (such as believing one has superhuman powers) and a potential to precipitate psychopathologies in predisposed, but not healthy, individuals. Unlike many of the social drugs, hallucinogens are not addictive, and they do not exert any direct effects on dopaminergic neurotransmission at normal doses and therefore do not interact with the "reward" pathways in the brain.[371] The one exception here is the semisynthetic LSD, which has specific affinity for dopamine receptors.[592] However, there is little evidence of addiction or reinforcement in humans, while animal research suggests a mixed pattern of reinforcing effects, with monkeys developing negative reinforcement to LSD, so actively avoiding it where possible,[593] and rats exhibiting either weak avoidance or reinforcement depending on factors such as strain and gender.[371]

■ ECOLOGICAL ROLES OF HALLUCINOGENS—SOME GENERAL OBSERVATIONS

The ecological roles of some of the hallucinogenic secondary chemicals have received comparatively little attention. However, the evidence of ecological roles for the ergot alkaloids and the plant-derived tryptamines are summarized in the relevant sections below. As a general observation, while it is intuitively pleasing to suppose that the primary role of these chemicals is to produce deterrent hallucinogenic effects in vertebrate herbivores, it is rather more likely that, as with most defensive secondary metabolites, their intended targets are the insects and invertebrates that represent the immediate neighbors of the plants and fungi that express these chemicals. In this respect it is notable that insects express four distinct serotonin receptors orthologous to the mammalian equivalents: two 5-HT_1-like receptors, one 5-HT_2-like receptor, and one 5-HT_7-like receptor.[294] Both of the insect 5-HT_1 and 5-HT_2 receptors are involved in circadian processes and sleep,[594] and all three subtypes of receptors play a role in memory processes.[319] The key 5-HT_2 receptor is particularly richly expressed in areas of the insect brain, for instance the proto-cerebrum and ellipsoid body, that participate in higher-order behaviors, including perception.[294]

We can be sure that LSD has direct insect receptor-binding properties as this compound was used to characterize the existence and nature of all three serotonin receptor subtypes in flies, bees, and locusts.[406,451–453] It seems safe to assume that the other hallucinogens exert similar receptor-binding effects; indeed, direct binding of LSD, DMT, and bufotenin, with a similar affinity to serotonin itself, has also been demonstrated at the insect 5-HT_1 receptor.[405] We can also be relatively confident on the basis of some early insect studies that this receptor binding modulates brain function. For instance, in ants, the consumption of LSD occasioned outstretched forelegs and antennae and fumbling movements that suggested perceptual distortions.[456] It also led to an initial increase in social activities such as food sharing,[456] followed by a decrease in the same social activities along with reduced foraging behavior,[456,458] decreased aggression and the modulation of brain electrical activity,[457] with impaired locomotion and death at higher doses.[449] In the most recent series of studies, Nichols et al.[455] showed that genetically modified flies with increased sensitivity to serotonergic modulation were partially or fully immobilized by LSD. The researchers also harnessed the tendency of wild-type flies to follow visual stimuli to show that LSD interfered with the visual processing of drugged flies, with this effect reduced by both 5-HT_1 and 5-HT_2 receptor antagonists, confirming that interactions with these receptors were responsible for the visual effects. Brain metabolism was also increased by LSD, as assessed by activation of *Dfos*, the protein product of a gene related to neuronal activity, in the fly brain. These results paralleled previous findings with regard to the mammalian equivalent of the same gene, suggesting that comparable signal-transduction mechanisms are involved in both species.[455]

■ THE PHENETHYLAMINES

Mescaline (3,4,5-Trimethoxy-phenethylamine)

The peyote cactus (*Lophophora williamsii*) and San Pedro cactus (*Trichocereus pachanoi*) have separate geographical ranges but share hallucinogenic properties, a common principal bioactive component, and a long history of use as shamanic entheogens. Peyote cactus is indigenous to areas north of the American isthmus, growing in the desert regions of Mexico and Texas. It comprises a small spineless crown above the ground and a tuberous root below. Both the spineless top of the cactus, dried into a "button" and then rehydrated in the mouth or in a beverage, and the freshly severed crown of the plant made into an infusion are consumed. South of the isthmus the San Pedro cactus, a columnar cactus native to the Andes, grows at elevations above 2,000 meters and is widely consumed for its entheogenic properties in the form of a sacred intoxicating drink made from slices of cactus, sometimes mixed with material from members of the tropane alkaloid-bearing *Brugmansia* genus of *Solanaceae* (see Chapter 7).

Both cacti have a long history of use in religion and ritual. Archaeological evidence of peyote consumption includes cactus material excavated from cave dwellings with an 8,000-year history of occupation in Coahuila in Mexico[16] and peyote buttons recovered from the Shumla caves on the edge of the Chihuahan desert, Texas. These were radiocarbon dated to 4000 BC and were found to still contain mescaline.[595] The ritual use of peyote is also confirmed by rock art found in the Lower Pecos River region of Southwest Texas dated to between 750 and 2200 BC.[595] Oral histories also date peyote use by the indigenous tribes of Mexico to as early as 300 BC. Certainly, by the early 16th century, the Franciscan friar and anthropologist Bernardino de Sahagún described the widespread historic use of peyote by the seminomadic Chichimecas. He wrote that those who imbibed "peyotl" experienced "terrible or ludicrous visions. . . . It gives them strength, incites them to battle, alleviates fear, and they feel neither hunger nor thirst".[578] Within the last 500 years the use of peyote by indigenous peoples has also spread throughout the northern part of the American continent and is now legally sanctioned as a sacrament of the 300,000-strong Native American Church, whose all-night ceremonies take place in tepees or other traditional structures under the guidance of a "Road Chief" priest and feature long periods of eating or imbibing peyote in order to attain a state of religious exaltation.[596] While there is less archaeology supporting the longevity of the use of San Pedro cactus, the evidence does include stone carvings from around 1300 BC showing slices of cactus in the hands of mythological beings[16] and depictions of the cactus in the ceramics of the Chavin culture of the Peruvian highlands (circa 300 BC).[43] Certainly by the time of the Spanish conquests the consumption of the cactus was ubiquitous across its home range, and its use was subsequently suppressed along with the other psychotropic plants used in ritual.

The modern era for hallucinogens dawned with the rediscovery of mescaline by the Western world in the closing years of the 19th century. The Swedish explorer, anthropologist, and ethnographer Carl Lumholtz chronicled the mythology and the medicinal and shamanic uses of the peyote cactus among widely distributed tribes such as the Huichol and Tarahumare in the journal articles and books describing his years of travels in Middle America. In the first volume of his 1902 book *Unknown Mexico* he described the exhilaration and color visions he experienced after chewing peyote buttons. Lumholtz also described trees dancing before his eyes, and fluctuations in mood during peyote ceremonies, with attendees alternately weeping and laughing. In 1898, soon after Lumholtz's first articles, the German chemist Arthur Heffter isolated mescaline, naming it after the Mescalero Apache tribe who had provided him with the plant material for his studies. Careful animal studies and self-experimentation identified mescaline as peyote's active component. It was another two decades before Ernst Späth finally succeeded in chemically synthesizing mescaline. Louis Lewin, in *Phantastica*,[578] his seminal book on psychoactive plants, had described the sensory hallucinations, "colour-symphonies," and spiritual experiences after consuming peyote buttons with their complex cocktail of alkaloids, but now it was possible to investigate the effects of the single active molecule.

The specific pharmacology of mescaline and the resultant chemically induced changes in visual perception, emotional changes, states of ecstasy, and altered states of consciousness were systematically investigated and described in books by K. Beringer (*Der Meskalinrausch*) and Heinrich Kluver (*Mescal and Mechanisms of Hallucinations*) published in 1927/1928. However, with no apparent medical applications, research into the effects of mescaline waned. It was 1952 before the English psychiatrists Humphry Osmond and John Smythies reignited interest with their observation that mescaline mimicked some of the symptoms of schizophrenia. They noted, for the first time, the close similarity between the chemical structure of mescaline and adrenaline, tying these observations together to theorize that schizophrenia might be caused by an endogenous "failure of metabolism" leading to the production of a mind-altering adrenaline/mescaline-like chemical.[597] The English writer and intellectual Aldous Huxley was so taken with their theorizing that he contacted Osmond, offering himself as an experimental subject. Huxley's own opinion, expressed in his correspondence with Osmond, was that the brain acted as a "reducing valve," preventing certain perceptions, memories, thoughts, and emotions from ever reaching the conscious, and that mescaline might loosen this constriction, leading to greater awareness. Osmond duly visited Huxley on May 3, 1953, and administered the author 400 mg of mescaline. The narrative of Huxley's subsequent book *The Doors of Perception* was a semi-autobiographical account of that day, describing his experience of a profound change in his perceptions, the heightening of visual impressions and his intellectual response to art and music, and ultimately his observations on the utility of mescaline in terms of its ability to act as a "door" to self-transcendence. Osmond and Huxley went on to enjoy a prolonged correspondence, during which they toyed with names to describe the psychological effects of hallucinogens. Osmond finally proposed "psychedelic" (in the playful couplet "To fathom Hell or soar angelic, just take a pinch of psychedelic"). Although it would be several years before Huxley would first self-experiment with LSD, his book went on to have a profound effect on the 1960s drug counter-culture, influencing seminal figures such as the rogue Harvard academic Timothy Leary, who coined the phrase "turn on, tune in, drop out." In its own way Huxley's book left the door slightly ajar in readiness for the psychedelic decade.

Osmond and Smythies' observation that the structure of mescaline resembled that of adrenaline was, of course, absolutely correct. Both mescaline and adrenaline are derived from dopamine (see Fig. 4.3 in Chapter 4), and indeed peyote and San Pedro cactus also contain significant levels of dopamine, adrenaline, and noradrenaline. Both cacti also synthesize a wide range of structurally related alkaloids from L-tyrosine, with total alkaloids representing some 8% to 9% of the dry weight of peyote buttons. Of these alkaloids mescaline is the most abundant and prevalent alkaloid, but the chemical cocktail includes a range of non-hallucinogenic alkaloids such as anhalamine and anhalonine.[195]

Mescaline itself is the least potent of all the classic hallucinogens, with an active dose of 200 to 400 mg; this is some 2,000 times the quantity of LSD required to generate comparative effects.[371] This lack of potency, and a delayed onset of action, is due both to mescaline's comparatively low affinity for serotonin receptors and its poor ability to penetrate the blood–brain barrier because of its polar molecular characteristics. Levels in the brain peak at about a third of the concentration seen in the blood,[598] but the hallucinogenic effects are comparatively long-lived, at 10 to 12 hours. Ultimately, mescaline's bioactivity is predicated in part on its broad affinity and excitatory effects at serotonin receptors, including 5-HT_{2A}, 5-HT_{2B}, 5-HT_{2C}, and 5-HT_{1A}.[371,599] The comparative contribution of each of these receptor subtypes is unclear, as is the contribution to mescaline's net effect of dopaminergic and noradrenergic receptor binding.[372,598] The close structural relationship of mescaline to these catecholamine neurotransmitters may account for a number of adrenergic symptoms following ingestion, including increased heart rate, perspiration, pupillary dilation, dry mouth, anxiety, heightened temperature, and nausea.[600]

Slight modifications of the structure of mescaline result in synthetic hallucinogenic amphetamine compounds with far greater receptor-binding specificity and potency, including 2,5-dimethoxy-4-methylamphetamine (DOM), 2,5-dimethoxy-4-iodoamphetamine (DOI), and 4-bromo-2,5-dimethoxyamphetamine (DOB). The active dose of these compounds in humans is in the region of 100 times less than mescaline itself, and their increased receptor specificity has led to their being used as neuropharmacological probes in a plethora of animal and in vitro studies.[581]

Very little empirical research has been undertaken assessing the psychotropic effects of mescaline in humans, beyond the self-experiments of the early hallucinogen pioneers. The few small studies have assessed the extent to which mescaline mimics the symptoms of acute psychosis and have included studies that have demonstrated that mescaline aggravates the psychotic symptoms of sufferers from schizophrenia. In the three published reports involving healthy participants, mescaline was investigated to determine whether its effects were similar to the symptoms of acute psychosis in healthy males. Across the studies, 3.5 to 4 hours after taking mescaline, participants' ratings on the Brief Psychiatric Rating Scale (BPRS) and Paranoid Depression Scale (PDS) had degenerated to resemble an "acute psychotic state",[601–603] with a pattern of increased blood flow in the right hemisphere of the frontal cortex, as measured with SPECT, which was correlated with mescaline-induced psychotic psychopathology.[603] The ecological roles of mescaline have not been investigated to date.

■ ERGOT ALKALOIDS: LYSERGIC ACID DERIVATIVES

Throughout recorded history there have been outbreaks of a mysterious disease that came to be known in Roman times as *ignis sacer* (holy fire) and in the Middle Ages as "Saint Anthony's fire" (after its patron saint).[604] We now know that these outbreaks, some of which affected thousands of people, were caused by the alkaloids synthesized by a seemingly innocuous fungus colloquially known as ergot (*Clavicipitaceae* family, typically *Claviceps purpurea*) that grows symbiotically on cereal. Ergot-contaminated flour is associated with two separate disease syndromes, depending on the particular profile of alkaloids present on the cereal. Gangrenous ergotism occurs as a result of the extreme vasoconstriction associated with ergotamine and related alkaloids and results in restricted blood flow to the extremities, which become hypoxic and develop gangrene. Convulsive ergotism, on the other hand, is likely to be related to the additional presence of clavines and features a range of symptoms related to involuntary muscle contractions and a number of neurological symptoms, including vertigo, double vision, headaches, hallucinations, and delirium.[605] The symptoms of communal ergotism would often have been attributed to witchcraft, and it has been theorized that this condition was responsible for the apparent bewitchments that triggered the Salem witch trials.[606] Although the prevalence of ergotism declined with the realization that ergot on cereals was the culprit, the disease did survive into more recent times, with outbreaks seen in Russia in 1926–7 and India and Ethiopia at the turn of the 21st century.[605,607]

If we look back slightly further in time we find that ergot-infected cereal was also used in ritual and religion in Ancient Greece and subsequently Rome. The *kykeón*, a mysterious flour-based potion consumed during the "Eleusinian Mysteries" initiation ceremony, may well have owed its hallucinogenic and entheogenic properties to water-soluble lysergic acid amides from ergot.[14,26,35,36] Ergot also has a long history, extending back as far as Mesopotamia in 600 BC, of medicinal use in labor, both to precipitate childbirth and staunch blood flow.[605] It continued to be used and appears in European pharmacopeia in the 16th century, although its use fell away due to its lack of predictability. Eventually its only remaining use was in preventing potentially deadly postpartum bleeding.[607]

Lysergic Acid Diethylamide (LSD)

The ergot alkaloids eventually leapt into the public's consciousness due to the work of the Swiss chemist Albert Hofmann. The identification, synthesis, and modification of the ergot alkaloids was of considerable interest to chemists as they offered great potential for the development of pharmacological compounds due to their chemical structure, which has commonalities with both the catecholamine neurotransmitters and serotonin (see Fig. 4.3 in Chapter 4).[195,605] In 1938 Hofmann was hard at work isolating, synthesizing, and modifying a number of lysergic acid derivatives. His research had already generated molecules such as ergometrine, which would have obstetric applications that echo the crude (but more dangerous) effects of ergot; lysergic acid amide, which he named ergine; and ergoloid mesylate (Hydergine), a nootropic that would be used to treat the cognitive deficits of dementia.[607] The 25th among a long series of derivatives that Hofmann produced was lysergic acid diethylamide, abbreviated in his notes as LSD-25. Hofmann put this "relatively uninteresting" compound to one side. Five years later a "peculiar presentiment" that LSD may have unexplored pharmacological properties led him to resynthesize it, and on April 16, 1943, as he worked in the laboratory, Hofmann accidentally absorbed some of the crystalized LSD through his skin. Restless and dizzy, he retired home to lie down and experienced "a not unpleasant intoxicated-like condition, characterized by an extremely stimulated imagination. . . . a dreamlike state" in which he saw "an uninterrupted stream of fantastic pictures, extraordinary shapes with intense, kaleidoscopic play of colors." After 2 hours the effects faded away. Three days later Hofmann confirmed the psychedelic effects of LSD by taking 0.25 mg of the drug in his laboratory. Later, already experiencing hallucinations, he endured a surreal and distorted bicycle ride home and, bedridden, he suffered the first "bad trip" of the psychedelic age. As he recalled: "My surroundings had now transformed themselves in more terrifying ways. Everything in the room spun around, and the familiar objects and pieces of furniture assumed grotesque, threatening forms. They were in continuous motion, animated, as if driven by an inner restlessness. The lady next door. . . . [became] a malevolent, insidious witch with a colored mask. Even worse than these demonic transformations of the outer world, were the alterations that I perceived in myself, in my inner being".[607]

Hofmann's creation, LSD, is both the most potent and most pharmacologically complex of all of the hallucinogens, although its mechanisms of action are still poorly understood. Unlike mescaline, LSD easily passes the blood–brain barrier.[459] In terms of receptor interactions it can be most aptly described as a partial agonist at the 5-HT_2/5-HT_1 receptors that have both excitatory and inhibitory effects on neuronal function respectively. LSD is specifically a partial agonist at excitatory 5-HT_{2A} receptors, including those on the apical dendrites of the richly connected neocortical pyramidal cells. As with all of the hallucinogens, the activation of 5-HT_{2A} receptors also leads to increased levels of the excitatory neurotransmitter glutamate, possibly mediated by activity originating in the thalamus, with resultant increases in focal activity across wide regions of the brain. LSD is also an agonist at inhibitory 5-HT_{1A} receptors, particularly in the cortex and the key areas of the brain that modulate serotonergic function, most notably the locus coeruleus and raphe nuclei, and these interactions have the potential to modulate downstream activity throughout other neurotransmitter systems.[459] This picture is further complicated by LSD's ability to bind and modulate the activity of most of the other 5-HT_1 and 5-HT_2 receptor subtypes, several adrenergic receptors, and the five subtypes of dopamine receptor (D_{1-5}).[372]

These receptor interactions and their multiple potential downstream effects underlie a range of psychological effects. Across a raft of laboratory-based studies participants have been shown to be impaired in terms of psychomotor function, distortions in the perception of time, and in most aspects of cognitive functioning, although these apparent neurocognitive impairments may simply be due to a lack of cooperation because of the intense physical and perceptual effects of the drug.[459] These include a range of subjective effects: perceptual distortions and hallucinations, such as an intensification of color perception and visual imagery; the metamorphosis of objects and faces; changes in or an intensification of mood; alterations in thinking, such as less abstract and more imaginative thoughts; distorted perceptions of the body; changes in memory function, including re-experiencing biographical memories and mass recall of memories; and mystical experiences.[459] LSD can also be associated with somatic symptoms related to activation of both branches of the autonomic nervous system—so, on one hand sympathetic symptoms such as pupillary dilation and increased heart rate and blood pressure, and on the other hand, parasympathetic symptoms such as sweating, increased salivation, and occasional nausea.[459]

Lysergic Acid Amide

The seeds from several members of the *Convolvulaceae* (morning glory) family, such as *Ipomoea violacea* and *Turbinia corymbosa*, have been used as entheogens in Mesoamerica from pre-Aztec times until the present and are collectively known as ololiuqui. The Spanish physician Francisco Hernandez, writing circa 1570, noted: "When the priests of the Indians wanted to visit with the gods and obtain information from them, they ate of this plant in order to become inebriated. Thousands of fantastic images and demons then appeared to them. . .".[607] Mexican tribes from isolated regions of the mountainous state of Oaxaca continue to use the seeds to induce a trance-like state typified by sedation, lethargy and apathy, increased perceptual sensitivity, and a stimulation of imagination that inculcates a divinatory state.[59] In some cases *piuleros*, professional soothsayers or medicine men, take the crushed seeds in water or in an alcoholic beverage and then give their clients advice under the influence of the drink. Alternatively they might give the ololiuqui to the client or patient and then question him or her to divine the best course of action, or appropriate treatment for his or her ailment.[63]

Interest in ololiuqui was piqued by a 1941 monograph by the ethnobotanist Richard Schultes[608] that detailed the botany, history, and use of ololiuqui from before the times of the Aztecs. The mescaline enthusiasts soon picked up on this potential hallucinogen. Humphry Osmond, in the preamble of his *British Medical Journal* paper describing his own self-experiment with ololiuqui,[609] noted anecdotally that, whereas Aldous Huxley experienced little effect from eating six seeds, Mrs. Huxley "had some delightful visions depicting some new adventure in the life of Wu-Cheng-Ens' heroic and archetypal ape, Monkey." He then recounted his own self-experiments, in which he consumed increasing numbers of seeds (14, 26, 60, 100) crushed up with water on different occasions. At the lower doses the effects were modest and consisted largely of apathy and lethargy, accompanied by alert thought processes, and this increased along with the dose until they were accompanied by a sharpening of visual perception, distortion of the perception of time, and hypnagogic visions when the eyes were closed. Osmond also noted that these effects were eventually "replaced by a period of alert, calm, relaxed well-being lasting many hours." Frank hallucinations were not included in the experience.

Naturally, Albert Hofmann became interested and, indeed, his involvement in elucidating the chemistry of the indole alkaloid hallucinogens came full circle when he isolated the main active ergot alkaloid in ololiuqui and found that it was lysergic acid amide. This was the same chemical that Hofmann himself had synthesized from lysergic acid, calling it ergine, during the series of studies that had thrown up LSD. The difference here was that it was occurring naturally in the ololiuqui seeds. Of course, Hofmann's new discovery was closely related structurally to his semisynthetic creation LSD and shares many of the psychotropic effects, albeit at 20 to 40 times less potency. Hofmann's sample of ololiuqui also contained a number of other ergot alkaloids, including d-isolysergic acid amide and another of the molecules he had previously synthesized, the vasoconstrictor ergometrine.[63]

The news that ergot alkaloids were present in two very dissimilar taxa, the seeds of an angiosperm plant and a lower fungus, was initially met with some incredulity by the scientific community.[607] However, the conundrum of how a plant came to synthesize fungal alkaloids was finally resolved a few years ago when the source of the plant alkaloids was traced to the *Periglandula* genus of fungi, whose symbiotic relationship with *Turbina corymbosa* and other members of the *Convolvulacea* involves the fungus that lives on the leaves of the plant synthesizing the ergot alkaloids, which are then actively sequestered by the plant for its own ecological purposes.[610,611]

Naturally Hofmann ran his customary self-experiments with both lysergic acid amide (LSA) and isolysergic acid amide (isoLSA) and described them both as having similar depressive and sedative effects; LSA led to "indifference, a decrease in psychomotor activity, the feeling of sinking into nothingness and a desire to sleep," and isoLSA created an experience of "tiredness, apathy, a feeling of mental emptiness and of the unreality and complete meaninglessness of the outside world".[63] A few, small, laboratory-based controlled trials of the ololiuqui alkaloids were also subsequently conducted. Perhaps the most thorough involved the administration of the total alkaloids from ololiuqui, and each of the two principal alkaloids alone, to healthy adults, who then underwent psychometric, behavioral, and subjective assessments.[612] The researchers found that all three treatments reduced the ability to concentrate and that the subjective effects of LSA and isoLSA differed in that the former featured intoxication, reduced consciousness, and autonomic side effects and the latter engendered more euphoria, synesthesia, and distorted perceptions of time. Smaller doses of the total ololuiqui alkaloids were then similar to isoLSA but became more LSA-like as the dose ascended. Overall they found little evidence of hallucinogenic effects per se.

There is comparatively little information on the pharmacological properties of LSA and isoLSA, with the exception that 5-HT_{2A} receptor binding has been established.[384]

Investigations of the effects of ergot alkaloids from fescue grasses, carried out in an agricultural context, also show that LSA also binds to dopamine D_2 receptors with a similar potency to dopamine (but much less potency than several other ergot alkaloids), reducing cyclic AMP production, offering a mechanism for the reduced prolactin levels seen in cattle that graze on infected grasses.[385]

Of course, no good hallucinogen ever goes to waste, and both the seeds from the traditional ololiuqui-producing plants and those of Hawaiian baby woodrose (*Argyreia nervosa*), which is reported to express the highest concentrations of ergot alkaloids of any of the *Convolvulaceae*, are increasingly consumed around the world for their psychoactive effects.[613]

Ecological Roles of Ergot Alkaloids and Brain Function

One hint as to the ecological roles of these hallucinogens can be garnered from research focusing on some of the lower fungi that survive in a symbiotic relationship with their host plant.[614] *Claviceps purpurea*, or ergot, which enjoys a parasitic/symbiotic relationship with rye, has received a great deal of attention because it grows on a crop consumed by humans and has the potential to cause ergotism. However, several of ergot's close relatives within the fungal family *Clavicipitaceae* have been investigated because of their significance in agricultural terms. The *Epichloë* species, and their asexual derivatives, the *Neotyphodium* species of fungi, are transmitted via airborne spores or via the host's seeds and grow ubiquitously on turfs and grasses. These fungi synthesize a range of secondary metabolites that confer a number of fitness advantages on their host plants, including increased resistance to disease and herbivory, especially by insects, with a resultant increase in the growth and the geographical range of the host plant.[614] To illustrate the advantages to the plant, the sustainability of perennial ryegrass in New Zealand is poor unless a high percentage of the plants within a population are infected with *Neotyphodium lolii*.[615] The specific secondary metabolites involved in this relationship include lolitrems, a neurotoxic indole-terpene alkaloid that causes "ryegrass staggers" in grazing animals, and insecticidal chemicals such as the "loline" pyrrolizidine alkaloids (e.g., N-formylloline and N-acetylloline), and peramine. The fungi also synthesize a range of ergot alkaloids,[616] including ergovaline, clavines, lysergic acid, and lysergic acid amide. In terms of deterring insect herbivores, aphids have been shown to prefer noninfected grasses and suffer reduced population growth when feeding on infected grasses.[617] The direct insecticidal effects of the ergot alkaloids in particular has been assessed previously using grasses infected with strains of "genetic knockout" fungi that differed from wild-type fungi only in that they lacked the ability to synthesize either all of the ergot alkaloids, or alternatively the more complex ergovaline and lysergic acid amide alkaloids. Neonate generalist caterpillars fed grass infected with the normal "wild-type" fungi ate less leaf and suffered stunted growth, delayed development, and increased mortality. However, these effects were significantly reduced in caterpillars fed grass bearing either of the two genetically modified fungi that lacked the ergot alkaloids. The fact that there were no differences between the two genetically modified fungi suggests that the more complex ergovaline and lysergic acid amide were directly responsible for the insecticidal effects of the fungi.[616]

As noted above, until very recently the ergot alkaloids responsible for the psychotropic effects of ololiuqui, including lysergic acid amide, were assumed to be synthesized by the plant itself. In reality, these plants simply enjoy another symbiotic relationship with members of the recently described *Periglandula* genus of the *Clavicipitaceae* family of fungi.[610] These fungi colonize the plant, living in and around the oil secretory glands on the leaf surface, and synthesize ergot alkaloids, which are then actively transported into the plant and distributed throughout the plant's tissues, where they perform defensive functions for the plant. The fungus is then transmitted to the next generation of plants via the seeds, which have now also become hallucinogenic as a consequence. The symbiotic nature of this entire process is confirmed by the observation that the fungus itself is largely devoid of the alkaloids that it synthesizes.[611]

The ergot alkaloids synthesized by fungi, and employed as defense compounds by plants, do have well-established negative effects on large grazing animals; for example, the descriptively named "drunken-horse grass" (*Achnatherum inebrians*) and "sleepygrass" (*Achnatherum robustum*) both owe their effects to the presence of fungal lysergic acid derivatives. However, it is unlikely that deterring large herbivores is their primary ecological role; the psychotropic effects require the consumption of quantities of infected grass that were unlikely to ever be achieved by any grazing animal prior to the relatively recent introduction of large expanses of monoculture crops. It is also notable that there is little evidence that grass fungal infections are detected by grazers or that they deter consumption, rendering them somewhat ineffective. It is far more likely that the ergot alkaloids offer protection against invertebrates, including insects and nematodes, and other pathogens such as bacteria.[605] Any effects on insects and other invertebrates would not be entirely surprising, as many of the extended fungal family of *Clavicipitaceae* are specialist arthropod and invertebrate pathogens. Their life cycles involve infecting species from at least 10 arthropod orders, with any one fungus typically being specialized to a single species or a narrow range of species of arthropods.[618] Intriguingly, recent phylogenetic analyses have shown that the fungi living in symbiotic relationships with grasses, including *Claviceps purpurea, Epichloë*, and *Neotyphodium*, themselves originated as specific arthropod pathogens. Their closest relatives are the fungi specializing in infecting *Hymenoptera, Lepidoptera, Coleoptera*, nematodes, and mollusks.[618] Presumably, the close, mutually beneficial relationship between the grasses and these fungi was originally forged on the basis of the anti-invertebrate nature of the fungus, with its toxic ergot alkaloids, after fungus-bearing insects, worms, or mollusks had brought their deadly cargo unwittingly to the grass.

■ SIMPLE TRYPTAMINES

Psilocybin/Psilocin

Psilocybe is a genus comprising approximately 250 species of small mushrooms that are distributed across all of the world's continents. About 150 of these species have hallucinogenic properties, with 53 of these found in Mexico alone.[62] The earliest evidence of the divinatory or ritual use of *Psilocybe* mushrooms comes from representations in cave rock art. The oldest example is a mural depicting the local *Psilocybe mairei* in a cave in the Tassili-N-Ajjer region of the Sahara, in Algeria. The estimated date, 5000 to 7000 BC, would have coincided with an era in which today's desert was forested and fertile.[6] Slightly later glyphs from the area include many depictions that could be interpreted as including mushrooms, including a bee figure, complete with mushroom shapes projecting from its body, and a human figure that appears to be defecating two types of mushrooms, possibly representing *Psilocybe* and *Amanita muscaria*.[15] Cave murals in the Spanish Sierra de las Cuerdas mountains dating to 4000–6000 BC similarly appear to depict *Psilocybe hispanica*, a hallucinogenic mushroom unique to the Pyrenees.[5] Representations of mushrooms crop up as far afield as the easternmost tip of Siberia, where the Chukchi Eskimos' word for intoxication literally translates as "bemushroomed".[22] A plethora of indirect evidence drawn from sculptures, reliefs, hieroglyphs, jewelry, clothing, and burial practices also suggests the widespread use of mushrooms such as *Psilocybe cubensis* in the religious practices of the early Egyptians. The hallucinogenic experience would have allowed the user to commune with their gods and may have been believed to confer immortality and divinity. Berlant[15] suggests that the "Eye of Horus" and the god Osiris may both have represented hallucinogenic mushrooms.

The inhabitants of Mexico and Mesoamerica, a region with a plethora of indigenous *Psilocybe* species, have a long history of mushroom use, with depictions on ceramics and the sculpted "mushroom stones" of the Mayan civilization dating back 2,500 years.[5] The use of mushrooms in this region encompassed entheogenic communication with the gods, in which a shamanic figure would enter a visionary state in order to divine the answer to specific questions from the community. Mushrooms would also be consumed during festivities or in more mundane social settings. As an example, the Dominican Friar Diego Duran reported that the mushrooms called *teonanácatl*, or "flesh of the gods," were eaten during the festivities marking the accession of Moctezuma II to the Aztec throne in 1502. Slightly later, Friar Bernardino de Sahagún described the use of sacred mushrooms in his *Historia General*. In his account he describes how merchants celebrated a successful trip with chocolate drinks and mushrooms, which led to dancing and weeping, and an assortment of visions both good and bad. As he noted: "Some saw in a vision that they would die in war. Some saw in a vision that they would be devoured by wild beasts. . . . Some saw in a vision that they would become rich, wealthy".[59,607] Naturally, to the fervently Catholic Spaniards, inebriation, visions, and hallucinations were all works of the devil, so the Conquistadors viciously suppressed mushroom use from the 16th century onwards. Indigenous use only survived the attentions of the Spanish in higher, less accessible mountainous regions of Mexico. As mushroom use had fallen away in other areas of the globe, it was now restricted to these mountainous areas, and some regions of Siberia, for the next 400 years.[62]

Psilocybe mushrooms were, in effect, rediscovered by the Western world in 1955 when the wealthy amateur ethnomycologist R. Gordon Wasson and the photographer Allan Richardson were given sacred mushrooms for the first time while travelling in Mexico. They reported the experience of perceptual distortions, waking dreams, and mystical awakening. Naturally, the rediscovered psychedelic mushrooms eventually found their way into the laboratory of Albert Hofmann, who had to personally start the process of identifying the bioactive principals due to an antipathy toward hallucinogenics by his employers, Sandoz, in the wake of the growing misuse of LSD. Eventually, in 1957, following more of his trademark self-experimentation, Hofmann and colleagues isolated the active components underlying the psychoactive effects of *Psilocybe mexicana*, noting that the hallucinogenic effects were very similar to those of LSD, but at 100 times the dose.[607]

It transpired that the active constituents of the hallucinogenic *Psilocybe* mushrooms were the indole alkaloids psilocybin and psilocin, both of which are derived from the amino acid L-tryptophan via the monoamine alkaloid tryptamine. Psilocybin is rapidly dephosphorylated to the active metabolite psilocin in the body. Synthesis of psilocin itself in the mushroom is via a similar, if slightly reordered, pathway as to that which leads to the plant signaling molecule, and animal neurotransmitter, serotonin.[195] The close structural similarities between L-tryptophan, serotonin, and psilocin are shown in Figure 4.3 in Chapter 4.

As with the other hallucinogens, psilocin's structural similarity to the endogenous neurotransmitter endows it with serotonin receptor-binding properties. In this regard it is relatively nonselective, displaying moderate to high affinity for all mammalian 5-HT_1 and 5-HT_2 subtypes, with hallucinations and visual distortions related to agonist binding at the 5-HT_{2A} receptor.[588] This relationship has been confirmed by a positron emission tomography (PET) imaging study that demonstrated a direct correlation between the perceptual effects of psilocybin and its occupation of 5-HT_{2A} receptors in the anterior cingulate cortex and medial prefrontal cortex.[619] Other effects, in particular those evident across a substantial animal experimentation literature, are likely to be mediated by activation of 5-HT_{1A} and other serotonin receptors, and a plethora of potential downstream neurochemical events,[588] including, for instance, demonstrations of an increase in striatal dopamine as assessed by PET.[620]

In the 1960s a number of early investigations probed the utility of psilocybin in enhancing the psychotherapeutic process. Medium oral doses (12 to 20 mg) were found to engender an altered state of consciousness that included increased emotionality and introspection, and perceptual distortions and illusions. The overall effect following higher doses was often likened to a mystical experience.[17,621,622] Indeed, in the most famous study, on Good Friday 1962, Walter Pahnke, as part of a PhD dissertation co-supervised by Timothy Leary, investigated whether psilocybin could facilitate an entheogenic religious experience in a group of spiritually predisposed theology students. The group received either 30 mg psilocybin or a placebo in a church service context. The psilocybin group, with one exception, reported religious or mystical experiences, whereas the placebo group, again with one exception, did not. The experience also left members of the psilocybin group reporting positive changes in their attitudes and behavior 6 months later.[622] A follow-up study, while critiquing the methodology of the original study, relocated the majority of the participants after 24 years and found that the psilocybin group still reported that the mystical experience following their single exposure to the drug had resulted in persisting spiritual benefits.[623]

During a more recent resurgence in psilocybin research, Griffiths et al.[20] revisited the research question underlying the "Good Friday experiment." In their study, 36 hallucinogen-naïve, religiously or spiritually active volunteers were administered a high dose of psilocybin (30 mg/70 kg) and a non-hallucinatory active comparison treatment (methylphenidate [Ritalin]). The results showed that psilocybin engendered mystical-type experiences and, for a minority, significant fear. Reassessments carried out 2 and 14 months later showed that participants rated the experience as having long-term spiritual significance and a positive effect on their outlook and behavior.[20,624] Several years later the same research team expanded on their earlier findings with a dose-ranging study[625] in which 18 healthy adults were given a placebo and ~5, 10, 20, and 30 mg psilocybin. The majority of measures showed orderly time- and dose-related patterns of modulation, with ascending doses of psilocybin associated with increased hallucinatory symptoms, increased ratings of "altered states of consciousness" and "mystical experience," and, again for a sizeable minority, increased fear and anxiety. These results have been replicated consistently within a growing literature. For instance, Studerus et al.[577] conducted a meta-analysis of data from eight double-blind, placebo-controlled psilocybin studies involving 110 healthy participants. These individuals had not been recruited on the basis of spirituality or religious tendencies. The analysis demonstrated consistent, dose-related changes in affective state (such as increased emotional excitation, and dreaminess), visual perceptual distortions and pseudo-hallucinations, and altered states of consciousness (as assessed by the Altered States of Consciousness Rating Scale) that could be described as mystical. There were few negative effects, and no later flashbacks as assessed by clinical criteria. During follow-up interviews 8 to 16 months after taking part, participants rated the psilocybin experience as having beneficial effects on their attitude or personality and as having had an "enriching" effect upon them.[577]

Psilocybin has also been used in several brain imaging studies. For instance, in a PET study comparing its effects to those of the entactogen 3,4-methylenedioxyethylam phetamine (MDE) and methamphetamine, the three drugs engendered different patterns of neurometabolic changes, with psilocybin being associated with frontal cortex overactivity and a blunted response in the prefrontal cortex in response to cognitive tasks. These effects were directly analogous to those seen in schizophrenia.[626] As noted above, in the most recent brain imaging study, which used two variants of fMRI, psilocybin led to reduced blood flow within a number of regions of the brain that act as communication hubs for activity, including the thalamus. The reduced activity in this structure correlated negatively with the intensity of the experience for the participant, enabling what the authors described as "a state of unconstrained cognition".[587]

Dimethyltryptamine and Derivatives

The hallucinogenic tryptamines N,N-dimethyltryptamine (DMT) and its derivatives, 5-methoxy-DMT (5-MeO-DMT) and 5-hydroxy-dimethyltryptamine (5-HO-DMT or bufotenin), are synthesized by plants from a number of unrelated families. Historically they have been consumed in South America/Mesoamerica in a variety of forms, depending on the geographical area. The most common and ubiquitous mode of consumption was, and remains, the drinking of a decoction called ayahuasca, which is made from the jungle vine *Banisteriopsis caapi* and the shrub *Psychotria* viridis (or less often *Diplopterys cabrerana*). Another popular method was the snuffing or smoking of bufotenin-rich powders made from the toasted seeds of the two members of the *Anadenanthera* genus, *A. colubrine* and *A. peregrine*, which produce the snuffs most commonly known as vilca/cebil and yopo/cohoba respectively. While there is little archaeology surrounding ayahuasca, primarily because it is a liquid and does not require any specialized paraphernalia, the earliest archaeological evidence of the use of the *Anadenanthera* genus comes in the form of snuff paraphernalia, often in the form of a kit comprising a wooden snuff tray, a tube made of bone or wood, a spatula, and sometimes a pestle and mortar, often contained in a woolen bag. The earliest examples, snuffing trays in the form of birds and fish, have been found in southern Brazil and eastern Paraguay and have been dated to circa 3000 BC. The archaeological evidence then gradually tracks north over the next two millennia, presumably as the use of the snuff spread through indigenous groups.[43,627] Certainly by the time of the Spanish Conquests both snuffing and ayahuasca were ubiquitous throughout the accessible parts of South America. Many of the more recently dated finds, grave goods from the early part of the first millennium AD, contain identifiable plant material, and analysis of this powder shows the presence of DMT, 5-MeO-DMT, and bufotenin, the latter of which confirms that the material comes from a member of the *Anadenanthera* genus.[627] Analysis of a snuffing pipe dated 2100 BC found in northwest Argentina also confirmed the presence of DMT.[628] The act of snuffing or smoking these plant powders is significant in that orally consumed tryptamines are metabolized in the liver and gut by the ubiquitous enzyme monoamine oxidase A (MAO-A) before they can reach the brain, rendering them inactive. Absorption by the mucous membranes of the nose or lungs bypasses this process.

The bioavailability of the tryptamines in ayahuasca, on the other hand, relies on a different mechanism. The drink itself is made by boiling the scraped-off bark of the jungle vine *Banisteriopsis caapi* (ayahuasca means "vine of the soul" in the widely spoken Andean Quechua language) with the leaves of *P. viridis* or *D. cabrerana*, which contribute the DMT. The reason that the two plants were combined, presumably as a consequence of an unlikely but fortuitous co-consumption in prehistory, is that *B. caapi* contains a range of β-carboline alkaloids, including harmine, tetrahydroharmine, and harmaline, which inhibit the activity of monoamine oxidase in the consumer's digestive system and liver, leaving the DMT unmetabolized and free to cross the blood–brain barrier.[393,629]

Ayahuasca continues to play a central role in the religious, ceremonial, and medicinal practices of South American shamanism.[630] Alternative names include daime or hoasca in Brazil, yajé in Colombia, and natem in Ecuador. Ayahuasca has also spread alongside the syncretic religions that originated in Brazil, such as Santo Daime and the Uniao do Vegetal, and is now used globally both in this religious context and as a social drug.[631]

In terms of structure, DMT, along with serotonin and the other hallucinogens (except mescaline), is synthesized from the amino acid L-tryptophan. Synthesis in this case is via tryptamine, and DMT is then modified to give 5-MeO-DMT and bufotenin. The relevant structures are shown in Figure 4.3 in Chapter 4. As to its psychopharmacology, the hallucinogenic effects of DMT are described typically as being a consequence of its affinity for binding to 5-HT_{2A}, 5-HT_{1A}, and 5-HT_{2C} receptors.[379,632,633] However, early receptor-binding studies also suggested a modest affinity for α-adrenergic receptors.[379] A more recent analysis demonstrated that DMT is actually the most highly promiscuous naturally occurring hallucinogenic alkaloid in terms of receptor interactions and rivals the semisynthetic LSD for the broadest binding profile of any classic hallucinogen. Its favored binding sites (in descending order of affinity) from a comprehensive screening study were 5-HT_7, 5-HT_{2B}, 5-HT_{1D}, α_{2b}, α_{2c}, D_1, 5-HT_{2C}, 5-HT_{1E}, 5-HT_6, 5-HT_{5A}, α_{1B}, α_{2A}, $\alpha 1_A$, 5-HT_{2A}, and σ_1 receptors.[372] DMT also modulates the uptake of monoamine neurotransmitters and competes for their metabolic enzymes, leading to complex effects unrelated to receptor binding.[634]

5-MeO-DMT is also a nonselective serotonin receptor agonist, with greatest affinity for the 5-HT_{1A} receptor and comparatively low affinity for the 5-HT_{2A} receptor.[635] It also exhibits more pronounced binding activity at 5-HT_7, 5-HT_{1D}, 5-HT_6, 5-HT_{1B}, 5-HT_{5A}, 5-HT_{1E}, and 5-HT_{2C} receptors and interacts with dopamine ($D_{1,2,3}$) and adrenergic $\alpha_{2A,B,C}$ receptors,[372] and it inhibits serotonin reuptake (with no effect on dopamine) to a similar extent as cocaine and methamphetamine.[635] However, once consumed 5-MeO-DMT is also readily metabolized by the cytochrome P450 "2D6" to its active metabolite bufotenin, which exhibits a far greater affinity for the 5-HT_{2A} receptor and approximately three times the potency of 5-MeO-DMT when they are both present in the brain. Although bufotenin's ability to cross the blood–brain barrier was initially questioned, it readily produces hallucinogenic psychoactive effects if injected or snorted. Bufotenin therefore joined DMT and 5-MeO-DMT as a Schedule I controlled substance in the United States some decades ago.[635]

In a modern context, but harking back to ayuahasca, 5-MeO-DMT is often taken with an MAO inhibitor, typically harmaline, and this serves to sharply increase and extend the duration of brain levels of both 5-MeO-DMT and bufotenin.[635] The MAO-A inhibitory properties of harmaline or the other β-carboline alkaloids from *B. caapi* may also lead to generally increased levels of the monoamine neurotransmitters, including serotonin, dopamine, and noradrenaline, due to an inhibition of their metabolism[392,393] and direct interactions with dopamine, 5-HT, GABA, and NMDA receptors.[635] This may lend ayahuasca and 5-MeO-DMT a much more complicated palette of pharmacological effects than those attributable to DMT alone.

Beyond the 5-HT_{2A} receptor another complementary mechanisms that might contribute to the hallucinatory effects of plant/fungal secondary metabolites is the shared affinity of hallucinogens for the group of TAARs, which were first identified as recently as 2001.[268] These receptors, as well as playing roles in cardiac and vascular function, play key neuromodulatory roles and are implicated in a host of mood and neurological disorders,[269] including putative roles in the hallucinogenic and delusional phenomena experienced by schizophrenics.[270] They are also expressed throughout areas of the brain that underpin sensory perception, such as the prefrontal cortex, hippocampus, substantia nigra, amygdala, and basal ganglia.[267] The particular relevance here is that DMT, 5-methoxy-DMT, and bufotenin were identified several decades ago as endogenous mammalian neurochemicals, and very recently as trace amines that bind to members of the TAAR family.[267] TAAR_1 is a particularly promiscuous binder of monoamines, and its ligands include the above endogenous hallucinogenic trace amines and most of the "classic hallucinogens" described above. Intriguingly, the group of exogenous ligands also includes semisynthetic compounds such as amphetamine and MDMA, both of which are reported to have perception-distorting or hallucinatory properties at high doses.[317,401] These observations may suggest a further common pathway for the hallucinatory effects of all of these compounds.[581] In this respect Wallach[267] notes that the tryptamine trace amines must play an endogenous role in the creation of waking awareness, or consciousness, which he suggests is an internally created (psychedelic) state that normally correlates with events or stimuli from the "real" world. Sensory distortions attributable to the consumption of hallucinogens are then the result of their coopting the TAAR pathways associated with the endogenous trace amines within perceptually relevant areas of the brain. In a similar vein, the sigma receptors σ_1 and σ_2, which are widely distributed in the brain, are activated by a number of social drugs, including cocaine, MDMA, and methamphetamine.[636] However, these receptors had "orphan" status, in that no endogenous ligand for the receptor had been identified, until this role was recently attributed to endogenous DMT, at least with regard to the σ_1 receptor. Agonist properties of both DMT and bufotenin at this receptor may contribute to their psychotropic properties.[637]

In terms of isolating the effects of individual tryptamines, DMT, administered intravenously to avoid metabolism of the drug, has been shown to engender a near-instantaneous, dose-related anxiety-provoking physical and mental "rush," followed for higher doses (0.2 to 0.4 mg/kg) by the onset of visual hallucinations comprising complex concrete imagery, a subjective feeling of loss of control and bodily dissociation, and extreme shifts in mood. Auditory hallucinations were experienced by 50% of participants. These effects paralleled the pharmacokinetics of the drug, resolving after 30 minutes.[638] These dose-related effects have subsequently been replicated.[638,639]

Bufotenin, after an inauspicious start in unethical experimentation on convicts and the mentally ill in the 1950s, has also received only limited interest. However, Ott[640] reviews the small number of relevant studies and describes his own dose-ranging self-experiments, ultimately concluding that consuming bufotenin, by inhaling an evaporated freebase or taking capsules orally, had similar, but qualitatively distinguishable, psychoactive and perception-distorting effects as a similar dose of 5-MeO-DMT, whereas DMT required a higher dose or co-consumption of an MAO inhibitor to attain the same level of psychoactivity.

Ayahuasca has received much more research attention than the single compounds, with interest galvanized by the global expansion in its use. So, for instance, in two dose-ranging studies participants were administered doses of the drink containing 0.5, 0.75, and 1 mg DMT/kg[393] and 0.6 and 0.85 mg/kg DMT[641] (along with about three times this quantity of β-carbolines) respectively. The subjective dose-related effects experienced by the participants included a "rush," euphoria and increased well-being, dream-like imagery when the eyes were closed, perceptual modifications, changes in thought content, swings in emotion, and increased physical bodily awareness. In the later study[641] the pharmacokinetic profile of the DMT administered in the form of ayahuasca was found to be much slower than following intravenous administration of DMT alone, with levels peaking in plasma at 90 minutes post-dose, as opposed to almost immediately after injection. Overall, the subjective effects of ayahuasca were found to be somewhat milder than those following intravenously administered DMT,[393] and they tracked the bioavailability of DMT, starting between 30 and 45 minutes post-dose, peaking between 90 and 120 minutes post-dose, and resolving by the end of 4 hours.[641] The subjective effects and pharmacokinetics of ayahuasca have subsequently been consistently replicated.[632,642,643] A brain imaging study using single photon emission computed tomography (SPECT) later demonstrated that these subjective effects were accompanied by increased regional cerebral blood flow in the neo-cortex and paralimbic regions, including the right anterior insula, and amygdala.[643] The authors suggest that these areas are key to interoception (self-awareness of your own, entire, physical body) and emotion respectively, and note that similar frontal cortex activation has been seen following ingestion of mescaline[602] and psilocybin[644] using different imaging techniques. Given its widespread use in South America, several studies have assessed the effects of the long-term use of ayahuasca on neuropsychological functioning. These studies have shown no deficits; on the contrary, the ayahuasca group in one study that included 127 users and 115 matched controls exhibited better scores in terms of ratings of spirituality and psychopathology and performed better on a number of cognitive tasks.[645] There is also no evidence of ayahuasca having addictive properties.[646]

Ibogaine

The root bark of the shrub *Tabernanthe iboga* has a long tradition of use in Congo and equatorial Africa in ritual, in medicine, for reducing fatigue, and as an aphrodisiac.[195] In particular, it is used by the Bwiti cult as a sacrament and in a rites-of-passage ceremony in which inductees drink the preparation, vomit, and then fall into a trance during which they typically have entheogenic spiritual hallucinations.[73] It owes its effects to a group of alkaloids, the most prevalent of which is ibogaine, a terpene-indole alkaloid comprising a tryptamine portion and a C9 terpene fragment derived from the C10 iridoid terpene secologanin[195] (for structure see Fig. 4.3 in Chapter 4).

The African use of Iboga root preparations was described by travelers in the late 19th century, and ibogaine was isolated in purified form in 1901. It emerged onto the French market in the late 1930s as Lambarène, a "neuromuscular stimulant," recommended for the treatment of fatigue and depression. It remained on sale until ibogaine became an internationally controlled illicit drug in 1970. In the meantime, from the 1950s onwards, ibogaine began to be used in the treatment of drug dependence.[647] In this role it is still informally used to treat addiction to a variety of drugs, including cocaine, methamphetamine, and alcohol. Animal research shows that ibogaine reduces self-administration of a range of potentially addictive drugs, reducing the place-preference elicited by amphetamine and morphine and decreasing the symptoms of opiate withdrawal.[647] In humans, case studies and an open-label trial that included 33 patients suggest that ibogaine significantly reduces both the symptoms of opiate withdrawal and their resolution, as well as decreasing post-treatment drug craving.[648]

In terms of psychopharmacology, ibogaine noncompetitively binds to and inhibits both serotonin and dopamine reuptake transporters[380]; it acts as an antagonist at $\alpha3\beta4$ nicotinic acetylcholine[381] and NMDA glutamate receptors[647] and binds to the serotonin 5-$HT_{2A,}$ σ_1 and σ_2, and κ-opioid and μ-opioid receptors.[372] Interactions with these receptors, in particular the σ and opioid receptors, may underpin the attenuation of withdrawal and behavioral effects of opiates.[649] These effects may also, in part, be related to the observation from an animal study that coadministration with ibogaine reduced the release of dopamine seen in the nucleus accumbens following the administration of addictive drugs.[647]

In terms of behavior, animal studies suggest improved spatial memory at lower doses, but with decrements appearing at higher doses.[650,651] In humans, low doses of ibogaine have a stimulant effect. However, the best-documented effects, drawn from case reports, relate to ibogaine's hallucinogenic effects. These usually commence 1 to 3 hours after ingestion, with the eyes closed, and involve "waking-dream"-like visual hallucinations, often involving "panoramic" visual recall of past events or the experience of entering into complex landscapes. As this phase subsides, consumers often enter a period in which they passively evaluate their experience while remaining stationary. As the psychoactive effects resolve consumers then often experience a lengthy period of arousal or increased vigilance and often report poor sleep for several days.[647] However, a host of negative side effects, including vomiting and tremors, make ibogaine unattractive as a social drug in its own right, and its use as an "official" medicinal treatment for drug cravings and withdrawal symptoms has been stymied by reports of neurotoxicity in rats at high doses.[647] Ibogaine may also generate life-threatening cardiac arrhythmias by interfering with the functioning of cardiac potassium ion channels.[394]

Ecological Roles of Simple Tryptamines and Brain Function

Direct research does suggest that tryptamine and its derivates are expressed by plants as deterrents to herbivory, reducing feeding and inhibiting the growth of insects.[652,653] As an example, the 5-MeO-DMT synthesized by the Amazonian snuff plant *Virola calophylla* acts as an anti-feedant against cotton boll weevil[654] and, when applied to plants at naturally occurring concentrations, is a feeding deterrent to aphid nymphs[655] and reduces aphid nymph survival.[655] Applied to seeds it is also lethal to seed-eating beetle larvae.[526] A simple ecological role in defense against herbivory is also strongly suggested by the secretion of the tryptamines, in particular 5-MeO-DMT and bufotenin, in the toxic skin secretions of toads from the *Bufo* genus. This phenomenon represents an example of alkaloids ascending through several trophic levels, with the *Bufo* toads sequestering the tryptamines from specialist herbivore arthropods that in turn had ingested them in their diet of plant material and incorporated them into their own cuticle as a defense against predation.[656,657] Naturally, this chain of events is predicated on the tryptamines being toxic to the insect in the first place, and therefore requiring safe sequestration by the insect. Interestingly, it is suggested that the excretions of *Bufo* and other genera of toads have been smoked for their psychoactive properties in Mesoamerican societies from pre-Colombian times.[658]

Research using DMT and bufotenin as probes to characterize the insect serotonergic system show that both of the tryptamines bind to the insect 5-HT_1 receptor, suggesting that this is their major intended target (although other targets have not been investigated in the insect).[405] It should also be remembered that octopamine and tyramine function endogenously as trace amines in mammals, binding to TAARs, but that these same chemicals function as major neurotransmitters in insects via G-protein-coupled receptors analogous to TAARs.[267,269] Given the reciprocal effects of many mammalian monoamine receptor ligands at insect octopamine and tyramine receptors[269] and the affinity of hallucinogens for binding mammalian TAARs, it seems likely that hallucinogens may also bind to and disrupt the functioning of insect octopamine and tyramine receptors.

7 The Deliriants—The Nightshade (*Solanaceae*) Family

The deliriants are a group of secondary metabolites with a venerable history of use as entheogens. They can be clearly distinguished from the hallucinogens in that their predominant effect, when taken at a high enough dose, is the induction of a state of delirium, characterized by confusion, disorientation, and a trance-like state. Although the subjective experience of high doses can include hallucinations, these are typically confused with reality, as opposed to the pseudo-hallucinations of the classic hallucinogens.

At first glance, the history of shamanism, witchcraft, potions, and poisons would seem to be very different between the Americas and Eurasia. However, one family of plants with deliriant properties play a key role on both continents. The *Solanaceae* or nightshade family comprises around 21 tribes, 100 genera, and 2,500 species[659] and includes a number of psychoactive plants and a huge range of important food crops, including the potato, tomato, and eggplant (aubergine).[51]

The *Solanaceae* has a center of origin and diversity firmly fixed in the crucible of biodiversity of South America.[659] On the American continent several genera would play a pivotal role in the lives of humans as they spread across the landmass: the *Nicotiana* and *Datura* genera, whose origins can be traced to the region below the narrow isthmus of land connecting North and South America,[660,661] and *Datura*'s close relative, the tree/shrub genus *Brugmansia*, which originated in the humid, highland areas of the more northern reaches of South America.[660] By a twist of fate ancestors of these nightshade genera also escaped from their continent on several occasions. Members of the *Nicotiana* genus traveled to Australia and Africa, carried by birds or more likely ocean currents.[662] Once established in Australia the immigrant plant prospered and radiated into a multitude of species, including several members of the *Nicotiana* and *Duboisia* genera of shrubs/trees, the leaves of which provided the indigenous population with the psychotropic "pituri" leaves.[663] Similarly, cousins of *Datura/Brugmansia* were distributed to the Eurasian continent on two separate occasions.[664] Molecular phylogenetics suggests that the first of these two distributions took place around 20 million years ago, as the ancestor that would foster the *Mandragora* family washed up in the Mediterranean/Turanian region of Europe. This was followed 2 to 3 million years later by the relative that was the progenitor of the Eurasian *Hyoscyameae* tribe that includes the *Atropa, Hyoscyamus,* and *Scopolia* genera. These immigrant plants were lucky in that they arrived in an area of the world that, like their South American homeland, was to be one of the most fertile regions for biodiversity and speciation. They were unlucky in that the continent was also enduring a long period of geological turmoil, including the raising of mountain ranges and the Tibetan plateau. Members of the two families subsequently became isolated on either side of these new geological barriers, with speciation between the Tibetan plateau and environs and the Mediterranean region, eventually leading to, for instance, five geographically separated species of *Mandragora*, five species of *Atropa*, and 11 species of *Hyoscyamus*.[664]

Interestingly, taxonomic evidence strongly suggests that all of the current species of the *Datura* genus originated in South America. Yet, copious written evidence predating Columbus's rediscovery of the Americas describes the use of fruits in the Middle East and India that were originally identified in the modern era as being those of *Datura stramonium*. This species, and eventually several cousins, would proliferate across Europe and into Asia during the Middle Ages.[665] The circumstances of the arrival of the *Datura* genus in Eurasia are therefore somewhat mysterious, and this may represent another, very recent, unaided migration, or even a pre-Columbian human introduction of a *Solanaceae* from the New World. However and whenever they arrived, the deliriant *Solanaceae* species went on to play a reciprocal role in Eurasian witchcraft, religion, and spirituality as their American cousins.

The common key factor that would help these *Solanaceae* species survive in the new environments they migrated to, and would eventually make them indispensable to the rituals, shamanism, witchcraft, and medicine of American, Australasian, and Eurasian humans, was the evolution of synthetic pathways that would allow them to synthesize a cocktail of secondary metabolite chemicals peculiar to the nightshades. As noted previously (Chapter 4), the genetic event that ultimately fostered the synthesis of the psychoactive *Solanaceae* alkaloids was the duplication of a gene involved in the primary metabolism pathways that synthesized polyamines. The duplicate, free from selective pressure, mutated and evolved a new function, coding for a slightly modified enzyme that, in turn, catalyzed the synthesis of a novel compound (N-methylputrescine) that is the crucial precursor for the pyrrolidine ring system that features in this entire group of alkaloids.[114] This new pathway led to the elaboration of the synthesis of nicotine and related alkaloids in the *Nicotiana* genus, and tropane alkaloids, including hyoscyamine, scopolamine, and atropine, in the South American *Datura*, *Brugmansia*, and *Solandra* genera and the Eurasian *Hyoscamus*, *Atropa*, *Scopolia*, and *Mandragora* genera.

■ NICOTINE

It might come as a surprise to many contemporary smokers, but tobacco (usually *Nicotiana tabacum* or *Nicotiana attenuata*) became the single most common plant to be employed as an entheogen in shamanism across the American continent as a consequence of its deliriant properties when consumed at high doses.[666] The amount of tobacco consumed, often smoked over many hours or consumed as a juice, would have disturbed color vision, leaving the consumer experiencing a world of white, yellow, and black; it would have caused visual hallucinations, including moving flashes that could be interpreted as birds; and it would have precipitated the user into delirium and an immobile stupor, including a feeling of disembodiment and the experience of "out-of-body" flight.[666,667] These subjective effects are represented in the many images of flying birds depicted in the continent's rock art and archaeology.[668] Hallucinations may also have been engendered due to the knock-on effect of nicotine increasing neural activity associated with numerous neurotransmitters, including serotonin, along with the monoamine oxidase inhibitory properties of β-carboline alkaloids in the smoke.[669] Of course, tobacco would often also be consumed alongside other psychoactive plants.[666]

By the time Europeans arrived, tobacco had long since pervaded the entire American continent. Archaeological evidence in the form of tobacco seeds or pipe residues suggests that it was in use in New Mexico well before 1000 BC and confirms that its use had spread to the northeast portion of the continent before 200 BC and the northwest portion by the mid-9th century AD.[670] By that time it was grown ubiquitously as a cultivated crop, even among societies that survived as hunter-gatherers. Methods of consumption differed and included snuffing dried leaves, drinking juices, and making the leaves into suppositories. However, the most common mode of administration was smoking: by pipe in North America and rolled into cigars in South America.[61] In Australia, the aboriginal term "pituri" refers historically to the leaves of either local immigrant members of the *Nicotiana* genus (e.g., *N. rosulata* or *N. gossei*) or to their close relative *Duboisia hopwoodi*. The leaves of both plants were chewed as a source of nicotine, either habitually or in order to engender a trance-like state for ritual purposes. It is notable that *Duboisia hopwoodi* combines the synthesis of nicotine, and the related but more potent alkaloid nornicotine, with modest levels of hyoscyamine and scopolamine.[195,671] The other two members of the Australian *Duboisia* genus, *D. myoporoides* and *D. leichhardtii*, synthesize only the tropane alkaloids and are currently grown commercially in Australia as the primary global source of medicinal hyoscyamine, scopolamine, and atropine.[195]

As Alexander von Humboldt noted with regard to South America, "The continent has given us one great blessing and one great curse: the blessing is the potato and the curse is tobacco!".[51] The egregious contribution that the human's addiction to tobacco has made to the turbulent history of mankind is summarized in Chapter 1. Suffice it to say that globally some 1.3 billion adults currently smoke tobacco. Just as the prevalence of smoking has peaked and started to decline in developed countries, so the consumption of tobacco in developing countries has continued to rise inexorably. Projections suggest that by 2025 up to 1.9 billion adults will smoke worldwide, with the preponderance of these smokers in developing countries.[672] The World Health Organization estimates that smoking eventually kills up to 50% of smokers in one way or another. Setting aside the longer-term health issues, it is also notable that nicotine is highly toxic to mammals. While very low doses are delivered by smoking, nicotine can be administered at lethal doses transdermally or orally. Some of the symptoms of nicotine poisoning are manifested in "green tobacco sickness," a condition associated with the absorption of nicotine through the skin as a consequence of picking wet tobacco; these include dizziness, nausea, headache, and indications of parasympathetic activation such as breathing difficulty, fluctuations in blood pressure or heart rate, and increased perspiration and salivation.[472]

The key active component of tobacco smoke is, of course, nicotine. This alkaloid combines the pyrrolidine ring common to the *Solanaceae* alkaloids with pyridine, derived from nicotinic acid.[195] It then owes its pharmacological activity to the spatial configuration of components that it has in common with the neurotransmitter acetylcholine, which allows it to bind to nicotinic receptors. It does this with such specificity that these receptors were identified, and subsequently named, on the basis of this binding. Nicotine continues to be used as an essential probe in neuropharmacological studies investigating the distribution and function of nicotinic receptors. The structure of nicotine and the elements it shares in common with acetylcholine are shown in Figure 7.1.

Both the toxicity and addictive properties of nicotine are predicated on its ability to bind to and activate all three of the principal central nervous system ionotropic

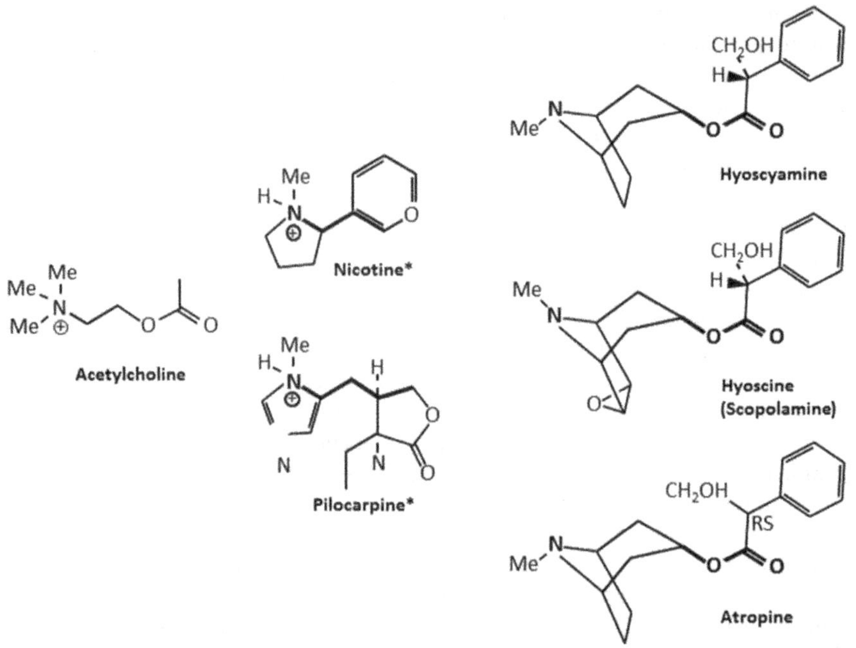

Figure 7.1 Structural similarity between the neurotransmitter acetylcholine and nicotine and the tropane alkaloids derived from the *Solanaceae*. The structural elements that each compound has in common with acetylcholine are shown in bold within each structure. The muscarinic agonist pilocarpine is also included.

nicotinic receptors: the $\alpha4\beta2$, $\alpha3\beta4$, and $\alpha7$ receptors.[a] Nicotine opens the receptor complex's central channel, allowing a flow of positively charged ions into the neuron; the change in voltage then further activates voltage-dependent calcium channels, allowing more calcium ions to enter. The downstream consequence of this change in voltage is the release of other neurotransmitters from the neuron. Nicotine then has different effects depending on the receptor type: the $\alpha3\beta4$ subtype mediates cardiovascular effects, $\alpha7$ underpins cognitive effects (particularly memory, via long-term potentiation), and $\alpha4\beta2$ is largely responsible for nicotine's addictive properties. These are due to nicotine mediating the release of dopamine in the corpus striatum, frontal cortex, and mesolimbic area. The latter includes the brain's "reward center." In particular the release of dopamine in the ventral tegmental area and nucleus accumbens is the critical event that unintentionally makes nicotine consumption, and indeed consumption of the other rewarding drugs, addictive (see Chapter 5). Nicotine also increases glutamate and GABA release, which have opposite effects on dopamine release; however, desensitization of some, but not all, nicotinic receptors after extended exposure

a. Each ionotropic nicotinic receptor consists of five components arranged symmetrically around a central channel. In the brain each component of the receptor can be either an α subunit, of which there are nine ($\alpha2$ to $\alpha10$), or a β subunit, of which there are three ($\beta2$ to $\beta4$). The most abundant receptors in the central nervous system are $\alpha4\beta2$ (i.e., composed of these two types of subunit), $\alpha2\beta4$, and $\alpha7$ (which is "homomeric," composed of only this one type of subunit).

leads to a net increase in glutamate-mediated excitation.[390,673,674] The increased release of several other neurotransmitters contributes to the complex effects of nicotine. For instance, GABA and β-endorphins mediate anxiolytic properties, noradrenaline contributes to the arousing effects, serotonin affects mood, and serotonin, dopamine, and noradrenaline all serve to reduce appetite.[673]

In terms of improved brain function, much of the early research was confounded by conducting research in smokers who had abstained from smoking, making the interpretation of any net effects of nicotine difficult to disentangle from alleviation of decrements due to being in a state of a withdrawal. However, a recent meta-analysis reviewed the data from 50 methodologically adequate, double-blind, placebo-controlled studies that had assessed the effects of nicotine administered via various methods in nondeprived smokers, minimally deprived smokers, or nonsmokers. The authors concluded that nicotine consistently improved cognitive performance in a number of domains, including attention and episodic/working memory.[675] These effects are mediated by cholinergic projections to the prefrontal cortex and direct binding to receptors in the amygdala and hippocampus, respectively.[676,677] An attendant increase in the release of glutamate is also associated with the memory effects.[673] Because of its cognition-enhancing properties, transdermal nicotine treatment has also been proposed as a method of attenuating the behavioral deficits and cognitive decrements associated with old age and a number of conditions, including attention-deficit/hyperactivity disorder, Alzheimer's disease, and schizophrenia.[678] Nicotine's potent addictive properties would argue against this particular usage. Interestingly, schizophrenics (and other groups suffering psychopathologies) typically smoke more cigarettes, and do so more intensively, than the general population, suggesting they are self-medicating with nicotine in order to attenuate some of the subjective symptoms of the disease.[679] Naturally, it is difficult to disentangle the cause and effect of antipsychotic drug treatment, smoking, and psychotic symptoms in schizophrenia sufferers. However, recent research does show that smoking in untreated, first-episode sufferers from schizophrenia is related to increased positive psychotic symptoms, such as auditory hallucinations.[680] Smoking is also positively related to levels of psychosis in bipolar disorder,[681] and in a longitudinal study of a cross-section of the UK population, smokers were significantly more likely to report psychotic symptoms, including hallucinations and delusions.[682] This research hints at the mind-altering properties of nicotine that made smoking such an essential weapon in the entheogenic arsenal of the shaman.

Ecological Roles of Nicotine and Brain Function

Nicotine is the archetypal insecticidal secondary metabolite, and tobacco extracts were the first insecticides to be used in agriculture. They currently remain a permitted pesticide for organic farming in the United States, and nicotine-containing products such as Black Leaf 40 also continue to be used as household insecticides.

In the plant, nicotine is synthesized in the roots and transported to the aerial parts of the plant, where it can be stored in glandular trichomes (see Chapter 11).[683] Synthesis is both constitutive, providing background levels of nicotine, and induced, via the jasmonic acid pathway, in response to leaf wounding, microbial attack, and insect damage.[684] As with many alkaloids, the *Nicotiana* genus of plants have their own specialist herbivores, such as the larvae of *Manduca sexta* (tobacco hornworm moth), which have evolved tolerance to nicotine and which sequester the alkaloid within their own body as a protection against predators.[364] Fatty acids and amino acids in the oral secretions and regurgitants of the hornworm trigger a jasmonic acid-mediated increase both in the synthesis of nicotine and in the density and size of trichomes. This response is over and above that seen following simple mechanical damage and is accompanied by reduced root growth as resources are diverted from growth to defense.[685] In many cases herbivory by specialists is countered by reductions in defense chemical synthesis and increased growth, but in this case the nicotine is still having a detrimental, but nondeadly, effect on the hornworm. Indeed, when reared on tobacco with reduced nicotine synthesis (due to the transgenic silencing of the gene for *putrescine N-methyl transferase*), hornworm larvae grow significantly faster. The nicotine-free plants are also subject to significantly more herbivory, particularly from generalist herbivores, and they ultimately lose more than three times the leaf area as does unmodified tobacco.[363] Intriguingly, the pattern of herbivore-elicited gene transcription in the tobacco plant was shown to vary depending on the specific herbivore, with differences seen following an attack by the sap-feeding mirid bugs, *Tupiocoris notatus*, or the tobacco hornworm caterpillar, and following a combined attack by the two species.[686] Another common specialist, the earworm caterpillar, has evolved a mechanism for feeding on tobacco that does not involve secretion and sequestration. In this case it excretes a salivary regurgitant into the leaf that contains glucose oxidase, which decreases the level of nicotine in the leaf.[687]

As noted previously, acetylcholine is the principal excitatory neurotransmitter in the insect central nervous system and, naturally, nicotine owes its insecticidal and anti-herbivory properties to its ability to bind directly to nicotinic acetylcholine receptors.[363] In this regard, five nicotinic receptor subunits have been identified in insects.[688] These receptors are abundantly expressed throughout the nervous system, including a wide distribution in the synaptic neuropil region of many brain areas and a particularly high concentration in the sensory and integrative structures of the brain. These nicotinic receptors are therefore involved in sensory and motor functions and in aspects of insect memory, with nicotinic antagonists such as mecamylamine and α-bungarotoxin disturbing memory processes.[309] In the intact fly brain in vivo bioluminescence imaging of CA^{2+} signaling demonstrated increased activity in the mushroom bodies following nicotine administration, which was followed by a delayed secondary response suggestive of the synaptic plasticity associated with long-term potentiation and memory processes.[469] In terms of behavior, administration of nicotine improves insect olfactory memory.[468] Several studies have also shown that administering low doses of volatilized nicotine to flies leads to hyperactivity, aberrant movements, and interference with geotaxis (the natural tendency of flies to climb when startled). As the dose ascends, activity is reduced and eventually eliminated entirely.[298,439] These effects can be reduced by dopaminergic blockade, suggesting that they may be mediated by the downstream dopaminergic effects of increased nicotinic receptor activity.[439] They can also be increased by repeated exposure to nicotine. This sensitization to the effects of nicotine was as a consequence of the same cellular mechanisms (cAMP/protein kinase A cascade and CREB-dependent gene transcription) as those underlying the same phenomenon in mammals.[689]

Interestingly, when given the choice, bees were shown to actively prefer nectar containing nicotine at the very low concentrations naturally present in nectar but avoided higher doses that were closer to those found in plant material.[575,690] The preferred doses had no detrimental effects on bee survival and fecundity at naturally occurring levels but reduced larval hatching and survival at higher levels.[691] However, groups of weak bees that were assumed to have microbial infections survived longer following nicotine consumption, suggesting an antimicrobial benefit for the bee from nicotine consumption.[690]

■ HYOSCYAMINE, SCOPOLAMINE, AND ATROPINE

Members of the American *Datura* and *Brugmansia* genera, both of which are sometimes called "angel trumpets" due to the shape of their flowers, have a long history of use in ritual, medicine, and magic. They feature among the nine core psychotropic "shamanic" plants used across South America as sacred entheogenic plants.[660] *Datura* is a genus incorporating a number of plants with white/blue flowers that grow most happily in coastal valleys, as well as jungles and mountainous regions. Amongst these, *Datura stramonium* is the most widely used species, and its large, unpleasant-smelling leaves are smoked or made into infusions and decoctions along with the seeds. The *Brugmansia* genus incorporates a number of trees and shrubs that grow above 1,800 meters of elevation in warm and humid mountainous regions. The flowers, leaves, and seeds are used in shamanic rituals, typically as an infusion or drink, and often in combination with other psychoactive plants. The primary use of both genera was, and remains, to precipitate a delirious state in a ritual setting, in particular to allow divination and communication with ancestors. Both were also used medicinally, and *Brugmansia*, in particular, had a number of other more practical applications. These included the pacification of children, and, for the Chibcha tribe of Columbia, the drugging of wives and slaves prior to burying them alive with a recently deceased tribal chief.[660] The choice of genera employed by indigenous peoples would typically have depended on geography, with *Brugmansia* in use within mountainous regions such as the Andes[692] and *Datura*, having made the journey across the continental isthmus, being used in North America, where it was one of the three staple psychoactive plants (along with tobacco and peyote cactus).[667]

Oral histories suggest that the use of these species predated the arrival of Europeans by many hundreds if not thousands of years. However, as they don't require any specialized equipment, their impact on the archaeology of the region is limited. Nevertheless, *Brugmansia* is represented in the pottery produced by the Chavín culture (600–300 BC) and of the Nazca culture of Peru (300 BC to 700 AD).[660] It would seem a fair assumption that their use was as enduring as tobacco and ololiuqui, both of which left a durable archaeological trace extending 3,000 to 7,000 years into the past, in the form of snuff paraphernalia and, in North America, smoking pipes.

The recent history of the *Solanaceae* genera that emigrated to the Mediterranean is somewhat clearer. The members of three taxa in particular have played a notable role in ritual, medicine, and murder spanning four millennia. The *Hyoscyamus* genus of annual or biennial flowering plants comprises 11 species that attract pollinators by exuding a smell reminiscent of rotting flesh, which is attributable to the compound tetrahydroputrescine. These species include *Hyoscyamus niger*, also known descriptively as henbane.[34] The *Atropa* genus contains five perennial flowering plants, the most celebrated of which is *Atropa belladonna*, or deadly nightshade. The descriptive belladonna, or "beautiful lady," of the botanical name refers to the plant's cosmetic use as a pupil dilatant. The *Mandragora* genus comprises five species of perennial plants that have a large taproot, bell-shaped flowers, and yellow fruits. The genus includes

Mandragora officinarum, better known as mandrake. Of these three nightshade species mandrake has perhaps the most ancient recorded history. Theophrastus, in his Greek treatise on plants, noted that the plant's large taproot resembles a small figure of a man, and this correspondence was to lead to its name ("Man" "dragon," alluding to its shape and magical powers), as well as many of the strange beliefs and ceremonies that have revolved around the plant throughout its history. These include the reputed shrieks that emanate from the uprooted mandrake. Shakespeare describes the effects of the screams in Romeo and Juliet: "Living mortals hearing them run mad".[693] Mandrake is not only described in the book of Genesis, where its effects on the fecundity of Rachel led to the birth of Joseph,[693] but it also featured in ancient Egyptian religious practices, as evidenced by artistic, written, and paleo-ethnobotanical sources.[4] It was often used in conjunction with members of the *Nymphaea* genus of lotus flowers, which contain the aphrodisiac alkaloid apomorphine.[31] *Nymphaea* and mandrake feature in a number of frescos and wall scenes, and on ungent jars, from tombs spanning from 1550 to 1350 BC. The most famous example may be the several depictions of mandrake in the tomb of Tutankhamen, including a depiction of the ill pharaoh being administered mandrake, opium, and *Nymphaea* by his wife. The tomb also contained a gold-plated shrine decorated with a bas-relief showing the pharaoh holding two *Nymphaea* and a mandrake in his left hand and wearing a collar bearing a mandrake fruit.[4] Extracts from these three plants also feature as vehicles to ecstasy in the papyri that were intended only for the higher castes of Egyptian society, including the "book of the dead".[4]

All three nightshades were also entwined in Greek mythology. Mandrake, the fruits of which were known as "apples of love," provided Aphrodite, the goddess of love, with one of her names, Mandragoritis, translated as "She of the Mandrake".[693] In the Greek myths, the souls of the dead that roamed the banks of the river Styx were garlanded with henbane,[34] and the genus name *Atropa* was taken from the name of the Greek Fate Atropos (literally meaning inexorable or inevitable), who chose the manner of an individual's death and, at the moment of death, cut the thread of his or her life with her shears. Deadly nightshade was probably also the poison wielded by Circe against the crew of Odysseus' ship, for which he took the antidote "Moly," which may refer to the snowdrop, *Galanthus nivalis*, or snowflake (genus *Leucojum*), both of which might provide an antidote to the anticholinergic properties of this genus (see below).[32] Certainly the Greek physician Dioscorides in his pharmacopaeia "De Materia Medica", published ~50 AD, describes Moly as being from the latter genus and "good against poisonings and bewitchings" (1655 translation).[37] The later association of deadly nightshade with medieval witchcraft was presaged by the Greek association of *Atropa belladonna* with the cult of Dionysius (later the Roman god Bacchus). Devotees of the cult used a decoction of wine, mixed with deadly nightshade and other herbs, to enter a trance-like state in which they danced with libidinous and orgiastic abandon.[33]

The three *Solanaceae* also went on to appear throughout the herbal medicinal texts of the ancient Greeks and were later absorbed by the Romans. Mandrake root was prescribed at low doses by Hippocrates (400 BC) for depression and anxiety and by Theophrastus (230 BC) for gout and sleeplessness and as a love potion. All three of the nightshades were also well known to Dioscorides who attributed many properties to mandrake root, including as a sleep-inducing suppository and orally as a hypnotic, "love medicine," and surgical anesthetic. However, this came with the warning "being too much drank it drives out life".[37] Rather than risk the extreme oral toxicity of henbane, the plant was recommended for pain and insomnia in the form of an oily salve with local anesthetic properties.[34] Dioscorides also clarified the distinction between mandrake and deadly nightshade, which were often taken to be the same plant, with potentially disastrous consequences, as the latter was considerably more toxic than mandrake.[33] The Romans also started a tradition of using a sponge soaked in opium, henbane, and mandrake, the "spongia somnifera," to induce anesthesia, although the dangers of the practice were also luridly described.[34]

The use of the three *Solanaceae* survived into the Dark Ages in the transcribed works of the Greek and Roman physicians, but they emerged into the later Middle Ages in Europe largely in the hands of herbalists and traditional healers. All three plants could also be found as key ingredients in the pharmacopeia of European medieval witchcraft and were closely associated with Hecate, the pre-Olympian goddess of magic, witchcraft, and necromancy. Hecate herself had an illustrious history, having originated before recorded history in Anatolia before being absorbed first into Greek mythology, then being transferred wholesale to the Romans, and finally appearing as the primary goddess of medieval witchcraft. The mandrake itself might have been rare, particularly in Northern Europe, but would have been imported from its indigenous area. Often the root would not be consumed but retained for its talismanic homunculus-like properties and passed through the generations as an heirloom.[693] It could also be replaced with a local alternative. For instance, in Britain, black and white brionies, resembling mandrake and replete with their own narcotic poisons, were passed off as the real thing.[693] As well as playing a role in medicinal herbalism, the three genera of nightshades featured as key ingredients, in various combinations, in the induction of trances and as love potions and "flying ointments." In the case of the latter the ointment, which usually featured one or two nightshades in combination with other ingredients, was smeared over the body, including the anus and genitals. It might have been smeared on the handle of the witch's broom, which was then used as a phallus. This ointment, absorbed through the skin and mucous membranes, would then induce the experience of flying and delusions that the witch was engaging in coitus with the devil, and, as with the ancient Greeks and Romans, it would induce an orgiastic state of abandon.[33,34,693]

Knowledge of the anesthetic properties of the nightshades was also preserved within monastic traditions of apothecary. In particular, henbane was used alone as a potent sedative. As Gerard's herbal from the late 16th century described, henbane's "leaves seed and juice taken inwardly causeth an unquiet sleepe, like unto the sleepe of drunkennesse, which continueth long and is deadly to the party." Various combinations of nightshades, with opium and other extracts, continued to be used as a "spongia somnifera" well into the 19th century.[694] As a mark of their ubiquity, various *Solanaceae* "drowsy syrups of the world" appear in many of the plays of Shakespeare. References include the likely somnolent that triggers the tragic end in *Romeo and Juliet*, wherein it was described with the words "Within the infant rind of this small flower, Poison hath residence, and medicine power."

One factor that must have helped the migrating nightshades to survive when they washed up in the Mediterranean was their family's cocktail of secondary metabolites. The same chemicals underlie the psychoactive properties that have made them so attractive in ritual, religion, witchcraft, and medicine across continents. In the case of the *Datura/Brugmansia*, *Atropa*, *Mandragora*, and *Hysocamus* genera, the most prominent secondary metabolite tropane alkaloids that underpin their various effects are hyoscyamine, scopolamine (also called hyoscine), and atropine, the racemic form of hyoscyamine. These compounds combine the *Solanaceae* alkaloids' pyrrolidine ring system, in the form of tropine, with tropic acid derived from L-phenylalanine. Due to a structural similarity to the neurotransmitter acetylcholine (i.e., the spatial relationship between the nitrogen atom and ester linkage), these alkaloids can bind to all of the five subtypes of muscarinic postsynaptic receptor for acetylcholine, disrupting its function.[195] High doses of these tropane alkaloids can also block nicotinic receptors. The structures of the three alkaloids, illustrating the common structural elements with acetylcholine, are shown in Figure 7.1.

The differences in the binding properties, and therefore pattern of cholinergic effects, of the three chemicals are then related to slight structural differences.[195] The sensitivity of these differences can be seen in the inability of anisodamine, a tropane alkaloid from *Scopolia tangutica* that differs structurally from scopolamine in only one minor aspect (substitution of an OH group instead of an oxygen bridge), to cross the mammalian blood–brain barrier due to its reduced lipophilic solubility.[695]

The pattern of effects of scopolamine and atropine in humans are of course directly related to muscarinic receptor blockade. This has two notable consequences related to the areas that the receptors populate. In the brain muscarinic receptors are concentrated in the basal forebrain/septum, cortex, hippocampus, amygdala, and striatum, and in the periphery they are found in the heart, lung, ileum, and exocrine glands. Subtypes M_2, M_3, and M_4 are fairly uniformly distributed across these central and peripheral areas, and the M_1 and M_5 subtypes occur in the brain, with the M_5 subtype most prevalent in the hippocampus.[391] The effects of these alkaloids then predominantly feature disturbances of brain function, for instance in cognitive function, motor control, and respiratory and cardiovascular activity, while peripheral disturbances include heart rate control, smooth muscle contraction, vasodilation, and a host of other effects related to parasympathetic deactivation.[391,696] The ability of scopolamine to engender a state of altered consciousness, including delirium and the experience of psychotic hallucinations, may well be related to muscarinic M_1 receptor-mediated disruption of cholinergic and GABAergic control of the functioning of the thalamus—in effect disturbing the filtering function of this brain structure and allowing inappropriate intrinsic and sensory information into conscious awareness.[697,698] Atropine is considerably less potent than scopolamine, by a factor of about 10, but otherwise its effects are largely the same.[391]

The deleterious effects of these nightshades on central nervous system function are most often seen in cases of deliberate intoxication, typically by adolescents. Consumers of high doses of any of the deliriant nightshades will typically conform to the mnemonic used to identify the symptoms of anticholinergic poisoning: "hot as a hare, blind as a bat, dry as a bone, red as a beet, and mad as a hatter." For instance, the symptoms of those admitted to the hospital after consuming one of these plants are typically confusion and disorientation, delirium, visual and auditory hallucinations, bizarre behavior and combativeness, restlessness and agitation, dilated pupils, dry mouth, and blurred vision. Severe cases can result in seizures, coma, and death.[699]

In terms of the direct effects of the individual alkaloids on brain function, scopolamine has received the lion's share of attention. The potential for scopolamine to interfere with memory processes was identified in the opening years of the 20th century.[700] It went on to be employed during childbirth to induce a state of "twilight sleep"; in anesthesia as a secondary treatment; and even to facilitate interrogation as a supposed "truth drug".[701] Scopolamine also found a role in countering the symptoms of motion sickness, which continues today, despite a preponderance of centrally and peripherally mediated symptoms of parasympathetic nervous system depression.[702] Scopolamine administered as a medicinal treatment, including in the form of eye drops, transdermal patches for travel sickness, and as an anticholinergic treatment in the elderly, can on occasion engender elements of delirium, including hallucinations and confusion.[703,704]

In terms of brain function, psychopharmacological studies using low doses (in comparison to "deliriant" doses) confirmed that scopolamine leads to rapid, wide-ranging dose-related decrements in memory, learning, psychomotor function, attention, and executive function.[481,705,706] It also reduces electrical activity in the brain, primarily in the alpha waveband,[481] and decreases alertness while engendering psychotomimetic effects such as changes in visual perception, psychedelic effects measured in terms of subjective internal and external perceptions, and feelings of being "high".[481] Concomitant peripheral parasympathetic symptoms include pupil dilation and decrements in eye focus, decreased blood pressure and heart rate, constipation, and difficulty urinating.[481]

Scopolamine has also been employed as an indirect tool in two expansive areas of research. In the first of these, it has been used as a pharmacological probe to disentangle muscarinic and nicotinic receptor function as a consequence of its ability to unselectively bind to, and block, the activity of acetylcholine at all of the subtypes of muscarinic receptors. The results of rodent studies into the effects of central administration of scopolamine, in agreement with human studies, have generally suggested a disruption of short-term memory function, although these effects may simply be predicated on more general impairments in attention. More focused studies involving the injection of scopolamine into discrete brain areas have confirmed muscarinic involvement in hippocampal memory processes and in the prefrontal cortex mediation of attention.[391] In human brain imaging studies, muscarinic blockade with scopolamine causes a reduction in activity in regions rich in muscarinic receptors, including the frontal cortex, with concomitant decrements in attention and executive function.[391,707] Scopolamine was also shown to decrease activity in areas of the frontal lobe during an "n-back" working memory task[708] and in task-specific brain regions including the hippocampus while concomitantly reducing performance on a "delayed matching to sample" working memory task.[709] With regard to longer-term memory, scopolamine decreased activation in the fusiform cortex, hippocampus, and inferior prefrontal cortex[710]and reduced retrieval-related activity in the fusiform cortex,[711] with decrements in memory performance also evident.

In the second major area of research scopolamine has been used to model the cognitive decrements of aging and dementia. According to the "cholinergic hypothesis," the memory deficits seen with increasing age and dementia are due to downregulation of the acetylcholine neurotransmission system. A "scopolamine challenge"[712] has therefore been used to mimic these effects, effectively providing a model for assessing the potential of an intervention to attenuate the cognitive decrements associated with dementia and age. This approach has been adopted in a plethora of animal and human studies. However, where treatments that counteract the effects of scopolamine have been identified, the results have rarely translated into benefits in humans. This may be due to a wide range of confounding factors, including the possibility that scopolamine primarily targets attention rather than memory, with degraded attention then affecting other cognitive domains. The nonselective blockade may also lead to nonspecific effects on other behaviors, such as locomotor activity or anxiety. Similarly, it is necessarily difficult to interpret the effects of a drug with such wide-ranging effects across the brain and periphery.[391]

Ecological Roles of the Tropane Alkaloids and Brain Function

Hyoscyamine, scopolamine, and atropine are proposed to play a specific role in the plant's defense against insect herbivores,[713–715] and their synthesis is directly induced in response to insect herbivory.[716] Constitutively synthesized tropane alkaloids are also stored at higher concentrations in more valuable plant parts, again suggesting a specific defensive function. As an example, in *Brugmansia suaveolens,* scopolamine is more concentrated in the unripe fruit, seeds, shoots and young leaves, and flowers. However, physical damage induces higher concentrations of scopolamine in the leaves, with concentrations peaking after 24 hours. This comes at a cost; for instance, dramatically increasing the synthesis of scopolamine by applying methyl jasmonate to the roots was balanced by the rate of leaf growth being more than halved.[714] As to the fate of the insects, both undamaged leaves of *B. suaveolens* and scopolamine added to the diet of the larvae of the generalist herbivore moth larvae *Spodoptera frugiperda* negatively affected its life course and growth, and this effect was magnified when the larvae were fed damaged leaves that were expressing higher concentrations of scopolamine.[714] Similarly, herbivory by generalist but not specialist herbivores was negatively correlated with hyoscyamine content in *Datura stramonium*, and transgenically manipulated levels of scopolamine in *Atropa belladonna* were negatively associated with beetle damage.[717]

The sophistication of the *Solanaceae* alkaloid defense system can be gauged from the response to the larvae of the specialist herbivore *Placidina euryanassa.* This species of butterfly tolerates and actively sequesters scopolamine to defend itself against predation by birds. It therefore exhibits a preference for scopolamine-rich tissue. In response to species-specific feeding cues, such as saliva from the larva of this particular herbivore, *B. suaveolens* reduces its synthesis of scopolamine in favor of alternative defenses along with increased tissue growth, thereby compensating for leaf area lost to the larvae.[362] In a further interesting example of co-evolution, the winter cherry bug (*Acanthocoris sordidus*) also preferentially feeds on tropane alkaloid-bearing *Solanaceae*, but in this case its regurgitant enters the leaf and reverses the synthetic pathway that culminates in scopolamine, thereby reconverting it to the less toxic atropine. The atropine is then sequestered by the winter cherry bug as a defense against its own predators.[718]

In terms of the mode of action in insects, it's clear that acetylcholine is the most abundant neurochemical and the primary excitatory neurotransmitter in the insect central nervous system and that insects possess orthologues of the mammalian muscarinic receptors.[294] Evidence clearly shows that scopolamine and atropine modify both mammalian and insect behavior, most likely via the same muscarinic mechanism. Grant and Grant[719] famously observed that hawkmoths (*Manduca* genus) apparently became intoxicated, becoming immobile or flying erratically, after feeding on nectar from *Datura meteloides* and proposed that the tropane alkaloids represented a psychoactive reward for the insect. Whether this is the case or not (bearing in mind that these alkaloids do not have rewarding properties in mammals), both scopolamine and atropine certainly kill nonherbivorous insects. Both chemicals showed high affinity for *Drosophila* muscarinic receptors and led to suppressed movement, followed by knock-down or death in up to half of houseflies within an hour.[415] We can also gauge some of the effects of the tropane alkaloids from studies using scopolamine and atropine as muscarinic antagonists in studies employing insects as simple models of the animal central nervous system. Examples include the inhibition of flight motor-pattern generators in locusts by scopolamine, with abolition of this effect following administration of the muscarinic agonist pilocarpine,[720] and inhibition of grasshopper song generation.[721] A number of studies have also probed memory processes. For instance, atropine and scopolamine have been used as muscarinic receptor antagonists in electrophysiological studies of the insect mushroom body, the brain structure implicated in olfactory learning and memory.[722] Direct injections of scopolamine into some but not all regions of the mushroom body have been shown to disturb memory processes.[414] Administration of the muscarinic agonist pilocarpine to worker bees has also been shown to mimic the effect of foraging in terms of increasing the size of the bee's mushroom body neuropil and increasing the complexity of its neuronal dendritic connections. These effects were abolished by administration of scopolamine.[723,724] Atropine also engenders a concentration-dependent inhibition of mushroom body kenyon cell nicotinic receptor currents.[725] In terms of direct modulation of behavior, scopolamine has been shown to increase attacks on nest-mates in honey bees, suggesting decrements in recognition memory,[343] and both atropine[726] and scopolamine[311] degrade memory retrieval (using a proboscis extension reflex paradigm) in honey bees. These studies hint at the effects that the tropane alkaloids must have on the brain function of herbivorous insects and bear several striking similarities to the effects seen in humans.

THE ANTIDOTE TO HYOSCYAMINE, SCOPOLAMINE, AND ATROPINE: THE CHOLINESTERASE INHIBITORS

Typically the antidote for poisoning with one of the anticholinergic *Solanaceae* plants is the administration of physostigmine, a reversible acetylcholinesterase (AChE) inhibitor, which reduces the enzymatic breakdown in acetylcholine leading to a net increase in synaptic levels of the neurotransmitter. This pyrroloindole alkaloid was originally derived from the Calabar bean (*Physostigma venenosum*), which had a long history of use by the indigenous peoples of West Africa as an "ordeal poison." This was administered to the accused in witchcraft trials, and if the victim survived the resultant convulsions without suffocating he or she was proclaimed innocent. The extract of Calabar bean was subsequently first used in 1864 as an antidote for accidental atropine poisoning when five prisoners interred in Prague prison managed to get hold of a bottle of medicinal atropine. Since that time it has continued to be used in cases of atropine and scopolamine poisoning, as well as serving as a treatment for glaucoma and as a neuropharmacological probe in studies investigating the cholinergic neurotransmission system.[727] Just as the observation that the cholinergic deficits associated with scopolamine mimicked the cognitive impairments seen in Alzheimer's disease, contributing to the "cholinergic hypothesis" of dementia, it was also suggested that drugs that had an opposite effect to that of scopolamine might make suitable treatments for the symptoms of dementia. The thinking underlying this approach is that the devastating loss of cholinergic neurons in the brain associated with dementia may be offset, in part, by any treatment that increases cholinergic neurotransmission in what remains of the cholinergic system. Physostigmine certainly fits the bill, and small doses not only reverse the cognitive deficits associated with scopolamine in animal studies but also improve cognitive function, in particular memory, in aged primates and humans.[419] Unfortunately, the use of physostigmine in dementia treatment was discontinued across much of the world due to its short half-life and high incidence of side effects. Indeed, a Cochrane review in 2001[728] found that it was associated with few benefits and high levels of side effects. However, it has been replaced by rivastigmine, a semisynthetic derivative with a better pharmacokinetic and side-effect profile. Two further plant-derived cholinesterase inhibitors are now also widely used for treating the cognitive and behavioral deficits of Alzheimer's disease: galantamine and huperzine. The first of these is an alkaloid derived from members of the *Amaryllidaceae* family, including snowdrops (*Galanthus* genus), daffodils (*Narcissus pseudonarcissus*), and snowflakes (*Leucojum* genus). Although its history is uncertain, the "Moly" that saved Odysseus from Circe's poisoning with deadly nightshade is believed to be a reference to one of the galantamine-containing plants. It also has a putative history of traditional use in the Caucasus, although this is largely anecdotal.[729] Galantamine represents a potential improvement over the other cholinesterase-inhibiting treatments for Alzheimer's disease in that it is both a competitive inhibitor of AChE and an allosteric modulator at nicotinic cholinergic receptor sites.[730] The remaining plant-derived treatment is huperzine (huperzine A), from the firmoss *Huperzia serrata*, which is an alkaloid with a sesquiterpene portion that primarily exerts competitive, reversible inhibition of AChE, with additional neuroprotective properties related to the attenuation of hydrogen peroxide, glutamate, and β-amyloid-related neurotoxicity.[731]

In terms of efficacy, the most widely used treatment for the symptoms of dementia, rivastigmine, has been investigated in a number of controlled trials. For instance a recent Cochrane review included nine trials involving 4,775 Alzheimer's disease patients and concluded there was evidence of slight, but significant, improvements in the rate of decline of cognitive function, activities of daily living, and dementia severity following higher doses. However, its effectiveness has to be balanced with its linear dose-related cholinergic side effects, which can include nausea, vomiting, and diarrhea, headache, transient loss of consciousness, loss of appetite, and abdominal pain.[732] Similarly, data from 10 trials incorporating a total of 6,805 patients with Alzheimer's disease or mild cognitive impairment (MCI) also showed that galantamine, across the dose range administered, attenuated declines in cognitive function, with some evidence of improved activities of daily living and behavior. However, it was also associated with the same pattern of dose-related cholinergic side effects as seen following rivastigmine. Interestingly, the two studies included in this review that incorporated MCI patients showed that there was evidence of marginally improved cognitive function following galantamine in this group but an unexplained increase in the death rate.[730] These findings with regard to efficacy were echoed in a very recent review assessing the cholinesterase inhibitors rivastigmine, galantamine, and the synthetic donepezil; the question of the cost-effectiveness of these treatments was unresolved.[733] Whereas the efficacy of these treatments has been subject to substantial research, that of the huperzine A is less certain. In another Cochrane review, Li et al.[731] managed to find only six studies, which included a total of 465 participants. They concluded there was provisional evidence of beneficial effects in terms of an attenuation of the gross cognitive impairment, behavioral disturbance, and degraded functional performance associated with dementia, but that the evidence base was too small to reach any firm conclusions. A recent study involving 210 patients with mild to moderate Alzheimer's disease failed to answer this question, finding no benefits on any measure, including in terms of cognitive function, following huperzine A.[734]

In terms of the ecological roles of the cholinesterase inhibitors, there is simply very little relevant research published. The exception are indirect studies showing that physostigmine is a cholinesterase inhibitor within the insect nervous system and that it binds agonistically to nicotinic receptors,[473] along with demonstrations of an increase in the number of trials required for habituation of the proboscis extension reflex.[408]

PART THREE
The Phenolics

8 Phenolics and the Lives of Plants and Animals

The phenolics represent a large group of phytochemicals that incorporate within their structure at least one phenyl aromatic hydrocarbon ring with one or more hydroxyl groups attached. This motif lends the phenolics the property of being able to reduce reactive oxygen species and other organic/nonorganic substrates. As a group, phenolics are impossible to avoid as they are ubiquitous in plant tissue and therefore form an intrinsic part of the human diet. This is just as well, as animals, unlike plants and microorganisms, do not possess the "shikimate" pathway that contributes to the synthesis of this group of chemicals. The same pathway also synthesizes a number of essential amino acids, and indeed L-phenylalanine (and occasionally L-tyrosine) goes on to provide the starting substrate for most of the phenolics. These amino acids are termed "essential" because, while they are essential to physiological functioning, mammals do not synthesize them and therefore have to derive them from their diet. A reasonable argument could be made that plant-derived phenolics as a group are also, to a certain extent, essential to human health. This topic will be returned to later in the chapter.

In terms of synthesis, the common starting point for all phenolics is shikimic acid derived via the shikimate pathway, although other pathways such as the acetate and malonate pathways contribute components of the more complex phenolic structures. Several simple, low-molecular-weight phenolic acids can be synthesized directly from shikimic acid, including gallic acid, which constitutes the base unit of hydrolysable tannins such as the polymeric gallotannins and ellagitannins. Otherwise phenolic synthesis typically proceeds in plants via chorismic acid and its derivative amino acid, L-phenylalanine, which is first deaminated by phenylalanine ammonia-lyase to form cinnamic acid.[195] A series of hydroxylation and methylation reactions then create the other "cinnamic acids." These include p-coumaric, caffeic, ferulic, and sinapic acids, and this group of acids then form the substrates of a wide range of structures comprising molecules of greatly varying complexity. At the simplest end of the spectrum, these include low-molecular-weight volatile aromatic phenolic acids that are formed by a number of operations that remove two carbon atoms from the cinnamic acids. These include benzoic acid and its derivatives, such as the plant hormone salicylic acid. Similarly the coumarins and phenylpropenes are simple derivatives of cinnamic acids. The phenylpropenes include a number of compounds associated with odors and flavors, such as cinnamaldahyde from cinnamon, myristicin from nutmeg, and eugenol from cloves. Increasing in complexity the cinnamic acids can also be esterified, and the most prevalent dietary products here are the chlorogenic acids, found in abundance in coffee, which are esters of either p-coumaric, caffeic, or ferulic acids with quinic acid. Similarly, p-coumaric acid is esterified with 3,4-dihydroxyphenyl–lactic acid to give rosmarinic acid, a trademark chemical constituent of the *Lamiaceae* family of plants. Alcohols of p-coumaric, ferulic, and sinapic acids also provide the building blocks needed to construct lignans (composed of two units of the alcohol) and lignins

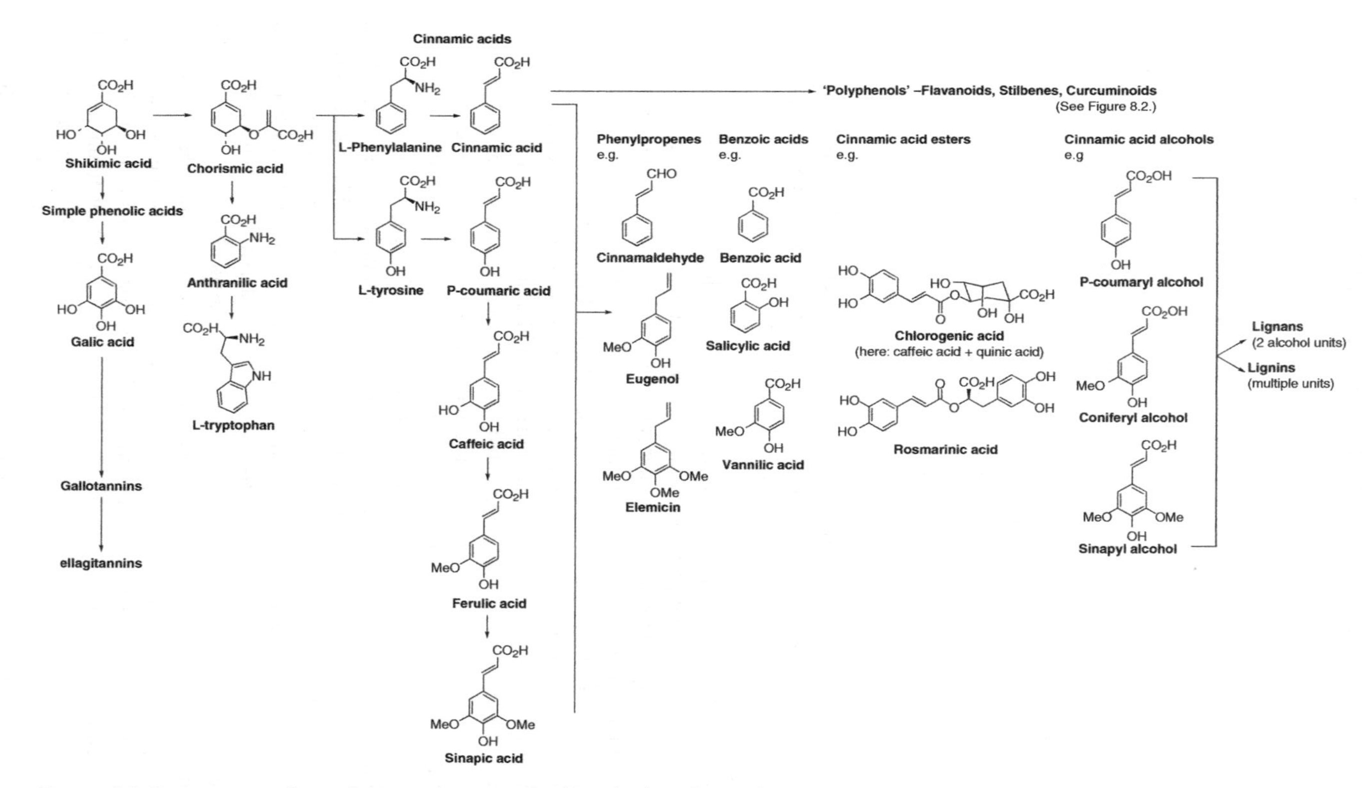

Figure 8.1 Structures and synthetic pathways of selected phenolics. The arrows indicate synthetic pathways. The amino acids L-phenylalanine, L-tyrosine, and L-tryptophan are shown where they occur in the synthetic pathways.

(polymers of multiple units of the alcohols). This last group includes structures of varying complexity that, along with cellulose, provide plants with structural strength and represent the bulk of wood. The structures of a number of the above phenolics and their simplified synthetic pathways are shown in Figure 8.1.

Cinnamic acid also goes on to provide the substrate for a wide range of more complex phenolic structures, which are ubiquitous in the human diet as a group and are of particular relevance here. These "polyphenols"[a] combine a shikimate pathway-derived

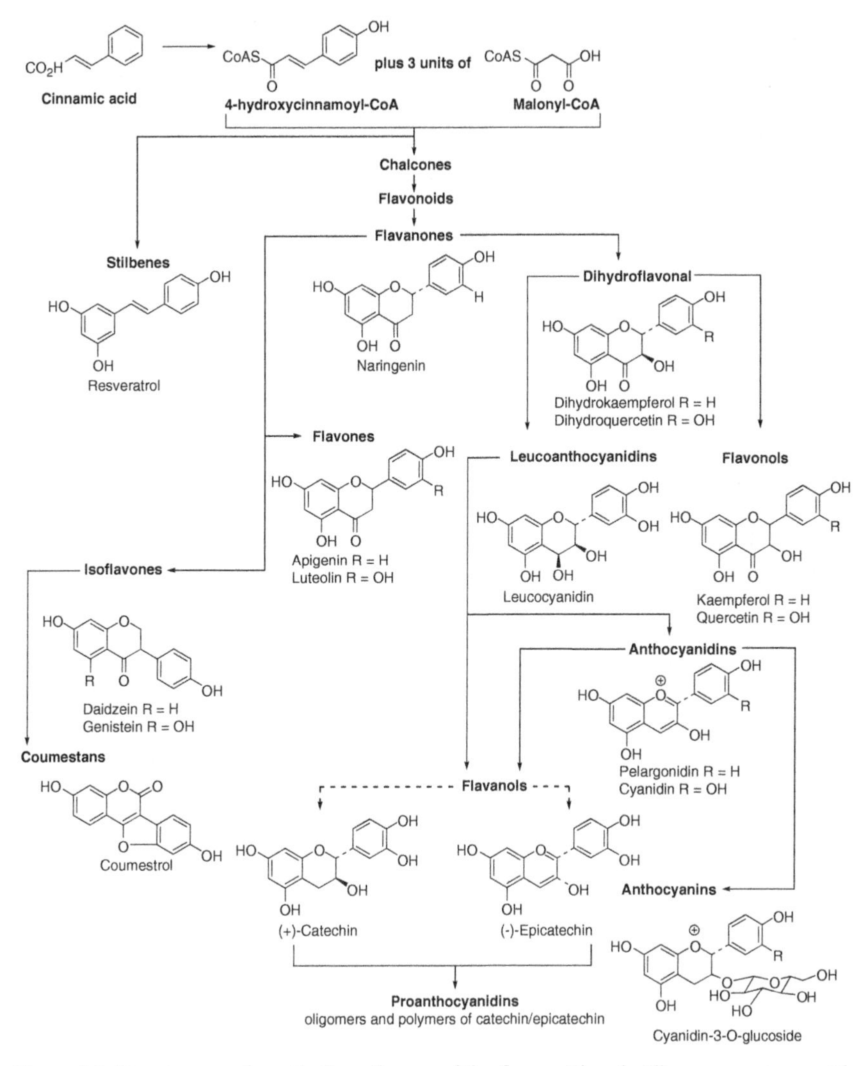

Figure 8.2 Structures and synthetic pathways of the flavonoid and stilbene groupings, with selected example structures. Arrows indicate synthetic pathways.

a. Strictly speaking, "polyphenol" refers to any molecule that features more than one phenolic ring, although in practice it is often, as here, used to encompass the large group of dietary phytochemicals composed of cinnamic acid starter units combined with malonates.

cinnamic acid starter unit (cinnamoyl-CoA) with malonyl-CoA, which is derived via the acetate pathway. This class includes a number of compounds with restricted distributions, such as the curcuminoids and kavalactones, which encompass one and two malonyl-CoA units respectively. However, the majority incorporate three malonyl-CoA units. The resultant group starts with an identical polyketide and then differentiates into stilbenes and chalcones on the basis of the nature of the first unique enzymes (stilbene synthase vs. chalcone synthase) in their respective pathways. The stilbene products of one of these two pathways have a comparatively restricted distribution, but the chalcones go on to act as the precursors for the entire flavonoid group of structures. The flavonoids are ubiquitous, appearing in tissue throughout the plant and throughout the plant kingdom, and they represent the largest, most diverse single group of secondary metabolites, encompassing more than 8,000 compounds. All of these share a common underlying structure of two six-carbon rings, with a three-carbon bridge, which usually forms a third ring.[195] Flavonoids can be subdivided into the chalcones, their derivatives the flavanones, and their derivatives the flavones, flavonols, isoflavones, flavanols, and anthocyanins.[735] Naturally some of these subgroups can themselves be subdivided. So, for instance, the flavanols are the most complex and varied subclass of flavonoid structures, which range from simple monomers, such as catechin and epicatechin, to their oligomers and polymers, or proanthocyanidins (also known as condensed tannins).[736] The relationships among the stilbene and flavonoid subgroups, along with structural examples, are shown in Figure 8.2. Before leaving the topic of synthetic pathways it is also worth mentioning the quinones, another group of complex phenolics, which can be derived by oxidation of phenolic compounds. Examples of this group include the tocopherols and the naphthoquinone juglone.

PHENOLICS IN THE LIFE OF PLANTS—ECOLOGICAL ROLES

The best way to gauge the breadth of ecological roles played by phenolics of varying levels of complexity is to take a flying insect's view of the plant. From a distance the first phenolic signal that our insect would detect would be the splashes of color provided by flowers, fruits, and some leaves against the background of chlorophyll-green photosynthetic foliage. The use of these colors evolved to attract pollinators and fruit-dispersing animals, as well as to deter herbivores, and they are provided primarily by a wide range of flavonoids, most notably the anthocyanins and anthocyanidins (the terpene group of carotenoids also provide some yellow, red, and orange colors). These flavonoids occur throughout plant tissue, but they are more concentrated in the epidermis of leaves or cuticle of flowers and fruits. Here three key chemicals, cyanidin (magenta), delphinidin (mauve), and pelargonidin (orange-red), and a handful of less widespread anthocyanin pigments combine to produce rich colors that range from oranges, through reds to blues and every shade between. Other flavonoids, such as the flavones and flavonols, contribute whites and yellows. The eventual color of the specific flower or plant will have evolved as a consequence of the method of pollination and will have been matched to the portion of the spectrum most readily detected by the chosen pollinator. As our insect draws closer, nectar guides will become discernible on the flower's petals, advertising the direction to the nectar reward. Many of these are drawn in flavonols that reflect ultraviolet light and are therefore invisible to the mammalian eye.[84] Of course, in the world of the phenolics there are few compounds that serve only one function, and all of these pigments also function as sunscreens and antioxidants, absorbing dangerous wavelengths of ultraviolet light as they strike the plant tissue and mopping up reactive oxygen species.[146]

As the insect approaches closer to the stand of plants, it begins to enter a complex world of chemicals. The air will be rich with volatile compounds, including many low-molecular-weight phenolics that are being released by the plants to attract and deter pollinators and herbivores. Closer still, the air will be thick with plant signaling molecules, including phenolic aromatic compounds such as methyl salicylate, the volatile ester of the hormone salicylic acid, and phenylpropenes such as eugenol and estragole.[737] These chemicals are being released into the air around the plant as a rapid signaling system designed to deliver information about stressors and physical integrity rapidly to the plant's distant parts. The airborne signaling chemicals will be absorbed by leaves and other tissues, triggering a range of responses.[162] However, these airborne chemical signals are also being detected by neighboring plants, priming a range of potential responses, including similar stress-related responses in communities of plants, or the reallocation of growth resources in competitor species.[738]

If our insect is a pollinator, it will alight on one of the plant's flowers and collect its reward in the form of pollen and nectar. This sugar-rich liquid usually contains a range of secondary metabolites, including phenolics, and these chemicals help manage the flower's relationships by attracting and managing visits by symbiotic pollinators and by deterring "nectar robbers"—for instance, by endowing the nectar with acetylcholinesterase inhibitory properties.[739] Phenolics also protect the nectar from bacterial and fungal infection and provide antioxidant protection.[151,739,740] If our insect is a herbivore intent on feeding, it will first encounter basal, constitutive levels of flavonoids in the plant tissue that act as general-purpose antifeedants, digestibility reducers, and toxins.[147] A number of flavones and flavonols may also interfere with the insect's endocrine function, for instance having an antagonistic effect on its ecdysteroid hormone system and thereby modulating its growth, metamorphosis, molting, and life course.[741] The direct reaction of the plant to the herbivorous assault will be dictated by whether the insect has mandibles for tearing tissue or stylet mouth parts for less intrusive piercing and sucking. The physical act of tissue wounding might release stored precursors that combine in the damaged tissue and trigger a chemical stress response, or alternatively enzymes within the insect's saliva, such as beta-glucosidases or fatty acid/amino acid conjugates, might trigger a specific "immune-like" reaction.[742] In general, the response to chewing is driven by jasmonic acid signaling, whereas the response to sucking includes salicylic acid resistance pathways as well (see below).[743] These hormonal signals will trigger a cascade of biochemical events that might result in the synthesis and eventual polymerization of phenolic cinnamic acid alcohol monomers that will interlock to form woody lignin as a physical defense.[123] It will also inevitably result in an increase in the synthesis of defensive secondary metabolites, including the many stress-induced phenolic compounds that directly protect plant tissue. The low-molecular-weight phenolics will also contribute to the volatile emissions that prime distant parts of the plant for similar attacks[162] and attract the predatory enemies of the herbivorous insects.[154]

As a secondary result of the herbivorous attack, microbial pathogens may penetrate the plant via the wound, where they will immediately encounter antimicrobial chemical defenses in the form, for instance, of the anthocyanidins that are distributed throughout plant tissue.[147] If the fungus or bacteria survives this initial encounter, its molecular fingerprint will be identified by transmembrane "pattern recognition receptors," leading to a chain of cellular events that will lead, via activation of the jasmonate hormonal pathway, to several days during which "phytoalexin" secondary metabolites will be synthesized throughout the plant. These will accumulate around the site of infection.[177] The exact nature, in terms of chemical group, of this phytoalexin response is specific to the plant division or family, and it is dictated largely by the nature of the secondary metabolites that the plant uses as a constitutive first defense. As this role most often falls to phenolics, they are also the most commonly employed response to infection, playing this role in all gymnosperms and a multitude of families of angiosperm plants.[175]

Any above-ground induction of defense mechanisms will also have a knock-on systemic effect that reaches below ground into the root system. This might involve the reallocation of resources from the roots to the attacked area of the plant, or vice versa. Alternatively it may involve a tandem, systemic increase in the synthesis of both direct and indirect defense compounds in the roots.[744] If we drop down below the soil we are entering the "rhizosphere" surrounding the root system. This is a world that is rich in secondary metabolites, with phenolics predominating.[171,745] These are often exuded into the soil in response to specific signals, for instance from symbiotic and pathogenic microbes, and a range of abiotic stresses, such as nutrients and the mineral status of the soil, temperature, and water stress.[746] The full range of structures from simple phenolic, gallic acid and tannin derivates, and every level of cinnamic acid derivative (benzoic acids, coumarins, flavonoids) can be detected in the rhizosphere. Many of the flavonoids are excreted into the rhizosphere to play benign roles in the management of microbial symbionts, attracting mutualist bacteria and triggering the germination of fungal spores. These microbes then colonize the roots, facilitating the uptake of nitrogen, water, nutrients, and minerals. In return they are paid with photosynthates such as carbohydrate by the plant.[158,159] These relationships can be quite specific; in the legume (*Fabaceae*) family the interplay between plant and bacteria includes the growth by the plant of root nodules for the bacteria to inhabit, with individual flavonoids such as luteolin and apigenin managing relationships with individual species of *Rhizobium* bacteria.[147,746] Alternatively they can be more general; for instance, ubiquitous flavonoids such as quercetin and rutin manage the plants' relationship with the arbuscular mycorrhizal fungi that contribute to plant phosphorus uptake by colonizing the surface of roots in the vast majority of plants.[147,159] Naturally, the plant's protective reaction to infection by underground pathogenic fungi and bacteria is identical to that seen above ground and often also involves the synthesis of phenolics.

As we move away out of the immediate rhizosphere we find that the soil surrounding the plant still harbors a variety of secondary metabolites that have been exuded by roots or leached from leaves and fallen plant tissue. Even at a distance from the plant flavanols and tannins protect against nematodes[147] and mollusks.[747] Many of the phenolics in soil are also allelopaths, inhibiting the growth and germination of competitor plants in their immediate surroundings. For instance, wheat plants exude a range of simple phenolics, including vanillic, p-coumaric, syringic, and ferulic acids, which function as both above-ground defense chemicals and below-ground allelopaths for wheat plants. They accomplish the latter by influencing nutrient uptake, plant morphology, enzyme function, and protein synthesis in competitor plants, potentially via a modulation of hormone levels and subsequent gene expression in the encroaching plant.[157,171] Many flavonoids also have allelopathic properties, and these compounds may function by interfering with the signaling and transport of the hormone auxin in the roots of competitors, or they may act as pro-oxidants, among many other potential mechanisms of action.[746] The classic example of an allelopathic plant was described in the first century AD by Pliny the Elder, who noted that "the shadow of walnut trees is poison to all plants within its compass." The active allelopathic chemical exuded by walnut is juglone (5-hydroxy-l,4-naphthoquinone), a naphthoquinone phenolic derivative that prevents root elongation in competitors by interfering with gene expression relating to protein kinase signaling pathways and jasmonic acid, abscisic acid, and gibberellic acid hormonal pathways. Juglone can persist in the soil for more than a year.[165] Unfortunately, many flavonoid-excreting plant species also suffer from autotoxicity, whereby flavonoids and other allelopathic chemicals build up in the soil and reduce the growth and survival of the emitting plant over a period of years.[746]

In all of these many and varied interactions, individual phenolics are often multifunctional, multitalented compounds. As an example, the simple flavanol catechin can play a host of differing roles. In the rhizosphere, catechin, secreted into the surrounding soil by the invasive weed *Centaurea maculosa*, acts allelopathically, inhibiting the growth of competitor species by inducing oxidative stress and cell death in their encroaching roots. Catechin also functions as an auto-inhibitor, ensuring optimal spread of plants by preventing the germination of its own seeds within its immediate vicinity; it acts as an induced antimicrobial agent and as an antioxidant; it chelates metals, increasing their uptake by the plant; and it acts as a complex modulator of relationships with mutualistic and pathogenic microbes and nematodes.[170] Of course, catechin, like the other flavonoids, also plays a host of other protective and defensive roles above ground. Indeed, phylogenetic analyses show that flavonoid biosynthesis genes were already present at the point that plants started to colonize the land. The very balancing act of offering ultraviolet-absorbing properties, while providing cellular antioxidant protection by these evolutionarily ancient molecules, along with a role as developmental regulators, was a key factor that allowed the first plants to migrate out of the sea onto the harsh, sundrenched shores of our sterile planet.[118] Of course, these molecules also act against microbial infection and reduce the palatability of plant tissue to insects, and it is interesting to note that exposure to light will both increase anthocyanidin concentrations in irradiated tissue and also the synthesis of other flavonoids throughout the plant. This increases the plant's resistance to all stressors, and vice versa.[146,174]

■ PHENOLICS IN THE LIFE OF HUMANS

As phenolics are expressed ubiquitously in plant tissue, they are also ever-present and unavoidable in the human diet. We obtain them from fruit, vegetables, cereals, seeds and beans, spices and herbs, and oils, and all of the food products made from these basic components. The concentration of phenolics varies hugely within these food sources, with the highest concentrations often seen in food sources that we eat in very small amounts, such as herbs and spices, or foods that we might only eat occasionally, such as cocoa and highly colored berries.[748] However, overall, and unsurprisingly, evidence suggests that we derive the bulk of our daily intake of phenolics from the plant-based food products we consume most often and in the greatest quantities, particularly non-alcoholic beverages such as coffee and tea, fruit and fruit juices, vegetables, and alcoholic beverages. Both red wine and beer are excellent sources of phenolics.

In terms of chemical diversity, all of the structural groups of phenolics are present in foods, albeit at differing concentrations; so, where they occur in a food, anthocyanins and flavanols typically appear at high concentrations, whereas flavones and stilbenes are usually present at low concentrations. The most recent and comprehensive database describing the phenolic contents of food, "Phenol-explorer," contains information on over 500 individual phenolics that appear in more than 450 regularly consumed foods.[749]

While we can measure the phenolic content of a given food, quantifying the actual amounts eaten by humans is more problematic. One way to estimate actual consumption is to use a food diary or questionnaire in which participants note the types and amounts of foods they consume over a period of time. With the help of a database that describes the nutritional components making up individual foods, a very rough approximation for the individual's total consumption of phenolics can be calculated. As an example, in the first systematic assessment of flavonoid consumption across the United States, Chun et al.[750] collected diary information from nearly 9,000 adults and calculated flavonoid contents using the U.S. Department of Agriculture's "Database for the Flavonoid Content of Selected Foods." The estimated mean daily total flavonoid intake across their sample was 189.7 mg flavonoids per day, with flavanols making up a rather striking 80% of the total. The flavonoids were most often consumed in tea, citrus fruit juices, fruit, and wine. Using a similar method Zamora-Ros et al.[751] assessed flavonoid consumption in a Spanish cohort of more than 40,000 adults aged 35 to 64 and demonstrated a somewhat higher average consumption of 313.26 mg/day, with proanthocyanidins and other flavanols representing about 70% of their sample's total consumption. Similarly, 24-hour dietary recall data from nearly 14,000 Australian adults showed total flavonoid consumption levels of 454 mg/day, with 92% of this composed of flavanols, predominantly from tea. This single source provided 70% of the total flavonoids for the average participant.[752] Of course flavonoids are not the whole story, and when a study conducted in Finland also included an assessment of the consumption of phenolics other than flavonoids, the researchers found that the mean consumption of 222 mg flavonoids per day (58% of which were flavanols) was dwarfed by the 640 mg phenolic acids consumed per day. Caffeic acid alone, consumed in its esterified chlorogenic acid form found in coffee, represented 417 mg of the average daily total of phenolics.[753] A recent French study,[754] involving nearly 5,000 adults and using the "Phenol-explorer" database, demonstrated that a total of 337 separate phenolics were consumed by their cohort, with 258 of these consumed regularly by at least 50% of the cohort. The total average phenolic intake for the sample was estimated at 1,193 mg/day, of which flavonoids contributed 506 mg and phenolic acids 639 mg. Several things are notable from these studies. The first is that the results are highly variable, probably because of a number of factors, including differences in eating habits between countries, differences in measurement techniques and the accuracy and comprehensiveness of the reference database, and differences in the scope of the studies (i.e., concentrating just on flavonoids or all phenolics). The second is that tea in particular, followed by fruit and red wine, make large contributions to flavonoid consumption, and that coffee contributes the lion's share of the phenolic acids in the form of esterified chlorogenic acids. The big discrepancy between the flavonoid consumption of the U.S. sample (189 mg/day) and the French sample (506 mg/day) may also illustrate another key possibility: that many developed countries, often adopting eating habits originating in the United States, are continuing a drift away from the high-phenolic diet rich in fruit and vegetables that formed the nutritional backdrop for human evolution. We'll return to this subject a bit later.

It is evident that phenolics are an unavoidable part of our daily diet, and epidemiological evidence suggests that they confer a wide range of health benefits on the consumer. For example, relatively recent epidemiological research shows that the consumption of polyphenol-rich foods and drinks such as chocolate[755] and tea[756] is inversely related to cardiovascular disease. Overall consumption levels of flavonoids have also consistently been shown to be associated with reduced mortality due to cardiovascular disease.[757] As an example, in a recent longitudinal study demographic and dietary data were collected from nearly 100,000 participants aged 69 or 70 years at the outset of the study. Seven years later participants with higher total overall flavonoid consumption and higher consumption of individual groups of flavonoids—including anthocyanidins, flavanols, flavones, flavonols, and proanthocyanidins—were significantly less likely to have died from cardiovascular disease.[758] While dietary recall data are inherently inaccurate, the relationship between the direct consumption of phenolics and lower blood pressure and reduced hypertension has been confirmed by assessing phenolic consumption by measuring urinary excretion of the phenolics and their metabolites.[759] Direct evidence in the form of controlled supplementation trials in humans have also confirmed the beneficial cardiovascular effects of a number of specific sources of flavonoids, although the vast majority of these studies have investigated the effects of flavanols, often derived from cocoa.[760] Taken as a whole, these trials suggest a consistent beneficial effect of cocoa-derived flavanols on cardiovascular parameters, including inflammatory biomarkers related to atherosclerosis,[761] blood pressure, and endothelial function.[757] To give an indication of the size of this literature, a recent review identified 24 individual controlled trials (incorporating 1,106 participants) that had assessed the effects of short-term administration of flavonoid-rich cocoa. Meta-analysis of the data showed that short-term consumption of cocoa products significantly improved blood pressure, insulin resistance, lipid profiles, and peripheral blood flow (i.e., endothelial function), with peak effects seen at 500 mg flavonoids per day.[762] Similarly, another meta-analysis, this time including the data from 42 studies, showed that endothelial function was significantly improved within two hours of taking cocoa-based treatments, and that this effect was most closely related to the level of epicatechin administered in the studies. Chronic administration also resulted in improved blood flow, decreased insulin levels, and improved insulin resistance, diastolic blood pressure, and mean arterial pressure.[760]

Given the close relationship between cardiovascular and cerebrovascular health, these relationships must also extend to brain function. So, for instance, the consumption of tea[763,764] and fruits and vegetables and total levels of flavonoids[765] have been shown to be associated with protection against or slowed progression of cerebrovascular diseases such as strokes and neurological disorders such as Alzheimer's disease and other dementias.[766–770] Similarly, in elderly populations, cognitive impairment has also been shown to be inversely associated with tea consumption,[763,767] while better cognitive function was shown to be associated with the consumption of polyphenol-rich foods such as chocolate, red wine, and tea.[771] Two recently reported longitudinal studies also focused specifically on cognitive function. In one study 16,000 nurses over 70 years of age, who had provided food consumption data every 4 years for the previous 15 to 20 years, were assessed using a telephone cognitive assessment battery three times over a period of 6 years. The results showed that total flavonoid consumption was associated with reduced rates of cognitive decline over the 6 years, as assessed by a global cognition outcome incorporating data from the six cognitive tasks. The participants in the highest third in terms of blueberry and strawberry consumption also had significantly lower rates of cognitive decline than the lowest third of consumers.[772] In the second study,[773] the researchers used the "Phenol-explorer" database and dietary data collected between the years 1994 and 1996 from 2,574 middle-aged adults to quantify the participants' consumption of phenolics. The researchers split their group into quartiles on the basis of their total phenol consumption, and then they assessed how cognitive function differed between these groups 13 years later (2007–2009). The top-line result was a positive relationship between overall consumption of phenolics at the earlier assessment and cognitive function 13 years later. Higher intake was related to better performance in terms of a "language and verbal memory" factor extracted from the data from the four tasks the researchers had used. The authors went on to carry out a number of subanalyses in which they redivided their cohort into quartiles on the basis of individual structural groups of phenolics, ultimately creating different subsets of participants with differing dietary patterns for each analysis. Unsurprisingly, this generated somewhat contradictory results. The study did, however, illustrate the difficulty inherent in epidemiological research generally, in that phenol consumption was also related to a plethora of lifestyle factors (e.g., education, exercise, smoking, BMI) and other dietary factors (e.g., energy intake, protein, fiber, vitamins B_9, C, and A), each of which has the potential to have an independent impact on cognitive function. While the authors tried to control statistically for these confounding factors, it is still possible that either unidentified covariates continued to drive the relationships, or the removal of the influence of a variable that simply covaried with phenol consumption as an epiphenomenon also modified the real relationship between the phenolics and function. In this context, splitting the cohort multiple times may have confused the issue further by creating subsets of participants with wildly differing diets and lifestyles for each subanalysis.

Of course, phenolics do not exist in isolation in food, and a large body of cross-sectional and prospective epidemiological research has also assessed the health effects of the so-called "Mediterranean diet," which is typified by a high consumption of fruit, vegetables, legumes, complex rather than simple carbohydrates, olive oil, and red wine, and moderate consumption of fish. Naturally, this diet is also very high in phenolics, along with other potentially beneficial dietary factors, such as vitamins. Typically the adherence to the Mediterranean-style diet by the participants is measured on a nine-point scale (with 9 reflecting complete adherence). In the most recent and comprehensive review, Sofi et al.[774] identified a total of 18 prospective studies that had been conducted with a combined sample size in excess of 2 million participants. In each study the researchers assessed adherence to the ideal Mediterranean diet at the outset and then assessed disease and mortality at follow-up times ranging from 4 to 20 years. The authors found that a 2-point (out of 9) increase in adherence to the Mediterranean diet was associated with significantly reduced mortality and specific protection against cardiovascular disease, tumors, and both cerebrovascular and neurodegenerative diseases such as dementia.

The first substantial epidemiological evidence of the associations between health parameters and the consumption of polyphenols only really emerged in the mid-1990s, but in the intervening years these observations have driven a huge research effort that encompasses hundreds of thousands of papers describing empirical investigations.

However, the vast majority of these studies have focused in vitro and in/ex vivo on delineating the molecular and physiological mechanisms underlying the putative benefits of polyphenols.[775,776] In other words, they have been almost exclusively concerned with "how" polyphenols might benefit health. Some of the evidence relating to brain function will be reviewed for the most promising molecules and groups of phenolics in the next chapter. However, despite the fevered interest in phenolics, and polyphenols in particular, almost no real attention has been spared for the question of "why" phenolics have an impact on human brain function. We'll return to this question after a small digression.

■ WHAT HAVE THE PLANTS EVER DONE FOR US? VITAMINS

We're all familiar with the notion that all of our naturally produced foods, either first-hand in the form of plants and plant-derived products or second-hand as animal products, derive from the plants and algae that turn sunlight into edible forms of energy by a process of photosynthesis. Ultimately, photosynthesizing organisms at the bottom of the food chain provide all of our macronutrients, the carbohydrates, fats, and proteins we require for our immediate survival. They also synthesize and provide the essential amino acids that form the building blocks of proteins and contribute a range of organic micronutrients or "vitamins" that are crucial to animal life. All of these components have to be sequestered from food.

In total, humans require adequate amounts of four fat-soluble vitamins (A, D, E, K) and nine water-soluble vitamins (B_1, B_2, B_3, B_5, B_6, B_7, B_9, B_{12}, C). Without adequate levels of each of these vitamins we become ill, and in extreme cases die. Most of these organic compounds[b] are synthesized by plants as primary metabolites. These play largely the same essential roles in plant metabolism as they will go on to play in the animals that consume them, acting as essential antioxidants and enzyme co-factors that carry out cellular chemical manipulations.[777] The big difference is that plants synthesize them, whereas animals have to consume them in their food, either directly in the form of plants or indirectly in the form of animal products such as meat, milk, or eggs from higher up the food chain than the plant (or in some cases microorganisms). Naturally, given their indispensable role in cellular chemistry, vitamins are essential for every aspect of human physiological functioning. This includes every facet of the functioning of the brain. Additionally, vitamins have adopted several brain-specific roles. For instance, vitamin C acts as the antioxidant of choice in the brain, and several vitamins play neuromodulatory roles as ligands, either at their own receptors or at the receptors for other signaling molecules, in cognition-relevant brain areas.[778]

It is interesting to note that the palette of vitamins required by differing animals can vary. This is because, paradoxically, despite the fact that they are critical for life, these compounds became "essential" only as animals, or their simpler forebears, lost the ability to synthesize the vitamin in sufficient quantities during evolution. The final requirement of a species therefore depends on the metabolic pathways that their ancestral family lost in the distant past. The clearest example of this is ascorbate, or vitamin

b. The exceptions being vitamin B12, which is synthesized by bacteria; vitamin C, which is synthesized by most eukaryotes, including plants, but not humans; and vitamin D, which can be consumed in food or synthesized in the skin in response to sunlight.

C, a monosaccharide that is structurally similar to glucose, which is synthesized by all eukaryotes as a part of their normal metabolic processes. The only exceptions are guinea pigs, bats, a few passerine birds, and the anthropoidea (tarsiers, monkeys, and apes, including humans). These few species have lost the ability to synthesize ascorbate endogenously. In the case of humans and our close primate relatives, this was due to a mutation in the gene for L-gulonolactone, an enzyme in the synthetic pathway of ascorbate, which was lost by a common ancestor some 35 to 55 million years ago.[779] Similarly, most, but not all, arms of the animal kingdom have lost one or more of the requisite genes expressing the enzymes required for the synthesis of vitamin B_6 (pyridoxal 5′-phosphate) at some point since the divergence of plants and animals.[780]

In many cases being able to utilize a vitamin is not necessarily straightforward. Vitamins often exist in foods in the form of non-bioactive precursors and have to be extensively tailored to arrive at their bioactive vitamin form. Naturally, the enzymatic pathways and processes for these bioconversions must have been in place prior to the ultimate loss of the ability to synthesize the individual vitamin. It's worth quickly looking at several of the many examples. For instance, in herbivores and omnivores the tetraterpene plant-derived carotenoids, predominantly β-carotene, represent the original plant source of vitamin A. These carotenoids are first converted to bioactive retinal in the intestinal mucosa and other tissues by β-carotene-monooxygenase, and then by other enzymatic reactions they are further converted to retinol and retinoic acid.[781] However, carnivores such as the cat obtain their vitamin A directly from meat in an already-converted form and have therefore never developed, or have lost, the carotenoid-converting enzymatic pathways. To the cat carotenoids are therefore not a form of "vitamin A," as they cannot utilize them.[782] Similarly, vitamin D is synthesized in the body by the hydroxylation of vitamin D_3 first to 25-hydroxyvitamin D3 and then to its bio-active form, 1,25-dihydroxyvitamin D_3,[195] and the active form of vitamin B6 is converted by enzymatic pathways, including pyridoxal kinase, from its non-bioactive forms of pyridoxal, pyridoxamine and pyridoxine, to its active form of pyridoxal 5′-phosphate.

The apparent evolutionary paradox of why an organism would benefit from losing the ability to synthesize a compound on which it relies so heavily for survival is resolved by the fact that, during evolution, vitamins have been in ubiquitous and plentiful supply within the food chain. The process of endogenous enzymatic de novo synthesis of these molecules would have entailed a disadvantageous cost in terms of energy expenditure, the need for cellular machinery, and the oxidative stress involved in metabolism. This places an organism that can simply sequester its "vitamins" from the environment at an evolutionary advantage.[783,784] This evolutionary hypothesis would suggest that human nutritional requirements have been shaped by evolution, culminating in approximately 7 million years[785] of hominin evolutionary development. During that time the general diet of the ancestors of today's humans will have changed relatively little, until the development of agriculture approximately 12,000 years ago.[786] This is a span of time that represents the blink of an eye in evolutionary terms. It has been suggested that many of the "diseases of civilization," such as dementia, diabetes, obesity, and cardiovascular disease, are partly predicated on the "collision of our ancient genome" with contemporary nutrition[787] as the diet of modern humans in developed societies has rapidly departed from our evolutionarily determined, largely herbivorous, micronutrient-rich diet, toward a high-energy, highly digestible, micronutrient-depleted diet.[786–788] Naturally, this proposition extends to vitamin consumption.[788] As a single example, reference to the rate of endogenous synthesis of vitamin C in other ascorbate-synthesizing mammals, the diet of gorillas,[784] and the vitamin constituents of what might be assumed to be a typical pre-agricultural diet[784,788,789] all suggest that our consumption of vitamin C should naturally be at least 10 times the recommended daily allowance of around 60 to 75 mg espoused by most governmental authorities. A similar if slightly less striking argument can be made for the discrepancy between our pre- and post-agricultural consumption of other vitamins.[784,788] Evidence certainly does suggest that large proportions of the populations of developed countries are suffering from levels of biochemical deficiency in one or more vitamins that would predispose them to disease, and a wealth of epidemiological evidence suggests a relationship between reduced vitamin levels and an increased incidence of the "diseases of civilization," including dementia, mood disorders, and poor cognitive function.[778] But of course vitamins are not the whole story.

The vitamins that function primarily as essential components of the human body's antioxidant defense system comprise a disparate group of plant chemicals: the monosaccharide ascorbate (vitamin C), a handful of tetraterpene carotenoids (vitamin A), and the quinone tocopherols derived from phenolics (vitamin E). However, it has also been suggested that humans have evolved to similarly take advantage of the antioxidant properties of other, broader classes of ubiquitous plant-based food components that would have formed part of our evolutionary diet. These include the many hundreds of carotenoids found in colored fruit and vegetables that do not contribute to vitamin A synthesis, and, most pertinently here, the phenolics that are found ubiquitously in all higher plants.[788] Naturally the shift in diet away from our paleolithic forebears has considerably reduced our consumption of these dietary components, in exactly the same way that our vitamin consumption has dropped off to a point where a lack of these micronutrients may be having detrimental effects on the health of a significant proportion of the populations of developed countries. The striking health-promoting, or disease-preventing, properties of the Mediterranean diet[774,790] may well be attributable to this dietary pattern conforming more closely to a typical paleolithic diet.[790] Indeed a more extreme "paleolithic" diet, more weighted toward fruit, vegetable, and tuber consumption, and lacking any refined or dairy products, has been shown to confer additional benefits over and above the classic Mediterranean diet.[791] Both of these diets, of course, provide a vastly increased quantity of both vitamins and phenolics.

While no single phenolic is essential in itself, and therefore escapes the definition of "vitamin," the evidence reviewed above does suggest that this entire class of phytochemicals confer a wide range of general and specific health benefits. Indeed, the first observation of the biological activities of polyphenols was made by Rusznyak and Szent-Györgyi in 1936[792] when they showed that compounds in lemon peel other than vitamin C could reduce the capillary permeability that caused subcutaneous bleeding in the skin-discoloration condition purpura. They named their discovery vitamin P (for permeability), although it later transpired that there were several active flavonoid compounds in their extract, and that these weren't "essential" as such. Despite this, flavonoids in general were commonly described as "vitamin P" for several decades, and, although the misnomer fell out of use, it still occasionally resurfaces in descriptions of rutin, a quercetin glycoside from citrus fruits.

While individual phenolics, or small groups of phenolics, do not conform to the required definition of "essential," it is an interesting proposition that this entire structural class of phytochemicals may play essential roles in human health.

■ WHY DO POLYPHENOLS HAVE BENEFICIAL EFFECTS ON HUMAN BRAIN FUNCTION?

In terms of brain function, as with general health, the working hypothesis underlying much of the interest in this area was originally the notion that polyphenols exert their benefits via their direct, potent antioxidant properties. However, emerging evidence, in particular showing low bioavailability, particularly in the brain, suggests that direct antioxidant properties are unlikely to be a key factor in the neuroprotective and neuroenhancing effects of these secondary metabolites.[793] Polyphenols may well, of course, indirectly upregulate endogenous antioxidant systems, but their utility to the consuming animal is more likely to come from their interactions with a number of cellular signaling pathways; many or all of these interactions may have their roots in the synthetic pathways and endogenous and ecological roles of the polyphenols in the plant.

The key to understanding why some classes of plant chemicals, including many polyphenols, interact with the human brain is likely to reside in great part with the many biochemical and biomolecular similarities between plants and mammals described in Chapter 3, and three factors in particular: the similarities in hormonal stress signaling between plants and animals; the plant signaling roles of polyphenols; and the potential for cross-kingdom polyphenol signaling between taxa. These factors will be considered individually below.

Similarities in Hormonal Stress Signaling Between Plants and Animals

Both plants and animals have a well-developed hormonal system that transduces signals either locally (autocrine and paracrine signaling) or over a longer distance (endocrine signaling). In plants the typical net effect of hormonal signaling will include modulation of the synthesis of numerous secondary metabolites, including phenolics. The key hormones upregulating defensive reactions, including secondary metabolite synthesis, are the jasmonates: jasmonic acid, and its conjugated and hydroxylated derivatives such as methyl-jasmonate and jasmonoyl–isoleucine. Jasmonic acid and its derivatives are oxylipins that are formed from the 18-carbon polyunsaturated fatty acid, α-linolenic acid, which is released from cellular glycerides by lipases in plant tissue as a consequence of biotic and abiotic stressors. The α-linolenic acid is then metabolized via the octadecanoid pathway, which commences with oxidation by lipoxygenase enzymes.[178,181] The jasmonate derivatives that eventually result travel to other cells and degrade cellular proteins (JASMONATE-ZIM DOMAIN proteins) that are suppressing the transcription factors, and therefore the gene expression, that dictate the many physical and chemical plant responses to stressors. This ultimately unleashes the appropriate defensive reactions.[182,183]

While jasmonates are therefore necessary for the induced synthesis of protective secondary metabolites, they represent only one strand of a complex interplay between hormones. These include salicylic acid, abscisic acid, auxin, and ethylene, along with ubiquitous signaling molecules such as nitric oxide. These chemicals are synthesized in varying quantities in response to differing stressors. So, for instance, jasmonic acid predominates in a plant's response to insect herbivory and abiotic stressors; jasmonic acid and ethylene combined are synthesized in response to necrotrophic pathogen attack; and salicylic acid is induced following infestation with biotrophic pathogens.[794]

The synergistic and antagonistic "cross-talk" between these hormones and signaling molecules, in effect, fine-tunes the plant's response to match the stressor,[185] leading to complex stress-specific patterns of gene transcription.[182,794] As an example, abscisic acid, a key promoter of the response to abiotic stressors, tends to work synergistically with the jasmonates.[178] In contrast, salicylic acid has an antagonistic relationship with the jasmonates in which one of its functions is to downregulate the jasmonate pathways, in effect switching the plant's response from a short-term synthesis of phytochemicals suited to abiotic stressors and herbivory and redirecting it toward a longer-term "immune" response that will be more suited to resisting biotrophic and viral pathogens. Naturally, the converse is also true, with the jasmonates working to downregulate the salicylates,[178,186,795,796] and salicylic acid downregulating the gene expression associated with other hormones, such as auxins.[797]

These plant hormonal systems bear some striking relationship to animal systems. The most straightforward example is the role that the "plant" hormone abscisic acid plays in the mammalian nervous system. This genetically conserved signaling molecule is also synthesized endogenously by mammals and operates via the same signaling pathways and conserved genes across plants and animals.[281,282] In both taxa it guides the cellular reactions to abiotic stressors such as heat and light. It also has wide-ranging functions within the mammalian immune system and modulates insulin release from the pancreas and the proliferation of stem cells. A number of these effects are related to its modulation, including upregulation, of the synthesis or function of prostaglandins.[281,284] No less intriguingly, the oxylipin pathways that synthesize jasmonates in plants and the prostaglandins in mammals are genetically conserved orthologues inherited from the common ancestors of plants and humans.[277] In mammals, the orthologous pathways lead to the synthesis of eicosanoids from the 20-carbon polyunsaturated fatty acid arachidonic acid, which is released from cellular phospholipids by phospholipases and metabolized via very similar pathways to those seen in the formation of jasmonates in the plant, but in this case featuring both lipoxygenase and the cyclooxygenase COX1 and COX2 enzymes.[798] In keeping with this shared heritage, the jasmonates in plants and the prostaglandins in animals are closely related in structural and functional terms.[277] The structural similarity is illustrated in Figure 8.3. In animals, prostaglandins and the COX enzymes contribute, for instance, to the modulation of blood flow via the dilation or constriction of blood vessels and by determining the aggregation, or stickiness, of blood platelets, and they contract or relax bronchial and smooth muscle. Directly in keeping with the plant roles of jasmonates, they also govern a number of responses to stressors, including the regulation of inflammation and immune function and the response to wounding.[275,798] The most striking difference in the response of the plant jasmonate and animal prostaglandin systems to stressors is therefore primarily seen in the end product, with the animal response most often associated with inflammation or immune system activation and the plant response, while including many analogous cellular responses, also being typified by the additional synthesis of secondary metabolite chemicals.[799]

A wealth of emerging evidence suggests that each of the key hormonal mediators of secondary metabolite synthesis in plants can have reciprocal effects within mammalian tissue. Modulation of endogenous abscisic acid function in humans is being investigated from a therapeutic point of view with respect to a variety of diseases, and both the jasmonates and salicylates exert multifarious, potentially beneficial, effects on mammalian cancerous cells in a manner that resembles their innate activity

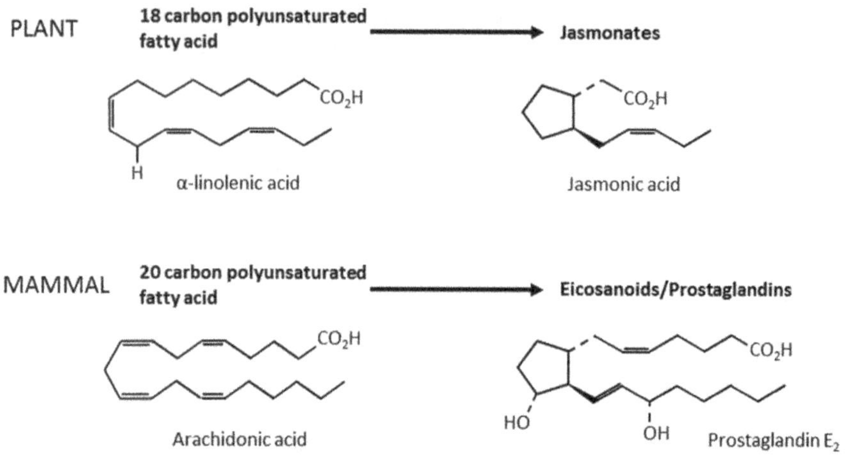

PLANT 18 carbon polyunsaturated fatty acid → Jasmonates

α-linolenic acid — Jasmonic acid

MAMMAL 20 carbon polyunsaturated fatty acid → Eicosanoids/Prostaglandins

Arachidonic acid — Prostaglandin E_2

Figure 8.3 Conserved stress signaling molecules, showing the fatty acid precursors and an example of the mammalian prostaglandins and the plant jasmonates.

in plant cells. The activity of salicylic acid in terms of inducing tumor cell apoptosis, in particular, resembles the hypersensitive response, or programmed cell death, that it coordinates in response to pathogens in plants.[799] Salicylic acid itself, particularly in the structurally close form of acetylsalicylic acid, better known as aspirin, has also attracted a huge amount of research over many decades. It has previously been noted that the activity of salicylates when introduced into mammalian systems is merely a natural extension of their roles within plants[800] and that many of their actions involve interactions with mammalian cellular processes that are directly conserved or very similar to those seen in plants. The most readily appreciable example is the inhibition of cyclooxygenase activity by salicylic acid, which leads to reduced prostaglandin synthesis. This effect mediates the celebrated anti-inflammatory properties of acetylsalicylic acid (aspirin) in humans.[799] Just as the upregulation of prostaglandin function by abscisic acid in the animal mirrors its synergistic relationship with the jasmonates in the plant, the inhibitory effect of salicylic acid on prostaglandin synthesis in mammals can also be seen as a direct reflection of salicylic acid's analogous antagonistic effect on the jasmonate pathways that induce secondary chemical synthesis in plants.[798] The above suggests some intriguing similarities between the stress signaling hormonal pathways in plants and mammals. We'll return to this topic shortly.

The Plant Signaling Roles of Polyphenols

We have some understanding, albeit patchy, of the complex endogenous signaling relationships enjoyed by plant hormones (summarized very briefly above). A huge amount of research has also been undertaken investigating the effects of polyphenols and other phenolics in mammalian cells, animal models, and even humans. However, we know very little about the endogenous signaling roles that polyphenols carry out in the life of the plant itself. With a few exceptions this is because this research has simply not been conducted to date. Nevertheless, a number of strands of evidence are beginning to point to the possibility that some phenolic secondary metabolite chemicals may play multiple "primary" signaling roles in the physiological functioning of the plant. For example, Pourcel and Grotewold[801] note that secondary chemicals are often synthesized locally by specialized cells or tissue at specific times dictated by developmental processes or stressors. They are often then actively transported long distances in the plant, becoming "integral components of the plant signaling machinery," functioning both as signaling molecules in their own right and by interfering with the signaling activity of other primary chemicals, such as hormones. As examples, a multitude of direct signaling roles have been attributed to the ubiquitous flavonoids. For instance, they play recently established signaling roles in pollen germination and dormancy, the sanctioning of the transmembrane movement of the hormone auxin, the nodulation process that allows the colonization of root systems by symbiotic bacteria, and the lignification process; they also partake in intra-plant alleochemical communications.[801,802] As an example of the latter, a recent study investigating gene expression showed that the allelopathic effects of juglone, a phenolic derivative that stunts the growth of plants in the vicinity of walnut trees by preventing root elongation, are related in major part to juglone's ability to interfere with MAPK (see below) cellular signal transduction and gene expression within the jasmonic acid, abscisic acid, and gibberellic acid hormonal pathways of the encroaching recipient plant.[165] A putative wide-ranging signaling role for flavonoids is also strongly supported by their detection in the nucleus of plant cells, suggesting that they may function in endogenous gene transcription.[803]

One of the most fascinating of these signaling roles, and one worth looking at in more detail, is the benign part that flavonoids play in the plant's symbiotic relationships with both the bacteria and fungi that colonize the plant's roots and facilitate the uptake of nitrogen, water, nutrients, and minerals. In return for their services these microbes are paid with nutrients generated by photosynthesis. The most ancient of these relationships, that between plants and arbuscular mycorrhizal fungi, dates back to the colonization of the terrestrial environment by the ancestors of all plants and is therefore ubiquitous across all plant lineages and controlled by equally ubiquitous flavonoids such as quercetin and rutin. There are only a few plants that appear to have severed these fungal relationships, but I'm sure they had their reasons. The more specialized relationships between nitrogen-fixing rhizobial bacteria and plants arose at a later date, and these are typically directed by other ubiquitous flavonoids such as naringenin. Both relationships work via the same mechanisms. Flavonoids released into the rhizosphere either attract bacteria, which move up their concentration gradient by chemotaxis, or stimulate the germination of fungal spores in the soil. The plant then perceives the presence of the symbiotic fungus or bacteria through the microbe's chemical emission of lipochitooligosaccharide "nod factors" or "myc factors" (standing for nodulation and mycorrhizal respectively), which bind to and activate "LysM" receptor kinases in the plant.[804] This leads to gene expression in the plant and an increase in the synthesis and emission of a range of phenolic acids and flavonoids from the roots, which are, in turn, detected by receptors such as the bacterial NodD protein in the microbes. This increases bacterial and fungal reproduction and the growth and attraction of hyphal branches that interconnect the fungi with the plant roots. In the large legume (*Fabaceae*) family the cross-talk between the bacteria and plant also results in the creation by the plant of specialized root nodules designed specifically to accommodate the bacterial colony, with the entire process dictated by flavonoids, such as luteolin and apigenin, that can be specific to the relationship with one species of *Rhizobium* bacteria.[147,746] This entire cross-kingdom signaling process, involving "nod" and "myc" factors and flavonoids, manages the plant/microbial relationship via multiple iterations of autoregulatory feedback.[147,159] In a similar manner a number of flavonoids, such as catechin and naringenin, are synthesized in response to the bacterial "quorum-sensing" signaling molecules that direct the population size and activity of bacteria in the rhizosphere. These flavonoids directly mimic the quorum-sensing chemicals by binding to their bacterial receptors in a positive feedback mechanism that directs both bacterial behavior and flavonoid synthesis.[159] To add a further level of complexity, "tripartite" symbiotic relationships involving the signaling molecules and respective receptors of plants, fungi, and bacteria also take place, and the net effect of any or all of these mutually beneficial interactions can be increased defensive capability for the various parties.[159]

All told, the slight evidence that we have to date suggests that secondary metabolites, and flavonoids in particular, can serve endogenous signaling roles both in the internal life of the plant and in the interactions of the plant with competitors and microbial symbionts.

Cross-Kingdom Signaling Between Plants and Humans by Polyphenols and Other Phenolics

The chemical interactions between plants and symbiotic bacteria and fungi are clearly cases of cross-kingdom signaling. The notion that polyphenols may have their beneficial effects in animals by a similar process of "cross-kingdom" communication, whereby the phytochemical interacts with cellular factors shared between taxa, has been advanced before.[801] For instance, Howitz and Sinclair[805] attempted to explain the molecular and health-benefiting effects of phytochemicals that have no apparent chemical relatives in animals with the concept of "xenohormesis." Straightforward "hormesis" in this context would refer to the notion that a small amount of a potential stressor such as a toxin, physical exercise, or starvation enhances functioning by triggering an adaptive stress response, thereby improving aspects of cellular and physical functioning in the organism, but with these benefits being reversed at higher doses of the stressor. The hypothesis underlying xenohormesis, on the other hand, expands this notion into an "inter-species hormesis" whereby animals and fungi, courtesy of the many similarities in their cellular signaling pathways, have evolved the ability to read the chemical evidence of environmental stress from the plants that they feed on, in the form of stress-induced secondary metabolites such as polyphenols. These chemical cues then allow consumers to mount their own preemptive, beneficial "hormetic" stress response that primes the organism to deal with the environmental stress that triggered the synthesis of the polyphenol in the plant. A favored example here is the ability of several polyphenols to interact in vitro with mammalian "sirtuin" genes (SIRT—Silent Information Regulator Two protein). These genes express enzymes that regulate cellular function by modifying (deacetylazing) histone and non-histone proteins, in effect regulating a number of transcription factors (e.g., p53, NF-κB, PPARγ, PGC-1α, and Foxo) that play key roles in stress responses, cellular differentiation, and metabolism.[806] Sirtuins therefore ultimately modulate a range of critical metabolic and physiological processes, including cellular metabolism, survival and ageing, stress resistance, inflammation, immune function, and endothelial function. Caloric restriction is proposed to owe its life-extending effects in several fungi and animals to the activation of sirtuin genes and their consequent downstream "hormetic" effects on several cellular stress responses. It is notable that much of the huge interest in the polyphenol resveratrol was sparked by the observation that it extends the lifespan of yeast, and animals such as the fruit fly (*Drosophila*) and roundworm (*C. elegans*), in a similar manner to caloric restriction, apparently by activating sirtuin genes.[807,808]

The notion of xenohormesis as an adaptive mechanism is intriguing, and it is possible to intuitively see how environmental stress information garnered from plants might be useful to an organism that survives alongside one species of plant (e.g., an endophytic bacteria or fungus) or an organism that consumes one or a few species of plant (e.g., the many specialist insect herbivores). However, as the organism and diet become more complex, xenohormetic information would become increasingly redundant. To give an example, the in vitro phenomena that support the concept of xenohormesis have been described for only a restricted range of polyphenols, rather than across broad chemical groups. These phytochemicals are indeed synthesized in response to stress but also exist in the plant at basal levels, either as a consequence of also being synthesized as a preemptive, constitutive defense mechanism or alternatively because, in reality, an "unstressed" plant probably does not exist.[175] An animal with a varied diet composed of a variety of plants would be unlikely to either consume an individual xenohormetic phytochemical frequently enough or be able to differentiate basal levels of the chemical in its plant-derived diet from increased levels due to environmental stress. Of course, it is possible that xenohormetic phenomena in mammals represent an echo of an adaptive mechanism from our much simpler, distant ancestors. However, as an explanatory tool of human/phytochemical interactions it falls afoul of the law of parsimony, which suggests that the simplest possible explanation is most likely to be the correct explanation.

The comparatively simple explanation as to why polyphenols, and indeed some other classes of phytochemical, might have beneficial effects on human health and brain function probably comes down to two factors: first, the similarities between signaling pathways in the taxa, and secondly, the function the phytochemicals are trying to fulfill for the plant, including their endogenous roles as signaling molecules within the plant itself. The plant/fungi/bacteria relationship described earlier illustrates one aspect of this quite nicely. The flavonoids and bacterial "nod factor" and fungal "myc factor" chemicals are all intentionally engaged in cross-kingdom signaling, which is clearly beneficial to all parties. However, as well as this intended role, the flavonoids involved also exert unintended cross-kingdom signaling effects in humans. They do this because the receptors to which the flavonoids bind in the bacteria and fungi are "estrogen-like receptors." They are described thus because they also bind mammalian estrogens such as 17β-estradiol; they function largely in the same manner as estrogen receptors in terms of the molecular mechanisms by which they modulate gene transcription.[809,810] In the case of the bacterial receptor, the NodD receptor protein, they have been reported in one paper to be partial homologues of the mammalian estrogen receptor,[811] which function alongside other homologous proteins within the respective signaling pathways.[812] The flavonoids in question, all of which are readily available from commonly eaten foods, all also function as agonists, partial agonists, or antagonists at mammalian estrogen receptors, with a range of attendant physiological effects. We'll look at these interactions and their knock-on effects in more detail, and in particular explore whether these phytoestrogens have an impact on human brain function, in the next chapter. However, for the present it will suffice to suggest that the estrogenic effects of these "phytoestrogens" are an unintended consequence of their intended cross-kingdom signaling role for the plant, predicated on the similarity in the cellular signal transduction equipment possessed by microbes and mammals. So, can the beneficial effect of other polyphenols and simpler phenolics be accommodated by their partaking in similar, unintended cross-kingdom signaling?

Unfortunately and surprisingly, as noted above, due to a striking lack of research we actually know very little about the signaling roles of secondary metabolites within their home plant. In contrast, in the case of polyphenols, we know a great deal about their interactions with signaling pathways in animals.[813] Emerging evidence shows that flavonoids owe their bioactivity in mammals to interactions with the cellular transduction and signaling pathways that mediate cellular responses to stressors. These pathways are typically initiated by the binding of extracellular signaling molecules, for instance hormones, cytokines, or growth factors, to proteinaceous receptors embedded in the cell membrane. Alternatively, "pattern recognition receptors" may identify proteins associated with microbial pathogens. In both cases the conformation of the receptor protein will be changed so that it transduces its activation across the membrane into the cell, triggering a signaling cascade involving multiple "protein kinases" within the cell. A pertinent example here is the ubiquitous mitogen-activated protein kinase (MAPK) cascade, in which a series of protein kinases (called MAPK kinase kinase, MAPK kinase, and MAPK) activate each other in turn by phosphorylation in a chain reaction that amplifies and transmits the signal within the cell. This chemical signal, along with interacting signals from other signaling cascades, reaches and activates transcription factors in the cytoplasm or cell nucleus. Typically they do this by removing inhibitory proteins from the transcription factors, allowing them to translocate to a specific sequence of DNA in the nucleus, which in turn regulates the activity of genes that lead to the synthesis of proteins in the cytoplasm and, ultimately, changed cellular function. In the case of stress signaling in the mammalian brain, for instance, protein kinase cascades modulate the activity of transcription factors such as NF-κB or CREB (cAMP response element binding protein). This in turn leads to a wide range of cellular responses, including cell proliferation, apoptosis, and the synthesis of growth factors, such as neurotrophins, and inflammatory molecules, such as iNOS, cytokines, and COX-2. Overactivity or dysregulation of these signaling pathways and transcription factors is implicated in the pathogenesis of many neurodegenerative diseases and cancers.[209,210] A simplified MAPK signaling cascade is shown in Figure 8.4.

Many of the key components of cellular signaling are highly conserved, having arisen in the unicellular common ancestor of plants and animals, and the same signaling cascades play often-identical roles in mediating the cellular responses of plants and animals to stressors.[206,207] Plants have a particularly rich complement of more than 1,000 protein kinase genes (known as their "kinome"), which is approximately double the number of kinase genes seen in the human genome and four times that seen in the fruit fly.[208] A striking 2.5% of the genome of Arabidopsis is given over solely to the 600-plus genes that express the kinase subfamily of "receptor-like kinases." These receptors bridge the cellular membrane and transduce extracellular signals into cellular secondary messenger MAPK signaling cascades as described above. Phylogenetic evidence shows that not only did the progenitors of the entire kinase superfamily evolve in a unicellular ancestor of both plants and animals, but that the specific "receptor-like kinases" also originated before the two clades split. Beyond this point the kinome of plants expanded greatly due to genetic duplication events, and the majority of plant

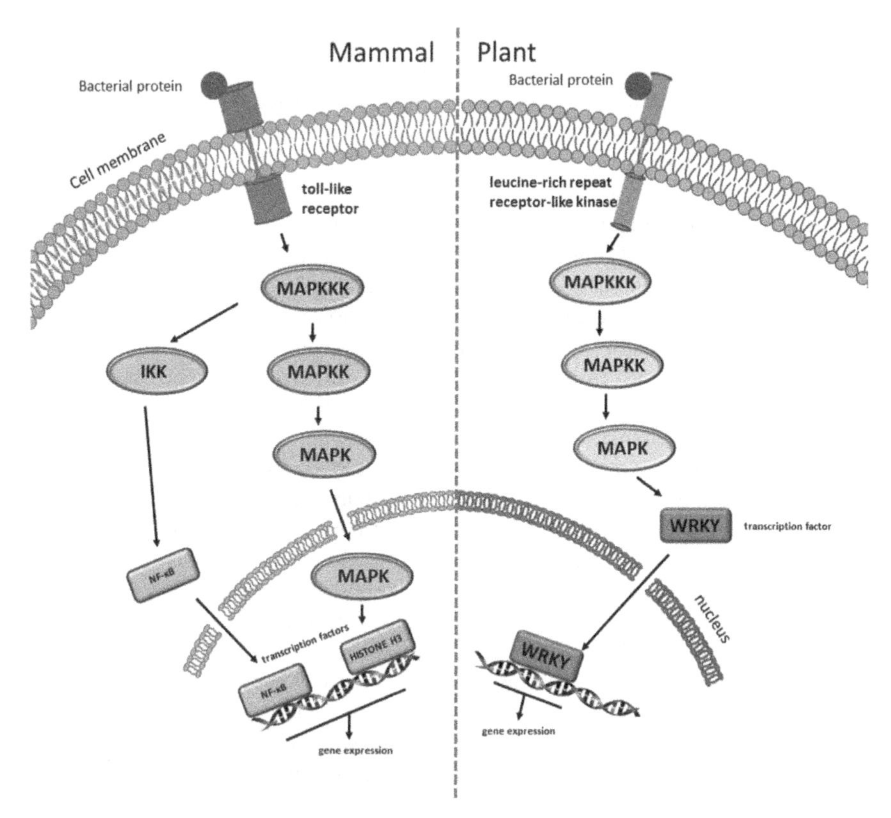

Figure 8.4 Simplified MAPK signaling cascade. The signaling cascade is triggered by the detection of signaling molecules or stressors, either by membrane-spanning receptors or receptors or sensors in the cytosol. These trigger a chain reaction in which a series of MAP kinases (MAPKKK, MAPKK, MAPK) activate each other in turn by phosphorylation, ultimately leading either to the removal of proteins attached to transcription factors in the cytosol, allowing them to translocate to the nucleus, or to direct interactions with transcription factors already in the nucleus. Typically, the ultimate gene expression will be the consequence of activity in multiple interacting signaling cascades. The cascades pictured are simplified versions of the highly conserved signal transduction pathways in animal and plant cells that are triggered by the detection of bacterial proteins by membrane "pattern recognition receptors." The animal cascade is shown with an additional NF-κB cascade which is mediated by IκB kinase (IKK). In animals the eventual cellular response will include the synthesis of inflammatory mediators, whereas the plant response will include the synthesis of secondary metabolites.

receptor-like kinases remained specific to serine/threonine, while the majority of animal receptors evolved specificity to tyrosine (i.e., "receptor tyrosine kinases").[205] As well as sharing these closely related protein kinase membrane receptors, both animals and plants also share the MAPK enzyme signaling pathways. Indeed, these are highly conserved across all eukaryotes,[206,207] with plant MAPKs being most closely homologous to the "extracellular signal-regulated kinase" (ERK) MAPK subfamily in mammals.[814] In addition, plants also possess conserved PI-3K (phosphatidylinositol 3-kinase) kinases and their interacting PKB (protein kinase B) kinases.[815] The Arabidopsis genome also encodes 39 distinct, conserved "animal" AGC kinases (named after the member kinases: cAMP-dependent protein kinase A, cGMP-dependent protein kinase G, and phospholipid-dependent protein kinase C). These kinases modulate the activity of other intracellular second messengers, including cAMP, cGMP, and phospholipids, and make their own independent (to MAPK) contribution to stress signaling as well as engaging in "cross-talk" interactions with the MAPK signaling pathways.[208,816]

This extended protein kinase family also plays analogous cellular transduction and signaling roles in plants to those in animals. One particularly pertinent example is the TOR (target of rapamycin) kinase signaling pathway, a crucial growth-regulating cellular system that simultaneously collects information on stressors, nutrients, and internal energy states from multiple extra- and intra-cellular inputs, including from the PI-3K/PKB pathways. In benign conditions of nutrient and energy availability, TOR upregulates a vast range of energy-consuming activities such as metabolism, cell proliferation, and the translation of mRNA, ultimately dictating a wealth of developmental and metabolic processes by regulating other key kinases.[817,818] The TOR kinase pathways are, again, highly conserved, having originated before the last common eukaryotic ancestor, and both their function and core molecular components, including the upstream PI-3K/PKB pathway, are shared across the cells of nearly all eukaryotic clades.[c] There are differences: for instance, signaling pathways involving growth factors such as insulin, along with the requisite transmembrane tyrosine kinase receptors, were connected up to the existing animal TOR system in response to its own evolutionary needs, whereas plants and microbes do not share this particular facility. However, in essence, the TOR kinase system functions largely the same across taxa.[819] This pathway comes sharply into focus for humans when we consider that activity or aberrant overactivity within the TOR pathway mediates tumorigenesis, insulin resistance in obesity, and the aging process. Conversely, TOR inhibitors may provide novel treatments for cancer, and the life-extending properties of caloric restriction are mediated by reduced TOR signaling.[818]

What Relevance Does This Have to the Effects of Polyphenols on Human Brain Function?

While the TOR and MAPK pathways and the other kinase pathways are heavily involved in directing gene transcription across taxa in response to a wide variety of stressors, the ultimate downstream products of this signaling effort differ between animals and plants. In animals, one consequence of activation will be the synthesis of cyclooxygenase and the prostaglandins, whereas in the plant the same signaling pathways will lead to the expression of lipoxygenase and the synthesis of jasmonates. This in turn will ultimately lead to the synthesis of a raft of protective secondary metabolite chemicals,[211,820] with cross-talk with other hormones modulating the response via their own interactions with the plant kinome.[814,821,822] This concordance between plants and animals becomes crucial when we consider that a plethora of in vitro, and in/ex vivo, research has identified a wide range of potentially beneficial effects of polyphenols on cellular and physiological parameters in mammals and in mammalian tissue. These

c. The only exceptions to date are protozoan parasites, such as the malarial *Plasmodium* species, which have lost this TOR system as they tap into the same information from the cell they are invading.

include beneficial effects on tumorigenesis, cardiovascular parameters, and insulin resistance and, specifically relevant to the brain, modulation of neuroinflammation, neuroprotection, cerebrovascular function, and the modulation of synaptic plasticity and neurogenesis. All of these effects can be related in one way or another to the interaction of polyphenols with the mammalian kinase signaling pathways, including, but not restricted to, the MAPK, PI-3K/PKB, and TOR pathways, and their resultant downstream effects on gene transcription.[209,210,803,823] For instance, the potentially beneficial effects of polyphenols, in terms of inhibiting tumorigenesis and in increasing longevity in animal models, seem to be predicated in great part on direct interactions with the nutrient- and stress-sensing TOR kinase signaling pathway.[818]

It has also been established that polyphenols may have these effects either via direct interactions with the components of the signaling pathways within the cell or alternatively via the direct binding of the polyphenol at nuclear receptors or neuronal membrane receptors. In this respect flavonoids have been shown to bind to mammalian membrane and nuclear receptors such as nicotinic, GABA, adenosine, and estrogen and androgen receptors, as well as binding to and modifying the function of the mammalian "receptor tyrosine kinases" that are homologues of the plant "receptor-like kinases".[824] All of these receptor interactions may feasibly trigger or modulate the activity of protein kinase signaling pathways.[209,210,803,823]

In general, across taxa, stress/immune signaling and gene transcription is a complex affair that is typified by an abundance of feedback and feed-forward loops that closely regulate the expression of signaling molecules and transcription factors. As an example, in plants, the expression of salicylic acid genes is essential for many stress responses but an overaccumulation of salicylic acid is toxic to the cell, so feedback and feed-forward loops between the components of pathways responsible for synthesizing salicylates, and other hormones and cellular factors, carefully regulate the ultimate expression of salicylic acid.[797] Polyphenols and other phenolics represent the ultimate downstream products of plant hormonal defense communications. They also travel within the plant, penetrate cells and cell nuclei,[101] and may exert independent signaling properties. They therefore represent ideal candidates both to feedback within their own synthetic pathways and to interact with the cellular signaling pathways in neighboring and distant cells.

An interesting analogy can be drawn here with the chemicals that play primary metabolism roles in plants and that represent "vitamins" when consumed by humans. Not only can the synthesis of several of these compounds be induced by a variety of stressors in the plant, but they can also self-regulate their own synthesis, preventing wasteful or damaging overexpression, via direct "feedback" interaction with their own biosynthetic pathways.[777] The possibility that polyphenols, and other secondary metabolites, interact with their own synthetic pathways in this manner has simply not been investigated in plants to date. However, it is clearly a possibility that flavonoid-related modulation of animal stress signaling pathways, such as MAPK and other protein kinase components, may simply echo the phytochemicals' role within the genetically conserved or analogous signaling pathways within their own plant. In this context it would make absolute sense, for instance, for the polyphenol products of stress signaling to provide one of the many endogenous signals that contribute in the plant to the activity of the global stress-, nutrition-, and energy-sensing TOR kinase pathways. This activity may then be transferred directly following consumption into modulation of the conserved PI-3K/PKB and TOR pathways in animals. Similarly, viewed in the context of the conserved nature of the jasmonate and prostaglandin synthetic pathways, the inhibition of the cyclooxygenase enzyme COX2 in animal tissue by a wide variety of flavonoids[209] can also be interpreted as the flavonoids attempting to inhibit lipoxygenase enzymes within the orthologous plant jasmonate pathways. This interaction within the plant may be part of the complex cross-talk between the antagonistic jasmonate and salicylate hormone systems (as suggested with regard to salicylic acid itself by Schultz[798]), or it may simply be feedback or feed-forward within and across the jasmonate system. It is also notable that the net effect of flavonoids within several animal signaling pathways is modulation of the synthesis of nitric oxide.[209] This ubiquitous signaling molecule, which represents a key downstream product of signaling cascades in both taxa, plays diverse modulatory roles in both the synthesis and function of jasmonates, salicylates, and ethylene.[825,826] It is therefore intrinsically tied to secondary metabolite synthesis pathways in plants. Any effects on nitric oxide synthesis in animal tissue may well also represent an attempt by the phytochemicals to interact with the various pathways dictating nitric oxide synthesis in its home plant.

If we also briefly return to the yeast/animal sirtuin genes that fostered so much interest in the polyphenol resveratrol, we find that while mammals express seven distinct, structurally similar sirtuin genes (SIRT 1 to 7), of which SIRT1 is the best characterized, plants also possess two conserved sirtuin genes that are homologues of the human equivalents.[806] These are expressed in the reproductive tissue (SRT1) and throughout plant tissue (SRT2) respectively.[827] The specific roles of these sirtuins in plants are poorly understood, but evidence does suggest that they play roles in plant development, via hormonal signaling, and in the hypersensitive immune response.[827] Intriguingly, the SRT2 plant homologue has been shown to suppress salicylic acid-dependent defense against pathogens, suggesting that it may play a role in the mutually inhibitory cross-talk between salicylic acid and the jasmonates.[828] Once again this would seem to suggest that any effects of polyphenols on sirtuin-mediated stress responses in animals may simply reflect an echo of polyphenol/sirtuin interactions in the plant.

Finally, just to confirm that polyphenols do exert direct cross-kingdom signaling effects as a consequence of the role they are trying to play in their home plant, it is interesting to note that one of the few plant signaling roles of flavonoids that has attracted any research attention is their modulation of transmembrane auxin movement. Flavonoids accomplish this by binding to multidrug resistance permeability glycoproteins (MDRP glycoproteins), with the consequence that they inhibit auxin efflux from the cell. The same transporters are genetically conserved in plants and animals, and they form the established target underlying the potential medicinal utility of flavonoids in terms of reducing drug efflux in mammalian multidrug-resistant cells.[802,829] Intriguingly, recent research also shows that members of the same family of MDRP glycoproteins transport prostaglandins out of mammalian cells[830] and that they can be inhibited in this role by a number of flavonoids, including quercetin and silymarin.[831]

■ SOME CONCLUSIONS

It is possible to start to piece together a reasonable argument that the potentially beneficial effects of polyphenols and other phenolics in mammals could be directly related to their interactions with cellular signaling systems that are largely conserved in both plants and animals. In turn these interactions are not necessarily random, but rather they may represent an echo of signaling roles that the phytochemicals may be trying to fulfill in their home plant, including interactions as part of the cross-talk between hormonal systems in the plant. Alternatively, in some instances they may reflect the role the polyphenol is trying to play in relationships with microbes for their home plant. Of course, the notion that many of the effects that polyphenols have in the animals that consume them are simply reflections of the effects they are trying to exert in their home plant requires confirmation of similar signaling effects within the plant.

Of course, one further factor also needs to be taken into account. Phenolics as a class of chemicals are ubiquitous in plant-derived foods such as fruit and vegetables and therefore would have figured daily in the human diet throughout our evolution. The human body is not a passive receiver of food components. It would be entirely expected that we would have evolved mechanisms that allowed the transformation of regularly ingested edible compounds into usable forms with functional properties related both to their structure and the microbiological systems shared between plants and animals. This is exactly what we see with many of those other ubiquitous micronutrients, the "vitamins." These are synthesized by plants and microorganisms but not animals, and they reach humans either first-hand, in plant-derived foods, or second-hand in animal products. Once consumed, they fulfill the same cellular roles for the consumer as they had been playing for the plant that synthesized them. However, vitamins are rarely utilized in the form in which they are eaten but rather are operated on by a raft of metabolic pathways, carried out in a few cases by symbiotic bacteria in the gut, that eventually turn them into their bioactive forms. Polyphenols, similarly, are often acted on by a plethora of enzymes in the gut and body and are often largely bioavailable only in the form of the metabolites from these processes. It would be surprising, given the above, if these metabolites were not also being tailored for bioactivity.

9 Phenolics and the Human Brain

In the previous chapter it was noted that the evidence from cross-sectional and prospective epidemiological studies suggests a relationship between the consumption of greater quantities of flavonoids (and occasionally phenolics in general) and improved cardiovascular function and a reduced incidence of cardiovascular disease (CVD) and CVD-related mortality.[755–759] Given that cardiovascular function is intimately linked to brain function, it is hardly surprising that the same protective relationships have been observed with regard to stroke and dementia in the elderly[763–770] and cognitive function in elderly[763,767,771,772] and middle-aged adults.[773] The protective effects of direct acute and chronic administration of flavonoids in terms of multiple cardiovascular parameters have also been confirmed in a wealth of clinical trials,[757,760–762,832] although it is notable that this research, and therefore the evidence, has accumulated primarily around the flavanols derived from cocoa. The evidence from controlled trials directly assessing the effects of polyphenols on brain function will be reviewed below, but first it will be useful to briefly summarize the potential mechanisms of action underlying putative improvements in brain function.

MECHANISMS OF ACTION—FLAVONOIDS

The bulk of research in this area has focused on the flavonoids as a group. Several notable non-flavonoid polyphenols, such as resveratrol and curcumin, will also be reviewed individually below. The basic synthetic pathways and structures of the flavonoids and stilbenes are shown in Figure 8.2 in Chapter 8.

One key observation worth noting is that, following oral consumption, a range of flavonoids or their potentially active metabolites can cross the mammalian blood–brain barrier and access the brain, where they have been detected in a range of rodent brain structures. This ability seems to be dictated both by the lipophilicity of the individual molecule and its interactions with the transmembrane proteins, such as the multidrug resistance permeability glycoproteins (MDR P-glycoproteins) that are conserved between plants and animals and that function in mammals as transporters ferrying molecules across cellular membranes and the blood–brain barrier.[210,833] Following dietary supplementation, flavonoids and their metabolites are present in the brain at the concentrations (10 to 300 nM) that would be necessary for them to exert pharmacological effects at receptors and within signaling pathways.[834] Interestingly, several classes of flavonoids, for example anthocyanins, become more concentrated and are retained in neural tissue longer than in plasma.[835] Naturally, access to the brain is a precondition for exerting direct effects on brain function.

Cognitive Function

A wealth of evidence in animals has shown that flavonoid-rich foods, extracts, and individual polyphenols can beneficially modulate cognitive function. The focus of this research has typically been on older and impaired rats that are suffering cognitive decrements as a consequence of age, brain insults, or induced pathologies. For instance, flavanol- and anthocyanin-rich fruits and berries and single flavonoids, such as the flavonol quercetin, consistently improve aspects of memory or prevent declines in memory performance.[834–837] These beneficial effects are seen across all aspects of memory, including memory acquisition, consolidation and retrieval, short-term working memory, and spatial memory. They are also accompanied by morphological changes to structures relevant to memory, for example by promoting or protecting neurogenesis and synaptic plasticity in the hippocampus.[210,834,835,838]

As described in the previous chapter, the early hypothesis that the beneficial effects of flavonoids on multiple physiological parameters were attributable to their direct antioxidant properties has given way to the proposal that they interact with a range of intracellular signaling pathways. The most direct way that flavonoids can modify these signaling pathways is by interacting with a range of receptors to which neurotransmitters and other signaling molecules bind. A clear example here is the ability of a wide range of flavonoids, including ubiquitous compounds such as the flavone luteolin, and widely consumed flavanols such as epigallocatechin gallate (EGCG), to exert anxiolytic effects in rodent models that are predicated on their allosteric modulation of $GABA_A$ receptor activity.[839] Flavonoids that are commonly encountered in the diet have also been shown to interact with adenosine,[840] opioid,[841] nicotinic,[842] and estrogen receptors (see below). They can also bind to and modulate the functioning of receptor tyrosine kinases such as tyrosine-related kinase B (TKB), which responds to key neurotrophins such as brain-derived neurotrophic factor (BDNF).[843] Evidence also suggests the existence of brain plasma membrane and nuclear binding sites that have yet to be fully characterized but for which flavonoids have a high affinity.[844] Of potentially even greater relevance, polyphenols can also interact directly within diverse neuronal and glial protein kinase signaling cascades, such as the ubiquitous mitogen-activated protein kinase (MAPK) and PI-3K/PKB signaling cascades described briefly in the previous chapter.[210,823] For instance, recent research demonstrates that a range of flavonoids bind directly to individual protein kinases within these cascades, modulating their phosphorylation state and thereby modifying the activity and outcome of the signaling pathway.[845] As an example, flavonoids may exert beneficial effects on cognitive function by activating components of the ERK signaling cascade that leads to increased activity of transcription factors such as CREB, with a resultant increase in the expression of neurotrophins such as BDNF. This can, ultimately, lead to an increase in the synaptic plasticity and long-term potentiation that underlies long-term memory consolidation. Alternatively, up-regulatory interactions within the PI-3K/PKB pathways may lead to increased activity in the nutrient-sensing TOR signaling pathways or the increased expression of endothelial nitric oxide synthase (eNOS) and therefore nitric oxide synthesis.

This modulation of "good" nitric oxide synthase activity leads to a brain-specific increase in local blood flow, angiogenesis, and neurogenesis, all of which may contribute to neuroprotection and neuronal repair in the face of aging and insults.[210,823] Modulation of eNOS synthesis also underpins the ability of flavonoids to improve peripheral vascular parameters such as endothelial function, blood pressure, and platelet aggregation, and in turn beneficially modulate gross cerebral blood flow and hemodynamic responses to neural activity. Notably, local increases in vasodilation, cerebral blood flow, and nitric oxide synthesis are also implicated in hippocampal angiogenesis and neurogenesis, processes that are, again, implicated in learning, memory, and neuroprotection.[210]

Neuroinflammation and Neuroprotection

As well as activating the normal cellular signaling cascades underpinning cerebral blood flow, hemodynamic responses, neurogenesis, and synaptic plasticity, flavonoids can also selectively inhibit deleterious over-activity in the same, and other, signaling pathways. The beneficial effects of flavonoids on cardiovascular health have been attributed to an attenuation of the inflammatory cascades implicated in the pathogenesis of CVD. Similarly, interference with the aberrant cellular pro-inflammatory signaling cascades that underpin carcinogenesis may underlie demonstrations of protection against cancer. In terms of brain function, the potential to modulate the same cellular signaling pathways endows many polyphenols with potential anti-inflammatory, and therefore neuroprotective, properties in brain tissue. As with other tissues in the body, short-term inflammation of brain tissue can be a beneficial, natural, defensive reaction to injury, infection, stroke, and toxins. However, sustained neuro-inflammation as a consequence of continued activation of microglia, the primary immune cells of the nervous system, and their subsequent sustained release of damaging pro-inflammatory mediators, may contribute to the neuronal damage associated with neurodegenerative diseases such as multiple sclerosis, Parkinson's disease, and various dementias, as well as the deterioration in cognitive function seen with aging.[846]

The inflammatory cascade within microglia is in large part dictated by activity in the ERK and stress-responsive JNK and p38 MAPK pathways and the resultant activation of transcription factors such as STAT-1, NF-κB, and AP-1. These lead to the expression of inducible nitric oxide synthase (iNOS), cytokines such as TNF-α and IL-1β, and enzymes such as COX-2 and LOX, and, in turn, the synthesis of prostaglandins and leukotrienes. One critical event on activation of microglia cells is the increased expression of potentially "bad" iNOS, which can result in a possibly damaging overproduction of nitric oxide. This molecule can go on to inhibit mitochondrial respiration and increase the generation of reactive oxygen species. Nitric oxide can, in turn, then react with superoxide created during this oxidative burst to form extremely damaging and neurotoxic peroxynitrite.

Evidence suggests that disparate flavonoids, including a number that feature as integral parts of the human diet, such as flavanols and isoflavones and their metabolites, can suppress the neuroinflammatory activity of microglia by inhibiting each stage of the inflammatory signaling process in activated glia: inhibiting the activity of pro-inflammatory transcription factors, the release of cytokines, the generation of reactive oxygen species, and the synthesis or activity of iNOS, COX2, and LOX and the resultant production of nitric oxide, prostaglandins, and leukotrienes. These effects in turn may be related to the ability of polyphenols to modulate the activity of multiple components of a range of neuronal and glial signaling pathways, including the individual components of the MAPK pathways involved in inflammatory and apoptotic responses described above.[209,847]

Evidence does suggest that interactions with the components of these signaling pathways is dependent on the exact structure of the polyphenol molecule in question, for instance in terms of the number and location of hydroxyl groups or the nature of specific bonds. This means that different polyphenols will exert markedly different cellular effects.[210] On top of this, many in vitro studies have investigated the effects of concentrations of flavonoids that are unlikely to be achieved in tissue in vivo, or concentrated on the parent molecules, rather than the many metabolites that are typically the potentially bioavailable molecules following digestion.[209] Having said this, while much of the evidence with regard to the anti-inflammatory effects of flavonoids comes from in vitro research exploring mechanisms of action, a recent meta-analysis of 25 human intervention studies that included an assessment of the effects of flavonoids on inflammatory biomarkers noted a significant reduction in TNF-α associated with consuming flavonoid-rich foods or supplements and several flavonoid subgroups.[848] The attenuation of neuroinflammation could certainly underpin the epidemiological evidence of protected brain function in elderly humans.

■ FLAVONOIDS, OTHER POLYPHENOLS, AND HUMAN BRAIN FUNCTION

The above raises the question of the extent to which the in vitro and in/ex vivo demonstrations of the modulation of signaling pathways, the fairly consistent observations of modulated rodent behavior and cognitive function, and the suggestive epidemiological evidence showing relationships between polyphenol consumption and cognitive function and dementia will carry over into evidence from controlled trials assessing brain function in humans. In the following section the evidence, where it exists, is reviewed with regard to the flavonoids and other polyphenols that have attracted attention in terms of direct research into human brain function. Some of the flavonoid structures not represented in Figure 8.2 in Chapter 8 are presented in Figure 9.1.

Flavanols

Flavanols are among the most widespread polyphenols in plants and are found in more than half of the plant-based foods consumed by humans. High levels are found in fruits, grapes (and red wine), and leguminous seeds, with the richest sources of all being encountered in green tea and cocoa. Flavanols can be broadly divided into two groups: catechins, which are the monomer (single-molecule) forms, and proanthocyanidins, which are constructed from multiple catechins.[849] Proanthocyanidin dimers and trimers are at least as abundant as monomers in most foods, although it is more difficult to quantify the more complex polymers. These proanthocyanidins often

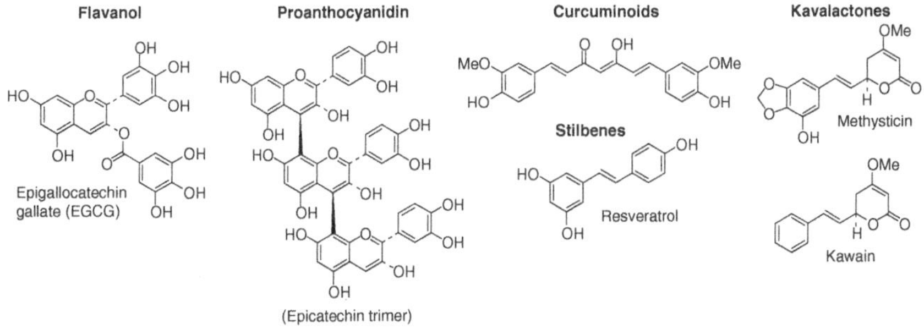

Figure 9.1 A few relevant phenolic structures not shown elsewhere.

provide the astringency of many foods and typically reduce in concentration during the ripening process.[849]

The catechins in general are highly bioavailable, being extensively absorbed and excreted,[850] although their metabolic fates are highly variable, and these ultimately depend on the molecule in question, inter-individual variability, food matrix, and dose.[851] Most are ultimately transformed into a range of potentially bioactive metabolites, typically by conjugation with glucuronic acid and/or sulfate groups. The major exception to this is EGCG, the most abundant catechin in tea, which is predominantly available in its original unmetabolized form in plasma after consumption.[852] The metabolism and absorption of the proanthocyanidins is less certain, as they are typically believed to be too large to be absorbed without major modification. Conflicting evidence suggests several possibilities: that they might be converted to smaller aromatic molecules including phenolic acids; that they may pass through the digestive system largely untouched; and that they may be degraded to monomers during transit through the digestive system and then absorbed.[850,852] As an example of the latter, the extended pharmacokinetic profile of catechin in human plasma over the 12 hours following the consumption of a pine bark extract (Pycnogenol™) composed mainly of proanthocyanidins suggested that the oligomers were being degraded to monomers and absorbed over time.[853] A recent study also confirmed several previous observations suggesting that a small percentage of proanthocyanidin oligomers may be absorbed and detected in their free state in plasma.[851] To a certain extent the questions over the bioavailability of proanthocyanidins may not be particularly relevant, as food sources always contain 5% to 25% monomers[852] and these may underlie any bioactivity.

Looking at individual sources of flavanols, cocoa primarily contains (−)-epicatechin, which typically accounts for 35% of the total phenolic content, and (+)-catechin,[a] along with their respective proanthocyanidins. Cocoa also contains flavonols, primarily quercetin.[854] Green tea, on the other hand, predominantly contains high levels of EGCG, gallocatechin, and epigallocatechin, all of which are very stable at high temperatures. Black tea then contains the same flavanols, but they are oxidized during fermentation to form dimeric theaflavins and polymeric thearubigins.[849] One final source of flavanols that will appear below are the commercial extracts made from the bark of pine trees, in particular *Pinus pinaster* and *Pinus radiata,* which provide extracts known as Pycnogenol™ and Enzogenol™ respectively. Both extracts are particularly rich in proanthocyanidins and catechins, along with much lower quantities of other flavonoids and phenolic acids.

The interest in the potential for flavanols, and indeed flavonoids, to mediate cardiovascular disease was originally piqued by the observation that the island-dwelling Panamanian Kuna Indians did not suffer from the typical rise in blood pressure and hypertension seen in aging human populations. However, once they relocated to the mainland their cardiovascular risk reverted to that of the mainland population. Eventually this effect was pinned down to a high consumption of flavanols in the form of cocoa, which represents the main drink of the Kuna and is incorporated into many foods.[855] As noted above, and in the previous chapter, epidemiological evidence has demonstrated a clear relationship between the dietary consumption of flavanols and cardiovascular risk factors.[755–757,758] These beneficial effects have also been captured

a.Catechin and epicatechin can exist in two stereoisomeric forms, + and −. However, they are typically seen in the form (−)-epicatechin and (+)-catechin, although they will be referred to without the +/−.

by controlled intervention studies showing that both acute and chronic administration of cocoa-flavanols can consistently improve cardiovascular parameters, including inflammatory biomarkers, blood pressure, insulin resistance, and endothelial function.[757,761,762,832] In contrast, the literature describing the effects of green tea and tea catechin extracts on cardiovascular parameters is smaller, methodologically heterogeneous, and of inadequate quantity and quality to arrive at any firm conclusion.[856] Likewise, Pycnogenol has attracted comparatively little attention in this respect, and therefore firm evidence is lacking for efficacy with regard to any indication.[857]

Naturally, peripheral blood pressure, blood flow, and insulin resistance, a condition that represents a failure of the body's normal glucoregulatory system, are inextricably linked to blood flow and metabolism in the brain. Consequently they are also inextricably linked to age-related cognitive decline, dementia, and mood disorders.[858–862] It is not surprising, then, that in controlled trials cocoa-flavanols have been shown to improve cerebral blood flow. This effect has been seen in healthy elderly adults using transcranial Doppler following 2 weeks' administration of 900 mg flavanols.[863] An increase in cerebral blood flow during task performance, as measured by fMRI, was also seen in healthy young adults following five days' administration of 150 mg flavanols.[864] In contrast, the lower of two doses of EGCG attenuated the task-related increase in blood flow in the frontal cortex of healthy young adults during demanding cognitive tasks, as measured by near-infrared spectroscopy (NIRS).[865]

In terms of direct modulation of cognitive function, a number of randomized, placebo-controlled studies have assessed both the acute and chronic effects of cocoa-flavanols. With regard to single doses, Field et al.[866] used a crossover design and assessed the effects of single doses of dark chocolate containing 720 mg flavanols, in comparison to a low-flavanol white chocolate control. They demonstrated improved cognitive performance on a spatial memory task and the predictable responses during a choice reaction time task following the high-flavanol chocolate. They also ran two visual tests and found that the consumption of the high-flavanol treatment increased the speed at which the participants detected the direction of motion of stimuli and led to an improvement in visual contrast sensitivity, which the authors attributed to a potential improvement in retinal blood flow. A further study,[867] also using a crossover design, assessed the effects of single doses of drinks containing either high (990 mg), medium (520 mg), or low (45 mg) levels of cocoa-flavanols during an hour of cognitively demanding task performance, which commenced 90 minutes after consumption of the treatments. The results showed that following both the high- and medium-flavanol treatments participants performed better than the control treatment on one of two demanding mental arithmetic tasks, while the high dose led to improved speed of performing a focused attention task but more errors on another mental arithmetic task. Following the medium dose of flavanols participants also reported consistently lower subjective mental fatigue as a consequence of task performance.

One of several studies that assessed chronic administration of cocoa-flavanols nicely illustrated the relationship between cardiovascular function, insulin resistance, cognitive function, and flavonoid consumption.[868] In this study, 90 adults suffering mild age-related cognitive impairment consumed daily drinks containing high (990 mg), medium (520 mg), or low (45 mg) doses of cocoa-flavanols for 8 weeks. By the end of the study period, in comparison to the low-flavanol group, participants in the medium- and high-flavanol groups had significantly reduced insulin resistance, blood pressure, and lipid peroxidation. Cognitive function was also improved in terms of performance of two of the three "paper and pencil" cognitive tasks that were used: a task (trail making) that measures visual attention and a verbal fluency task that measures executive functioning. Interestingly, the extent to which insulin resistance was improved correlated with improved performance on the cognitive tasks. In contrast, a previous study had assessed the effects of dark chocolate and cocoa on similar parameters and found no significant effects, although the active treatment was readily identified by the participants, potentially skewing the results.[869] Finally, a recently reported study also assessed the effects of both single doses and 30 days' administration of 500 mg and 250 mg of cocoa-flavanols and a low-flavanol control condition in healthy middle-aged participants using a comprehensive computerized cognitive assessment. There were no beneficial effects on cognitive function, although the subjective psychological state of the participants was improved in terms of "calmness" and "contentedness".[870]

To date the effects of catechins from tea on brain function have received very little research interest. Aside from the study that demonstrated a modified cerebral blood flow response to task performance noted above,[865] a single electroencephalography (EEG) study demonstrated that a single dose of 300 mg EGCG modified cerebro-electrical activity in healthy young participants, with a concomitant increase in self-ratings of calmness and reduced subjective stress. Any changes in cognitive function were not reported.[867] Two studies have also assessed the effects of green tea extracts and demonstrated some evidence, following a single dose of green tea, of increased brain activation (or cerebral blood flow, depending on your interpretation) during a working memory task, although this effect was restricted to the prefrontal cortex and was not accompanied by improved performance.[871] Finally, the effects of 16 weeks' administration of green tea extract was assessed in sufferers from mild cognitive impairment. In this study cognitive function improved marginally in a small subsection of the sample, with a modest increase in cerebro-electrical activity (EEG) restricted to one waveband.[872] Naturally the findings from these last two studies are particularly difficult to interpret as the treatments would also have contained the amino acid L-theanine, which has been shown to exert independent effects on brain function, along with other potentially bioactive components.

The effects of the proanthocyanidin-rich pine bark extracts Pycnogenol and Enzogenol have also been assessed in several studies. Two of these placebo-controlled trials concentrated on climacteric symptoms in menopausal females. They demonstrated attenuated sleep problems and improved subjective climacteric symptoms in 154 women receiving 60 mg Pycnogenol per day for 3 months[873] and improved subjective climacteric symptoms, accompanied by improved antioxidant and cholesterol status, in 101 women receiving 200 mg Pycnogenol for 6 months.[874] Pycnogenol has also been the subject of a single study assessing cognitive function and mood. In this instance 101 elderly participants (60 to 85 years) received Pycnogenol or placebo for three months and underwent a comprehensive cognitive and biological parameter assessment after 1, 2, and 3 months. The results demonstrated cognitive improvements that were restricted to working memory, along with improvements in markers of oxidative stress.[875] If anything, the other pine bark extract, Enzogenol, fares slightly worse in terms of brain function. An initial study[876] investigated the effects of 5 weeks' supplementation with 960 mg Enzogenol and placebo in a comparatively small sample of 42 healthy middle-aged men using a comprehensive battery of cognitive tasks, and incorporating cardiovascular and hematological parameters. There were no significant differences between conditions on any measure. Similarly, a recent study in a group of 60 adults who were suffering from cognitive deficits related to a mild traumatic brain injury sustained in the 12 months prior to taking part also showed that while treatment with 1,000 mg/day Enzogenol for 6 weeks resulted in fewer reported cognitive failures on a subjective questionnaire measure, there were no treatment-related differences on any of the objective cognitive tasks included in the assessment.[877]

Taken together, the above research seems to offer some mild, preliminary support to the notion that cocoa-flavanols may improve brain function. However, the evidence with regard to the other sources of flavanols is less promising, although all of these flavanol-containing interventions would benefit from more methodologically adequate research in this domain.

Anthocyanins

Anthocyanins, the glycosides of anthocyanidins, are found throughout plant tissue. However, their most notable role is providing the vibrant color to flowers and fruits. As such they are typically consumed by humans in brightly colored berries, grapes, and fruits, including eggplant (aubergine), and vegetables such as red cabbage. They are unusual among polyphenols in that they tend to exist at fairly high concentrations; for instance, a serving of 200 g colored plant food such as eggplant or black grapes can contain 1,500 mg anthocyanins.[852] Unlike the other polyphenols, the parent glycosides do not appear to undergo extensive metabolism. Evidence does, however, suggest that anthocyanins can cross the blood–brain barrier, and they have been found to accumulate in several of the key areas of the mammalian brain underpinning cognitive function.[878,879] Anthocyanin-rich foods have also consistently been shown to improve cognitive function in animal models of aging and neuropathology.[880]

Human research into the effects of anthocyanins on brain function has used two potential sources: berry juices that contained high levels of anthocyanins and phenolic acids (for instance, 877 mg and 734 mg per liter, respectively, in the juice used in the study by Krikorian et al.[881]) and Concord grape juice, which has a phenolic content comprising nearly 50% anthocyanins with smaller quantities of phenolic acids (~30%) and flavanols (~10%), plus minor constituents including flavonols and resveratrol.[882] However, the promising animal research has, by and large, failed to translate into human trials of anything approaching adequate size. For instance, Krikorian et al.[881,883] conducted a pair of studies in very small samples of elderly adults who were suffering from age-associated mild cognitive impairment. In one study,[883] a small sample of 12 participants received either ~500 mL Concord grape juice per day or a matched placebo drink for 12 weeks in a randomized, double-blind manner. In the other study by the same researchers,[881] a further nine participants were recruited and all received ~500 mL blueberry juice per day for the same period, using the placebo group from the Concord grape juice study as a comparator. The studies used the same two tasks. In the Concord grape juice study participants allocated to the grape juice condition performed better than placebo on one measure derived from the California Verbal Learning Task (CVLT); in the blueberry study the participants performed better on the other task (Verbal Paired Associates Learning). However, the sample size, design, lack of blinding, and unusual use of the placebo data from one study as a comparator condition for another study render the results largely uninterpretable. In a further study, Krikorian et al.[884] went on to assess the effects of 16 weeks' consumption of ~500 mL

Concord grape juice in another small group of participants, this time 21 elderly adults suffering from age-associated cognitive impairment. Once again there was a significant improvement seen on one measure from the CVLT, but no effects were seen in terms of mood, blood pressure, or glucoregulatory parameters. A small subsection of participants (four per group) also underwent fMRI during a working memory task, and an increase in activation was reported in two out of 12 "regions of interest" that are implicated in working memory. In contrast, a crossover study that used a much healthier sample of 36 younger adults, and a more powerful crossover design, reported no effects of a single dose of Concord grape juice (10 mL/kg bodyweight) on mood, an implicit memory task, or appetite and food consumption.[885] Taken together, the positive results from several very small studies, one of which was unblinded, and the negative results from one larger study, don't as yet lend any real support to the use of anthocyanins for improving brain function in humans.

Isoflavones (and Other) Phytoestrogens

Plants synthesize a range of flavonoid and polyphenol compounds that, among their various signaling pathway effects (as outlined above), have estrogenic or anti-estrogenic properties. These include the isoflavones genistein and daidzein and the structurally related coumestans, such as coumestrol, all of which are particularly richly expressed in the legume (*Fabaceae*) family. Other phytoestrogenic flavonoids include, in order of descending potency, a number of ubiquitous flavones, including kaempferol, quercetin, apigenin, and luteolin, and several flavanones, including naringenin.[886] The structurally related stilbenes also contribute several phytoestrogens, including resveratrol. The final group are members of the lignans, which are found in grains, fruits, and vegetables. This class of compounds are not notably estrogenic in themselves, but they become so after they have been converted by intestinal bacteria to biologically active estrogenic metabolites such as enterodiol, enterolactone, and secoisolariciresinol.[887]

All of these compounds owe their estrogenic properties to their direct binding affinity at the nuclear estrogen receptors α and β (ERα, ERβ), whose endogenous ligands are the mammalian estrogen sex hormones, and in particular 17β-estradiol. The estrogens (also including estrone and estriol) play a specific role in female reproduction, including regulating the menstrual cycle. Alongside this they play a general role in the mitosis, growth, and function of a variety of tissues across the sexes. The two receptor types are differentially expressed throughout tissues, with ERα expressed at higher concentrations in the liver, kidney, heart, and uterus and ERβ in the ovary and prostate, gastrointestinal tract, lung, bladder, and central nervous system. Where they are co-expressed, for instance in some brain regions and the thyroid gland, prostate, mammary, adrenal, and bone tissue, ERβ inhibits the gene expression associated with ERα.[888]

The phytoestrogens typically exhibit an affinity that is at least 1,000 times less than the receptors' primary endogenous ligand, 17β-estradiol. This weak binding is typically compensated for by the comparatively high concentrations of the phytoestrogens.[889] In many cases these compounds also show a marked preference for ERβ as opposed to ERα receptor. They can also act as agonists or partial agonists, or alternatively they function as partial antagonists, as their presence on the receptor blocks the activity of the more potent endogenous ligands. They can also bind to and inhibit the activity of androgen- or estrogen-metabolizing enzymes such as aromatase[887] and exert non-receptor binding properties, potentially via modulation of signaling cascades related to estrogenic functioning.[890] This multiplicity of effects means that the pro-estrogenic or anti-estrogenic activity of the phytoestrogen can be highly variable and can depend on a number of factors, including tissue type, receptor subtype, dose, and the amount of the endogenous ligand present.[887] This complex interplay of factors and mechanisms of action explains the potential utility of phytochemicals such as genistein both as pro-estrogens in osteoporosis and cardiovascular disease and as anti-estrogens in breast cancer cells.[890]

By and large, the ligand binding affinity of the various phenolic phytoestrogens is related to the extent to which they incorporate several structural features of 17β-estradiol. A small selection of phytoestrogens of varying potency and their structural similarity to 17β-estradiol are illustrated in Figure 9.2.

The soya bean plant (*Glycine max*) is the richest source of phytoestrogens and contains high concentrations of coumestrol, genistein, and daidzein. Given that soya is also an important food crop, the vast majority of human research in this area has concentrated on this source of phytoestrogens. While these tend to be described as "soy-isoflavones," these compounds are actually expressed in significant amounts by a wide range of plants, including the majority of the legume (*Fabaceae*) family, and other food crops, such as the *Caffea* genus that gives us coffee.

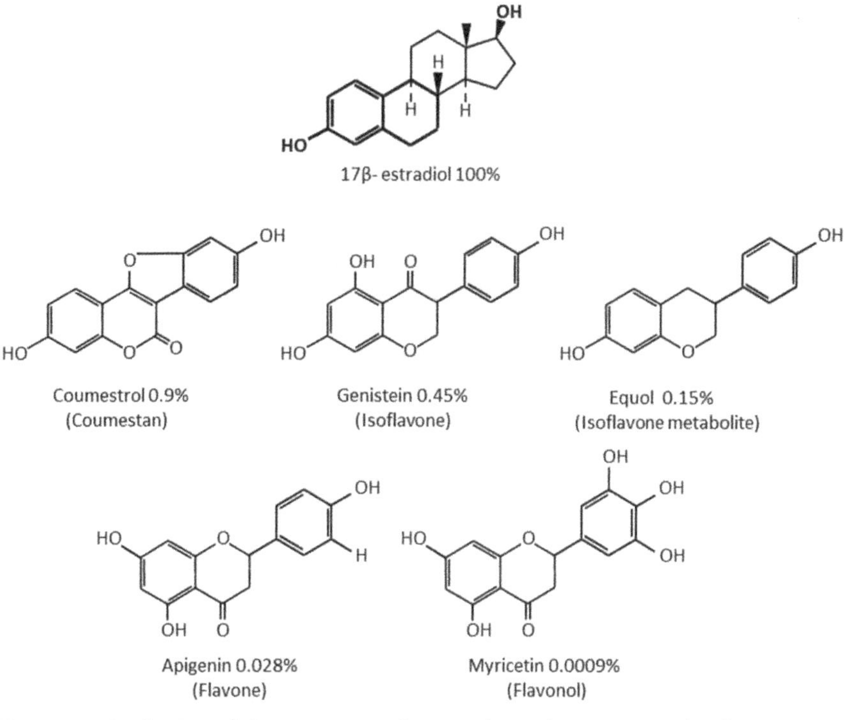

17β- estradiol 100%

Coumestrol 0.9% (Coumestan)

Genistein 0.45% (Isoflavone)

Equol 0.15% (Isoflavone metabolite)

Apigenin 0.028% (Flavone)

Myricetin 0.0009% (Flavonol)

Figure 9.2 A selection of phytoestrogens, showing their relative receptor binding potency (%) in comparison to 17β-estradiol.[886] The potency of the individual molecule is dictated by its resemblance to the structure of 17β-estradiol, in particular the possession of the two rings and the comparative position of the two hydroxyl groups depicted in bold.[810,886]

Daidzein itself can be metabolized by intestinal bacteria into the more potent estrogenic compound equol, although only 30% to 50% of individuals are equipped with the requisite intestinal flora to accomplish this, with the percentage rising in cultures that eat more daidzein-containing plant products. Whether an individual is an "equol producer" or not may markedly modify his or her sensitivity to isoflavones, as equol is more bioavailable, and has greater affinity for estrogen receptors than daidzein. Like genistein, equol also binds to the androgen dihydrotestosterone, preventing it from activating androgen receptors, potentially leading, among other things, to beneficial effects on the prostate gland. Equol may well underpin many of the health benefits of soy isoflavones, and inadequate definition of equol producers might be one factor obscuring the effects of these phytoestrogens in clinical research.[891]

Isoflavones from soya have been shown to exert a number of effects relevant to general health, including estrogenic modulation of cardiovascular and immune function parameters and modulation of the mechanisms underlying carcinogenesis, including the inhibition of aberrant mitogenic activity related to ERα activity. They also upregulate endogenous antioxidant activity.[892] These factors may underlie the tentative epidemiological observations of a relationship between soy-isoflavone consumption and breast cancer.[893] As an example, adolescent phytoestrogen intake and the consumption of soy-isoflavone supplements have been shown to be associated with a reduced incidence of postmenopausal breast cancer.[894,895] This appears to be particularly the case with regard to the large proportion of breast cancer cases that feature the overexpression of estrogen and progesterone receptors within tumor cells.[894] A recent meta-analyses of the sizeable literature describing prospective studies involving soy-isoflavone supplementation also confirmed a reduced risk of breast cancer incidence in Asian but not Western populations.[896] Soy-isoflavone supplementation has also been associated with decreased bodyweight and improved glucoregulation in postmenopausal non-Asian women,[897] decreased menopausal symptoms such as hot flushes,[898] and improved vascular parameters such as endothelial function[899] and blood pressure.[900]

With regard to the effects of soy-isoflavones on brain function, evidence from animal models suggests that the beneficial effects of 17β-estradiol on cognition may be related to enhanced cholinergic function and hippocampal plasticity,[901] and this is borne out by demonstrations of improved memory performance following soy-isoflavone consumption in intact[902] and ovariectomized rodents.[903] Similarly, memory was improved in middle-aged or older rats as a consequence of supplementation with both genistein[904] and a range of soy-isoflavones.[905] In humans, supplementation with soy-isoflavones in females has been shown to significantly improve the physical but not psychological symptoms of premenstrual syndrome in comparison to placebo,[906] and in postmenopausal women isoflavones improve ratings of quality of life,[907] decrease both follicle-stimulating hormone and luteinizing hormone levels, and increase circulating levels of estradiol.[908] With regard to neurocognitive function and mood, a recent review by Lamport et al.[909] identified 13 methodologically adequate studies that had assessed the effects of isoflavone treatments on cognitive function. Seven of these had demonstrated modest beneficial effects of supplementation. The authors rightly interpreted the findings as being inconclusive, although it is interesting to note that most of the studies involved postmenopausal women or groups of older women (who by implication should have been well past the menopause).[909] Two recent studies also seem to confirm the lack of clarity in the literature. In a cross-sectional study involving a cohort of women undergoing the menopause, Greendale et al.[910] found that dietary isoflavone consumption was related to either cognitive benefits or decrements, depending on the stage of menopause, while lignan consumption was associated with benefits, but only during the early stages during which the menopause develops. Coumestrol had no relationship to cognition either way. These results suggested that the cognitive effects of phytoestrogens were dependent on a number of variables, including the specific group to which the phytoestrogens belong and the phase of menopause.[910] In a similar vein, 2.5 years of supplementation with soy-isoflavones in postmenopausal women had no effect on a global measure of cognition, although benefits were evident on a "visual memory" sub-factor derived from the cognitive data.[911] These varied results are interesting because recent research has shown that, several years following ovariectomy, primates lose the ability to respond to either 17β-estradiol or equol in terms of hippocampal receptor binding and activity.[912] This would seem to confirm a theorized "window of opportunity" for hormone replacement therapy and suggests that a corresponding loss of responsiveness to phytoestrogens in the postmenopausal women involved in most of these studies may be a factor in the equivocal nature of the results. Interestingly, in the review by Lamport et al. three of the four studies that included men in their samples demonstrated cognitive benefits.[909] These studies included an intriguing investigation in which soy-isoflavone supplementation for 6 weeks in men was associated with selective improvement on a spatial working memory task in which women usually outperform men.[913] Evidence actually suggests both cognitive benefits and deficits in male rats following phytoestrogen administration,[902] so it would be interesting to see more data from male humans, and indeed younger premenopausal women. All told, to date, the promise of phytoestrogens as cognition enhancers has yet to be realized.

■ NON-FLAVONOID POLYPHENOLS

While the flavonoids tend to be expressed by a wide range of plants, and therefore feature as a ubiquitous component of the typical human diet, a range of structurally related phytochemicals are synthesized by more restricted ranges of plants. Several of those that have attracted research interest with regard to their potential psychotropic effects are described below.

Resveratrol

The phytoalexin resveratrol (3, 4,′ 5 trihydroxystilbene) is a produced within a range of plants with edible fruits and nuts in response to tissue damage and environmental stressors such as fungal/viral attack.[914,915] While it is a stilbene rather than a flavonoid, the two groups are closely related and derived from the same cinnamic acid and malonyl-CoA unit (X 3) precursors (see Fig. 8.2).

The consumption of resveratrol is associated with numerous protective health benefits in animal models. These include increased longevity,[916] anti-inflammatory[917] and antiviral properties,[918] and protection against cancer and tumorigenesis,[919] CVD,[920] and atherosclerosis.[921] With regard to cognitive performance, a number of in vivo studies in rodents have demonstrated preserved behavior and cognitive performance in aged rats[922] and following laboratory-induced brain insults.[923–926]

As with the flavonoids, these effects are potentially mediated by a plethora of cellular signaling effects. These include interactions with receptors and the modulation of multiple aspects of a number of signaling cascades. For instance, resveratrol's anti-inflammatory properties, which include a reduction in pro-inflammatory cytokines and mediators and a specific inhibition of COX2 expression and prostaglandin synthesis, are related to modulation of MAPK and protein kinase signaling cascades. Its modulation of antioxidant status is related to upregulation of the Nrf2 transcription factor (along with modulation of sirtuins) that dictates antioxidant enzyme expression, and improved endothelial function and cardiovascular parameters are related to modulation of MAPK (ERK) signaling pathways and the resultant modulation of eNOS and nitric oxide levels.[927]

The literature surrounding resveratrol is substantial and expanding, comprising more than 20,000 papers at the time of writing. However, this huge research effort has typically concentrated on in vitro/ex vivo studies that often have unknown relevance to humans due to questions over resveratrol's bioavailability and metabolism profile following oral consumption. Unfortunately, while the literature also includes many animal studies, the huge research effort lavished in laboratory models has not translated into a great deal of research involving humans.[836]

However, the few examples of studies in humans reported to date include a study demonstrating that oral administration of 40 mg resveratrol for 6 weeks resulted in reductions in oxidative stress parameters, reduced expression of pro-inflammatory kinases (including JNK) in mononuclear cells, inhibited activity of the pro-inflammatory transcription factor NF-κB, and reduced levels of related downstream pro-inflammatory cytokines.[928] In a subsequent study, the oxidative and inflammatory effects of a high-fat meal were offset by co-administration of resveratrol, with the beneficial effects including increased Nrf-2 activity and consequent upregulation of endogenous antioxidant parameters.[929] In contrast, 12 weeks' administration of 75 mg resveratrol a day to postmenopausal women failed to affect metabolic rate or inflammatory markers or to modulate molecular targets such as SIRT1.[930] While red wine has been shown to improve endothelial function in several human studies, only one study has directly investigated the effects of resveratrol on vascular function. Wong et al.[931] found that all three doses of resveratrol (30, 90, 270 mg) that they administered to obese individuals improved flow-mediated dilatation, with a modest relationship observed between plasma levels of resveratrol and the endothelial benefit. In terms of brain function, a single paper has been published describing a crossover study in which 24 healthy young participants were assessed on three separate occasions after taking either a placebo or doses of 250 mg and 500 mg resveratrol. At each assessment cerebral blood flow was monitored in the prefrontal cortex using NIRS during a 45-minute absorption period and then during 45 minutes of performing cognitively demanding mental arithmetic and attention tasks. These tasks have previously been shown to activate the frontal cortex. While there was no evidence of improved cognitive function, resveratrol did increase the cerebral blood flow response to task performance in a dose-dependent manner, and levels of deoxygenated hemoglobin, reflective of oxygen utilization, also increased.[932] The increases in cerebral blood flow during task performance following single doses of resveratrol, but the lack of any benefits to cognitive function following acute or chronic (4 weeks) dosage, have been replicated in two further studies that are either currently submitted to journals or in press.

Curcumin

Curcumin, a curcuminoid polyphenol responsible for the bright-yellow color of the Indian spice turmeric (*Curcuma longa*), has been used for centuries within the Ayurvedic system of medicine in the treatment of a whole host of ailments, including inflammation.[933] Its structure is formed from one cinnamic acid starter unit with two malonyl-CoA units, as opposed to the three incorporated into the flavonoids.[195] However, as with the flavonoids, the bioactive properties of curcumin in mammals can be attributed to interactions with multiple components of a variety of intercellular signaling cascades, including kinases and MAP kinases, transcription factors (e.g., NF-κB, AP-1, Nrf-2), and their multiple downstream products such as apoptotic proteins, cytokines, and other pro-inflammatory mediators (e.g., TNF-α, IL-1β, IL-6, COX2/prostaglandins), and endogenous antioxidant enzymes.[934,935]

To date over 100 clinical trials, usually of modest size, either have been completed or are under way, and curcumin has demonstrated potential efficacy as a monotreatment or adjunct treatment in treating several cancers, at least in terms of attenuating molecular etiological factors and markers of the disease. Curcumin has also shown promise in terms of attenuating the inflammation associated with a number of diseases. Likewise, early, if limited, evidence suggests potential benefits in CVD and diabetes. The most consistent finding is that curcumin is safe, with no significant side effects, although its bioavailability has been questioned.[935]

With regard to the brain, curcumin is associated with a raft of neuroprotective properties in animal models of neurodegenerative diseases,[936–938] and a recent study demonstrated antidepressant effects in mice as a result of MAP kinase (ERK) activation and a downstream increase in BDNF.[939] Curcumin is also associated with an attenuation of cognitive deficits in rodent models of Alzheimer's disease.[940,941] However, these promising findings in animals, and epidemiological data showing better cognitive performance related to curry consumption in humans,[942] have not translated into research focused on human brain function. To date the results of only two small studies in humans, both assessing the effects of curcumin in Alzheimer's sufferers, have been reported, and neither demonstrated symptomatic or biochemical efficacy.[943,944]

■ PHENOLIC ACIDS

To date the effects of simpler phenolics have received very little research attention with regard to brain function in humans. A single crossover study involving a sample of 39 older adults compared decaffeinated coffees with high and normal levels of chlorogenic acid on a number of cognitive tasks, mood measures, and EEG and found that the high chlorogenic acid treatment improved ratings on only one of two measures of alertness, with no interpretable differences on any other measure.[945] Similarly, one study assessed the effects of 6 weeks' administration of cranberry juice in 50 healthy adults over 60 years of age using a more comprehensive battery of cognitive tasks and mood measures and reported no significant differences between groups.[946] While this study failed to report the phenolic contents of its cranberry juice, previous analyses suggest that the phenolic contents are predominantly benzoic acid and derivatives.[947]

ATYPICAL PHENOLICS

Whereas the above phenolics are theorized to primarily owe their potential effects on brain function to what could be characterized as a fairly subtle modulation of intra- and inter-cellular signaling pathways, several plant extracts containing atypical, species-specific phenolics have more pronounced effects on nervous system parameters. The two that have attracted substantial human research attention are extracts from *Piper methysticum* (kava) and *Hypericum perforatum* (St. John's wort). The latter is included here on the basis that its many potentially bioactive constituents include a range of phenolics and a number of phloroglucinols, which can be derived via the oxidation of phenolic compounds.

Kavalactones (from *Piper methysticum*)

The South Pacific clade of the *Piper* genus (*Piperaceae* family) of wet forest lianas, shrubs, and herbs includes a number of kavalactone-producing species. The best known of these is *Piper methysticum*, also known as kava.[948] Drinks made from the rhizome and roots of kava have been used for millennia in the South Pacific as a part of important social occasions. It is also consumed as a recreational drink and medicinal treatment that reduces fatigue and induces relaxation and sleep. The effects of social kava consumption are reported to include muscle relaxant properties at low doses, with intoxication, sedation, analgesia, ataxia, and paralysis of the extremities as the dose ascends.[949] Kava is more commonly consumed today in the form of extracts, which are typically taken for their purported effects on anxiety, agitation, and insomnia,[195] although the sale of these supplements was restricted in many countries in the early 2000s following a number of cases of hepatotoxicity.

The major bioactive constituents of kava roots and rhizomes are a group of styrylpyrones generally referred to as kavalactones. These comprise the same cinnamic acid starter unit that contributes to the flavonoids but with only one additional malonyl-CoA unit incorporated in the structure.[195] The most prevalent kavalactones in the root and rhizomes are kavain and methysticin, with yangonin, dihydrokavain, demethoxyyangonin, and dihydromethysticin increasing in prevalence with ascent through the plant. The leaves and stem contain comparatively low levels of kavalactones but do contain potentially hepatotoxic alkaloids. Contemporary commercial kava extracts are generally standardized to contain approximately 30% to 70% of kavalactones.

The exact mechanisms underlying the central nervous system effects of kavalactones are poorly delineated to date. However, kavalactones have been shown, at the most fundamental level, to modulate neuronal excitability via inhibition of voltage-dependent calcium and sodium channels.[949,950] They also indirectly modulate activity in the serotonergic, glutamatergic, and dopaminergic neurotransmitter systems[949] and may inhibit the uptake of noradrenaline.[951] In addition, kava's anxiolytic properties have been attributed to interactions with NMDA, reversible inhibition of monoamine oxidase B (MAO-B), and modulation of GABA neurotransmission; however, kavalactones do not appear to bind to GABA receptors but may modulate the binding of other ligands.[950] Modulation of β-adrenergic activity may also play a role in the kavalactones' behavioral effects,[952] and individual kavalactones inhibit COX1 and COX2.[953] Recent research has also shown that yangonin is a ligand at cannabinoid CB1 receptors, albeit with over 100 times less potency than the cannabinoid tetrahydrocannabinol.

However, this property may be the upstream source of some of the other effects of kava on neurotransmission. In general, the serum concentrations of yangonin necessary for relevant cannabinoid receptor binding would be achievable in a traditional kava-drinking session but would be unlikely to be attained following the lower doses associated with supplementation.[954]

With regard to kava's anxiolytic effects, a 2003 Cochrane review[955] included a meta-analysis of the randomized controlled trials extant at that time. Twelve studies met their inclusion criteria, and the researchers concluded that, in comparison to placebo, kava extracts were associated with a significant reduction in anxiety as measured by the Hamilton Anxiety (HAM-A) scale. A subsequent meta-analysis of data from six studies that had used the same acetonic extract again found a significant treatment-related anxiolytic effect in comparison to placebo.[956] In the most recent controlled trial, and the only trial to date that assessed a theoretically safer aqueous extract (of a noble cultivar), 1 week's administration of a kava extract containing 250 mg kavalactones led to significantly reduced anxiety and depression in sufferers from generalized anxiety.[957] This was a potentially important study because kava had been mired in controversy for several years since the publication of a series of case reports of treatment-related liver failure, which were variously attributed to the use of ethanol/acetone extracts and the presence of aerial plant parts containing the alkaloid pipermethysticine. These reports led to the withdrawal of kava by many national authorities, and one recommendation from the World Health Organization was the adoption of theoretically safer aqueous extracts.

With regard to behavior, a small number of studies have assessed the cognitive effects of kava; the results are largely inconclusive, but this may be due to the heterogeneity and methodological limitations of the studies.[949,958] In a recent study assessing the effects of single doses of kava extract (180 mg kavalactones) on cognitive function, psychological state, and anxiety, kava had no significant effects, whereas 30 mg oxazepam had anxiolytic effects and reduced subjective alertness.[959]

The kavalactones are a structurally atypical group of phenolics, so it would be advantageous to quickly look at their ecological roles at this point. Kavalactones would seem to primarily have ecological functions related to the rhizosphere. They are most concentrated in the skin of the root and then decrease in concentration within the plant with increasing distance from the root tips.[960] They are also water-soluble, allowing distribution into the rhizosphere, where they have nematicidal properties[961] and inhibit the growth of pathogenic fungi[962] and competitor plants.[961,962] With regard to the latter, individual kavalactones reduce the germination, shoot length, and root length of competitor plants.[963] It is also notable that kavalactones are also synthesized by other species, including passion fruit (*Passiflora edulis*), which tend to have bare earth or reduced weed coverage below their canopy.[201]

Hypericum perforatum (St. John's Wort)

Extracts of *Hypericum perforatum* have been in recorded medicinal use from the time of the ancient Greeks.[964] *H. perforatum* extracts contain a wide variety of potentially bioactive constituents, including phenolic acids (e.g., chlorogenic acid) and a wide range of flavonoids (quercetin, quercitrin, isoquercitrin, rutin, hyperoside, epigenanin); structurally related phloroglucinol derivates such as hyperforin, which can be derived by oxidation of phenolic compounds[965]; and naphthodianthrones such as hypericin. The antidepressant and anti-inflammatory effects of *H. perforatum* were initially attributed to the naphthodianthrones[966] and more recently to hyperforin[967,968] and the range of flavonoid constituents.[969] It is now widely accepted that the various potential actives act synergistically.[970,971]

Beyond its pronounced anti-inflammatory and antibiotic properties,[968] *H. perforatum* also exerts a number of effects directly relevant to mammalian brain function, including inhibition of the neuronal reuptake of serotonin, dopamine, norepinephrine, GABA, and glutamate and increased neurotransmitter sensitivity and receptor binding.[972] Functional effects include neuroprotective properties, an attenuation of cognitive impairment, and improved cognitive performance in rodents.[973]

H. perforatum extracts have attracted some of the most methodologically rigorous research of any herbal products. As a consequence, they have been shown to have well-established, beneficial effects on brain function. In humans the vast majority of research has focused on the antidepressant effects of *H. perforatum* extracts. In this domain a number of reviews and meta-analyses have confirmed the efficacy of *H. perforatum* in the treatment of mild to moderate depression.[974] The most recent Cochrane review by Linde et al.[975] included a total of 29 methodologically adequate controlled trials involving a total of 5,489 depression sufferers. The studies compared *H. perforatum* extracts to placebo and/or the standard, prescribed antidepressants. In common with all previous reviews on this topic, the authors concluded that *H. perforatum* extracts seem to be more effective than placebo and as effective as standard antidepressants for treating major depression. *H. perforatum* also engendered significantly fewer side effects than synthetic antidepressants.

With regard to ecological roles, *H. perforatum* is generally considered an invasive weed outside of its native European range, and away from home its success can be partly attributed to reduced tolerance in local insects to its range of secondary metabolites.[122] Hypericin and related naphthodianthrones may well be the primary deterrents to insects, due to their potent phototoxicity. While generalist herbivores may eat the plant's tissue they avoid the plant parts with the highest levels of hypericin, whereas specialist herbivores have adapted by avoiding sunlight after consuming hypericin-containing tissue.[976]

■ INTERVENTION STUDIES WITH PHENOLICS: SOME CONCLUSIONS

Leaving aside the atypical phenolics found in *P. methysticum* and *H. perforatum*, it is noticeable that the promise suggested for the phenolics, and in particular the flavonoids and other polyphenols, by epidemiological evidence suggesting relationships between their consumption and a raft of benefits to health (e.g., cardiovascular disease/function, cognitive function, incidence or severity of dementia and strokes) has not generally translated into evidence from controlled intervention trials demonstrating benefits to brain function in humans. Even the cocoa-flavanols, which have been shown to reliably improve multiple aspects of human cardiovascular function in controlled trials, have noticeably failed to garner the same level of evidence with regard to any effects within the human nervous system. Overall, the evidence can best be described as promising; there is evidence of modulated cerebro-electrical activity and cerebral blood flow as a consequence of taking several of these compounds or groups of compounds. Cerebral blood flow, in particular, is a key physiological factor underpinning all aspects of brain function. It is intrinsically related to cardiovascular function, and it reduces in step with cognitive function as a consequence of aging and all of the neuropathologies. It would seem likely that reversing this trend in the long term must have an impact on brain health and function. Looking at the research from intervention trials summarized above, it is also notable that this entire area has simply failed to attract enough adequately sized trials to come to any firm conclusion about efficacy. There are a number of small trials that show benefits, but often as many that do not. This is likely to be a function of low sample sizes, short intervention periods, and heterogeneous methodology rather than any indication of the actual efficacy of the treatments in question. In general, the vast research effort in terms of basic laboratory-based science, and we are talking here about many tens of thousands of papers describing in vitro and in/ex vivo research related to the potential mechanisms of action of polyphenols alone, has not translated into human research, despite the fact that these compounds are typically very safe for human consumption. It may be time that this pattern was reversed.

■ THE ECOLOGICAL (OR PRIMARY) ROLES OF PHENOLICS, AND THE HUMAN BRAIN

The polyphenols are multitalented molecules that play a wide range of simultaneous roles in the plant; typically these include acting as sunscreens and antioxidants, as antifeedants in the face of herbivory, as allelopaths, and as antimicrobials. Emerging evidence also suggests that they act as signaling molecules in the plant in their own right.

I concluded the previous chapter with the argument that the potentially beneficial effects of polyphenols on health and brain function (for instance, their protective effect in the face of cancer and CVD and their neuroprotective, cerebrovascular, and synaptic plasticity-fostering effects in the brain) may be directly related to their "cross-kingdom" interactions within the cellular signaling pathways that are largely conserved in both plants and animals. These effects, in essence, may represent an unintended transfer of the polyphenol's endogenous cellular signaling role in the plant into the orthologous, or analogous, signaling pathways in the consuming animal. Alternatively, in some cases the effects may be predicated on an unintended transfer of the polyphenol's intended cross-kingdom signaling function into a different clade altogether—for instance, signaling by a flavonoid that is intended to manage the plant's relationship with symbiotic microbes transferring into the consuming animal. It might be worth briefly revisiting these propositions to add further meat to the bones of some specific but interrelated examples.

Unintentional Cross-Kingdom Signaling?

Revisiting an example given in the previous chapter illustrates how unintentional cross-kingdom signaling might have diverse effects on mammalian cellular functioning. The TOR signaling pathway and its afferent kinase signaling pathways such as PI-3K/PKB are conserved across all eukaryotes, including plants.[819] TOR plays a crucial role in all cells, simultaneously collecting internal and external cellular information on stressors and nutrient/energy availability and then directing growth, metabolism, and differentiation.[817,818] In mammals TOR signaling is aberrantly upregulated in cancer[977] and dysregulated in obesity and diabetes, contributing to insulin resistance and CVD. It also plays a key role in the process of cellular aging.[818] In mammals, the functioning of the TOR signaling pathway is also modulated by a wide range of flavonoids and other polyphenols. For instance, ubiquitous catechins such as EGCG and flavonols such as quercetin attenuate cell proliferation and tumorigenesis by inhibiting both PI-3K/PKB and TOR signaling in mammalian cells, potentially by binding to the kinase proteins' ATP binding sites.[978,979] Similarly, quercetin, curcumin, and resveratrol have all been shown to promote protective autophagy-mediated cell death in mammalian cancer cells via TOR inhibition.[980] Resveratrol may also have its beneficial effects on cellular senescence, cell growth, glucose homeostasis, and cardiovascular function through its inhibition of TOR signaling via both sirtuin-dependent and -independent mechanisms. In the latter case, established mechanisms include modulation by resveratrol of PI3-K/PKB signaling upstream of TOR as well as direct interactions with the TOR kinase itself.[981,982] Similarly, anthocyanin-rich berries protect the hippocampus in an irradiation model of aging by inhibiting TOR and attenuating inflammation, oxidative stress, and a loss of protective autophagy.[983] Conversely, anthocyanin and flavanol supplementation in rodents activates TOR, promoting "growth" in terms of synaptic plasticity in cognition-relevant brain regions.[823] This last finding suggests that flavonoids can exert biphasic, homeostatic effects within this single system.

In keeping with their nutrient-, energy-, and stress-sensing roles across eukaryote cells, TOR signaling pathways play a large part in regulating growth in plants, with the TOR gene expressed at high levels in rapidly growing and dividing tissues. As with mammalian cells, TOR also mediates cellular responses to stressors and directs autophagy, possibly via interactions with plant hormones such as abscisic acid.[984]Naturally, a role in stress regulation also encompasses the mediation of secondary metabolite synthesis. Therefore, given their emerging role as plant signaling molecules in their own right, it would make absolute sense for polyphenols to interact with the PI-3K/TOR pathways in plant cells, providing valuable information on the current level of secondary metabolite synthesis and stressors, in order to titrate the phytochemical's own synthesis and restrain or promote TOR's activity on, for instance, growth and autophagy. Any effect on mammalian TOR signaling would then most likely be a simple transfer of the endogenous role of the phenolic compounds within the plant TOR system to the orthologous system within the consuming animal.

As noted in the previous chapter, TOR is just one of many potential interactions by phenolic compounds with the conserved cellular signaling systems that dictate cellular responses to stressors and nutrients in both plants and animals.

Unintended Targets of Intentional Cross-Kingdom Signaling?

The Phytoestrogens

Three possible ecological roles have been proposed for polyphenolic phytoestrogens: protection against herbivores, the management of relationships with mutualist microbes, and protection against microbial pathogens.

Although intuitively pleasing, there is actually little evidence that these polyphenols, as a group, negatively affect the life course and mortality of insect herbivores via mechanisms related to their estrogenicity.[985] There is rather more evidence that phytoestrogens disturb vertebrate hormonal function. Extreme examples include uterotrophic responses in sheep to dietary isoflavones such as genistein and biochanin A and "clover disease," a condition featuring infertility, genital abnormalities, and reduced lambing rate, caused by the estrogenic isoflavone formononetin.[887] Birds also often encounter phytoestrogens in their diet, and the direct administration of estrogens and phytoestrogens can negatively affect a range of reproductive parameters. It has therefore been suggested that the phytoestrogen content of the diet of some bird species effectively titrates their reproductive success. For instance, during periods of environmental stress such as drought, phytoestrogen concentrations in some plants increase, and this may serve as a natural contraceptive to the bird population, reducing their overall consumption of the plant in question.[887] However, the opposite relationship has also been observed in red colobus monkeys, who eat more of the particularly phytoestrogenic young leaves of the leguminous *Millettia dura* plant during weeks of higher rainfall. Examination of their droppings shows that consumption of the leaves correlates with both estradiol and cortisol levels (via knock-on effects within the various hormonal systems) and that these hormone levels are, in turn, related to increased aggression and copulation and reduced grooming.[986] Unfortunately, reproductive success as a consequence of phytoestrogen consumption has not yet been reported in these monkeys.

In the absence of clear evidence of an anti-herbivore role for phytoestrogens, it seems more likely that any effects in mammals are related to the benign roles that these polyphenols play in the management of the mutualist microbial symbionts that colonize the roots of most plants.[158,159] As described in the previous chapter, this relationship is managed by three-way cross-kingdom signaling in which flavonoids excreted by plants and chemical "myc" and "nod" factors excreted by fungi and bacteria are detected by receptors within the other parties, thereby promoting the gene expression that fosters the mutualist relationships.[804] A key point is that both the bacterial and fungal receptors are "estrogen-like receptors" that function in much the same way as mammalian ERα and ERβ estrogen receptors in terms of the palette of chemicals and ligands that they recognize and respond to. They also exhibit the same ligand concentration-dependent activity, they co-occur with chaperone proteins in their unactivated state, and the nature of their gene transcription effects are similar.[809,810] These microbial receptors also bind mammalian estrogens such as 17β-estradiol.[810] Indeed, it has been suggested that many of the proteins within the respective signaling pathways are orthologous[812] and that the bacterial "estrogen-like receptor," known as the NodD protein, is a partial orthologue of the mammalian estrogen receptor[811] (although this is disputed).[987] Certainly the key structural elements that dictate the binding of phytoestrogens to mammalian estrogen receptors, for instance the aromatic ring with a hydroxyl group corresponding to the C3 position in the 17β-estradiol ring system,[810,886] also dictate the binding of the same compounds to microbial estrogen-like receptors.[810] It is also notable that the majority of all of the key flavonoids reported to exert estrogenic effects in mammals also take part in the plant/microbe estrogen-like receptor interactions described above. These include daidzein, genistein, coumestrol, myricetin, luteolin, quercetin, kaempferol, naringenin, apigenin, biochanin A, and chrysin.[809,810,988]

Given that the evolution of both flavonoid synthesis and microbial estrogen-like receptors predates the colonization of the land by plants, it seems likely that the selective pressure for the synthesis of phytoestrogenic flavonoids as rhizosphere agents of microbial/plant communication was predicated on the fact that microbes already possessed the receptors to detect them. These receptors, in turn, must already have been functioning in some capacity as environmental sensors.

Evidence suggests that the vertebrate ERα and ERβ estrogen receptors, and indeed all of the steroid receptors, are the descendants of the orphan nuclear estrogen-related receptors still found across vertebrates and invertebrates.[989] These ancestral estrogen-related receptors have no known endogenous ligand, but their transcriptional activity is stimulated, like ERα and ERβ, by the binding of a number of phytoestrogenic flavonoids.[990] They may well also have originated as receptors for sensing the many exogenous estrogenic signals in the environment, including structures similar to the phytoestrogens.[809,991] The estrogen-related receptor eventually paired up with its 17β-estradiol ligand in vertebrates, becoming the ERα and ERβ estrogen receptors, only after the divergence of vertebrates and invertebrates.[989] Presumably the factor that dictated the adoption of 17β-estradiol, and other estrogens, as ligands was the fact that the ancestral receptor could already respond to structurally similar molecules, such as phytoestrogens, in the first place.

This suggests that both the estrogen-like and estrogen-related receptors of bacteria/fungi and mammals respectively appear to have an early provenance as ancestral sensors of exogenous estrogenic molecules. If this is the case, the estrogenic/anti-estrogenic effects of phytoestrogens in mammalian tissue simply reflect an unintended transfer of an intended cross-kingdom signaling role due to the functional (rather than strictly orthologous) origins of the signaling equipment possessed by all of the parties.

Defense Against Microbial Pathogens

Of course, many of the above phytoestrogenic molecules also function in the defense against pathogenic bacteria and fungi.[988] Several antimicrobial phytoalexins with more restricted distributions, such as kievitone, phaseollin, and resveratrol,[992] which do not have an established role in microbial symbiosis, do have estrogenic properties. For all of these defensive compounds the ability to disrupt quorum-sensing and estrogen-like receptor signaling in pathogenic microbes may be one useful mode of action. However, the TOR pathways may also provide a better example of defensive disruption of microbial signaling. The TOR kinase signaling pathway itself was originally identified in the 1970s when researchers were trying to discover the mechanism of action of a potent antifungal chemical produced by bacteria found in a soil sample from Easter Island (also known as Rapa Nui). They named the antifungal macrolide molecule rapamycin, and after many years of research, culminating in the 1990s with genetic screening studies in yeast, they finally identified rapamycin's targets as a then-novel protein kinase they named "target of rapamycin" or TOR, as well as a specific binding protein (FKBP12) within the TOR pathway.[993] To the fungus, rapamycin's inhibition of the TOR pathway signals a lack of environmental nutrients and prevents protein synthesis and cellular proliferation, thereby inhibiting growth and handing an advantage to the rapamycin-synthesizing bacteria.

It also transpired, following further research, that all of the components of the TOR pathway targeted by rapamycin are conserved in fungi, animals, and plants.[819,994] If we accept that the bacteria had evolved the synthesis of rapamycin as an antifungal strategy (and of course it may have an alternative, unexplored role in bacterial/plant relationships), then any effects of rapamycin in mammals would represent a cross-kingdom transfer of the bacteria's intended inhibition of TOR signaling in fungi. Rapamycin certainly has numerous useful medicinal applications for humans, all of which are predicated on TOR inhibition. It started life in the late 1990s as an antifungal treatment, but it soon became apparent that its major use was as an immune suppressant and antiproliferative that could be used to prevent organ rejection after transplant surgery. Recent research has also demonstrated that rapamycin has a number of properties in common with TOR-inhibiting polyphenols such as EGCG, quercetin, and resveratrol, including the ability to increase longevity and protect against cancer in mammals.[995] Conversely, the synthesis of the same polyphenols is upregulated by biotic stressors, including fungi, and they all exhibit antifungal properties.[996,997] Unfortunately, as yet, we don't know if these antifungal properties are predicated on interference with TOR signaling. However, this does raise the possibility that the cross-kingdom, kinase signaling roles of many polyphenols may simply reflect an unintended transfer of their intended role as antimicrobials, in the same manner as rapamycin, due to the conserved nature of the TOR pathways.

■ AND FINALLY, A FEW MORE CONCLUSIONS

All of this is not to say that the effects of polyphenols within human signal transduction pathways are solely predicated on either their endogenous role in plant signaling or their potential roles in the management of symbiotic and pathogenic microbes. If one thing is clear, it is that the phenolics are all multipurpose molecules and, more than likely, their synthesis will have evolved to play both roles simultaneously, and potentially many more. Unfortunately, research into the potential endogenous signaling roles of phenolics (and other secondary metabolites) within the plant is in its infancy, so signaling roles within the plant analogous to those seen in the conserved mammalian cellular signaling pathways can't be confirmed as yet. I suspect, however, that they will emerge in due course, hopefully at about the same time that the direct effects of the phenolics on human brain function attract a more substantial research effort.

Of course, one final point should be borne in mind: the flavonoids, in particular, have been a ubiquitous component of the human ancestral diet for hundreds of millions of years. We can expect that the eventual role of these molecules in mammalian physiological systems has been the product of evolution throughout this time span.

PART FOUR
The Terpenes

10 Terpenes and the Lives of Plants and Animals

The terpenes comprise a large, structurally diverse family of compounds, with over 35,000 members identified to date across taxa. The key feature of this group of chemicals is that they are composed of units of the volatile, organic, five-carbon (C_5) compound isoprene. These units are notionally joined head to tail, although this configuration can be hard to discern in more complex compounds, and they are classified according to the number of isoprene units they contain. So isoprene, which itself is synthesized and released as a gas by plants, comprises one unit and is classified as a hemiterpene (C_5); monoterpenes incorporate two isoprene units (C_{10}), sesquiterpenes three units (C_{15}), diterpenes four units (C_{20}), sesterpenes five units (C_{25}), triterpenes six units (C_{30}), and tetraterpenes eight units (C_{40}).[195] In the environmental conditions and temperatures associated with plant life the less complex hemiterpenes, monoterpenes, sesquiterpenes, and some diterpenes are volatile compounds that evaporate or sublimate into the air as a vapor.[115]

Two pathways for terpene biosynthesis have been identified. The first, the mevalonate pathway, is ancestrally conserved and common to bacteria, fungi, animals, and plants, whereas the more recently discovered "non-mevalonate" or methylerythritol phosphate pathway has been found only in bacteria, protozoa, and plants to date. Within plants the mevalonate pathway generally operates in the cytosol, the intracellular compartmentalized fluid, and synthesizes the C_{15} sesquiterpenes and their multiples (i.e., C_{30} triterpenes) whereas the non-mevalonate pathway operates in plastids, chemical-synthesizing organelles found within cells. This pathway is responsible for producing multiples of C_{10}—thus, monoterpenes, diterpenes, and triterpenes.[116] The functionally diverse products of these pathways include many compounds that play a range of integral roles in plant primary metabolism. These include hormones such as abscisic acid, gibberellins, and brassinosteroids; components of the electron transfer chain such as ubiquinone (coenzyme Q10); components of membrane structures such as phytosterols; and photosynthetic pigments such as the carotenoids.[116] Terpenes also play a host of secondary metabolite roles, which will be reviewed briefly below.

Typically a plant will have a relatively small complement of full-length terpene synthase genes, ranging from a single gene in the moss *Physcomitrella patens*, through a modest 14 for the evolutionarily ancient lycophyte *Selaginella moellendorffii*, to 69 for the grape vine *Vitis vinifera*. Within these plants the expression of individual genes is specific to the tissue in question. For instance, among *Arabidopsis*' complement of 40 terpene synthase genes, the four genes that are expressed in the reproductive structures of the plant differ from the six genes expressed in the root. A single terpene synthase metabolizing a single substrate can also generate multiple terpenes. For instance, again in *Arabidopsis*, a single enzyme leads to the synthesis of ten separate monoterpenes.[115] Across the entire spectrum of plants the terpene synthases can also be subdivided into seven phylogenetic families (a, b, c, d, e/f, g, h) that illustrate the expansion and diversification of terpene synthesis over time. This journey has seen the spectrum of

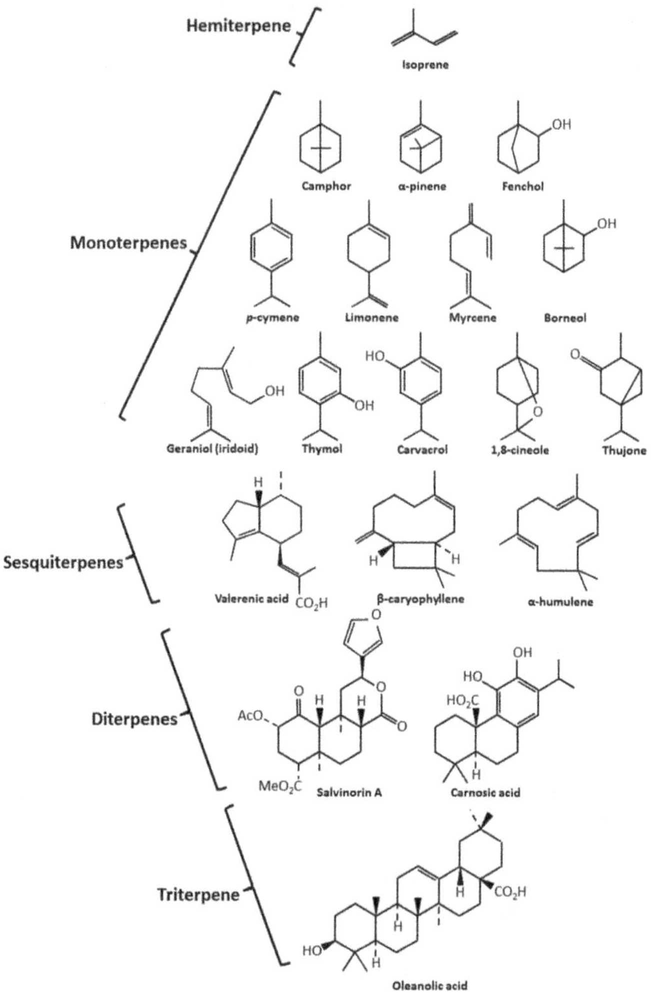

Figure 10.1 Assorted terpene structures.

plant terpenes in the plant evolve from a range of compounds solely involved in primary metabolism to include a wealth of specialized plant secondary metabolites. As an illustration, Chen et al.[115] note that all land plants and all vascular plants possess evolutionarily ancient terpene synthases from two of the synthase subfamilies (c and e/f) and that these produce only terpenes that originated as primary metabolites; for instance, performing roles as structural components such as phytosterols, photosynthetic pigments such as the carotenoids, and hormones, including abscisic acid and the ubiquitous gibberellin family. In addition to these primary metabolite synthases, gymnosperms then possess a more recently diverged, clade-specific d subfamily that synthesizes a range of secondary metabolite terpenes that they employ only in ecological roles. Similarly, the a, b, and g synthase subfamilies are specific to angiosperms and contribute to the synthesis of the monoterpenes, sesquiterpenes, or diterpenes that are solely deployed by flowering plants in ecological roles. This phylogenetic analysis suggests that although the origins of the terpene synthases involved in secondary metabolism were present before the differentiation of gymnosperms and angiosperms, the specific synthases involved in specialized secondary metabolite synthesis evolved convergently after the divergence of the gymnosperms and angiosperms.[115] However, the products of these pathways are often the same. As an example, the synthesis of the same single monoterpene, limonene, which is found in conifers and the angiosperm mint family (*Lamiaceae*), is the product of entirely separate, convergent evolutionary events.[116] Similarly, the synthetic pathways that produce the many triterpene saponins arose several times in different plant taxa by a process of convergent evolution.[998] Chen et al.'s[115] genomic analyses of terpene gene expression across species does illustrate nicely the varied divergent and convergent evolutionary processes that have led to the expansion and diversification of terpene synthases alongside the speciation of plants, and it also elucidates how a relatively small number of genes can lead to the expression of the 25,000 terpene structures synthesized by plants.

■ TERPENES AND THE LIVES OF PLANTS:— ECOLOGICAL ROLES

Where the alkaloids fulfill primarily toxic defensive roles and the phenolics are more typified by relatively benign protective, symbiotic, and attractant roles, the terpenes combine these extremes and fill areas of the middle ground. The most obvious example is in the dual roles played by many monoterpenes and sesquiterpenes as both toxic deterrents and attractants to insects. As an example of the multipurpose nature of this group, it is notable that the highly volatile monoterpenes and sesquiterpenes, comprising a range of hydrocarbons, aldehydes, ketones, alcohols, and esters, contribute the largest part to the plant's "volatilome".[737] This mixture of volatile airborne compounds can be synthesized and emitted by the plant in response to stressors as a means of rapidly communicating with its own distant parts, or indeed with nearby "eavesdropping" plants.[161] Alternatively they can function in direct defense, deterring or harming herbivores, or in indirect defense, whereby the emissions attract the natural predators of the herbivore. The same chemicals can also be synthesized constitutively as attractants for pollinators and other symbiotic microorganisms and animals.[737] One way or another, most of the functions of terpenes can be grouped together as offering either defense or attraction to the plant, although, as noted, many individual chemicals perform both roles.

Defense

When they are fulfilling defensive roles, the synthesis of terpenes can be induced by biotic and abiotic stressors. However, they are also often presynthesized constitutively and stored in discrete structures that are most concentrated on comparatively attack-prone or valuable tissue, such as young tissue or reproductive organs. These structures typically express the full complement of genes necessary for terpene synthesis and act as self-contained "chemical factories" that synthesize their final chemical products from a few simple metabolic inputs and precursors.[134] These products can then be stored in self-contained secretory structures that are technically outside the vascular system of the plant, reducing the dangers of autotoxicity. For example, in conifers complex mixtures of monoterpenes, sesquiterpenes, and diterpenes serve as first-line defenses against bark beetles and fungal pathogens, and the storage and delivery systems include branching canal systems, pressurized resin ducts, or laticifers, all of which ensure that the chemical is delivered to the herbivore or area of infection in liquid or volatile form at the time of first contact. In angiosperms terpene-containing glandular trichomes, pellucid glands, or secretory ducts are liable to be the first structures encountered by any herbivore intent on feeding.[116] These structures often excrete terpenes in the form of resins or latexes composed mainly of monoterpenes and sesquiterpenes, which coagulate after secretion. These sticky liquids can immobilize or gum up the mouth parts of herbivores. Alternatively, the same structures can emit a bouquet of volatile organic compounds including many monoterpenes, sesquiterpenes, and diterpenes released in a stressor-specific recipe to evaporate or sublimate into the air when the structure is stimulated or ruptured.

In defensive terms, volatile emissions that contain terpenes can be toxic or repellent to a wide range of insects[87,999–1001] and other invertebrates such as mollusks[1002] and nematodes.[1003] They can also directly deter feeding and exert multifarious direct effects within the herbivore's nervous system, modifying the growth and life course of insects and decreasing the oviposition rates of the feeding herbivores.[116] Terpenes are lipophilic and hydrophobic, and many of their defensive properties are predicated on the ability of these molecules to interact with and incorporate themselves into the selectively permeable cell biomembranes that regulate the movement of ions and organic molecules into and out of the cell and act as a scaffold in which the proteins that regulate the transfer of chemicals and information (in the form of ion channels, proton pumps, and G-protein-coupled receptors) are anchored. Terpenes can disturb the fluidity and permeability of the membrane and modify the function of these membrane proteins. This can result in the leakage of small molecules such as ions and cellular metabolites and disturbance of the close structural interaction between proteins and the surrounding membrane lipids, leading to loss of functionality. These effects are generally nonspecific and can affect organisms of any level of complexity from microbes to vertebrates.[137,998] As an example, the triterpene saponins have allelopathic, antifungal, antibacterial, insecticidal, and antiparasitic properties as a consequence of their ability to incorporate themselves into cellular membranes, thus compromising their integrity. In animals this activity is largely restricted to cold-blooded vertebrates and invertebrates, with low absorption and metabolism protecting warm-blooded animals following oral ingestion.[998,1004] Similarly, the irregular monoterpene pyrethrins, found in *Chrysanthemum cinerariaefolium*, are potent insecticides that bind to and block the voltage-gated sodium channels that are essential for the normal transmission of impulses in neurons, leading to the overactivity and nervous system dysfunction that ultimately causes paralysis and death in herbivores.[1005]

Terpenes and the invertebrate nervous system

Given that lipophilic volatile terpenes are readily absorbed by insects,[999] they may owe their overall toxicity in insects to interactions with multiple targets.[1006] However, as with the alkaloids, evidence does suggest that terpenes owe their defensive properties in the face of insect herbivory in part to interactions with specific functional proteins in the central nervous system. These include a range of receptors and enzymes. These properties typically also underlie the psychoactive properties of plant chemicals and extracts consumed by humans. It is certainly the case that very little research has been conducted elucidating the modes of action of terpenes in insects, certainly in comparison to that focusing on mammals, but several insect targets have attracted some attention.

GABA receptors

One established target for many terpenes is the insect GABA receptor. GABA functions as the major inhibitory neurotransmitter in both the invertebrate's central and peripheral nervous systems. It binds to ionotropic transmembrane receptors, allowing the ingress of negatively charged chloride ions and thereby reducing the electrical membrane potential of the neuron. This inhibits the generation of new action potentials, depressing the activity of the neuron. Many synthetic insecticides such as dieldrin and fipronil target GABA receptors, causing either an inflow or outflow of chloride ions, resulting in inhibition or excitation of individual neurons and the nervous system.[404,1007] A significant subsection of volatile terpenes also bind allosterically to insect ionotropic GABA receptors, increasing their inhibitory activity.[1006,1008] For instance, Tong and Coats[1006] found that nine of the 22 most widely distributed monoterpenes, including 1,8-cineole, thymol, camphor, and menthol, had significant GABA receptor binding properties in the housefly, and this property was related to several specific chemical and structural properties of the compounds. Volatile terpenes are readily absorbed by insects, and their ability to modulate insect behavior has been confirmed in tethered blowflies. Following topical application of the monoterpene thymol, or injections of GABA, electrical flight muscle impulses and wing beat frequency gradually reduced in a pattern suggesting central rather than peripheral interference with motor behavior.[1009] The neurotoxic effects of several other monoterpenes (eugenol, pulegone, citronellal, α-terpineol) were subsequently confirmed in the same experimental paradigm, again with modulation suggestive of increased GABA activity, although the pattern of modification differed between monoterpenes.[1010] More complex terpenes have also been shown to have direct GABAergic properties in insects, although these are typically inhibitory in nature, therefore leading to neural overactivity. For instance, the insecticidal activity of the picrotoxane group of diterpenes, which include picrodendrins from *Picrodendrum baccatum* and picrotoxinin from *Anamirta cocculus*, is directly related to their noncompetitive antagonistic binding at insect GABA receptors.[1011,1012] Similarly, the diterpene/sesquiterpene lactone ginkgolides and bilobalide owe their potent insecticidal properties to the same mechanism.[1013]

Octopamine/Tyramine

These invertebrate neurotransmitters, neurohormones, and neuromodulators play diverse roles, including in the regulation of sensory inputs, rhythmic processes, learning and memory, and the "fight-or-flight" modulation of energy expenditure. Activation of the G-protein-coupled receptors for octopamine and tyramine results in a rapid but short-term rise in intracellular Ca^{2+} and cyclic adenosine monophosphate (cAMP) levels,[1014] and this same effect has been detected following application of volatile terpenes to insect abdominal epidermal tissue, with the effect abolished by an octopamine antagonist.[1015] Similarly, both eugenol and α-terpineol have been shown to bind to octopamine receptors with resultant increases in cAMP levels seen at lower concentrations and decreased cAMP levels seen at higher concentrations. This biphasic effect was matched with behavior in the American cockroach, which exhibited hyperactivity at lower doses followed by immobilization and death as the dosage rose.[1016] In a later study, geraniol and citral evinced a similar biphasic, potentially octopaminergic effect on foregut contractions and electrical activity in American cockroach neurons.[1017] Similar binding by monoterpenes to *Drosophila* tyramine receptors has also been demonstrated, with differential compound-related effects on cellular cAMP and Ca^{2+}. The most potent monoterpenes, thymol and carvacrol, exhibited potent insecticidal effects that were abolished in tyramine receptor mutant flies.[1018] The nematicidal activity of thymol and carvacrol has also been related to tyramine receptor binding.[1019]

Acetylcholinesterase (AChE) Inhibition

As well as direct receptor binding effects, the volatile terpenes can interfere with other aspects of neurotransmission. For instance, a wide range of monoterpenes[87,1020,1021] exert neurotoxic effects in insects, in a similar manner to organophosphorus and carbamate pesticides, by reversibly inhibiting the acetylcholine hydrolyzing enzyme AChE. This results in the accumulation of acetylcholine in the synaptic cleft and neuronal overactivity, which can lead to modified behavior, paralysis, and death.[1022] As an example, Abdelgaleil et al.[1023] found that most of the 11 common monoterpenes they assessed were toxic to one or both of two insects (*Sitophilus oryzae* and *Tribolium castaneum*), with 1,8-cineole vapor having the highest toxicity and most potent AChE inhibitory properties across insects. However, although seven of the compounds (limonene, geraniol, linalool, carvone, fenchone, cuminaldehyde, 1,8-cineole) had pronounced AChE inhibitory properties, this factor was not necessarily related to their overall toxicity in terms of mortality. The AChE inhibitory properties of citral, pulegone, linalool, bornyl acetate, and 1,8-cineole were also confirmed using moth AChE, with the results showing dose-related escalation of inhibition.[1024] A number of other terpenes have established cholinesterase inhibitory properties in mammals but have not to date been investigated directly in insects (see below).

Ecdysteroids

Insects synthesize a number of endogenous modified triterpene steroidal hormones. These ecdysteroids (or ecdysones) bind to nuclear receptors and modulate every aspect of the insect's life course, including its embryogenesis, development, progression through life stages, metamorphosis, diapause, reproduction, and social interactions.[139,1025] However, due to an inability to synthesize the steroidal nucleus of these hormones, insects have to sequester their cholesterol and sitoserol precursors from their diet. Many plant species, for their part, also synthesize ecdysteroids, which are then conventionally called phytoecdysteroids. These compounds, of which more than 300 have been identified, are structurally identical or closely related to the insect hormones. While an unidentified endogenous role for these phytoecdysteroids in the life of the plant cannot be ruled out, it is generally accepted that these triterpene compounds are synthesized as a defense against herbivory.[1026] One way that they accomplish this is by simply mimicking the insect ecdysteroid hormones, binding directly to ecdysteroid nuclear receptors. This leads to disruption of endocrine function and the development, life course, and fertility of the insect.[123,140] The most ubiquitous endogenous insect hormone, 20-hydroxyecdysone, is also the most commonly synthesized plant phytoecdysteroid. In the plant the synthesis of 20-hydroxyecdysone is induced by herbivore-related damage via the jasmonate pathway,[179] and its inclusion in the diet of insects impairs numerous developmental processes, such as the ability of pupae to cast off their old cuticle, leading to death.[142] As a countermeasure, many insects are very sensitive to phytoecdysteroids in food and avoid them at all costs, while the physiological response to the ingestion of these compounds is also variable across species.[1026]

A number of other triterpenes, including triterpene acids, and steroidal and non-steroidal triterpene saponins also disrupt the life course and mortality of insects by interfering with the ecdysteroid hormonal system. They do this via two mechanisms: they can interact directly with insect ecdysteroid receptors by binding to the ecdysteroid binding site in the same manner as the "classic" phytoecdysteroids, and they can form a complex with non-membrane sterols, including cholesterol, thereby making them unavailable to the insect for the synthesis of ecdysteroids.[998,1027,1028] Examples of triterpenes with ecdysteroid-modulating properties include the ubiquitous ursolic acid, steroidal saponins such as the cucurbicatins from the cucumber/melon/marrow family, nonsteroidal saponins such as the ginsenosides from the *Panax* genus, and triterpene steroidal lactones such as the withanolides from *Withania somnifera*.[741,1027,1029–1031]

The key factor underlying the effects of all of these plant-synthesized chemicals in insects is that they are all structurally related, or identical, to the insect triterpene phytoecdysteroid hormones. The structural similarity will be illustrated in Figure 13.1 in Chapter 13.

Insect Counteradaptations

Of course, a wide variety of insects have developed counter-adaptations to make use of the defensive properties of plant terpenes. The classic example is the monarch butterfly (*Danaus plexippus*), which feeds on members of the milkweed (*Asclepias*) genus of plants and sequesters the plant's highly toxic, modified triterpene, cardenolides, within its own body as a defense against predation. In turn, one species of bird that co-habits the monarch's wintering ground in Mexico, the black-headed grosbeak, has developed immunity to the cardenolides. A number of weevils and beetles also recycle terpenes without directly sequestering them. For instance, a number of beetle genera either coat themselves in a layer of regurgitant and anal secretion composed of the defensive terpenes of their host plant, or alternatively, construct a solid fecal shield around their own body from terpenes found in the plant on which their eggs hatch.[116]

Other Defensive Roles: Allelopathy and Antimicrobial Properties

A further defensive role for a number of terpenes is in allelopathic interactions that constrain seed germination or the growth of neighboring plants. For instance, sagebrush (*Salvia leucophylla*) roots exude a range of monoterpenes that have been shown to inhibit germination and constrain the growth of other competing plant species. This creates the characteristic pattern of vegetation in the Californian chaparral, featuring sagebrushes in a sea of barren earth.[166] Similarly, triterpene ginsenosides from ginseng roots have direct concentration-related allelopathic effects on any surrounding vegetation that lead eventually to an autotoxic poisoning of the soil in which the plant itself is growing. This curtails the life expectancy of the plant and contributes to the period of several decades during which ginseng cannot be successfully regrown in the same earth.[1032] Terpenes are also the second most frequently cited chemical group in terms of possessing fungicidal effects (after phenolics). While this property extends to terpenes at all levels of complexity, it is a particularly common property of monoterpenes in particular, and the compounds with fungicidal properties include the majority of monoterpenes mentioned in this chapter.[367] The majority of terpenes also have antibacterial properties, at least when assessed with respect to bacteria with relevance to humans, and these properties are related to specific aspects of their molecular structure.[1033]

Attraction

Paradoxically, in contrast to their toxic and deterrent roles, many terpenes also function as attractants, fulfilling a range of functions for the plant. Typically, when functioning in this role terpenes are either synthesized at the time of emission or alternatively stored in structures, including specific types of trichomes or osmophores, that release them at the most propitious time. For instance, trichomes on flowering parts of the mint family differ in morphology from defensive trichomes elsewhere on the plant, and they synthesize and emit a cocktail of terpenes with a greater accent on typically attractive monoterpenes (such as borneol and limonene), releasing them as required.[1034,1035]

Pollination

Eighty-five percent of flowering plants, and therefore more than 70% of all species of plants, are pollinated by insects,[91] and the floral scents emitted by flowers provide the primary communication channel, along with color, between flowering plants and insect pollinators. Charles Darwin, having read the relatively obscure 1793 book *The Secret of Nature in the Form and Fertilisation of Flowers Discovered*, in which Christian Sprengel first proposed that plants could be pollinated by insects, wrote his own treatise *On the Various Contrivances by Which British and Foreign Orchids are Fertilised by Insects*. In this book Darwin extended his discussion of the co-evolutionary interactions between plants and symbiotic pollinating insects that he had mentioned briefly in *On the Origin of Species* and made what might have been the first observation of an ecological role for plant chemicals with his observations about the scent of flowers attracting pollinating insects in the absence of elaborated flowers.

Floral volatiles are typically synthesized within either floral organs themselves or nearby tissue, and all of the requisite biosynthetic genes and enzymes are expressed locally.[125] The resulting floral scents are usually complex, and the specificity of the attractive properties of the scent to specialist or generalist pollinators can be dictated by a single unusual compound, or the mixtures and ratios of more common constituents.[128] It is possible to gauge the range of compounds with potentially attractive properties from a study by Knudsen et al.,[125] who identified more than 1,700 compounds found in the "headspace" scents of 990 different plants. While these scents contained compounds from a wide range of chemical groups, monoterpenes were by far the most common components and featured in the scents of all of the orders and families of plants investigated. A number of monoterpenes, including limonene, β-ocimene, myrcene, linalool, and β-pinene, occurred in more than two thirds of all of the headspace floral bouquets. Sesquiterpenes were somewhat less common as a group but still occurred in the scent of 85% of plant orders, with caryophyllene appearing as a component in more than 50% of scents.

These monoterpenes and sesquiterpenes are not ubiquitous components of floral volatile emissions by chance. Many are also synthesized endogenously and excreted by insects, and they represent a key component of the chemical communication systems that insects employ to communicate with their own species (pheromones) and other species (allomones). These chemicals can be used to attract, repulse, or communicate information to other insects. For instance, they facilitate sexual contact, aggregation, and recruitment; mark trails and territorial limits; signal alarm or danger; and mark food sources.[126] To give an example, foraging honey bees synthesize and excrete a mixture of monoterpenes and sesquiterpenes, including geraniol and farnesol, from the tergal glands located on their abdomen onto nectar-bearing flowers and at the entrance to their hives. This pheromone scent complements their waggle dance in guiding co-workers to food sources and then back to the hive. Similarly, bumblebees return from a successful foraging trip and emit a scent composed primarily of monoterpenes and sesquiterpenes, including 1,8-cineole, ocimene, and farnesol, in order to recruit hive-mates to exploit the resource.[1036]

Schiestl,[126] using the data from Knudsen's headspace paper, identified 71 volatile compounds that were synthesized by at least 15 families of plants. Of these, well over half were either monoterpenes or sesquiterpenes. He then compared these common plant chemicals to the pheromones and allomones synthesized by insects and found that more than 90% of the terpenes were also synthesized by at least one family of insects, and they were often synthesized by many more. Schiestl also found that the chemicals most commonly synthesized by plants correlated with those most commonly produced by insects. Looking in more detail, there was a stronger correlation between the monoterpenes most commonly produced by plants as a whole and those synthesized by herbivorous insects, whereas the same relationship held for the cocktail of aromatic compounds produced by insect-pollinated plants and those synthesized and emitted by their pollinating insects. Similarly, the specific cocktail of volatiles produced by one order of pollinating insects was more likely to be more similar to those of its partner plants than to those of other plants.[126]

Given that terpene synthesis is an evolutionarily ancient ability that was enjoyed by both early plants and arthropods well before the radiation of the angiosperms, it seems likely that the selection of monoterpenes as defense chemicals in early plants arose because these compounds were already in use as insect semio-chemicals. This meant that they could interact deleteriously with existing sensory and neuronal mechanisms in herbivorous insects. This is supported by the observation that the wind-pollinated gymnosperms, a lineage that is far more ancient than the insect-pollinated angiosperms, have a volatile chemistry that is correlated to that of herbivores and not pollinators, suggesting that volatile monoterpenes originally functioned as defense chemicals. In this scenario the terpenes assumed roles as attractants only much later, with the advent of the angiosperms.[126] As Theis and Lerdau[116] note, evidence suggests that the first emergence of flowerlike structures and insects capable of pollination probably predated the radiation of the angiosperms by many tens of millions of years. Early herbivorous or predatory insects may inadvertently have assumed the role of pollinators as they went about their business, increasing the selection pressure for existing plant primary metabolites or defense volatiles that could also attract insects. The explosion in the number and subfamilies of terpene synthases then went hand in hand with the divergence of the angiosperms,[115] while the ranks of herbivorous and predatory insects were swollen by the evolution of floral specialist pollinator insects living on pollen and nectar.

This scenario does explain the apparently contradictory dual roles of volatile terpenes as both toxic deterrents to herbivores and attractants to pollinators. Dose is the only factor that differentiates the function of an insect pheromone as a semio-chemical rather than a neurotoxin, with many terpene pheromones demonstrating lethal toxicity to the very insects that synthesize them when they are administered at a high or sustained dose.[1037] It is also notable that the mechanisms of action of many of the terpenes that insects and plants have in common are not necessarily detrimental to insect brain function at low doses. For instance, many monoterpene pheromone compounds that are synthesized by both plants and insects, as well as a number that are restricted solely to insects, inhibit AChE in a dose-dependent manner.[1024,1038] The low doses of these compounds associated with attractant or pheromone volatile emissions would be unlikely to engender any negative effects at the concentrations required for chemical communication.[1038] Indeed, given the symbiotic relationship between plants and many insects, the effects of attractant terpenes within the insect nervous system must, at the very least, be neutral in nature. Indeed, mild AChE inhibition could well be beneficial to the cognitive function of the insect. As noted previously, both nicotinic and muscarinic receptors are expressed within insect nervous systems, and low doses of cholinergic agonists and antagonists have been shown to upregulate and downregulate insect memory function respectively.[302,343] Indeed, toxic synthetic cholinesterase inhibitors such as methyl parathion have been shown to specifically enhance visual and olfactory learning in insects at low doses.[310] It would certainly be advantageous for a pollinating insect to have improved memory for a food source as a consequence of increased cholinergic neurotransmission. Unfortunately, the possibility that volatile attractant terpenes may benefit insect cognitive function remains unexplored at present.

In general, the neurotoxicity of AChE-inhibiting monoterpenes depends both on dose[1037,1038] and the mode of delivery of the compound.[1039] This suggests that the specific function that a monoterpene or sesquiterpene plays must be predicated on whether it is encountered in food, absorbed directly through the cuticle, or delivered as an airborne volatile, and ultimately the final dose absorbed by the insect. It is notable that attractants are often synthesized and released immediately at relatively low doses. In contrast, defense compounds are often presynthesized and stored in structures such as trichomes ready for release in liquid, latex, resin, or volatile form in much higher concentrations.[116]

Indirect Defense

The release of volatile attractant chemicals, most often comprising terpenes, can also be a form of defense. In this case it would be an example of "indirect defense" in which the release of airborne, volatile compounds is intended to attract the natural predators of the herbivore that is threatening the plant. This mechanism can be activated in any type of plant tissue, from tip to root, and examples identified to date include the attraction of predators from five different invertebrate orders (nematodes, *Arachnida, Heteroptera, Coleoptera, Hymenoptera*) as well as vertebrates, although in the latter case this is generally restricted to a few species of birds.[153] The specific mix of volatiles can provide information to the predator on the presence and nature of the herbivore. To give a single example, when tobacco plants are attacked by tomato hornworm larvae they synthesize and release a cocktail of chemicals, including the monoterpenes α-ocimene and linalool and the sesquiterpenes β-farnesene and α-bergamotene, in response to specific amino acid and fatty acid conjugates in the oral secretions of their specialist herbivores. The volatile terpenes attract the hornworm's natural predators such as the parasitoid wasp *Cotesia congregata*, which lays its eggs within the larva's body, increasing the fitness of the plant. Applying the individual terpenes linalool and α-bergamotene, or the defensive plant hormone methyl-jasmonate to the stem of the plant also has the same effect, attracting the generalist, predatory big-eyed bug (*Geocoris pallens*), which then feeds on the hornworm eggs.[152]

Many terpenes are also synthesized as components of the extra-floral nectar that is excreted by various nonreproductive plant tissues in order to attract ants and other predatory insects. These symbionts effectively guard the plant against herbivores in return for the nectar.[737] However, volatiles emitted by the flowers of the same plants often include components such as the sesquiterpene α-farnesene. This chemical functions as an ant alarm pheromone, and its inclusion in the floral bouquet effectively dissuades the guard ants from straying beyond their guard duties. This prevents them from foraging valuable reproductive tissue or floral nectar. The same compound also happens to be an attractant for bees, as it forms part of their hive recruitment bouquet.[1040] In something of a twist, rather than employing the ants themselves as guards, a number of plants simply synthesize and release the terpenes that guard ants usually employ as pheromones. These emissions are then detected by herbivorous insects and warn them off the plant.[1041]

TERPENES AND THE LIVES OF HUMANS

Terpenes are integral to the lives of humans, both in terms of their many endogenous physiological roles and as components of a wide range of plant-derived products.

All animals synthesize a range of modified triterpene steroid structures via the mevalonate pathway. Principal among these are cholesterol, which forms an integral part of the cell membrane in all animal tissue and goes on to provide the substrate for a wide range of functional terpene derivatives. These include the human steroid hormones (glucocorticoids, mineralocorticoids, progesterone, estrogens, and androgens), vitamin D, bile acids, and lipoproteins. Vitamin D itself (calcitriol or 1α, 25-dihydroxyvitamin D_3) is unusual among the vitamins in that it can be synthesized endogenously, by hydroxylation of cholecalciferol (vitamin D_3), which in turn can be synthesized in the skin by sunlight-derived ultraviolet irradiation of the cholesterol derivative 7-dehydrocholesterol. Alternatively, cholecalciferol can be consumed presynthesized in animal-derived foods, or ergocalciferol (vitamin D_2) can be obtained direct from plants lower down the food chain and converted to vitamin D by the same hydroxylation process. The other terpene vitamin, vitamin A (retinol), has a diterpene structure that is derived in mammals by oxidative metabolism of plant-derived tetraterpene carotenoids, most notably β-carotene, to retinal and then to retinol. Alternatively it can also be consumed as premetabolized retinal in animal-derived foods.[195]

Terpenes are also key components in a multitude of products. Purely practical applications include the use of natural rubber and elastic, which are polymers of the hemiterpene isoprene, and the use of volatile monoterpenes and sesquiterpenes as antiseptics, antimicrobials, and disinfectants and as insect deterrents and insecticides. The latter class includes the irregular monoterpene pyrethin (from *Chrysanthemum cinerariaefolium*), which is one of the most widely used insecticides. The same anti-insect roles are often filled by essential oils. These are concentrated hydrophobic liquids containing volatile aroma compounds that are extracted, generally by distillation, from a limited number of plant genera. Typically monoterpenes are the major constituents, representing up to 90% of the oil's volume.[1042] Essential oils have a number of uses, and these include as widely used insect repellents. Currently there are more than 140 active patents for the use of various plant essential oils as mosquito deterrents alone.[1043] A great deal of research in this area has also concentrated on the efficacy of essential oils and individual and combined terpenes as toxic repellents to the larvae and adults of other insect parasites and disease vectors. The results often show that topical or fumigant application of terpene products can be more potent than synthetic insecticides/repellents. However, it is usually the case that the cocktail of terpenes in an essential oil is more effective than an individual component, with synergistic and suppressive interactions present among the many components.[1044] Essential oils are also used widely as perfumes, as mild sedatives, and in aromatherapy.

A vast range of food flavors and aromas are also provided by terpene constituents. Just to give a few examples, the core eucalyptus flavor is provided by 1,8-cineole, juniper by α-pinene, citrus fruits by limonene, coriander by linalool, saffron by safranal, and ginger by γ-bisabolene/zingiberene (plus other components). The mint family (*Lamiaceae*) also provides a wide range of culinary herbs, such as basil, rosemary, spearmint, peppermint, sage, marjoram, thyme, oregano, lavender, and perilla, all of which owe their distinctive odors and flavors to their volatile terpene constituents. More complex structures also feature in foods, for instance the triterpene acids such as the oleanolic acid found in olives, the glycyrrhizin triterpene saponins found in licorice, and the steroidal saponins such as diosgenin and yamogenin from the spice fenugreek. More recent, manufactured food additives include steviol, a diterpene from *Stevia rebaudiana*, which is increasingly used as a low-calorie sweetener, and plant sterol esters, found naturally in the diet in small quantities and manufactured in larger amounts by esterifying sitosterol from soya. These plant chemicals are more hydrophobic than cholesterol and therefore block its intestinal absorption, leading to lowered levels of circulating cholesterol and potential cardiovascular benefits.[195]

Terpenes also fulfill a wide range of medicinal roles. For instance, a range of steroidal drugs, including the endogenous mammalian steroid hormones, are typically manufactured from the triterpene steroidal saponin diosgenin. This compound is obtained in huge quantities by cultivation of the yam (edible tuber-producing species of the *Dioscorea* genus), which itself is reputed to have hormonal effects similar to progesterone. The related steroidal antibiotic, fusidic acid, is active against gram-positive bacteria.[195] Other medicinal roles are played by major drugs such as artemisinin, a sesquiterpene lactone from *Artemisia annua* that remains the most effective treatment for drug-resistant strains of malaria, and the major cancer drug taxol (or paclitaxel), a diterpene ester from the bark of the Pacific yew tree (*Taxus brevifolia*) that inhibits cell division by binding to microtubules in cells, interfering with their breakdown. An example of an old treatment incorporated into modern medicine is the use of cardiac glycosides such as the cardenolides. These modified triterpenoids are found as defense chemicals in a wide range of angiosperms, and several exist endogenously in small quantities in mammalian tissues. They function by blocking the activity of cellular sodium pumps, preventing efflux and thereby also raising intracellular calcium levels and generating a positive ionotropic effect in heart muscles. The earliest use of these particular secondary metabolites was as arrow poisons. In this role extracts of seeds of *Stropanthus kombe/gratus* containing their deadly component ouabain would be applied to arrow tips, and these would cause cardiac arrest in an animal wounded by the projectile. Luckily, the poor oral bioavailability of many terpenes in warm-blooded animals made the extracts safe to handle by the hunter. The more recent medicinal use of cardenolides dates back to the late 18th century, when digitalis, the dried leaf of the red foxglove (*Digitalis purpurea*), was used as the first effective treatment for heart complaints and dropsy, an accumulation of fluid caused by cardiac failure. The principal active component, digitoxin, along with a variety of semisynthetic derivatives, is now widely used as a drug.[195]

Terpenes and Human Brain Function

The terpenes that modulate human brain function tend to be encountered as components of plant extracts with a complex mix of chemical constituents. They can exert a broad range of neuropharmacological effects, many of which may be epiphenomena related to their primary functional properties. The effects of specific terpenes and terpene-containing extracts will be examined in more detail in subsequent chapters. However, it is possible to identify several common mechanisms of action across groups of terpene-containing plant extracts and chemicals. These may well represent their primary mode of action. These will be summarized below, along with some initial consideration of how these properties relate to the ecological roles of the terpenes.

Cholinesterase Inhibition

The insect cholinesterase-inhibiting properties of many terpenes carry over directly into mammals. For instance, the *Lamiaceae* (mint) family includes a wide range of plants with cholinesterase-inhibiting properties, including, to lesser or greater extents, the majority of the *Salvia* species investigated to date.[1045–1048] A number of studies assessing the cognitive effects of *Salvia* species in humans have demonstrated improved cognitive function commensurate with increased cholinergic neurotransmission following the administration of extracts with in vitro mammalian AChE inhibitory properties.[1049–1053] As noted above with regard to insects, a wide range of individual monoterpenes, many of which are characteristic components of the psychotropic *Lamiaceae*, also inhibit cholinesterase, and the overall potency of plant extracts has been shown to be due to synergistic interactions and antagonisms between the individual terpene components,[1054–1056] with 1,8-cineole being both the most abundant and active monoterpene in this respect. More complex terpenes also exhibit AChE inhibitory properties. Examples including a range of individual tanshinone diterpenes from *Salvia miltiorhizza*[1057–1060] and the triterpene withanolides from *Withania somnifera*.[1061] Terpene-rich extracts of both *Bacopa monnieri* and *Ginkgo biloba* have also been shown to inhibit AChE.[1062] Importantly, the cholinesterase-inhibiting properties of these and other terpene-containing extracts have been demonstrated in rodent brain tissue following oral administration, confirming the direct relevance of this mode of action in the mammalian brain.[1062–1064]

Direct Receptor Interactions

GABA Receptors

The most consistent terpene receptor ligand effects across both insects and mammals are reserved for GABA receptors. Unlike the other neurotransmitter receptor targets, the interactions of terpenes with insect GABA receptors have received a modicum of attention. As noted above, a wide range of terpenes are active functional ligands at insect GABA receptors, with many volatile monoterpenes and sesquiterpenes binding allosterically to receptors, increasing GABAergic activity, and thereby depressing insect neural activity.[1008,1009,1065] Conversely, some isolated sesquiterpenes, most notably bilobalide from *Ginkgo biloba*, and a range of diterpenes, including the pichrodendrins, picrotoxinin, and the the ginkgolides, exert antagonistic effects at the same insect receptors, leading to a deadly increase in neural activity.[404,1011–1013] The insect GABA receptors in question are homologues of the human $GABA_A$ receptors, which themselves are ubiquitous throughout mammalian neurons. It is notable that, as a general rule, positive allosteric modulators at $GABA_A$ receptors, whether synthetic or natural, have analogous effects at insect GABA receptors.[1008] In the case of mammals, these GABAergic effects may contribute to the central nervous system bioactivity of many terpene-containing extracts, and indeed the majority of herbal extracts with anxiolytic or sedative effects owe these properties to interactions with GABA receptors. Terpenes represent the largest group of bioactive compounds in this respect.[1066] Examples of single terpenes with direct or implied agonist properties at mammalian GABA receptors include monoterpenes such as menthol,[1067] thymol,[1008] borneol,[397] and the iridoid valepotriates from valerian[1068]; sesquiterpenes such as isocurcumenol[397] and valerenic acid from valerian[1069,1070]; diterpenes, including galdosol[1071] and miltirone[1072]; and the triterpene ursolic acid.[1073–1075] Ursolic acid and oleanolic acid also inhibit the enzyme GABA transaminase, leading to increased GABA activity.[1076]

In contrast to this, a number of terpenes also exert antagonistic effects at mammalian GABA receptors analogous to the effects seen in insects. The resultant increase in neuronal activity makes them potential convulsants.[1077] The most celebrated examples here are the monoterpene α-thujone, one of the primary psychoactive chemicals in absinthe, and the sesquiterpene picrotoxinin, the active component of picrotoxin (the other equimolar component is picrotin). This compound has such a pronounced activity in this respect that it is commonly used as a probe in neuropharmacological studies investigating the GABA system.[1078] However, several other potent GABA antagonists illustrate why the toxicity of some plant chemicals can be restricted to insects or other invertebrates and not mammals. Prime examples are the picrodendrins, a group of diterpene lactones from *Picrodendron baccatum*. Many of these phytochemicals are potent insecticides as a consequence of their noncompetitive antagonistic binding at insect GABA receptors. However, this group of compounds can also be subdivided on the basis of whether they are also potent ligands at mammalian GABA receptors. The difference in potency in this respect comes down to whether they possess either a single bond or double bond at one position in their structure, and the slight morphological differences between the insect and mammalian receptors.

These small differences mean that whereas the insecticidal picrodendrins with the double bond exhibit potentially disastrous excitatory effects in both insects and mammals, the compounds with the single bond are at least 30 times less potent at the mammalian GABA receptor than they are at the insect homologue.[1011,1012] The fact that some of the picrodendrins have very modest bioactivity at GABA receptors has little impact on human use of the plant material, as *P. baccatum* isn't eaten or used medicinally, presumably due to the presence of the more potent components. However, it is commonly used in the Dominican Republic as an insecticide, with the powdered leaves, called *mata becerro* ("calf killer"), used to kill lice and bedbugs.[1012] Similar examples, but this time with more relevance to human consumption, are the diterpene lactone ginkgolides and sesquiterpene lactone bilobalide, the major bioactive compounds in extracts from *Ginkgo biloba*. As with the structurally similar picrodendrins/picrotoxinin, these compounds are potent insecticides due to their noncompetitive antagonistic properties at insect GABA receptors, but they are only weak inhibitors of the homologous human $GABA_A$ receptor. This is due to a minor, single-amino-acid difference in the composition of the insect and vertebrate GABA receptor pores, which alters the docking of the phytochemical within the receptor and reduces the GABAergic potency of these terpenes by a factor of up to 10,000 in humans.[1013] Unlike *Picrodendron baccatum*, extracts made from *Ginkgo biloba* are consumed around the world in the form of popular herbal extracts that have very low toxicity in humans. They count mild GABAergic properties among their many potential mechanisms of action (see Chapter 13).

Other Receptors

The most striking examples of the receptor-binding properties of terpenes are the very specific, and somewhat atypical, interactions of the key components of the two social drugs, cannabis (*Cannabis sativa*) and *Salvia divinorum*, with their respective receptor targets. In the case of cannabis, its psychoactive effects are predominantly predicated on the partial agonist properties of the terpenophenolic Δ-9-tetrahydrocannabinol (THC) at G-protein-coupled CB_1 and CB_2 cannabinoid receptors, with subsequent activation of the ubiquitous endocannabinoid neuromodulatory system leading to downstream effects across multiple neurotransmitter systems.[1079] *Salvia divinorum*, on the other hand, contains the diterpene salvinorin A, which is a selective agonist at the widely distributed G-protein-coupled κ-opioid receptors that make up a key component of the endogenous opioid system, again driving multifarious downstream effects in other neurotransmitter systems.[1080] This mode of action may also be shared, albeit to a more modest extent, by the chemical constituents of other *Salvia* species such as *Salvia sclarea* and *Salvia dichroantha* and the individual monoterpene menthol.[1081,1082]

A number of terpenes also bind to and modulate the activity of acetylcholine receptors. At the volatile end of the spectrum both camphor and borneol are ligands at nicotinic receptors,[1056,1083] and extracts of terpene-rich *Melissa officinalis* beneficially modulate human cognitive function by binding to both human nicotinic and muscarinic cholinergic receptors.[1084–1086] At the other end of the spectrum, triterpene steroidal lactones from *Withania somnifera* have been shown to bind to muscarinic receptors in specific regions of the rat brain after oral administration.[1087]

A number of other more isolated terpene/receptor interactions have been established, and these will be described in subsequent chapters. One key point is that all of these receptor interactions would be equally prevalent to insects, although they have not yet been confirmed in this taxon. The one exception here is THC from cannabis, as the insect nervous systems lacks the CB_1 and CB_2 receptors at which it binds. However, a number of other invertebrate taxa do express these receptors, and other cannabinoids, in particular the equally prevalent cannabidiol, exert multiple effects within neurotransmitter systems that would be expected to have an impact on the analogous insect systems.

Ecdysteroids

The many plant-synthesized triterpene compounds that disturb the functioning of the ecdysteroid insect hormonal system also illustrate the close, if unintended, relationship between the ecological roles of terpenes and human brain function. 20-hydroxyecdysone, the compound that functions as the primary insect ecdysteroid hormone and that is also synthesized as a defensive secondary metabolite by a variety of plants, modulates a range of physiological processes when administered to mammals. These include having hormone-like anabolic effects within metabolic pathways, including those that underpin protein synthesis and lipid and carbohydrate metabolism. These effects lend this ecdysteroid the ability to increase muscle mass and physical performance, along with hypoglycemic, antioxidant, immune-protective, and hepatoprotective properties.[1025] Spinach (*Spinacia oleracea*) is a particularly rich source of 20-hydroxyecdysone, which is synthesized in response to insect herbivory via the jasmonate pathway.[1088] Spinach itself exhibits both in vitro and in vivo anabolic effects in mammals.[1089] Curiously, the anabolic effects of spinach had been noted by the time of Popeye the Sailor, and he consumed it with gusto when in need of superhuman strength. However, his creators didn't realize this was because of spinach's rich complement of 20-hydroxyecdysone, and they attributed its effects instead to its high iron and vitamin A content.

Although ecdysteroids exhibit a range of hormone-like effects in mammals, their mechanisms of action are poorly understood. Ecdysteroids do not seem to bind to classic vertebrate steroid nuclear receptors and therefore do not have the side effects associated with the activity of anabolic steroids at androgen receptors; rather, they may bind to as yet ill-defined, membrane-bound G-protein-coupled receptors, thereby influencing signal transduction.[1025,1089] This suggests a further potential mode of action for endogenous steroidal hormones.

Rather more intriguingly, a number of the other steroidal and nonsteroidal plant triterpenes with secondary metabolite roles that include interactions with the insect ecdysteroid system also have multifarious physiological effects in mammals. These include compounds such as ursolic acid, the ginsenosides, the withanolides, and the brassinosteroid plant hormones. In the case of these phytochemicals the mechanism of action may well be interactions with glucocorticoid and estrogen nuclear receptors.[1029,1090,1091] The ubiquitous glucocorticoid nuclear receptors, in particular, are present in every cell in the body and play a wide range of genomic and non-genomic roles. They typically modulate stress responses, including immune function and inflammatory processes, although their activity also encompasses widespread effects on brain function. Many of the wide-ranging effects of this subset of phytochemicals, including, for a number of this group, their "adaptogenic" anti-stress effects and their ability to modulate brain function, could well be explained solely by their interactions with these receptors. This would encompass their many in vitro/ex vivo modulatory effects on the transcription factors that drive inflammatory responses and contribute to carcinogenesis. The key point here is that insect ecdysteroid receptors and mammalian hormone receptors are all members of the same conserved metazoan nuclear receptor superfamily.[1092] Furthermore, the insect ecdysteroids, the mammalian steroid hormones (glucocorticoids, estrogens, androgens), the plant brassinosteroid hormones, and the phytoecdysteroids, ginsenosides, withanolides, and triterpene acids are all structurally related modified triterpenes. It seems entirely possible that the ability of the plant secondary metabolites to modify mammalian physiological parameters, including brain function, may be a byproduct of their ecological role as spoilers of the insect ecdysteroid hormonal system, with their more benign nature in mammalian systems predicated on the many differences in insect and animal hormonal systems. This possibility and some alternative hypotheses are discussed in more detail in Chapter 13. Representative structures of the hormones and secondary metabolites are shown in Figure 13.1 in Chapter 13.

■ SOME CONCLUSIONS

The ubiquitous terpenes play a host of primary and secondary metabolite roles for plants. The latter include a variety of interactions with the nervous systems of insect herbivores, pollinators, and symbionts. These insect nervous system effects seem to correspond broadly with the mechanisms of action of the same phytochemicals in humans and other mammals. This suggests that many of the effects on the human brain are likely to be predicated simply on the similarity between the insect and mammalian nervous systems. However, examples drawn from the literature on the comparative effects of terpenes with regard to insect and mammalian GABAergic and hormonal function suggest that differences in the morphology and functions of these systems in the respective taxa drastically modify the toxicity of the phytochemicals in question. The ecological roles and nervous system effects of a range of terpenes and terpene-containing extracts will be expanded on in the next three chapters.

11 The *Lamiaceae* Subtribe *Salviinae*—The *Salvia*, *Rosmarinus*, and *Melissa* Genera

The *Lamiaceae*, or mint family, is the sixth largest family of angiosperms, comprising 236 genera and 7,200 species of plant. These include a host of economically important food, culinary, and medicinal plants. The family can be subdivided into seven subfamilies. The *Nepetoideae* is the largest of these, comprising a third of the mint family's genera and half of its species. Phylogenetic evidence suggests that this subfamily arose some 57 million years ago in what was to become the modern-day Mediterranean region.[1093] The family history of the *Nepetoideae* took something of a turn when a single plant fostered the *Mentheae* tribe in the mid-Eocene (~46 million years ago), and one of its subsequent progeny gave rise in turn to the *Salviinae* subtribe 10 or more million years later, at the end of the Eocene (~34 million years ago). The *Salviinae*, still living around its home shores in the Mediterranean and extending into Southwest Asia, would go on to diversify to include the large *Salvia* genus, as well as smaller genera such as *Melissa* and *Rosmarinus*. All of these genera encompass psychotropic species.[1093]

While the ancestors of psychoactive species such as *Salvia officinalis, Salvia lavandulaefolia, Melissa officinalis*, and *Rosmarinus officinalis* stayed rooted in their family's home range, other members of the family were dispersed farther afield. They traveled, settled, and diversified into the 90-plus species of the *Salvia* genus that colonized Southeast Asia, a group that includes the neuroprotective *Salvia miltiorrhiza*. Likewise, one relatively recent emigrant arrived in Mexico less than 14 million years ago and fostered the *Calosphace* subgroup of 300 or more American *Salvia* species that include the hallucinogenic *Salvia divinorum*.[1094] One interesting feature of all of these various *Salviinae* species is that their members have been used independently by a number of distinct cultures, both across the millennia and around the globe, as psychotropics or cognition enhancers.

Naturally, the members of the extended *Nepetoideae* subfamily share a number of morphological and chemical similarities; for instance, unlike some of the other *Lamiaceae* subfamilies, the *Nepetoideae* rely heavily on volatile monoterpenes and sesquiterpenes to fulfil many of their ecological needs.[86] They therefore produce particularly high levels of terpene-rich essential oil. As the plants of the subfamily become more closely related, their complement of genetically determined terpene synthases becomes progressively more similar.[1095] So, as we progress up the *Nepetoideae* taxonomic tree into the *Mentheae* branch and *Salviinae* sub-branch, the similarities in their palette of terpene secondary metabolite chemicals increase.

One further commonality across many species of the *Lamiaceae* is their employment of glandular trichomes as the site of terpene synthesis and storage. Glandular trichomes are a variation of the many hair-like structures seen on plant tissue across vascular angiosperms, and they give the *Lamiaceae* leaves their familiar texture and stickiness and their distinctive aromas when damaged. In the *Mentheae* these trichomes come in both peltate and capitate guises and typically comprise a basal cell, in which terpene synthesis takes place; a stalk cell; and one, two, or four sack-like head cells that store the potentially toxic chemicals safely sealed off from the vasculature of the plant.[135] In many ways glandular trichomes operate as self-contained "chemical factories," receiving just a few simple inputs that fuel the primary metabolic pathways that generate energy and synthesize precursors. These are then translated by the internal synthetic pathways into the final chemical products, which, in turn, are safely stored externally to the plant's vasculature in the head cells to avoid autotoxicity.[134] These chemicals can, depending on the type of trichome, be emitted according to environmental factors such as heat, light, or time of day, or alternatively the sack-like head cells might be designed to rupture on contact, releasing a cloud of volatiles into the air or a sticky resin that unleashes both chemical and gummy physical defenses onto any small herbivore. The density of glandular trichomes is dictated by a number of environmental factors and controlled by the jasmonate family of plant hormones that play a crucial role in the synthesis of most defensive secondary metabolites.[1096] The chemical contents and the type of trichome then vary according to ecological necessity. So, for instance, *Salvia* glandular trichomes situated on the leaves and calyces that protect flowering parts, particularly in areas such as the underside of the leaf favored by insects and other herbivores, synthesize typically "defensive" monoterpenes and sesquiterpenes (such as camphor and beta-caryophyllene) and occasionally diterpenes, ready for deployment. In contrast, trichomes on flowering parts synthesize and emit a cocktail of terpenes with a greater accent of typically "attractive" monoterpenes (such as borneol and limonene), which can then be slowly emitted as required.[1034,1035] Observations of the behavior of pollinating insects visiting rosemary (*Rosmarinus officinalis*) show that they are adept at collecting nectar and pollen without touching the defensive trichome-bearing plant parts.[1097]

Naturally the functional nature of the individual chemicals also depends on the quantity and mode of release. So, for instance, attractant chemicals might be released in small quantities at the appropriate time, whereas defensive chemicals may be released only upon rupture of the head cell. It is notable that the *Lamiaceae* provide the largest proportion of reputedly insecticidal species of any plant family, with one paper reporting that 28% of plants reported to have insecticidal properties in the literature came from this clade.[367] In line with this, essential oils contained in the trichomes of plants from across the *Mentheae* taxon generally exhibit dose-dependent insecticidal properties against adult insects[1098] and kill and inhibit the growth of larvae from a number of species.[1099,1100]

The terpene-producing glandular trichome "chemical factories" lend the *Nepetoideae* subfamily a number of useful properties, other than as insecticides, from a human perspective. The taxon provides a wealth of culinary herbs, for instance oregano, basil, thyme, sage, lavender, spearmint, peppermint, and rosemary. The entire *Mentheae* family also provides a rich resource for the distillation of essential oils, which are employed, again, as insect repellents, as odors for the perfume and toiletry industries, as well as in aromatherapy. The group also affords a wide range of herbal medicinal treatments and probably represents a largely untapped resource in this respect. A number of the *Salviinae* subtribe in particular have notable effects on human brain function, and these will be reviewed below. The potential relationships between the ecological roles of these plants' terpene secondary metabolites and human brain function will also be explored at the end of the chapter.

■ THE *SALVIA* GENUS: SAGE

From its origins in the Mediterranean region, members of the *Salvia* genus traveled across the globe repeatedly, establishing two new regions of species diversification in Eastern Asia and Central/South America. These regions have added a further total of 90 and 500 species respectively to the contemporary roll call of some 250 European/ West Asian *Salvia* species.[1101] In each of these regions an entirely separate tradition of medicinal use has also grown up, driven by the differing pharmacological properties of the indigenous species. One commonality, though, is that the active chemical components of each species are drawn from the *Salvia* genus's arsenal of terpene secondary chemicals, which typically include a number of common terpenes in varying concentrations, such as 1,8-cineole, camphor, boreol, bornyl acetate, and β-caryophyllene.[1055]

European Sage: *Salvia officinalis* and *Salvia lavandulaefolia*

In Europe and the lands bordering the Mediterranean, the written history of the use of the *Salvia* genus stretches back to the Ancient Egyptians, although the genus owes its name to the Romans (from the Latin *salvare*, "to save"). The most commonly used European members of the genus are *Salvia officinalis* (garden sage) and *Salvia lavandulaefolia* (Spanish sage); cognition-enhancing properties have long been attributed to both. For instance, the Greeks considered *S. officinalis* to be "good for helping diminution of senses and loss of memory".[80] The Salerno medical school, the first in medieval Europe, considered an herb they named *Salvia salvatrix* to be sacred. It was described as a cure with a calming effect and featured in the proverb from the "Tabuli Salerni," "*Cur moritur, qui salvia crescit in horto*" (Why should he die who has sage in his garden?).[1102] This herb was later renamed in Carl Linnaeus' taxonomy as *S. officinalis* and was in common usage throughout Northern Europe, featuring in British herbal apothecaries from the 16th century onwards.[1103] Perry et al.,[1104] noting that during this epoch *S. officinalis* was recognized as an enhancer of memory, provide quotations from some of the foremost herbals of the day. For instance, Gerard, in the 16th century, suggests that "It is singularly good for the head and brain and quickeneth the nerves and memory"; Culpepper's "Complete Herbal" notes, in the mid-17th century, that "It also heals the memory, warming and quickening the senses"; and John Hill's "The Family Herbal" tells us, in the 18th century, that "Sage will retard the rapid progress of decay that treads upon our heels so fast in latter years of life, will preserve faculty and memory more valuable to the rational mind than life itself." One further curious indication for *S. officinalis*, which hints at potential anxiolytic effects, is in the mitigation of grief. As an example of this, Samuel Pepys in his diary notes that "Between Gosport and Southampton we observed a little churchyard where it was customary to sow all the graves with Sage".[1105] Interestingly, the medicinal use of indigenous varieties of sage in both Central America and China was supplemented with *S. officinalis* following contact with Europeans. In the case of China, *S. officinalis* became preferred above indigenous varieties, with an exchange rate in the Orient of 3:1 finest tea to European sage tea.[1105]

Specific contemporary indications for *S. officinalis/lavandulaefolia* include its use as a gargle or mouthwash for inflammation of the mouth, tongue, or throat; the alleviation of flatulent dyspepsia and loss of appetite; the reduction of blood sugar; as a treatment for cases of respiratory allergy, headache, anxiety, and nervousness in the elderly; and for the attenuation of poor memory, mental confusion, depression, and vertigo. Finally it is also suggested as a treatment for the symptoms of the menopause.[1106]

In terms of chemical components both *S. officinalis* and *S. lavandulaefolia* contain about 1.0% to 2.8% volatile oil,[1107] which is primarily composed of monoterpenes, most notably α-pinene, β-pinene, 1,8-cineole, camphor, geraniol, borneol, and camphene.[1034,1108] The oils also contain sesquiterpenes such as α-humulene and β-caryophyllene,[1034] diterpenes such as galdosol, carnosic acid, and carnosol,[1071] and triterpenes such as ursolic acid and oleanolic acid. The only really major difference in terpene constituents between the two species is that *S. lavandulaefolia* lacks the α- and β-thujone content of *S. officinalis*. Both α- and β-thujone are toxic convulsants, so *S. lavandulaefolia* is therefore theoretically more suitable for human consumption. However, it should be noted that *S. officinalis* does not exhibit the toxicity in humans that its thujone content might suggest.[1109] Both species also synthesize a number of polyphenolic compounds, including rosmarinic acid, methyl carnosate, caffeic acid, luteolin 7-0-glucoside, luteolin, apigenin, and hispidulin,[1110] and in the case of *S. officinalis* several salvianolic acid derivatives of rosmarinic acid.[1111,1112]

In terms of brain function, a number of individual *S. officinalis/lavandulaefolia* terpene components bind allosterically to inhibitory ionotropic GABA receptors. These include agonists such as the monoterpene borneol[397]; the diterpene galdosol, which binds to the benzodiazepine binding site[1071]; and carnosol and carnosic acid, which bind directly to the chloride channel of the $GABA_A$ benzodiazepine receptor complex.[1113] In contrast, the monoterpene α-thujone is a potent allosteric antagonist at GABA receptors.[1114] The resultant removal of GABA's inhibitory effects on neuronal activity lead to α-thujone's convulsant properties. However, despite a putative role as the psychoactive principal underlying the damaging effects of absinthe, α-thujone doesn't seem to have any damaging effects when consumed in the context of the complex components of *S. officinalis* extracts, potentially on account of the opposite effects of other terpene components on GABA function.

Essential oils and alcoholic extracts of both *S. officinalis* and *S. lavandulaefolia* have also been shown to inhibit acetylcholinesterase (AChE) in postmortem human brain homogenates or human erythrocytes[1049–1052,1054,1055,1085,1115,1116] and in vivo in the striatum and hippocampus of aged rats following oral administration.[1063] Several studies have also demonstrated inhibition of butyrylcholinesterase (BuChE), a cholinesterase found principally in the liver.[1049,1055] These cholinesterase-inhibiting properties have also been shown to be due to synergistic interactions and antagonisms between the terpene components.[1054–1056] In this respect 1,8-cineole is both the most abundant and the most active individual monoterpene; it works synergistically with α-pinene or caryophyllene but antagonistically with camphor. Components and extracts also exert nonenzymatic cholinergic properties, with camphor and borneol both binding to and modulating activity at nicotinic receptors.[1056,1083] In one rodent study that involved co-administration of receptor agonists and antagonists, the memory-improving properties of an ethanolic extract of *S. officinalis* appeared to be directly attributable to direct nicotinic and muscarinic receptor binding.[1117]

These cholinergic properties have also been matched with consistent demonstrations of improved cognitive function in humans following administration of single doses of *S. officinalis/lavandulaefolia* with in vitro cholinesterase-inhibiting properties. In the first of a series of double-blind, placebo-controlled, balanced crossover studies, Tildesley et al.[1051] demonstrated memory improvements (immediate and delayed recall) in healthy young humans within 1 hour and 2.5 hours of taking single oral doses of 50 μL and 100 μL of *S. lavandulaefolia* essential oil respectively. Similar mnemonic effects were subsequently confirmed following single doses of 25 μL and 50 μL of the same essential oil, along with improved performance on a mental arithmetic task and improved levels of subjective alertness, calmness, and contentment.[1052] A similar pattern of mood effects was also seen in comparison to placebo following the ingestion of 300 mg and 600 mg of encapsulated dried *S. officinalis* leaf, with reduced anxiety also evident following the lower dose, although these mood effects were subsequently abolished by completion of a laboratory psychological stressor.[1049] The memory-enhancing effects of sage were also confirmed following the two lowest of four single doses (167, 333, 666, 1,333 mg) of an ethanolic extract of *S. officinalis* administered to healthy elderly volunteers (>65 years). The 333-mg dose also improved attention task performance across all but one of the four post-dose assessments that were completed over the course of 6 hours after consuming the sage.[1050] Most recently the memory and attention-enhancing properties of *S. lavandulaefolia* were confirmed in healthy young adults.[1053] In this case the treatment was an encapsulated essential oil, exclusively containing monoterpenes, with a high concentration of 1,8-cineol and a particularly potent acetylcholinesterase inhibitory profile. These findings suggest that the monoterpene contents of sage extracts may underlie any acute beneficial effects on brain function and that these may, in turn, be related to cholinesterase inhibition. One study has also assessed the effects of aromatherapy with either *S. officinalis or S. lavandulaefolia* and a no-aroma control condition on cognitive function,[1118] finding that only *S. officinalis* improved memory performance in healthy young adults. Both aromas also improved mood, possibly reflecting the expectations of the participants.

While cholinergic properties may underlie the cognitive enhancement seen following sage, it is notable that extracts also exert antioxidant, anti-inflammatory, and mildly estrogenic properties,[1119] all of which may be relevant to brain function, both in healthy populations and sufferers from age-associated cognitive decline and dementia.[1106] Sage extracts are also well tolerated and exhibit few unwanted side effects at the levels likely to be consumed by humans, making them a potential replacement for the side-effect–prone, alkaloid-based cholinesterase inhibitors that are typically prescribed for the attenuation of the cognitive deficits related to Alzheimer's disease. To date the applications of sage in the latter groups has been investigated in only a single, double-blind, placebo-controlled trial in which a small cohort (N = 30) of Alzheimer's disease patients received a tincture of *S. officinalis* for 16 weeks. In comparison to the placebo, the sage group exhibited improvements in both cognitive (Alzheimer's Disease Assessment Scale—cognitive subscale) and behavioral function (Clinical Dementia Rating).[1120]

Salvia divinorum

The most notorious member of the *Calosphace* subgroup of 300 *Salvia* species that speciated in the Americas is *Salvia divinorum*. This plant has been used indigenously as a hallucinogen and as a treatment for a range of conditions, such as diarrhea, stomach conditions caused by "curses," headache, and rheumatism, for hundreds, if not thousands, of years.[1121] When *S. divinorum* was rediscovered in the 1960s its use had been restricted to the Sierra Mazateca mountains in Mexico. However, its closest relative, *Salvia venulosa*, is found exclusively in the Colombian Andes. As *S. divinorum* is propagated vegetatively by humans and appears to have been introduced into its current range, it is likely that it originated in the Andes.[1094] While few of the *Calosphace* sages have been subject to chemical characterization, it is notable that several other members, most notably *Salvia splendens* from Brazil, share the synthesis of the *S. divinorum*'s trademark neoclerodane diterpenes and may therefore share some of its psychoactive properties.[1122]

In a historical context *S. divinorum* is most often associated with the Mazatecs of northeastern Oaxaca in Mexico, who have used the leaves either chewed or in infusions along with teonanacatl and ololiuqui (*Psilocybe* mushrooms and *Turbina corymbosa* seeds respectively—see Chapter 6). The leaves have been consumed for their hallucinogenic properties in healing ceremonies, divination, and spiritual rites since well before they were first documented following the Spanish conquests.[1121] The current indigenous name, ska Maria pastora (the leaves of Mary the Shepherdess), and the belief that the plant represents the incarnation of the Virgin Mary reflect the incorporation of elements of Christianity into the belief system of the Mazatecs.[1121,1123] However, *S. divinorum* may also represent the plant known by the Aztecs as pipilzintzintli and the leaves used by the Cuicatec, Chinantec, and Otomı peoples as hallucinogens, and it may therefore have reached the area considered its current home via the trade routes of the ancient Nahua civilizations.[1094,1124]

The plant itself first came to the attention of Western science after an expedition in the early 1960s by the ethnomycologist R. Gordon Wasson, accompanied by the pharmacist Albert Hofmann and his wife, to search for the fabled hallucinogenic plant of the Mazatecs. Wasson and Anita Hofmann took *S. divinorum* during a local ceremony and experienced visual hallucinations, but on returning to Basel Hofmann was unable to identify the active ingredient in his (now ineffective) alcoholic tincture of *S. divinorum*. The elusive principal psychoactive component of *S. divinorum* was eventually identified in the early 1980s as salvinorin A, a neoclerodane diterpene.[1124,1125] This compound's mechanism of action, an ability to bind to and activate κ-opioid receptors, was only identified some two decades later.[1080] Salvinorin A itself is typically the most abundant diterpene in *S. divinorum* plant material, but its concentration is variable, and it usually constitutes less than 0.5% of plant material. It is also just one of a wide range of neoclerodane diterpenes, which also include salvinorins B to J, divinatorins A to F, salvinicins A and B, and salvinidins A to D, which generally lack substantial κ-opioid receptor activity.[1126] The plant also expresses a typical palette of monoterpenes and triterpenes. In the plant the synthesis and storage of these various terpenes is restricted to peltate trichomes.

It is interesting to note that salvinorin A is both structurally dissimilar to all of the other hallucinogens, being the only non-alkaloid structure, and that it exerts its effects by a unique mechanism, binding to and activating κ-opioid receptors rather than the $5HT_{2A}$ receptor that represents the principal common target of the other hallucinogens.[1080] In a recent "receptorome" screening program involving 35 putatively hallucinogenic natural and synthetic compounds, salvinorin A was shown to be the most selective compound in terms of receptor binding. Its affinity was restricted to the κ-opioid receptor, where it proved the single most potent binding ligand (followed by ibogaine and morphine).[372] Single studies have, however, also demonstrated allosteric interactions at the μ-opioid receptor, binding to dopamine D_2 receptors, while downstream effects within the noradrenaline, dopamine, and endocannabinoid systems have been reported.[1126] Several discriminative stimulus studies, which assess the extent to which an operant conditioned response to one drug translates to a another drug, have shown that conditioned responses to salvinorin A generalize to other κ-opioid receptor agonists in rats[1127] and monkeys.[1128] In the latter study the discrimination didn't translate to μ- and δ-opioid agonists, psilocybin, or ketamine, and was blocked by opioid but not 5-HT_2 receptor antagonists. These studies confirm the κ-opioid receptor as salvinorin A's site of action.

While the unique and potent κ-opioid receptor properties of salvinorin A and several close synthetic analogues have engendered a huge research effort, the other neoclerodane diterpenes have been comparatively overlooked. However, several other diterpenes have measureable, if weaker, κ-opioid receptor activity,[1129] including salvinicin A, salvinorin B, divinatorin D, and salvinorin G.[1130,1131] Salvinicin B also binds antagonistically to the μ-opioid receptor.[1131]

Animal models suggest that salvinorin A can reduce anxiety and that it has antidepressant properties at low doses but depressant properties at higher doses. It can also engender conditioned place preference at low doses but aversive behavior at higher doses. These properties may be a consequence of biphasic modulation of dopaminergic activity via an unidentified mechanism related to κ-opioid activation, and these dopaminergic effects can be blocked by antagonists of this receptor. Overall, the typical effect of salvinorin A is to decrease dopamine levels in the brain, including within the reward pathways; in line with this, high doses inhibit cocaine-seeking behavior in rats, suggesting a potential utility in the treatment of drug dependence.[1132]

Since its rediscovery by the West in the 1960s, the use of *S. divinorum* has been slowly working its way around the world. The leaves of the plant and salvinorin A are typically smoked or vaporized in a similar manner to freebase cocaine, with an active dose typically equating to 200 to 500 μg salvinorin A. By 2011 the sale or possession of *S. divinorum* has been regulated or banned in 20 countries and 23 states in America.[1126] Despite, or indeed because of, this, the results of a nationally representative survey published in the United States the same year, which included data from 70,000 individuals aged over 12 years, showed that 5% of adults had used *S. divinorum* at some point in their lives.[1133] The plant certainly seems to have made substantial inroads into drug culture, although this seems to have been driven largely by curiosity. More than half of the people who reported having used the drug in an online survey reported that they no longer did so, typically because they didn't particularly like the high or had simply lost interest.[1134]

In terms of *S. divinorum*'s psychoactive properties, a recent study trawled through several thousand potentially relevant video clips on YouTube and identified 34 films showing individuals' *S. divinorum* trips in their entirety. The researchers then analyzed overt behaviors such as immobility and facial and physical relaxation, emotion, and disordered speech. These video clips suggested a rapid onset and resolution of the drug's effects, with the strength of effects related to the number of inhalations and the length of time smoke was held in the lungs before exhalation.[1135] More conventional survey data also suggests that *S. divinorum* is typically used for recreational or spiritual reasons, occasionally rather than frequently; like the other hallucinogens, it seems to exert a low risk of dependence.[1136] Respondents typically report that the subjective effects of *S. divinorum* are almost immediate and that they peak relatively rapidly, in the region of 8 minutes after smoking, and are short-lived, typically lasting 20 minutes or less.[1137] A positron emission tomography (PET) study in baboons suggested that the transience of the experience is related to the very rapid penetration of the blood–brain barrier by salvinorin A,[1138] followed by its rapid metabolism to the much less active salvinorin B.[1126] The subjective experience is typically reported by users to include distortions of visual and auditory perception, mood and bodily sensations, and a highly modified perception of external reality, engendering a sense of unreality and depersonalization, bodily dissociation, and recall of past places and events, along with physical impairment.[111,1136,1137] This combination of hallucinogenic and dissociative effects is a profile unique to *S. divinorum* but has elements in common with the alkaloid hallucinogens and NMDA receptor antagonists such as ketamine.[1136] The PET study in baboons showed that the highest concentrations of radiolabeled salvinorin A were seen in the striatal cortex and cerebellum,[1138] areas of the brain responsible for visual processing and integrating sensory perception with motor control, respectively. Local effects in these areas may underlie some of salvinorin A's reported perception- and somatic-distorting effects.

The effects of *S. divinorum* and salvinorin A have been assessed in a small number of human trials. In the first of these, Siebert[111] assessed the comparative effects of differing modes of administration of *S. divinorum* and salvinorin A in studies involving six and 20 participants respectively. He found that swallowing the juice of the leaves or encapsulated salvinorin A was ineffective, whereas holding the juice in the mouth in liquid form, or spraying salvinorin A dissolved in alcohol into the mouth, prompted psychoactive effects that commenced 5 to 10 minutes after application. However, vaporization and inhalation of salvinorin A led to a more rapid and intense onset of effects that also dissipated more quickly. As *S. divinorum*'s mechanism of action was unknown at the time, Siebert also instigated a comprehensive in vitro screening of the potential receptor binding mechanisms underpinning salvinorin A's effects. However, he drew a blank, unfortunately having failed to include opioid receptor assays.

Two further studies have assessed the subjective effects of salvinorin A either smoked or inhaled as volatilized vapor. In one study Johnson et al.[1139] administered 16 ascending doses (0.375 μg/kg to 21μg/kg) of volatilized salvinorin A, along with four interspersed placebos, on separate days to four hallucinogen-experienced volunteers. The participants relaxed in reclining position with their eyes covered and gave verbal ratings of the strength of the drug's effects at 2-minute intervals and then completed the Hallucinogen Rating Scale and the Mysticism Scale 1 hour after taking the drug. The results showed a dose-dependent increase in ratings of drug strength and a consistent peak in the strength of effects that followed almost immediately after consumption. This was followed by a decrease to baseline over 30 minutes. The subjective effects, as assessed by the hallucinogen and mysticism rating scales and verbal reports, were similar to those seen following both psilocybin and dimethyltryptamine and included reports of physical and perceptual illusions and retrieved memories. In a subsequent study, on two separate occasions 30 hallucinogen-experienced participants smoked 25 mg *S. divinorum* leaf containing a dose of salvinorin A (100 μg) presumed to be inactive and the same quantity of leaf adulterated to give a total dose of 1,000 μg salvinorin A.[1140] The higher dose increased the movement, laughing, and physical contact of the participants and increased Hallucinogen Rating Scale scores. The majority of the participants described the effects of salvinorin A as being dissimilar to other drugs, and a sizeable minority found them similar to a state of dreaming.

To date, the research pertaining to *S. divinorum* has concentrated almost exclusively on salvinorin A, and no study has adequately attempted to disentangle the effects of salvinorin A from those of the other secondary metabolites expressed by the plant. This includes an absence of research investigating the effects of the other diterpenes with regard to other potential mechanisms of action.

Some Other *Salvia* Species

A number of other *Salvia* species show psychotropic potential that hasn't yet been matched to any direct evidence of their effect in humans.

The third distinct region in which the *Salvia* genus diversified was Eastern Asia, which counts some 90 local species of sage. A number of these species have been used in traditional Chinese medicine, with the genus being described as a "superior" herb in the Shen Nong Ben Cao Jing (25–220 AD). Medicinal species include *S. plebeia, S. chinensis,* and *S. miltiorrhiza.* The roots of the latter plant are highly valued as the versatile treatment called Danshen in traditional Chinese medicine.[1141] The early, historical use of the bright-red root (*miltiorrhiza* means "red juice extracted from a root") was dictated as a treatment for blood disorders by the Doctrine of Signatures, whereby the physical appearance of the plant part dictates the indication. Curiously, this use has survived into the modern era as it transpires that Danshen has well-established platelet-aggregation–inhibitory properties, leading to a thinning of the blood.[1141] Related to this, contemporary widespread use is also typically in the treatment or prevention of cardiovascular and cerebrovascular disease and hyperlipidemia.[1142]

S. miltiorrhiza is unusual within the genus in that the roots provide the medicinal extracts. The principal active constituents of extracts are a group of unique diterpenes including miltirone and up to 30 different tanshinones, which make up about 1% of the root and provide their bright-red color. Tanshinone I and II and cryptotanshinone are by far the most abundant. Their synthesis is controlled by the jasmonate hormones[1143,1144] and can be increased by a number of abiotic stressors and bacterial infection.[1145,1146] In vitro/ex vivo studies have shown that the tanshinones have a number of potentially beneficial properties, including inhibiting platelet aggregation and properties relevant to tumorigenesis, cardioprotection, reduction of inflammation, and vasodilation,[1141] along with antimicrobial properties.[1147] The root also contains high levels of triterpenes such as oleanolic acid, ursolic acid and phytosterols such as stigmasterol, plus a number of phenolic compounds, including flavonoids such as baicalin, phenolic acids, and rosmarinic acid and its derivatives, salvianolic acid A and B.[1141,1148]

In terms of brain function the beneficial effects of tanshinones are often attributed to their pronounced AChE inhibitory properties.[1057,1058] In this respect the individual diterpenes tend to be more potent than a complete extract, and this may be particularly relevant as the lipophilicity of the diterpenes suggests they will readily cross the blood–brain barrier.[1057] It is also notable that miltirone is a positive allosteric modulator at the $GABA_A$ benzodiazepine binding site.[1072] Animal behavioral studies have demonstrated that cryptotanshinone[1059,1060] and dihydrotanshinone[1060] have potent cholinesterase-inhibiting properties and attenuate the memory deficits related to scopolamine treatment in rats.[1059,1060] Tanshinone I and II have both also been shown to reverse diazepam-induced cognitive deficits in rodents,[1060] while an extract of

S. miltiorrhiza also increased cerebral blood flow and reduced levels of neuronal damage in rats with cerebral ischemia/reperfusion impairment,[1149] and tanshinone II attenuated neuronal damage in a rat model of oxidative stress.[1141] On the basis of these and a number of other studies it has been suggested that tanshinones may be a potential treatment for dementia,[1150] although to date there are no human data to support this contention.

Returning to the Mediterranean region, several studies have also assessed the cholinesterase-inhibitory properties of a range of different fractions extracted from *Salvia* species indigenous to Turkey.[1045–1047] These studies demonstrated that all of the *Salvia* species had antioxidant properties and that the majority also exhibited some anti-cholinesterase properties. However, this was typically restricted to the nonpolar, terpene-containing fractions. An investigation that included disparate essential oil extracts from 12 members of the *Nepetoideae* subtribe, including members of the *Melissa*, *Mentha*, *Lavandula*, and *Salvia* genera, found that almost all of the essential oils showed cholinesterase-inhibitory properties. Screening of the most abundant constituent chemicals suggested that this effect was due to the synergistic properties of the constituents of the total extracts.[1047] In a later study involving 55 *Salvia* species, *S. fruticosa* (also known as *Salvia triloba*) stood out as the most potent AChE inhibitor.[1048] This fits well with the history of this species as a memory enhancer in European folk medicine and demonstrations of both in vitro AChE inhibition and in vivo attenuation of scopolamine-related memory deficits in rodents.[1046] It may be notable that *S. fruticosa* also contains a high proportion of 1,8-cineole and a low proportion of camphor.[1054] In a separate study, *S. leriifolia* also exhibited both AChE and BuChE inhibitory properties, with this being most pronounced for a fraction containing a range of monoterpenes, sesquiterpenes, and diterpenes.[1151]

Interestingly, research also suggests that extracts from *Salvia* species other than *S. divinorum* exhibit κ-opioid receptor-binding properties. For instance, *S. sclarea* and *S. dichroantha*, species that are native to Turkey and that have local histories as analgesics, both partially displaced [(3)H] diprenorphine, an unselective opioid antagonist, from opioid receptors.[1082] A number of other *Salvia* species from the Calosphace subgroup, including *S. splendens*, also synthesize neoclerodane diterpenes that may possess opioid receptor-binding properties.[1122]

■ *MELISSA OFFICINALIS* (LEMON BALM)

Melissa officinalis is a cultivated, perennial, lemon-scented herb. Originating in Southern Europe, its cultivation and use had spread throughout Europe by the Middle Ages. Records concerning the medicinal use of *M. officinalis* date back over 2,000 years, with entries in the *Historia Plantarum* (approximately 300 BC) and the *Materia Medica* (approximately 50–80 BC). Medicinal use throughout earlier epochs included a recommendation by Paracelsus (1493–1541) that lemon balm would completely revivify a man and was indicated for "all complaints supposed to proceed from a disordered state of the nervous system"'.[1105] Several early herbal apothecaries also considered lemon balm tea to have not only general beneficial effects upon the brain, but also the ability to improve memory. Examples include, from the 17th century, John Evelyn, who writes that "Balm is sovereign for the brain, strengthening the memory and powerfully chasing away melancholy," and Nicholas Culpepper, who suggests that "Sage is of excellent use to help the Memory, warming and quickening the senses." Similarly, the authors of the

Pharmacopoeia Londinensis of 1618 recommended lemon balm for the "strengthening of the brain."

Contemporary indications include the administration of *M. officinalis* as a mild sedative, in disturbed sleep, and in the attenuation of the symptoms of nervous disorders, including the reduction of excitability, anxiety, and stress.[1106] *M. officinalis* is particularly widely used in the Iranian traditional system of medicine, where its uses include those noted above plus the treatment of headaches, flatulence, stomach disorders, vertigo, asthma, bronchitis, and epilepsy.[1152] *M. officinalis* has also been used in the enhancement of memory for more than two millennia.[1106] Its potentially active components primarily include monoterpenoids and sesquiterpenes, including 1,8-cineole, geranial, neral, 6-methyl-5-hepten-2-one, citronellal, geranyl-acetate, β-caryophyllene and β-caryophyllene oxide, and triterpenes such as ursolic acid and oleanolic acid.[1153]

The in vitro effects of *M. officinalis* extracts potentially relevant to the central nervous system include antioxidant properties,[1154] direct nicotinic and muscarinic cholinergic receptor-binding properties in human brain tissue,[1084–1086] and AChE-inhibitory properties[1085,1155] arising from synergies between components.[1054] An ethanolic extract of *M. officinalis* has also been shown to have an affinity for $GABA_A$ receptors,[1156] and a fraction containing rosmarinic acid and the triterpenes ursolic acid and oleanolic acid, as well as these individual compounds, inhibited the enzyme GABA transaminase, leading to increased GABA activity.[1076] These properties may underlie observations of a reduction in both inhibitory and excitatory neurotransmission, with a net depressant effect, in cultures of rat cortical neurons[1157] and the anxiolytic properties of extracts in terms of rodent behavior.[1158] Similarly, methanolic and to a lesser extent aqueous extracts have been shown to scavenge free radicals and inhibit MAO-A, potentially increasing the availability of monoamine neurotransmitters.[1159] It is notable that β-caryophyllene is also a cannabinoid CB_2 receptor ligand and thus offers potential for the prevention/treatment of inflammation.[1160] In an interesting study the effects of an *M. officinalis* infusion (1.5 g twice per day) were studied in radiology staff exposed to persistent low-dose radiation. Over the course of 30 days the individuals' endogenous antioxidant capacity was enhanced, while oxidative stress-related damage was reduced.[1161]

In humans two randomized, double-blind, placebo-controlled, balanced-crossover trials have demonstrated dose-dependent memory decrements[1162] and anxiolytic-like modulation of mood[1162,1163] following single doses of an ethanolic *M. officinalis* extract. However, the extract in question had no cholinergic receptor-binding properties. A subsequent trial used an encapsulated dried *M. officinalis* leaf with nicotinic and muscarinic-binding properties (in human *post mortem* brain tissue) and demonstrated a dose-related improvement in memory task performance and ratings of mood.[1084] These results suggest that *M. officinalis* extracts have consistent anxiolytic properties, which may be related to modulation of the inhibitory GABA system. However, it would appear that interactions with nicotinic and/or muscarinic receptor interactions are required for overall improvements in cognitive function.

Three double-blind, placebo-controlled studies have also assessed the effects of *M. officinalis* in dementia patients, with Ballard et al.[1164] finding improvements in agitation and quality of life following essential oil aromatherapy in 71 patients with severe dementia. A subsequent study in a slightly larger cohort, which included a donepezil condition, failed to replicate this effect, with all three conditions improving equally. The authors attribute this to beneficial effects of the simple application of sham and active essential oils.[1165] Akhondzadeh et al.[1166] also demonstrated reduced agitation and improved cognitive (Alzheimer's Disease Assessment Scale—cognitive subscale) and behavioral function (Clinical Dementia Rating) following 16 weeks' administration of a *M. officinalis* alcoholic tincture to a small cohort (N = 35) of patients with mild to moderate dementia.

ROSMARINUS OFFICINALIS

The *Rosmarinus* genus of four plants, of which *Rosmarinus officinalis*, or rosemary, is the most celebrated member, is closely related to *Salvia officinalis*. Indeed, phylogenetic analyses show that *R. officinalis* is a closer relative of *S. officinalis* than many of the globally distributed *Salvia* genus.[1093] Rosemary itself has a long and eminent history as a cognition enhancer and has a particular association with the remembrance of love and death. This facet of its history originates with the ancient Egyptians' use of the herb in the mummification process, and it spans the cultures of ancient Greece and the Romans. In these latter cultures one of its other, more curious uses was as a memory enhancer. Greek students wore sprigs or garlands of rosemary at times of educational demand, and Roman students massaged their temples and foreheads with rosemary oil prior to examinations. This notion of rosemary as a cognition enhancer survives to the present day, with references sprinkled throughout the historical and literary record. This is nicely illustrated in *Hamlet* by Shakespeare, who has Ophelia distribute a bouquet of symbolic flowers after learning of the death of her father, saying, "There's rosemary; that's for remembrance." Reference to the mnemonic potential of rosemary can also be found in the writings of most herbal apothecaries from the 15th century onwards. Examples from the 16th and 17th centuries include allusions in Culpepper's *Complete Herbal*: "It helps a weak memory, and quickens the senses," and in the work of John Gerard, who writes, "Rosemary comforteth the braine, the memorie, the inward senses." Treveris's *Grete Herball* of 1526 also includes an indication for "weyknesse of the brain".[1106]

The most widespread contemporary uses of rosemary are as an antimicrobial and antioxidant agent in foodstuffs and as a culinary flavoring. However, its herbal medicinal applications include its use as a tonic or stimulant and as a treatment for stomach complaints and dyspepsia, dysmenorrhea, rheumatic pain, headaches, circulatory disorders, and nervous tension, as an analgesic and antispasmodic. Finally, it retains its historic use as a treatment for memory problems.[1167,1168]

With regard to its constituents, essential oils consistently contain 1,8-cineole as the most abundant component, with this single compound representing up to 60% of the total volatile terpenes.[1169] Other monoterpenes include α-pinene, borneol, camphor, and camphene, and sesquiterpenes such as β-caryophyllene.[1169–1171] The components of the leaf include several nonvolatile diterpenes and triterpenes, most notably carnosol, carnosic acid and ursolic acid and oleanolic acid, and phenolics such as luteolin, caffeic acid, and rosmarinic acid.[1168,1171,1172]

Essential oil extracts of rosemary have been shown to have antibacterial and antifungal activity, although typically the potency of the extract is greater than the individual components, such as 1,8-cineole and α-pinene, alone.[1170] Extracts have also been shown to exert a wide range of in vivo effects following oral consumption in animal models. These include the bolstering of antioxidant status and the activity of endogenous antioxidant enzymes,[1168] anti-inflammatory effects,[1173] and analgesic properties.

The latter can be blocked by the opioid receptor antagonist naloxone, the $GABA_A$ receptor antagonist bicuculline, glutamate uptake inhibitors, and a $5\text{-}HT_{1A}$ receptor antagonist, suggesting the involvement of these neurotransmitter systems or receptors in the pain-relieving effects.[1174,1175] A recent fractionation study also suggested that the triterpene components alone were only marginally less potent than the whole extract in terms of analgesia.[1167] A potential opioidergic mechanism of action may also be suggested by the ability of aqueous and ethanol rosemary extracts to reduce morphine withdrawal symptoms in rodents.[1176] A recent ex vivo study also demonstrated that an ethanol extract abolished muscle spasms, and that this effect was mediated by the blockade of muscarinic receptors and calcium channels.[1177]

The cognition-enhancing effects of rosemary may relate to interactions with the cholinergic system. In an early study both fresh leaf and an essential oil demonstrated measurable in vitro cholinergic receptor-binding and AChE-inhibitory properties in human brain tissue.[1085] More recent studies have confirmed the AChE-inhibitory properties of an essential oil and demonstrated that multiple fractions of rosemary also inhibit BuChE.[1178] Importantly, the AChE-inhibitory properties have been confirmed in vivo in the mouse brain following oral administration.[1179]

These potentially relevant mechanisms are matched by rodent studies showing that rosemary extracts exert antidepressant effects,[1179] attenuate scopolamine-induced memory deficits,[1180] and improve short- and long-term memory.[1181] A handful of studies have also assessed the effects of rosemary on human brain function, either in the form of vaporized essential oil or dried leaf. In the first of these studies rosemary and lavender aromatherapy had opposite effects on cerebro-electrical activity (EEG) but improved mood and mathematical performance.[1182] However, the results of this study are difficult to interpret due to the lack of blinding or a control condition. However, in a more recent single-blind study, Moss et al.[1183] included a further "no-scent" condition and used a larger group of 140 participants. They found that those in the volatilized rosemary aromatherapy condition exhibited a significant enhancement of memory task performance, but with a corresponding significant reduction in the speed of memory task performance. Subjective ratings of alertness and contentedness were also increased in comparison to the no-scent condition. This study was followed by a correlational study in which 20 participants were exposed to vaporized rosemary essential oil for varying lengths of time before undertaking tasks assessing attention, working memory, and executive function. Intriguingly, not only was 1,8-cineole detectable in blood samples taken after exposure, confirming the bioavailability of the potentially active terpenes via absorption through mucous membranes, but performance of the two serial subtraction mental arithmetic tasks and a test of focused attention correlated with serum levels of 1,8-cineole in terms of faster performance and/or accuracy. Participants also felt less "content" with rising levels of the monoterpene.[1184] In contrast, in the only randomized, double-blind study to date, 28 older adults, with a mean age of 75, were administered single oral doses of 750, 1,500, 3,000, and 6,000 mg powdered rosemary leaf on separate occasions, after which their cognitive function was assessed over four time points spanning 6 hours. The results demonstrated improvements in function in terms of increased alertness and improved speed of performing memory tasks following the lowest dose but significant decrements on the same measures following the highest dose. The intermediate doses (and the lowest dose in one instance) were also associated with decrements, rather than improvements, in task performance.[1171] The last findings may illustrate the potential for even innocuous phytochemicals to have deleterious effects at high doses and may find parallels in the biphasic reward and deterrent effects of terpenes at low and high doses in insects.

ECOLOGICAL ROLES OF THE *SALVIINAE* TERPENES AND BRAIN FUNCTION

Given that the *Salviinae* are just a small subtribe within the much larger *Lamiaceae* (mint) family, it is no great surprise that all of the psychotropic plants described above employ similar secondary metabolite terpenes to fulfil the majority of their ecological needs. For instance, they typically express a similar complement of monoterpenes, typified by high levels of 1,8-cineole, α-pinene, camphor, and borneol. Their complement of sesquiterpenes typically include β-caryophyllene; their diterpenes feature carnosic acid and carnosol; and ursolic acid and oleanolic acid constitute their major triterpenes.

Many *Lamiaceae*, including all of the Mediterranean species described above, have established traditional roles as insecticides and insect repellents. Contemporary research has confirmed, for instance, the toxicity of rosemary as a fumigant in a variety of insect taxa[1185] and the insecticidal/nematicidal and anti-feedant effects of extracts or essential oils of *Salvia*, *Melissa*, and *Rosmarinus* species following volatile, oral, or topical application to insects, larvae, and nematodes.[1000,1098–1100,1186–1189] Alongside this they exert allelopathic properties against other plants.[1190]

The lipophilic volatile terpenes are readily absorbed by insects, and many of the individual monoterpenes found in abundance in these plant species have been assessed for their toxicity and repellency to adult and larval insect parasites and disease vectors. For instance, borneol, 1,8-cineole, α- and β-pinene, and camphene have all been shown to be deadly to the mosquito or its larvae and insecticidal or repellent to other insects. As an example, 1,8-cineole applied topically kills adult sandflies, blowflies, house flies, and louse eggs, while fumigation with vapor kills the blood-sucking "assassin bug" (*Rhodnius milesi*) and has greater toxicity than synthetic treatments for head lice. However, it is usually the case that the cocktail of terpenes in an essential oil is more effective than a single compound, with synergistic and suppressive interactions present among combined terpene components.[1044]

Many of the terpenes synthesized by the *Salviinae* also have a seemingly contradictory role, acting as airborne volatile attractants both for pollination, where they form a major part of the floral bouquets emitted by flowers, and as indirect defense chemicals. In this role their synthesis and release is induced by tissue wounding in order to attract the natural predators of attacking herbivores. The evolved role of these chemicals can clearly be seen when we consider that many of the most abundant monoterpenes and sesquiterpenes synthesized by these plant species are also endogenously synthesized and emitted by multiple insect families as pheromones or allomones, chemicals that regulate behaviors such as mating, aggregation, alarm signaling, and trail following.[126]

The complex and contradictory roles of these terpenes can best be illustrated by reference to a single compound that is particularly richly expressed across this subgroup of plants, wherein it often represents the single most abundant component in essential oils. The monoterpene 1,8-cineole is toxic to many insect orders, including flies, lice, and beetles.[1044,1191,1192] Indeed, it is so toxic when encountered in the diet that a number of insects have adapted countermeasures. For instance, larvae and adults of species of the Australian sawfly that feed on 1,8-cineole–rich eucalyptus leaves have developed a brush-like extension to their mandibles that is used to separate the toxic oil from the leaves and sequester it in a sack-like extension of their foregut, from where it is either excreted or used as a chemical defense by the larvae, earning them the common name "spitfire bugs".[1193] However, 1,8-cineole, when released in the form of natural volatile emissions, is harmless to many insect species, for instance bees.[1194] Indeed, it functions as a fragrant attractant for insect pollination[128] and in the indirect defense of the plant via attraction of the natural predators of herbivores.[1195] It is also synthesized by at least five insect families as a pheromone[126] and, for instance, forms a core component of the recruitment pheromone synthesized by bumblebees in order to summon their nest-mates following successful foraging.[1036] The comparatively small body of work conducted in insects shows that a number of mechanisms of action of terpenes are the same in insects and mammalian nervous systems. In the case of 1,8-cineole, this certainly holds true with regard to its agonistic allosteric binding at inhibitory insect and mammalian ionotropic GABA receptors,[1006,1065] which results in downregulation of the activity in the respective central nervous systems. In contrast, α-thujone owes its lethal effects to an opposite antagonist effect that leads to central nervous system overactivity in flies[1114] and convulsions in mammals.[1077] 1,8-cineole is also one of many monoterpene and sesquiterpene inhibitors of both insect and mammalian AChE,[1032–1034,1196] and it may modulate behavior in both insects and mammals accordingly. This does not appear to be the mechanism of its insecticidal/larvicidal effects[1020,1021]; in fact, when acting as an attractant it would be advantageous for 1,8-cineole to exert beneficial cholinergic effects within the nervous systems of symbiotic insects. Indeed, improved memory for the location of a deliberately attractive plant or a source of nectar or pollen could be one consequence of cholinergic upregulation via AChE inhibition and would benefit all parties in the mutualist exchange.

Naturally, these apparently contradictory roles in both attraction and defense must be a reflection of the overall exposure of the insect to 1,8-cineole (or other terpenes), as the mode of delivery, dose, and duration of exposure of the volatile terpenes dictate their toxicity.[1188] Hence, even endogenously synthesized insect allomones/pheromones become neurotoxic for the producing insect when they are administered at high enough doses.[1037] It is notable that the density of glandular trichomes, the comparative concentration of individual chemicals, and the mechanisms of chemical release of the trichomes differ according to age and the plant part in a similar manner throughout the *Salviinae*. So, for instance, chemicals functioning as attractants to pollinators will predominate in the slow-release trichomes in reproductive parts of the plant, whereas chemicals intended as toxins will be more concentrated in the trichomes on leaves and calyces that are designed to rupture on contact with a herbivore.[1034] This allows for the overall palette of terpenes to exert a wide range of ecological functions.

Naturally, the *Salviinae* also share a number of nonvolatile terpenes that seem to have more straightforward roles. For instance, ursolic acid and oleanolic acid are the principal triterpenes synthesized by all of the plants described above. These compounds are found in a wide variety of plants and are particularly concentrated in the skin of many fruits. They have specific insect anti-feedant properties[1197] that increase in direct proportion to their concentration.[1198] For instance, the life expectancy and fertility of aphids decrease in direct proportion to the amount of ursolic acid added to their diet.[1199] Similarly, armyworm moth (*Spodoptera frugiperda*) and *Drosophila* larvae suffer stunted growth and development along with increased mortality as increasing concentrations of ursolic acid are applied to their cuticles or added to their diet.[1200]

In mammals, ursolic acid has anxiolytic, antiepileptic, sedative, and anticonvulsant effects, which are most likely mediated via GABA neurotransmission.[1073–1075] It has analgesic effects that can be attenuated by naloxone[1075] and antidepressant effects mediated by agonist properties at dopamine D_1 and D_2 receptors.[1201] Ursolic acid also has a structure similar to the mammalian steroid hormones, which themselves are modified triterpenes, and it has been shown to bind to the ubiquitous vertebrate glucocorticoid receptors and other "orphan" nuclear receptors, endowing it with a wide range of potentially positive cellular properties (see Chapter 13).[1090] Ursolic acid appears to kill and disturb the growth of insects by impairing the functioning of the related insect (modified triterpene) ecdysteroid hormone system.[1200] Interactions with the related vertebrate/invertebrate hormonal systems may clearly be related, and naturally GABA, opioid, and dopamine receptors are expressed in insects and are equally prevalent as targets in both taxa.

Naturally, some terpenes unique to a particular species also function in ecological roles. The most striking example are the clerodane diterpenes synthesized by the hallucinogenic *S. divinorum*. These compounds are synthesized and stored in peltate trichomes with a defensive distribution across the plant.[1202] In line with this, the broad group of clerodane diterpenes have established roles as insect anti-feedants and deterrents, with the effects of an individual clerodane compound often restricted to a specific subset of the species.[1203] The resin from the plant also contains loliolide, a potent terpene ant repellent.[1204] It therefore seems likely that the ensemble of secondary chemicals perform defensive ecological roles. Whether the κ-opioid-receptor activity of salvinorin A is directed toward insects is difficult to divine. However, both μ- and κ-opioid receptors and the endogenous opioid receptor ligand enkephalin are widely expressed in the insect brain and nervous system,[325–328] and specific κ-opioid-receptor antagonists/agonists have been shown to affect insect feeding behavior, body mass,[328] and locomotor activity.[326] Given that morphine's wide-ranging ability to modify insect behavior is predicated on the same μ- and κ-opioid-receptor–binding mechanisms as those seen in humans (see Chapter 4), it seems likely that the feeding behavior or well-being of insects will also be modulated by salvinorin A's potent κ-opioid-receptor activity. It may also be relevant that the monoterpene menthol, an archetypal anti-insect secondary metabolite of the mint family, owes its analgesic effects in humans to κ-opioid-receptor agonism, albeit with this effect weaker than salvinorin A.[1081] Regardless of the insect question, the pharmacological properties of salvinorin A would also be liable to protect the plant from repetitive grazing by vertebrates, a useful trick in a region once heavily populated by deer and other larger mammals.[1202]

■ SOME CONCLUSIONS

This comparatively small subgroup of plants, nested within a ub-family of the *Lamiaceae* family of angiosperms, have provided us with a wide range of psychotropic plant species. All of these plants share many aspects of their secondary metabolite chemistry and employ relatively ubiquitous terpenes in ecological roles. The evidence for the efficacy of *S. officinalis/lavandulaefolia* and *M. officinalis* with regard to human brain function is consistent across studies, although, as always, it would be strengthened by larger-scale clinical trials. Along with *R. officinalis* and a host of other related *Salvia* species that have not yet benefitted from direct research in humans, the mechanisms of action of these plants would seem to be firmly rooted in the contradictory attraction/defense interactions between their secondary metabolites and insects, and, in turn, the commonalities between the insect and mammalian nervous systems. *S. divinorum*, on the other hand, owes its combined hallucinogenic/dissociative effects in humans to the expression of a slightly more specialized set of compounds—although again, the mechanism of action of the principal component would be equally prevalent in insects. One other notable factor shared by these plants is their safety and tolerability in terms of human consumption; on this basis their psychotropic effects deserve more research attention.

12 Cannabis and the Cannabinoids

The *Cannabis* genus belongs within the small *Cannabaceae* family, alongside its closest relatives, the members of the *Humulus* (hops) genus. The taxonomy of the *Cannabis* genus is disputed,[1205] with varying accounts describing either several species, two species (*C. sativa/indica*), or a single *C. sativa* species with up to five subspecies: *sativa, indica, ruderalis, spontanea*, and *kafiristanca*.[1206] Whichever account is correct, *C. sativa* is the most prominent psychoactive plant/species, with the species/subspecies *C. indica* also widely used.

Cannabis plants are herbaceous plants that exist as either males or females. They have a 4- to 8-month life cycle in the wild, with growth patterns, flowering, and senescence dictated by the lengthening and then shortening hours of darkness as the seasons progress. The female plants have flowers in dense clusters at the base of their leaves, which produce up to half a kilogram of seed in an individual plant. In the wild, cannabis is wind pollinated, with the male plant dying shortly after shedding its pollen. However, most cannabis cultivated for drug use is "sinsemilla," denoting female-only cultivation, with the unfertilized plants compensating for the setting of seed by producing additional flowers with increased trichomes containing higher levels of cannabinoids and other terpenes.[1205] The cannabis plant has manifold historical and contemporary uses, which include eating its leaves and seeds, using its fibrous stems for making ropes and cloth, and producing oil from the seeds. However, its most infamous use is as a source of psychoactive drug preparations, including marijuana, the dried leaves and female flowers of the hemp plant, and hashish, the resin exudate from the trichomes on the leaves and flowers of the plant.[1205]

The cannabis plant itself, commonly called hemp, has a point of origin somewhere in the valleys of Central Asia or the foothills of the Himalayas.[1205] It has an illustrious history that encompasses evidence of its use for cloth and rope making, in the form of imprints on pottery, dating back to 8000 BC in Taiwan and China. Archaeological evidence and Chinese historical records suggest that hemp has been domesticated and cultivated for up to 5,000 to 6,000 years, and the first written treatise on agricultural practices, the *Xia Xiao Zheng*, written in the 16th century BC, describes hemp as one of the main agricultural crops.[4] At some undefined point hemp crossed over from being the source of practical fibrous materials and edible seeds to being consumed for its psychoactive properties. Its oldest written description dates back to stonework from around 2350 BC, situated in the Egyptian Old Kingdom city of Memphis, after which point the writings of the ancient Egyptians suggest that it was used medicinally in oral, topical, smoke, and suppository forms. Physical evidence of its use includes hemp fibers in the tomb of Akhenaten (ca. 1350) and cannabis pollen grains inside the bound mummy of Ramesses II, who died ca. 1213 BC.[30] Its psychoactive use is also confirmed by preserved cannabis residues in bowls and religious paraphernalia from the Oxus city-dwelling civilization, which existed during the early part of the Egyptian epoch (circa 2300–1700 BC) at the intersection of modern-day Turkmenistan, Afghanistan, and Iran. Other pottery residues from the same region suggests that cannabis may have been one ingredient of the ritualistic and religious beverage known as soma or haoma that originated at about this time.[4] Other solid evidence of early cannabis use includes its mention on clay tablets found among the ruins of the Royal Library in the Sumerian city of Nineveh, sacked and destroyed in 612 BC,[30] and the account of the Greek historian Herodotus, who described (circa 446 BC) how the Scythians who had destroyed Nineveh would throw cannabis onto hot rocks in a sealed tent and "howl with pleasure." From Asia the psychoactive use of "drug-type" cannabis radiated outwards over the centuries, and it appears in the Northern European archaeological record in approximately 500 BC near Stuttgart, before reaching Africa and Madagascar along the trade routes from India in the first century AD.[4]

If we skip forward to the present time, cannabis is now the single most widely consumed illicit drug globally. According to the 2005 United Nations World Drug Report, the total value of the world's cannabis market was over $140 billion, with an estimated 164 million users around the globe. The steady increase in the use of illicit drugs over recent years can also be attributed largely to cannabis use, certainly according to data from the United States.[80] One reason for the increase in cannabis use is the changing nature of the product itself. Selective breeding, improved growing techniques, and modified processing techniques mean that cannabis has been becoming steadily more potent in psychotropic terms as the comparative ratios and overall contents of cannabinoids have changed. As an example, analysis of samples from over 46,000 drug seizures in the United States between 1993 and 2008 showed that the average content of Δ-9-tetrahydrocannabinol (THC), the principal psychoactive cannabinoid component, nearly tripled from 3.4% in 1993 to 8.8% in 2008.[1207]

Cannabinoids are unique to the *Cannabis* genus and are present in all aerial plant tissues, although they are most highly concentrated in the resin. This terpene-rich exudate is synthesized in capitate glandular trichomes and stored in spherical cells made up of a thin, delicate sack-like membrane. This is attached to the end of a trichome stalk, giving the whole structure the appearance of a glistening droplet. Cannabinoids are initially synthesized in the trichomes in the form of cannabinoid acids and, as these acidic forms are highly toxic to plant cells, the final stage of their synthesis is undertaken extracellularly in the trichome.[1208] The acid forms of the cannabinoids are eventually decarboxylated to give the key cannabinoids and their many derivatives, although much of this process takes place as the tissue dries or is heated after harvesting.[1209] The harvested resin is composed primarily of odorless cannabinoids (~80% or more), with a host of related terpenes giving the resin its distinctive aroma.[1205] In all, more than 60 cannabinoids and some 140 or more terpenes have been identified in resin, along with a plethora of minor constituents from many other chemical groups.[1210] The cannabinoids themselves are classed as terpeno-phenolics, in that their core structure includes a monoterpene C10 unit formed from geranyl pyrophosphate attached to a phenolic ring.[195] In "drug-type" cannabis plants the most abundant compound, and principal psychoactive component, is THC, which constitutes in the region of 0.5% to 3% of the leaves, 3% to 10% of the flowering tops and bracts, and 14% to 25% of the resin.[195] However, the wild cannabis plants that were first used by humans would have had equal or higher levels of cannabidiol (CBD) than THC, and indeed the majority of cannabis plants still conform to this chemotype.[1211] The difference between the two chemotypes is dictated by the comparative expression of two related enzymes that differentially modify the phenolic moiety of cannabigerolic acid to form either tetrahydrocannabinolic acid or cannabidiolic acid. The selection of plants and selective breeding mean that the levels of THC have been increasing, probably for millennia, as cultivated plants have shifted toward the drug chemotype. Although most research attention is focused on THC, cannabidiol exerts its own effects on brain function, as do a sizable minority of the other cannabinoids, including cannabinol, Δ-8-THC, Δ-9-tetrahydrocannabivarin, and cannabigerol.

Synthesis of the many other terpene constituents of cannabis is closely related to that of the cannabinoids, and they are formed via geranyl pyrophosphate, which is also the precursor for the C10 monoterpene unit in the cannabinoids.[1212] For the most part the terpene content comprises monoterpenes, with high levels of myrcene, limonene, α-pinene, and α-terpinolene (and trace amounts of many others), while β-caryophyllene and α-humulene are the most abundant sesquiterpenes.[1210] The structures of the most abundant cannabinoids and terpenes are shown in Figure 12.1).

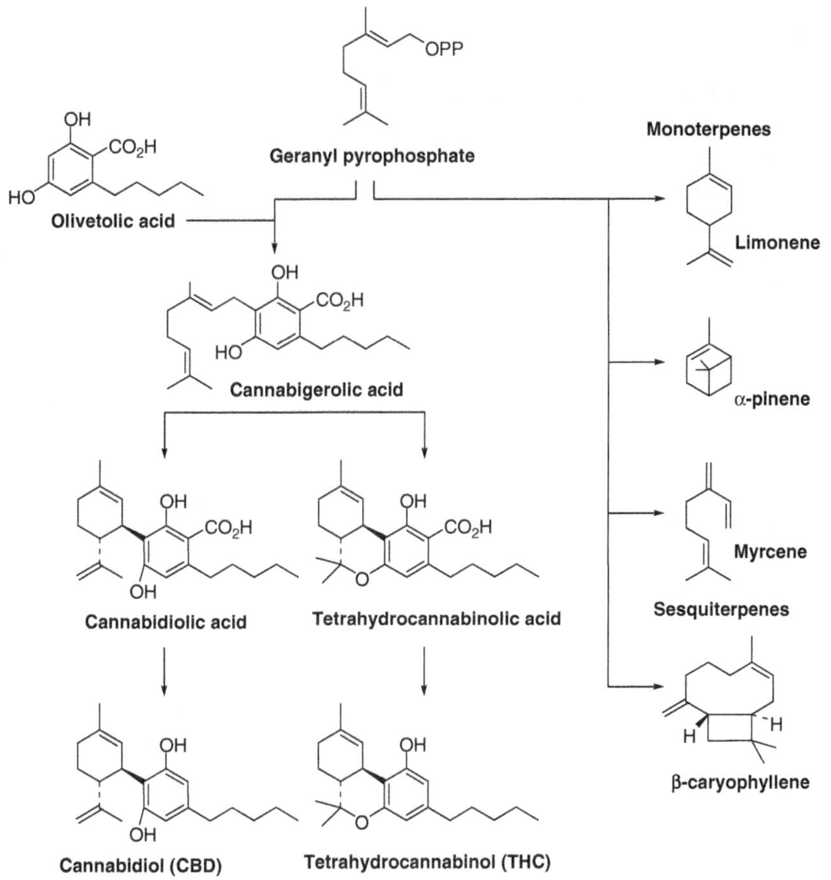

Figure12.1 Cannabinoids and other terpenes from cannabis. Geranyl pyrophosphate is the precursor for both the C10 monoterpene unit in the 60-plus cannabinoids and the 140-plus monoterpenes and sesquiterpenes found in cannabis resin. The most abundant representatives of each class are shown.

As noted above, the psychoactive properties of cannabis are largely attributed to THC. Its bioactivity is primarily attributed to partial agonist properties at cannabinoid receptors, which endow it with many central nervous system effects.[1079] Vertebrates as a group express two "classic" G-protein-coupled cannabinoid receptors, CB_1 and CB_2, although a further three potential receptors, GPR18, GPR55, and GPR119, have been recently identified.[1213] The receptors are called cannabinoid receptors because THC and the CB_1 receptor to which it binds were identified in 1964 and 1988 respectively,[1214,1215] some years before the first endogenous ligand for the CB_1 receptor was identified (in 1992).[1216] This neurotransmitter (N-arachidonoylethanolamide) was termed anandamide (from "anand," the Sanskrit for "bliss") and was later joined by another endogenous signaling molecule, 2-arachidonoylglycerol, the ligand associated with the CB_2 receptor.[1217] These compounds, and other ligands at the cannabinoid receptors, were termed endocannabinoids, a contraction of "endogenous cannabinoids."

The endocannabinoid system itself has been described as "a general intercellular signalling mechanism"[1218] in which the sustained activation of a wide range of postsynaptic G-protein-coupled receptors, including but not restricted to glutamate and acetylcholine receptors, trigger the release of endocannabinoids by the postsynaptic cell. These then cross back over the synaptic cleft and bind to cannabinoid receptors on the presynaptic cell, modulating neurotransmitter release and excitatory and inhibitory activity.[1218] In terms of distribution CB_1 and CB_2 receptors differ markedly, with CB_1 receptors more prevalent in the central nervous system, where they are richly expressed on presynaptic neuronal terminals throughout the cerebral cortex and brain structures such as the basal ganglia, cerebellum, and hippocampus.[1219] Although CB_2 receptors are also expressed by a number of non-neuronal brain cells, including astrocytes, microglia, and cerebrovascular endothelial cells, they are also expressed in the periphery, including in the cardiovascular, gastrointestinal, and immune systems, the liver, peripheral nerves, and the reproductive organs.[1218,1220] The ubiquitous distribution and broad functionality of the cannabinoid receptors mean that endocannabinoids are involved in most aspects of brain function along with a plethora of physiological processes. So, for instance, they play roles in the central nervous system control of pain, learning/memory, emotional functions, appetite (including milk ingestion in the newborn), reward, and motor regulation.[1221]

The reported subjective effects of cannabis consumption vary greatly across studies, although they typically include the experiencing of a state of relaxation, increased happiness or mild euphoria, perceptual and sensual distortions, a deepening of thought processes, cognitive alterations, increased appetite, and the intensification of sensory experiences.[1222,1223] On the potentially negative side, anecdotal and controlled-trial evidence suggests that acute use of cannabis can result in changed mood, increased or decreased anxiety, and transient, dose-related psychotic symptoms, potentially including paranoia, delusions, disorganized thinking, auditory and visual hallucinations, and impairments in attention and memory.[1222] Evidence also suggests that cannabis can exacerbate preexisting psychotic symptoms in sufferers from schizophrenia and may be a precipitating factor in the development of psychosis in predisposed individuals. Controlled studies suggest that the THC component alone can engender the negative anxiogenic, psychotomimetic, and psychosis-exacerbating effects of the whole extract, possibly due to modulation of dopaminergic, glutamatergic, or GABAergic activity throughout cannabinoid receptor-expressing regions of the brain. The schizophrenia-precipitating effects of extracts may be related to the disruption of the role of endocannabinoids in neurodevelopmental processes.[1222] To compound the downsides, the chronic use of cannabis leads to tolerance to the effects of THC and a "cannabis-withdrawal syndrome" that can feature such psychological disturbances as anxiety, insomnia and strange dreams, depression, and craving, along with chills, shakiness, stomach pain, and disturbance of appetite.[1224,1225]

While the majority of psychoactive effects can largely be attributed to THC, it is notable that several minor cannabinoids also interact with CB_1 and/or CB_2 receptors, including Δ-8-THC and Δ-9-tetrahydrocannabivarin, the latter of which exerts complex mixed effects as a CB_1 antagonist and CB_2 agonist.[1226] Conversely, other cannabinoids function via non-cannabinoid receptor mechanisms. CBD, the most prevalent cannabinoid in the resin from non–drug-type cannabis plants, and typically the second most abundant cannabinoid in drug chemotypes,[1227] is credited, in particular, with mitigating many of the more negative properties of THC. It has little binding affinity at CB_1 or CB_2 receptors, but nevertheless it counters THC's agonist activity at this receptor, with evidence suggesting that CBD attenuates the sedation, tachycardia, and anxiety associated with THC.[1228] To give an example, studies in humans in which THC and CBD have been directly compared show that they have opposite effects in terms of anxiety, with THC increasing and CBD attenuating ratings, and that CBD can reduce the psychotic symptoms associated with THC. The two cannabinoids also have non-contiguous or opposite effects in terms of brain activation. For instance, in an fMRI study, THC engendered anxiety and psychotic symptoms and modulated the activation in frontal and parietal areas of the brain during the presentation of fear-inducing stimuli. In contrast, CBD had no psychotomimetic effects and decreased anxiety and levels of neuronal activity in the key regions that respond to anxiety-provoking stimuli, including the amygdala and the anterior and posterior cingulate cortex.[1229] In a similar vein, CBD and THC had opposite effects on the activation of key brain areas during cognitive tasks and sensory processing,[a] while CBD alone attenuated the activation of the amygdala in response to fearful stimuli. Pretreatment with CBD also abolished the psychotic symptoms associated with THC.[1230] The bioactivity of CBD and its interactions with the effects of THC are attributable to a number of unique properties.[1228] For instance, CBD inhibits the reuptake and hydrolysis of anandamide and mimics its agonist activity at the physical and chemical heat-sensing vanilloid VR_1 receptor[1231]; it has antagonist activities at the recently identified novel cannabinoid GPR18 and GPR55 receptors[1227]; it inhibits the reuptake of adenosine and therefore increases signaling via adenosine receptors[1232]; and of particular interest, it is an agonist at $5\text{-}HT_1$ and to a lesser extent $5\text{-}HT_2$ receptors[1233] and exerts antidepressant-like properties in rodent models that are attributable to agonist properties at $5\text{-}HT_{1A}$ receptors.[1234] These receptor interactions alone may underlie a plethora of CBD-related effects, including its independent antipsychotic, cognition-enhancing, neuroprotective, anxiolytic, and potentially antidepressant effects (seen in rodent studies).[1227,1233] CBD also inhibits the activity of cytochrome P450 3A11, which is responsible for metabolizing THC to the significantly more psychoactive 11-hydroxy-THC, reducing THC's overall potency.[1210] Of course, the ability of CBD to attenuate many of the negative effects of THC becomes particularly relevant in light of the trend for THC concentrations to increase, while CBD concentrations decrease, in drug-type cannabis.[1222]

a. The striatum during verbal recall, the hippocampus during the response inhibition task, and the superior temporal cortex and occipital cortex respectively while participants processed speech and visual stimuli.

It is notable that while THC is responsible for many of the negative effects of cannabis it also exhibits a number of medicinal properties, including as a bronchodilator, antioxidant, neuroprotectant, and potent anti-inflammatory, with 20 times the potency of aspirin.[1227] Cannabinoids also exert a number of specific neuroprotective properties. These include THC-mediated reduction in excitotoxicity via its activity at CB_1 receptors, and, in the case of both THC and tetrahydrocannabivarin, activation of CB_2 receptors located on the microglia that regulate immune function in neural tissue, leading to an attenuation of potentially damaging inflammatory cascades. CBD also exerts anti-inflammatory effects via several receptor-independent mechanisms, including inhibition of inflammatory cytokines and the attenuation of nitric oxide over-synthesis. THC and CBD also exert antioxidant properties in vivo.[1226] It has been suggested, on the basis of anecdotal evidence, preclinical animal models, and very limited human trial data, that cannabis and selected cannabinoids may have some utility in treating some of the symptoms of and the neurodegeneration underlying nervous system diseases such as Parkinson's disease, multiple sclerosis, and Alzheimer's disease.[1226]

The medical utility of THC is, of course, somewhat limited by its psychoactivity. However, the interaction between THC and CBD is liable to have a multitude of practical medicinal applications, including the potential for CBD to increase the potential dose of THC whilst attenuating unwanted side effects[1228] and for the combination to have polyvalent, synergistic, or additive analgesic properties. As an example the combination of CBD and THC was found to significantly reduce pain in patients with intractable pain, whereas THC alone performed on a par with placebo.[1235]

To date the lion's share of research has focused on THC and its cannabinoid receptor-binding properties, with a much smaller body of work focusing on CBD. Many of the other cannabinoids also have interesting properties unrelated to any interactions with cannabinoid receptors. For instance, cannabigerol, the parent compound to the other cannabinoids, which is itself generally expressed at relatively low concentrations in resin, is a very weak agonist at CB receptors. However, it does exert a number of other effects relevant to the central nervous system, including inhibition of GABA and anadamide reuptake, moderate 5-HT_{1A} antagonism, and potent α_2 adrenergic receptor antagonism. These mechanisms may underlie the analgesic and antidepressant effects seen in animal models.[1227] While most of the other 60-plus cannabinoid structures have no known bioactivity to date,[1210] this may be because they simply have not been subjected to investigation. It may well be the case that cannabinoids represent "a neglected pharmacological treasure trove"[1236] or, indeed, that the psychoactive effects of cannabis are mediated by the independent and interactive effects of other constituents, including the many terpenes and flavonoids present in extracts. Certainly, many of the monoterpenes would have potential effects on brain function. For instance, limonene, one of the most abundant monoterpenes in cannabis resin, has been shown to have cholinesterase-inhibiting properties,[1023] to increase brain levels of GABA, and to reduce glutamate levels and attenuate stress-induced increases in corticosteroids and serotonin.[1237] It also exerts anxiolysis via GABA-receptor antagonism in rodents.[1238] It is also notable that the subjective effects of smoking pure THC differ from those following THC combined with small amounts of the other terpenes from cannabis.[1205] Interestingly, a synergy or polyvalency has also been noted between cannabis and other cholinergic plant extracts. The most notable example is tobacco, although historically, species of *Datura*, betel nut, and henbane have also been co-consumed with cannabis, and these extracts may contribute to the overall effects of the combination via mechanisms including interplay between and upregulation of receptors, and effects on third-party neurotransmitters.[1239] There is also a history of concomitant use of monoterpene-rich plant products, such as limonene-bearing citrus fruits, and calamnus (*Acorus calamus*) root or nuts, which may exert cholinesterase-inhibiting properties also.[1227]

■ CANNABIS AND REWARD

Data from 2005 suggest that a much larger percentage of the population of the United States, in the region of 1.7%, had met the DSM-IV criteria for cannabis abuse or dependence during the previous year than for any other drug of dependence.[1240] However, the percentage of people who have used cannabis at any time in their life who then transition to dependence is lower, at 9.2%, than for the other drugs of dependence. The comparative figures for lifetime users of heroin and tobacco who become clinically dependent are 23.1% and 31.9% respectively.[1241] This does suggest that cannabis exerts rewarding effects, but that they are mild in comparison to the key "addictive" drugs.

In animal models low doses of THC have been shown to have rewarding properties in conditioned place preference paradigms, but aversive properties at higher doses. Similarly, lower doses decrease intracranial self-stimulation thresholds in rodents, and low doses, comparable to the human equivalent of smoking cannabis, maintain self-administration behaviors in monkeys. These findings suggesting hedonic, rewarding properties.[1225] In terms of mechanisms, cannabinoid receptors are found throughout the brain regions involved in the reward process, including the prefrontal cortex, ventral tegmental area, and nucleus accumbens. They often co-occur with opioid receptors, for instance, potentially being expressed as heterodimers in the nucleus accumbens that control GABA and glutamate release.[1242] Animal studies show that acute administration of THC elevates dopamine levels and increases both tonic and phasic firing of dopaminergic neurons in the striatal dopamine reward pathways.[484,1224,1225] The increased dopaminergic function might reflect CB_1 receptor-mediated decreases in GABAergic inhibition of dopamine neural activity[1224] or alternatively GABAergic and glutamatergic excitatory/inhibitory inputs from the frontal cortex.[1243] On the other hand, withdrawal following chronic exposure to cannabis, as with other drugs of reward, leads to a decrease in mesolimbic dopaminergic neurotransmission, which in turn contributes to some of the psychological manifestations of the "cannabis-withdrawal syndrome" that negatively reinforces cannabis use.[1224,1244–1246] Evidence also suggests that cannabinoid receptors play multifarious roles in the development of addiction to the other drugs of reward[1247]; as well as being mildly rewarding in itself, THC also increases the rewarding properties of other drugs, such as morphine, in animal models.[1248]

The observation that THC, in common with the other rewarding drugs (see Chapter 5), activates the reward pathways in the brain has been confirmed in humans by a PET study showing increased dopamine release in the reward-mediating structures of the ventral striatum.[1246] However, a subsequent study using fMRI suggested that THC had fairly selective effects within the reward pathways, modulating activity related to the feedback of reward-related information.[1245] This is generally concordant with a previous study showing that chronic cannabis use blunted the activation of reward-related areas of the brain to anticipated monetary rewards, presumably via desensitization and downregulation of CB_1 receptors, suggesting that impaired reward processing may underpin the reduced ability to experience pleasure and the drug cravings that can result from long-term cannabis use.[1249]

Before leaving the issue of reward it is worth noting that in contrast to THC, CBD does not exhibit any rewarding properties; indeed, it has the opposite effects as THC in antagonizing the rewarding properties of opiates. This is a consequence of its ability to interact with $5\text{-}HT_{1A}$ receptors, and this property lends it some potential in the treatment of addiction.[1248]

■ THE ECOLOGICAL ROLES OF CANNABINOIDS AND BRAIN FUNCTION

The distribution of cannabinoids and associated terpenes throughout the aerial parts of the plant, with the highest concentrations seen in female plants and in the youngest and highest leaves, flowers, and reproductive tissue, suggests a specific role as chemical defense agents. Their deployment in delicate, easily ruptured vesicles at the end of the glandular trichome stalks, often on the insect-favored underside of the leaf, is also typical of one method of delivery employed for many terpenes that defend tissue specifically against insects and other small animals, as it combines both the physical threat of a sticky exudate with the delivery of anti-herbivore chemicals.[1250] The synthesis of resin, cannabinoids, and other terpenes can also be increased by a range of abiotic stressors. These include the aridity of the growing region (the more arid, the more resin), wounding, desiccation of leaves, and UV-B light exposure. Similarly, a number of biotic stressors induce synthesis, including physical damage related to insects and other herbivores and microbial attack.[1250] Indeed many individual cannabinoids, including THC, CBD, cannabichromene, and cannabigerol,[1251,1252] have antimicrobial properties. With regard to herbivores, cannabis has developed something of a reputation for being pest-free, but in reality the plants are simply "pest-tolerant".[1253] Hemp fields usually have a rich population of insects, including herbivores, their parasitoids and predators, and a range of indifferent insects, but they don't typically require pesticides.[1254] This suggests that a range of insects eat cannabis plant tissue, but in relatively small quantities. Historically, cannabis, either powdered or extracted using water or solvents, has been used as agricultural insecticide, molluscicide, and nematicide, and the plants have often been grown as a companion crop with the intention of protecting the main crop in a field.[1250,1253] Controlled studies show that crude cannabis extracts are highly toxic to mosquito larvae,[1255] termites,[1256] and nematodes.[1257] As with all of the best defensive chemicals, insects such as species of *Arctiid* moths have been shown to sequester the cannabinoids from high-THC cannabis plants into their exoskeleton, presumably for their own protection.[1258]

While the mode of storage and delivery of cannabinoids indicates defensive properties against insect herbivory, the identification of the intended active components in this respect is more problematic. It is tempting to assume that THC, as the component that has the most relevance to humans due to its psychoactive properties, is the key secondary metabolite in ecological terms. However, individual cannabis plants belong to one of three chemotypes that have very similar overall levels of cannabinoids. However, they vary on the basis of their expression of the enzymes that metabolize cannabigerol to THC and CBD respectively. The resultant type I chemotype has a ratio that favors THC over CBD and is typically described as the "drug-type" plant. Type

II has approximately the same levels of each compound. Type III has higher levels of CBD and is typically the "hemp-type" plant. Type I plants tend to be members of the *Indica* species/subspecies and to originate in hotter, more arid climes, and of course this group has been selected and bred by humans for its psychoactive properties for millennia.[1211] The division of cannabis plants into these three chemotypes does mean that the majority of naturally occurring plants have either the same levels or higher levels of CBD than THC.

THC itself would certainly prove an excellent deterrent to many herbivores. Its rewarding properties in mammals are restricted to low doses, and it has specifically aversive properties at the higher doses that would be encountered in nutritionally significant quantities of plant material. However, although cannabinoid receptors are ubiquitous across all vertebrate taxa, their pattern of expression across invertebrates is patchier, suggesting that cannabinoid receptors evolved in the last common ancestor of bilaterians but were then lost in a number of invertebrate phyla.[1217] For instance, a receptor homologous to CB_1 has been identified in annelids, mollusks, lobsters, and nematodes, but not in insects. This does suggest that the mechanism by which cannabis resin exerts any anti-insect effects is not related to the THC/CB_1 receptor mechanism that has such profound effects in humans. Indeed, by and large, THC fails to elicit a behavioral response in most investigations involving insects, with a few exceptions. For instance, THC-rich but not CBD-rich cannabis proved fatal to tiger moth (*Arctia caja*) larvae, and pure THC was repellent to the large white cabbage butterfly,[1250] whereas the moth *Pieris brassicae* exhibited identical changes in behavior when exposed to both THC and CBD.[1259] Of course, plants are prone to attack by a number of herbivores, and THC and the other CB_1/CB_2 receptor-binding cannabinoids may provide protection against the large proportion of herbivores, including all vertebrates, snails, slugs, and worms, that express CB receptors. It may be the case that the anti-insect effects of resins are adequately provided by the significant levels of monoterpenes and sesquiterpenes, or indeed the wide range of other minor components. Alternatively, CBD, which is equally abundant or more abundant than THC in the majority of plants, has a number of neurotransmitter effects that are relevant to insects. These include excitatory properties with respect to vanilloid and adenosine neurotransmission and agonistic binding properties at a range of serotonin receptors. It has also been established that many other cannabinoids, including cannabinol, have their effects via a range of non-CB receptors.

Having said this, neither THC nor CBD might be the relevant compounds in ecological terms. Cannabinoids are synthesized and stored extracellularly in trichomes in an acid form, which is then decarboxylated during storage or heating to form the non-acid forms prized by humans for their psychoactive properties; thus, THC originates as tetrahydrocannabinolic acid and CBD as cannabidiolic acid. The acidic forms may fulfil two, or indeed many more, ecological roles: they produce necrosis in plant cells, hence their extracellular storage, and this may provide an autopruning mechanism that disinfects and destroys the tissue surrounding the trichome when it ruptures as a consequence of age or infestation.[1227] They also induce apoptosis in insect cells, suggesting that the cannabinoid acids (along with hydrogen peroxide, a byproduct of the synthetic pathway concomitantly released from the ruptured trichome) may be the insecticidal components in trichomes in situ on the plant.[1208] Interestingly, while currently under-researched, the cannabinoid acids also exert similar or more potent effects on brain function parameters. As an example, cannabidiolic acid has greater effects on rodent nausea via 5-HT_{1A} receptor activation than cannabidiol,[1260] and consumption by humans of unheated cannabis extracts, which have higher levels of the acidic forms of CBD and THC, increases the bioavailability of CBD metabolites, more so than those of THC, in comparison to heated extracts.[1261]

Just returning briefly to the theme of the invertebrates that do express CB receptors, which include worms, it is interesting to note that cannabis has a history as an antihelminthic stretching back to the Ancient Egyptians.[30] Parkinson, in his evocatively named *Theatrum Botanicum* ("Theatre of Plants"), published in 1640, described various medicinal uses of indigenous hemp, including as an anti-inflammatory and analgesic, and noted also that "it is held very good to kill the worms in man or beast, as also the worms in the ears, or the juice dropped therein, or to draw forth another living creature that has crept therein." Rather more curiously he describes how "the decocotion thereof, powred into the holes of earthworms, will draw them forth, and that fishermen and anglers have used this feate to get wormes to baite their hookes." This certainly conjures an intriguing image of fishermen drugging worms to obtain their bait and may tell us something about the ecological roles of cannabinoids.

13 Some Miscellaneous Terpenes

THE ADAPTOGENS—MODIFIED TRITERPENES FROM THE *PANAX, WITHANIA*, AND *BACOPA* GENERA

The adaptogens are a class of phytochemicals and extracts that are theorized to function primarily by protecting the consuming organism from the negative impact of physical, biological, chemical, and psychological stress. Typically they are theorized to do this by regulating the homeostasis of physiological stress parameters related to the functioning of the hypothalamic-pituitary-adrenal (HPA) axis.[1262,1263] This system comprises the hypothalamus, the brain structure responsible for detecting external and internal stressors, the pituitary gland, and the adrenal glands on top of the kidneys. Glucocorticoid hormones released by the latter organ control reactions to stress and regulate a multitude of cellular and bodily processes—including immune function, mood and emotions, cognitive function, the storage and expenditure of energy, sexuality, and digestion. The overriding function of the system is to prepare the organism to deal appropriately with the stressor. The glucocorticoids, exemplified in humans by cortisol and corticosterone, bind to the glucocorticoid receptors that are present inside nearly every cell of the body, modulating the expression of some 10% of human genes[1264] and thus downregulating the potentially harmful cellular responses to stressors. Glucocorticoid receptors are members of the nuclear receptor family and are expressed in the cytosol of the cell, where they exist in an unactivated state complexed with a number of proteins such as the heat shock proteins. When activated by their ligands they shed these other proteins and translocate themselves to the nucleus of the cell, where they bind to response elements and directly modify gene expression, for example upregulating the expression of anti-inflammatory proteins in the nucleus. Alternatively, they form a complex with other transcription factors in the cytosol, for instance NF-κB and AP-1, preventing their own translocation to the nucleus and thereby preventing the transcription of genes involved, for instance, in modulating inflammatory and immune responses to stressors. Glucocorticoid receptor activation can also result in "non-genomic" effects, including, for instance, an increase in endothelial nitric oxide synthesis, causing vasodilation, and a decrease in potentially harmful inducible nitric oxide synthesis, via the activity of the freed-up heat shock proteins. The net effect of modulating glucocorticoid function includes the regulation of genes involved in inflammation and immunity; metabolism; endocrine function; the growth, differentiation, and apoptosis of cells; membrane transport and signal transduction; and multiple aspects of neurotransmission.[1264] HPA axis activation is also associated with increased release of adrenaline and noradrenaline from the adrenal medulla. All told, HPA axis dysregulation, possibly in response to extended exposure to stress, is implicated in a host of diseases, including diabetes, cancer, and psychiatric disorders. Given the above, an adaptogen that restores homeostasis to the HPA axis should theoretically be able to modulate a wide range of parameters, including many aspects of brain function.[1262–1264]

Broadly speaking, the group of plant secondary metabolites that have shown demonstrable evidence of adaptogenic effects in humans are triterpene saponins and related steroidal triterpene derivatives. These include the ciwujianosides from *Eleutherococcus senticosus* and the ginsenosides, withanolides, and bacosides from plants from the *Panax*, *Withania*, and *Bacopa* genera respectively. The adaptogens for which we have some evidence suggesting effects on human brain function will be considered in more detail below.

It has been noted that many of these structurally related triterpenes have the ability to bind to and modify the functioning of glucocorticoid receptors, and in some cases they also inhibit the enzymes that metabolize glucocorticoids.[998,1265] Any impact on this or any other aspect of HPA axis function is not merely coincidental. Both the phytochemicals and the glucocorticoids, and indeed the other mammalian hormones (progesterone, estrogens, androgens), fall into the same broad structural category as the animal steroid hormones; they are all modified triterpenes.[195] All of these mammalian

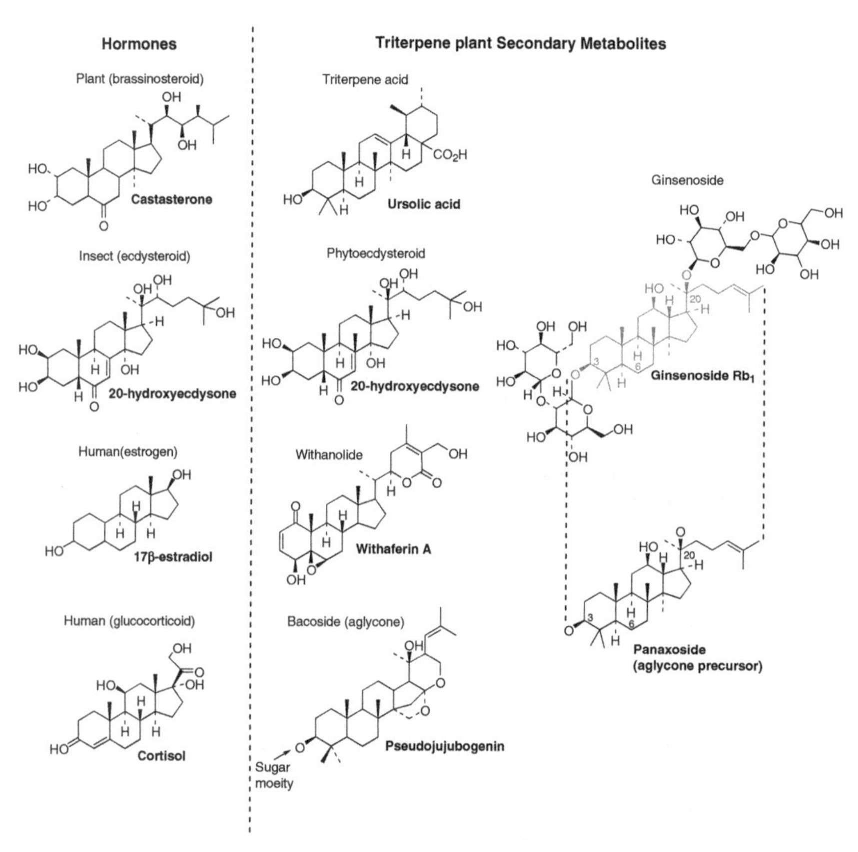

Figure 13.1 Structural similarity between modified triterpene hormones from plants, insects, and mammals, and steroidal and nonsteroidal triterpene secondary metabolites with potential cellular effects in animals due to their structural resemblance to these hormones. The bacosides are represented by the aglycone pseudojujubogenin. The glycoside ginsenoside Rb_1 is pictured with the aglycone panaxoside core of the ginsenosides. The two ginsenoside subgroups are differentiated by the position of the attached sugars (at the 3- and 20-hydroxy groups for the panaxadiols, and at the 6- and 20-hydroxy groups for the panaxatriols).

hormones also act via members of the same nuclear receptor family as the glucocorticoids, with several also activating membrane-bound G-protein-coupled receptors.

All of these modified triterpenes also start with the same precursor, squalene, before their synthetic pathways diverge. The nonsteroidal ginsenosides and bacosides are eventually based on a dammarane skeleton, the steroidal lactone withanolides are based on an ergostane skeleton, and animal hormones are synthesized from cholesterol via pregnenolone or progesterone. However, all of the compounds are broadly similar in structure. Representative structures from all of these classes are shown in Figure 13.1.

The *Panax* Genus—*Panax ginseng, Panax quinquefolius*

The *Panax* or "ginseng" genus comprises ten species distributed around Eastern Asia and two species that grow naturally in North America.[1266] The two most commonly cultivated species, *P. ginseng* and *P. quinquefolius* (American ginseng), along with *P. japonicus*, are generally thought of as representing a monophyletic group. The American emigrants diverged comparatively recently (in evolutionary terms) at some point after two genome duplication events, the more recent of which occurred in their common ancestor between 1.6 and 3.3 million years ago. After that point the immediate ancestor of *P. quinquefolius* crossed over into North America via the Bering land bridge that connected Alaska and Siberia during the Pleistocene ice ages.[1267]

Ginseng first appears in written history in the pharmacopeia (25–220 AD) that putatively described the verbal traditions passed down from Shen Nong, the first emperor and an agriculturalist/herbalist who reigned approximately 5,000 years ago. The Emperor was reputed to have tasted thousands of plants to discern their medicinal qualities, and when he chewed the root of a ginseng plant he experienced a warm and sexual feeling. This, along with a "doctrine of signatures" interpretation of the phallic or man-shaped root, led to ginseng root being used as an aphrodisiac, in the treatment for erectile dysfunction, and as a general panacea for the entire body.[1268] Ginseng's typical contemporary use is as an adaptogen, purportedly revitalizing the body and mind, increasing strength, and preventing aging.[1269,1270] Independently, *P. quinquefolius* also has a long history of use by indigenous North Americans,[1267] who themselves emigrated to North America via the Bering land bridge approximately 15,000 years ago.[574]

The putative major active components of ginseng comprise 40 or more species-specific triterpene saponins or "ginsenosides".[1271] Saponins are glycosides that produce a frothing in aqueous solutions due to their soap-like properties; they are hydrolyzed during digestion to their constituent sugars and their aglycone sapogenin. Most ginsenosides are built around a 17-carbon dammarane skeleton comprising a four-ring structure. Ginsenosides can then be differentiated into two groups on the basis of the position of the sugars attached to their common aglycone skeleton. For the panaxadiols (e.g., ginsenoside Rb1, Rb2, Rb3, Rc, Rd, Rg3, Rh2, Rs1) these are attached at the 3- and 20-hydroxy groups, and for the panaxatriols (e.g., Re, Rf, Rg1, Rg2, Rh1) at the 6- and 20-hydroxy groups. The individual ginsenosides within the two groups are then differentiated by the nature of the attached sugars.[195,1029] Figure 13.1 shows the basic aglycone ginsenoside skeleton within the structure of Rb1, the most abundant of all of the ginsenosides.

Across a huge and growing literature, ginsenosides have been shown to have well-established in vitro/ex vivo/in vivo properties that might make them suitable as treatments for a host of diseases, including cardiovascular disease, neurodegenerative diseases, and cancer.[1270,1272] However, it is notable that individual ginsenosides can have opposite effects on many parameters.[1272] Many of the cellular mechanisms underlying these effects, including the modulation of multifarious aspects of the inflammatory process, can be accommodated within a "glucocorticoid" framework. Indeed, many of the effects of ginseng have been attributed to modulation of nitric oxide synthesis,[1271,1273] which could be a consequence of either the genomic or non-genomic effects of modulating the glucocorticoid system. Ginsenosides certainly exert measurable in vivo effects on many aspects of HPA axis function. They can directly bind to and interact at multiple steroid hormone receptors; for instance, ginsenosides Rh1 and Rb1 are functional ligands of estrogen receptors, and Rg1 and Re are partial agonists at glucocorticoid receptors.[1029] Additionally, the panaxatriol and panaxadiol aglycone metabolites of ginsenosides bind to and activate both mammalian estrogen and glucocorticoid receptors. This finding may be particularly important as these aglycone forms may be the bioavailable end product of ginsenoside metabolism.[1029] Interactions with these ubiquitous receptors, leading to gene transcription or the inhibition of transcription factors such as NF-κB and AP-1, could theoretically underpin many of the reported effects of ginseng at a cellular and "whole-organism" level.[1265] These could include, for instance, adaptogenic and anti-stress effects and modulation of adrenaline/noradrenaline release and neuronal, cardiac, and nitric oxide synthesis/endothelial parameters.[1029] The rare reported side effects of ginseng in humans, including insomnia, anxiety, restlessness, and mania, are commensurate with overactivity in the neuroendocrine HPA axis that modulates stress responses via the activity of glucocorticoids. Naturally, as noted above, the interactions with animal hormonal systems may be related to the structural similarity between ginsenosides and estrogens and glucocorticoids.

Evidence of ginseng's efficacy with regard to its many indications is patchy. In China ginseng is often used in the treatment of ischemic heart disease due to its effects on vasodilation and nitric oxide synthesis. A recent meta-analysis of data from 18 methodologically adequate studies suggests that it has greater efficacy than nitrates.[1274] There is also promising evidence across a number of trials suggesting utility in the treatment of erectile dysfunction,[1275] and a recent study showed that 4 weeks' administration of ginseng decreased markers of oxidative stress and increased endogenous antioxidant defenses in healthy adults.[1276] However, the evidence with regard to several of its most popular over-the-counter supplement uses, for instance in improving physical or sporting performance or subjective "quality of life," is equivocal, although this may be attributable to numerous methodological inconsistencies in the literature.[1273,1277]

One area where evidence of efficacy is growing is in terms of brain function. Ginsenosides have multifarious effects relevant to brain function; they indirectly modulate the activity of $GABA_A$, serotonin, and nicotinic and glutamate NMDA receptors,[1029] and any modulation of nitric oxide-related activity could have far-reaching consequences within the brain. Animal behavioral models suggest that ginsenosides have anti-stress, antidepressant, and anxiolytic effects and that they moderate fatigue and improve memory in impaired rodents, potentially by fostering neurogenesis and modulating long-term potentiation in the hippocampus.[1271,1273,1278,1279]

In terms of direct modulation of human brain function, a number of randomized, controlled, balanced-crossover trials of single doses of standardized *Panax ginseng* extract have demonstrated consistent improvements in the accuracy of memory task performance,[1280–1282] improvements in the speed of performing attention tasks,[1282,1283] and decreased latency of evoked potentials along with topographical modulation of electrical activity as measured by EEG.[1284] Reay et al.[1285,1286] also demonstrated that ginseng extract improved performance on difficult mental arithmetic tasks and reduced the mental fatigue associated with task performance in fasted individuals. These effects were accompanied by concomitant reductions in fasted blood glucose levels but were abolished by the coadministration of glucose.[1286] A recent study[1287] also extended the findings of these single-dose studies by administering ginseng for 7 days and demonstrated improved mood, in terms of "calmness," following both doses (200/400 mg) investigated, and improved performance of the "3-back" working memory task following the higher dose but slower performance following the lower dose. A subsequent study assessing the effects of 8 weeks' administration of a Korean red ginseng extract also demonstrated improved working memory performance and modulation of ratings of "calmness".[1288] A single study has also extended these findings to single doses of a standardized *P. quinquefolius* extract, with demonstrations of improved working memory performance and dose-dependent increases in the speed of task performance.[1289] Two studies have also suggested the possibility that chronic administration of ginseng as an adjunct treatment improves cognitive function in Alzheimer's disease, although serious methodological limitations make interpretation of these findings difficult.[1290]

Withania somnifera

Withania somnifera is a shrubby bush with long tuberous roots that belongs to the nightshade (*Solanaceae*) family. The natural distribution of the *Withania* genus's 23 species in the Canary Islands, North Africa, the Mediterranean region, and South Asia suggests, as with the deliriant *Solanaceae* reviewed in Chapter 7, an initial sea-borne ancestral distribution from the *Solanaceae*'s center of diversity in South America.

W. somnifera, also known as ashwagandha, has a long history of medicinal use in India and the Middle East, appearing as a key plant in Ayurvedic, Siddha, and Unani traditional medicinal systems. Its written history dates back some 3,000 years, and it appears in the medical treatises *Charaka Samhita* (circa 900 BC) and *Sushruta Samhita* (circa 600 BC). Its contemporary ayurvedic uses include as a general "adaptogenic" and memory-enhancing tonic classified as a medharasayana, meaning "that which promotes learning and a good memory".[1291] Alongside these general uses it is also thought to have aphrodisiac, anti-inflammatory, diuretic, narcotic, and sedative properties. Additional contemporary indications include as an antibacterial, antihyperglycemic, and antitumoral agent, along with its recommended use in the treatment of dementia.[1292]

In terms of active constituents, the roots and aerial parts of *W. somnifera* contain comparatively low levels of a number of alkaloids, including tropine, nicotine, somniferin, and withanine, all of which are known to exhibit their own pharmacological activities. However, the beneficial effects of extracts are generally attributed to the presence of high levels of the withanolides, a group of 40 or more modified triterpene steroidal lactone derivatives of ergostane. The most abundant of the withanolides is withaferin A. The plant also synthesizes several sitoindosides (withanolides containing a glucose molecule).[1292,1293] The withanolides, which bear a close structural similarity to the triterpene ginsenosides, share many of their purported functional properties. Withanolides themselves are not unique to the *Withania* genus, being expressed by 15 *Solanaceae* genera, including *Datura*.[1292] The structure of withaferin A is shown in Figure 13.1.

Research in rodents shows that the oral administration of *W. somnifera* extracts can increase levels of endogenous antioxidants, decrease inflammation and the level of inflammatory markers, bolster immune responses, and inhibit carcinogenesis.[1292,1294,1295]

These effects can largely be attributed to the molecular effects of the withanolides, which include multifarious interactions with cellular signal transduction pathways and transcription factors, including NF-κB and the heat shock proteins that interact with glucocorticoid receptors.[1296] It is also notable that among the withanolides' anti-stress properties is a propensity to beneficially modulate glucocorticoid levels and function.[1295,1297] Other examples of hormonal effects following the administration of *W. somnifera* extracts include the modulation of sex hormone parameters, with, for instance, an increase seen in the levels of luteinizing hormone, progesterone, and testosterone and a decrease in follicle-stimulating hormone in rats following *W. somnifera* administration.[1298,1299]

In terms of direct effects on brain function, withanolides have been shown to have acetylcholinesterase (AChE)- and butyrylcholinesterase (BuChE)-inhibiting effects in vitro.[1061] *W. somnifera* extracts also inhibited both AChE and nitric oxide synthase in the brain of live rodents, with these effects co-localized in individual neurons.[1064] In vitro research demonstrating GABA displacement and electrophysiological effects of the extract suggested direct binding to and modulation of GABA receptors,[1300] and a methanol extract of *W. somnifera* also increased the activity of gonadotrophin-releasing hormone neurons in the hypothalamus via agonist properties at $GABA_A$ receptors.[1301] Extracts have also been shown to modulate GABA function[1302] and muscarinic receptor function in vivo.[1087]

Extracts of *W. somnifera* and withanolides have anxiolytic, antidepressant, and anti-stress properties in rodents.[1295,1303,1304] They also exert adaptogenic effects, including the suppression of glucocorticoids and stress-related brain oxidative stress parameters, along with the attenuation of impaired immune function and memory.[1295] As an example, hypobaric hypoxia has been shown to cause neurodegeneration and memory loss in rodents via a prolonged increase in corticosterone levels and glucocorticoid receptor activation, with glucocorticoid receptor inhibitors attenuating these effects..[1305] Chronic oral administration of a *W. somnifera* extract has been shown to ameliorate the neurodegenerative and cognition-impairing effects of hypobaric hypoxia-related stress. The mechanisms underlying these effects appear to be decreases in nitric oxide synthesis, plasma and hippocampal levels of corticosterone, and AChE overactivity, with reduction of the latter increasing acetylcholine levels.[1297] Both *W. somnifera* and its withanolide component have also been shown to reduce oxidative stress and ameliorate reductions in hippocampal glutamate NMDA receptor populations, and the cognitive deficits associated with epilepsy in rats.[1306]

The effects of *W. somnifera* on human brain function have been investigated in only a few, comparatively small placebo-controlled studies. In the first of these studies, which included the data from 33 participants, 6 weeks' administration of *W. somnifera* extract led to significantly more patients with anxiety disorders meeting clinical criteria for improvements in comparison to placebo, with a near-significant improvement also seen on the Hamilton Anxiety Rating Scale.[1307] A more recent study included 64 subjects with chronic stress and demonstrated improved ratings on clinical measures of stress, depression, and psychological well-being, along with significantly reduced cortisol levels, following daily doses of 300 mg standardized *W. somnifera* extract taken for 60 days.[1308] One study also compared 12 weeks of standard psychotherapy to a treatment combining a multivitamin and 300 mg *W. somnifera* extract in 81 subjects with moderate to severe anxiety. Both treatments resulted in improved anxiety as measured by the Beck Anxiety Inventory, but the improvement was significantly greater in the vitamin/*W. somnifera* group and was accompanied by improvements in self-ratings of quality of life and concentration.[1309]

Bacopa monnieri

Bacopa monnieri is a creeping annual plant found throughout the Indian subcontinent in wet, damp, and marshy areas. Its recorded ayurvedic medicinal use stretches back to the *Charaka Samhita* (circa 900 BC) and, as with *W. somnifera*, it is classified both as an adaptogen and as a medharasayana. In this respect its early historical indications included the treatment of anxiety, poor concentration, and impaired cognitive function.[1310]

Although *B. monnieri* extracts may contain a range of alkaloids and phenolics, the active constituents are thought to be a range of triterpene saponins based on the same dammarane skeleton as the ginsenosides. Typically, standardized extracts contain 50% or more of bacoside A and B, with A being the more prevalent. However, bacoside A actually represents a group of related structures including bacosides, bacopasides, and bacosaponins, all of which are glycosides of the sapogenins jujubogenin and pseudojujubogenin.[a,1310,1311]

Mechanistic studies suggest that *B. monnieri* extracts have in vivo anti-inflammatory properties predicated on the downregulation of pro-inflammatory mediators (TNF-α, IL-6, and iNOS).[1312] They also possess a number of properties directly relevant to the central nervous system. These include the ability of *B. monnieri* extracts to attenuate the cognitive decrements caused by the muscarinic antagonist scopolamine, potentially via AChE inhibition,[1062,1313] and analgesic properties that can be blocked by the opioid receptor antagonist naloxone, suggesting interactions with the opioid system.[1314] Extracts also attenuate the anterograde amnesia caused by diazepam[1315] and reverse the epilepsy associated with GABAergic dysregulation.[1316] Both of the latter effects suggest the modulation of GABAergic function. However, despite these indications of potential modes of action, the precise mechanism remain unknown. For instance, as yet, no direct receptor interactions have been identified, and all of the above phenomena may be downstream consequences of modulation of an unidentified system. Of course, one possibility is that they all reflect adaptogenic effects. In this regard *B. monnieri* has been shown to normalize markers of HPA axis dysfunction in rat models—for instance, attenuating stress-induced changes in plasma glucose levels, adrenal gland weight, corticosterone levels, and the decrease seen in the levels of the monoamine neurotransmitters serotonin, dopamine, and noradrenaline.[1317] Whatever the mechanisms, *B. monnieri* and its constituent triterpene saponins have been shown to exert consistent anxiolytic, antidepressant, cognition-enhancing, and anticonvulsant properties in rodent behavior models.[1303,1310]

In humans, one study assessed the effects of *B. monnieri* on general health parameters, demonstrating reduced cholesterol and triglyceride levels along with improvements in ratings of quality of life.[1318] A number of studies have also assessed the effects of between 300 and 450 mg of a standardized *B. monnieri* extract containing a minimum of 40% of bacoside A and B on psychological functioning. Two of these studies have assessed the acute effects of single doses of *B. monnieri*. The first[1319] found no cognitive effects and the second found only modest improvements restricted to the performance of one task within a cognitively demanding battery of tasks following

a. Jujubogenin glycosides are the active components of several medicinal plants, but pseudojujubogenin glycosides are seemingly unique to *B. monnieri*.

the lower of two doses of *B. monnieri* extract.[1320] However, a number of double-blind, placebo-controlled studies have assessed the cognitive and mood effects of supplementation with *B. monnieri* extracts for approximately 3 months. Two of these studies recruited healthy younger adults. In the first of these studies, self-ratings of "state anxiety" and performance on cognitive tasks assessing speed of early information processing, verbal learning rate, and memory consolidation were improved.[1321] In the second study, working memory performance was augmented.[1322] A number of studies have also focused on older adults (over 55 years). In this group cognitive benefits of *B. monnieri* supplementation have been seen in terms of attention and working memory,[1323] tasks assessing aspects of executive function and memory,[1324] word recall and executive function[1325], and memory acquisition and retention.[1326] In one study benefits were also seen in terms of improved mood and decreased levels of anxiety.[1325] One study also failed to demonstrate any good evidence of cognitive improvements.[1327] Although all of these studies were relatively small, with sample sizes ranging from 35 to 98 (average 62), the picture seems to be one of relatively consistent improvements in aspects of cognitive function. Indeed, a recent meta-analysis that included data from six of these studies concluded that the combined results did demonstrate evidence of efficacy in terms of memory enhancement. However, the authors noted that the studies were biased toward assessing memory, and, as yet, there was no evidence with regard to other aspects of cognitive function.[1328] One further study has been published since this meta-analysis. In this study the effects of two doses (300/600 mg) of an ethanolic extract of *B. monnieri* were investigated in 60 healthy adults. The measurements included an assessment of cognitive function, along with cerebro-electrical activity parameters (EEG) and both AChE and monoamine oxidase inhibitory activity. The results showed that working memory performance was improved following both doses of *B. monnieri*, along with a decreased latency of evoked potentials (i.e., faster processing). Consuming the extract also led to AChE inhibition and therefore may have increased cholinergic function, offering one potential mechanism for the modulation of brain function.[1329]

Ecological Roles of the Terpene Adaptogens and Brain Function

Triterpene saponins and triterpene-derived phytosteroids, as a group, have presumed ecological roles as antifungal, antiviral, insecticidal, feeding deterrent, and molluscicidal agents.[1265] In line with this, both ginsenosides and withanolides have been shown to exert anti-feedant, insecticidal, and larvicidal properties against a variety of insect taxa.[1330–1334] As an example, withanolides have been shown to accumulate in the most valuable plant tissue and reduce the survival and weight of feeding herbivores.[1335] This broad group of phytochemicals are typically produced constitutively and their levels vary according to need, season, and the plant's life cycle. Their synthesis can also be induced by a range of biotic and abiotic stressors and by the application of jasmonates, again suggesting a defensive provenance.[998,1004,1029,1336] The mechanism of action of this group in cold-blooded animals is assumed to be largely related to their ability to form a complex with cholesterol and other sterols within cellular membranes, causing pores or curvature in the membrane and thereby affecting its permeability.[998] However, these triterpenes also have the ability to disrupt the life course and mortality of insects by interfering with their ecdysteroid hormonal system. They can do this via two potential mechanisms: by once again forming a complex, but this time with nonmembrane sterols, including cholesterol, thereby making them unavailable for the synthesis of ecdysteroids, or by interacting with ecdysteroid receptors by binding directly to the receptor's ecdysteroid-binding site.[998,1027,1028] Triterpenes and derivatives with identified effects at insect ecdysteroid receptors include a range of steroidal saponins such as the cucurbicatins synthesized by members of the *Cucurbitaceae* cucumber/melon/marrow family,[741] the protopanaxatriol ginsenosides from ginseng,[1029] and the withanolides from *W. somnifera*.[1027,1030,1031]

Predictably, the insect ecdysteroids themselves are also modified triterpenes that are structurally closely related to the mammalian steroid hormones (see Fig. 13.1). In turn, their receptors are members of the same, highly conserved, nuclear receptor superfamily that transduce glucocorticoid and other steroid signals in mammals. Indeed, insects express members of each of the six identified nuclear receptor subfamilies, and their variants include homologues of mammalian receptors, including members of the estrogen-related receptor family.[1337] These insect nuclear receptors function in the same manner as their mammalian counterparts. This includes their inhibition of similar transcription factors and their complexation with heat shock proteins prior to the ligand binding. Taken as a group, insect nuclear receptors modulate as wide a variety of functions as in vertebrates. However, for insects the majority of nuclear receptor ligands identified to date are juvenile hormones and ecdysteroids, although many nuclear receptors from both taxa, including the estrogen-related receptors (described in Chapter 8 and 9), remain orphan receptors in that their endogenous ligand has not yet been identified.[1092]

As a consequence of these similarities, many triterpene saponins and phytosteroids also have the ability to bind to and modify the activity of the mammalian nuclear receptors for the structurally related steroid hormones.[998] As an example, as noted previously (Chapter 11), the ubiquitous triterpene ursolic acid, which is found in the skin of fruits and a range of plants, binds to mammalian glucocorticoid receptors and other orphan nuclear receptors[1090] and impairs the functioning of the insect (triterpene) ecdysteroid hormone system.[1200] Of particular relevance, the small amount of research in this area has shown that ginsenosides are also functional ligands at both estrogen and glucocorticoid receptors,[1029] while withanolides have antagonistic properties at the estrogen receptor-α.[1338] Alongside this, other saponins, such as gycyrrhetinic acid from licorice, also inhibit the enzymes that metabolize glucocorticoids, potentially modulating endogenous levels of these key hormones.[1265] It has been noted that interactions with glucocorticoid receptors alone could explain many of the cellular and wider physiological effects of adaptogenic compounds in mammals, via their multifarious effects on signal transduction pathways, such as those associated with heat shock proteins, direct gene transcription, and indirect inhibition of transcription factors such as NF-κB and AP-1.[1265] It may well therefore be the case that any such effects in humans simply reflect the evolved ecological role of the saponins as spoilers of the herbivorous insect's ecdysteroid system.

However, to add another layer of complexity, the modified triterpene brassinosteroids are steroidal plant hormones, structurally related to insect and mammalian steroid hormones, which regulate many aspects of plant growth and mediate responses to biotic and abiotic stressors.[1339] When consumed by insects brassinosteroids act as binding inhibitors and antagonists at ecdysteroid receptors, with disastrous developmental consequences.[1340] In mammalian tissue they exert antiproliferative, neuroprotective, immunomodulatory,[1341] and anabolic properties,[1342] although the mode of action is unclear to date. Conversely, plants have been shown to synthesize "mammalian" androgens, estrogens, and progesterone by partially conserved synthetic pathways,[243,248–250] and progesterone, in particular, exhibits many of the characteristics of an endogenous plant hormone that functions via membrane-bound and cytosolic receptors in plant tissue.[1343] These receptors may include partially conserved plant homologues of mammalian progesterone receptor proteins that bind both endogenous plant progesterone and, with lower affinity, brassinosteroids.[243,244]

This suggests another two unexplored possibilities. The first is that evolution has simply selected or retained modified triterpenes as a convenient vector for hormonal signaling across taxa, and the various cross-kingdom effects of the insect, mammal, and plant steroid hormones and the steroidal and nonsteroidal triterpene secondary metabolites are simply predicated on structural similarity. This would not preclude the selection of triterpenes such as the ginsenosides and withanolides, or indeed the many other phytoecdysteroids produced by plants, on the basis that they could interfere with the insect's ecdysteroid system. The second possibility is that, given their structural similarity to the endogenous plant brassinosteroid hormones, the ginsenosides, the withanolides, and the many other bioactive triterpenes and derivatives may also play stress-related signaling roles in the plant, for instance regulating their own production by feeding back and suppressing their own stress-activated synthetic pathways by interacting with the plant brassinosteroid signaling system. If this were the case any bioactive effects in mammals could represent a cross-kingdom transfer of these signaling properties to the analogous component of the mammalian system in a similar manner to that suggested for many polyphenols (Chapter 8).

Interestingly, the triterpenes under discussion here are also bioactive in the rhizosphere. Indeed, the cultivated ginseng plant has something of a problem in that over the 4 to 6 years that it typically grows, the rhizosphere and soil surrounding the plant becomes progressively incapable of supporting its survival. Once harvested, the soil cannot then be replanted with ginseng for as long as 30 years. This extreme autotoxicity has been attributed to a number of factors, including soil deterioration and disease. Research does show that ginseng roots are a particularly rich domain for beneficial endophytic fungi, which help protect the plant against phytopathogenic microbes. However, the diversity of these endophytic species peaks and then decreases after several years.[1344] Ginsenosides appear to play a constitutive role as defenders against a broad range of phytopathogens, in particular fungi. However, it may be the case that the increasing concentration of ginsenosides in the soil, released by the shedding of roots into the rhizosphere, has a detrimental effect on endophytes, while stimulating the growth of some phytopathogenic fungi[1345] such as the specialist fungi *Pythium irregulare*, which has developed an acquired resistance to ginsenosides.[1346] The resultant autotoxicity may partly represent an imbalance between symbiotic endophytic fungi and the phytopathogenic microbes in the soil, which is exacerbated as ginsenosides build up over time. Certainly, ginseng roots are targeted by an unusually rich palette of fungal pathogens.[1344] Looking on the bright side, ginseng extracts have antibacterial and anti-yeast properties against human microbial pathogens.[1347]

Of course, ginseng's autotoxicity may also be attributable to simple allelopathy, as ginsenosides at low concentrations may be stimulatory to the growth of successive ginseng plants, but at higher concentrations they may become allelopathic as they foster increased oxidative stress in encroaching roots.[1032,1348] The withanolides also possess similar properties, inhibiting the germination and growth of other plants.[1349]

In line with this, *W. somnifera* extracts inhibit the germination and radicle growth of competitor seeds,[1350] including when leaf residue is simply added to the soil surrounding the plant.[1351] These rhizosphere effects raise two further unexplored possibilities: that the allelopathic ecological role of these triterpenes may be predicated on an ability to interfere with the hormonal regulation of the encroaching competitor plant via the mechanisms outlined above, or alternatively that their intended targets are soil-dwelling microbes. In this respect it is notable that bacteria possess analogues of the estrogen-related receptor, which has an established role in plant–bacteria signaling.[811] The role of phenolic secondary metabolites in similar rhizosphere–microbial interactions are reviewed in Chapters 8 and 9.

■ GINKGO BILOBA

Ginkgo biloba is a dioecious tree often described as a "living fossil" that can have a lifespan of 1,500 years or more. It is the only living representative of the *Ginkgoaceae* family, whose early ancestors can be traced back some 280 million years. The *Ginkgo* genus itself dates back 180 million years to the early Jurassic period.[1352] Although the survival of the ancient lineage of *G. biloba* is often attributed to some unidentified property possessed by the tree, it has actually had something of a checkered history. From a peak of speciation approximately 120 million years ago, at which point the family comprised some 16 separate genera and had a worldwide distribution, the *Ginkgoaceae* family was successively challenged by the spread of angiosperms and environmental factors such as the ebb and flow of glaciation. It eventually disappeared from North America 10 million years ago and Europe some 2 million years ago. By that point the family had been reduced to the single species that we know today, which clung on to survival in two environmental refuges in the forests of southwest China and on Mt. Tianmu in eastern China.[1352] From there, the future of the ginkgo tree was entirely in the hands of humans. The cultivation of the trees first spread throughout China, typically in association with monasteries and nunneries, possibly because of Mt. Tianmu's association with multiple religions, including Buddhism, and the traditional herbal medicinal expertise of religious acolytes. They subsequently reached Japan ~1,000 years ago and Korea ~800 years ago alongside the spread of Buddhism. The original immigrant trees still survive today. The first plants to reach Europe arrived courtesy of the botanist Engelbert Kaempfer, and the oldest example, planted in Utrecht in 1731, may be from one of the seeds that he sent from Japan.[1352]

Ginkgo fruit and seeds have a 5,000-year history as medicinal treatments in China, originally as a treatment for asthma and coughs. The first recorded oral use of the leaves dates back to the early 16th century.[1353] Today ginkgo extracts are one of the most commonly taken herbal products globally, with indications for circulatory disorders, tinnitus, and cognitive function. They are prescribed routinely in many European countries as a treatment for age-related cognitive deterioration and dementia.[1354]

The biologically active components of ginkgo extracts are thought to include species-specific terpene lactones. These comprise the diterpenes—ginkgolide A, B, C, and J and M—and the sesquiterpene bilobalide. The latter may simply represent a degraded ginkgolide. The other major class of active components are a range of flavonoids, consisting mainly of mono-, di-, and tri-glycosides of the ubiquitous flavonols kaempferol and quercitin. Extracts are typically standardized to contain 26%/27% flavonoids and 6%/7% terpenes. They must also have less than 5 parts per million of

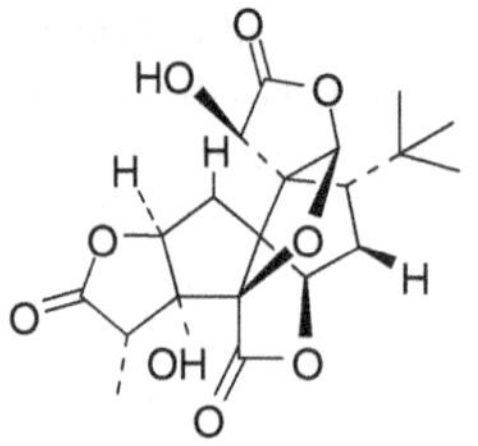

Ginkgolide A

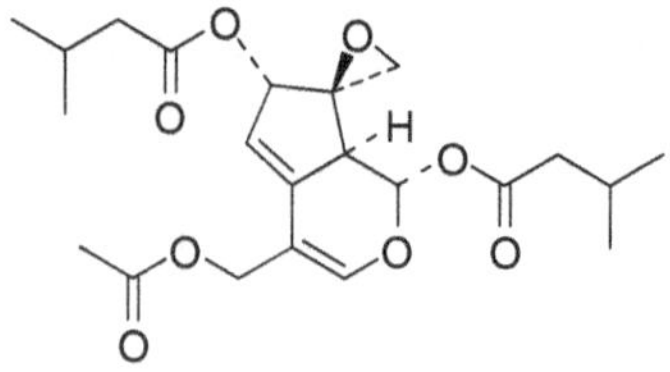

Valtrate

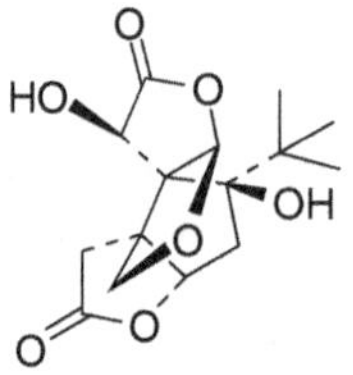

Bilobalide

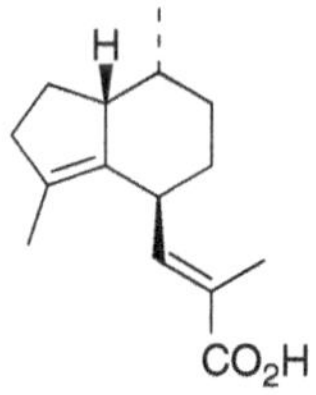

Valerenic acid

Figure 13.2 Structures of ginkgolide A and bilobalide from *Ginkgo biloba* and valerenic acid and valtrate from *Valeriana officinalis*.

alkyl phenols because of their potentially allergenic, cytotoxic, and mutagenic properties.[195] Example terpene structures are shown in Figure 13.2.

The potential neurologically relevant mechanisms of action of ginkgo extract include reduced blood viscosity via antagonism of platelet-activating factor, enhanced constitutive nitric oxide bioavailability and consequent beneficial effects on peripheral and cerebral blood flow parameters, scavenging and inhibition of free radicals, improved mitochondrial function, and both in vitro and in vivo neuroprotective properties, including inhibition of amyloid-β neurotoxicity and protection against hypoxic challenges and increased oxidative stress.[1355-1358] Ginkgo components also have direct effects on a variety of aspects of neurotransmission, including inhibitory binding properties at a variety of ionotropic "Cys-loop" receptors such as GABA, glycine,[b] and 5-HT_3 receptors by bilobalide and ginkgolide B,[1359-1361] and inhibition of the synaptosomal reuptake of dopamine and noradrenaline by ginkgo extract and both its flavonol glycoside and ginkgolide fractions.[1362] The latter highlights the difficulty of attributing the effects of ginkgo extract to any one class of compounds. Notably, in a recent study ginkgo extract administered to rats for 14 days led to a significant and dose-dependent increase in extracellular dopamine and noradrenaline levels in the prefrontal cortex. This effect was partially replicated by administering the flavonoid fraction alone, suggesting that the full effect was a consequence of either an additive or synergistic effect of both the flavonoids and terpene constituents.[1363]

Standardized *G. biloba* extracts have a number of effects on brain function parameters in humans. For instance, a single dose of ginkgo extract modulated cerebroelectrical activity (as measured by EEG) in healthy, young volunteers,[1284] and 2 weeks

b. Glycine is an amino acid and its receptors are mainly found in the brain stem and spinal cord.

of administration increased the amplitude and latency of evoked potentials during a working memory task, with these changes interpreted as indicating more efficient processing.[1364] Cerebral blood flow, as assessed by magnetic resonance imaging (MRI) and single photon emission computed tomography (SPECT), has also been shown to be increased in older adults following 4 weeks[1365] and 8 months[1366] of supplementation with ginkgo extract. These physiological effects are matched by evidence suggesting improved cognitive function. A number of randomized control trials have demonstrated cognitive enhancement in young adults following single doses of ginkgo extract[1367–1371] and in both younger[1372] and older[1373–1375] "cognitively intact" populations administered ginkgo for 7 days or longer. Having said this, evidence in this respect is not unequivocal, with two studies failing to find similar effects.[1376,1377] Reviews of these earlier studies have reached starkly differing conclusions as to whether ginkgo has any significant effect on cognitive function,[1378–1380] although they agree that the literature in this area suffers from methodological heterogeneity that makes interpretation difficult. However, recent studies employing larger samples of healthy middle-aged participants have shown improved memory performance after 6 weeks of supplementation with ginkgo in 188 healthy middle-aged participants[1381] and improved attention, memory, and subjective ratings of physical health in 300 middle-aged participants administered ginkgo for 12 weeks.[1382] A recent EEG study also showed improved working memory task performance after 14 days of supplementation along with modulation in cerebro-electrical activity.[1364]

While the effects of ginkgo in healthy humans look promising, the bulk of the research in this area has concentrated on the effects of ginkgo in dementia and, less often, age-associated cognitive impairment. In terms of these studies, a comprehensive Cochrane review in 2002[1383] included meta-analyses of 33 studies that involved cohorts suffering from dementia or age-related cognitive impairment. The authors concluded, "Overall there is promising evidence of improvement in cognition and function associated with ginkgo." However, in a more recent update and reanalysis published in 2009, the same authors[1384] modified their inclusion criteria and analyses and concluded that the evidence was "inconsistent and unconvincing." This conclusion seemed to be supported by two more recent studies that assessed the protective effects of administering 240 mg *G. biloba* extract every day for a number of years to large cohorts (3,069 and 2,854 participants respectively) of participants over 70 years of age who were healthy at the study outset. In the first study,[1385] 6 years of administration of ginkgo failed to reduce declines in memory, attention, visuospatial abilities, and language and executive functions any more than a placebo. In the second study,[1386] there was a similar incidence of decline into dementia following both ginkgo and placebo taken for 5 years. However, it was notable that in the first of these studies the cognitive function of participants failed to decline as expected, and in the second far fewer elderly participants were diagnosed with dementia over the 5 years than would be expected. In contrast, a number of recent meta-analyses of studies assessing the efficacy of ginkgo in treating the cognitive and neuropsychiatric sequelae of dementia have shown more promising results. Wang et al.[1387] included six studies that involved 6 months of administration of ginkgo in their meta-analysis and found significant improvements on the cognitive subscale of the Alzheimer's Disease Assessment Scale when baseline risk was taken into account. Weinmann et al.[1388] conducted a meta-analysis of the data from nine studies that involved the administration of ginkgo extract to patients with Alzheimer's disease and vascular and mixed dementia for 12 to 52 weeks. Results across the 2,372 patients included in the analysis showed that ginkgo attenuated declines in cognitive performance across the dementia groups and that there were additional improvements in terms of activities of daily living in the Alzheimer's group. Finally, Janssen et al.,[1389] in a meta-analysis and systematic review, included six Alzheimer's disease studies involving more than 1,800 patients and found that 16 weeks or more of ginkgo treatment led to improvements in activities of daily living and cognitive function and the amelioration of neuropsychiatric symptoms (such as psychosis, agitation, aggression, anxiety, euphoria/dysphoria, or disordered motor behavior), with these effects seen most strikingly in patients who had significant levels of neuropsychiatric symptoms. Despite the generally favorable results with regard to ginkgo, the authors of these meta-analyses concur that the results overall are somewhat heterogeneous and the improvements are moderate.[1388,1389] However, since the publication of these reviews more evidence has emerged of the efficacy of ginkgo in the subset of dementia sufferers with neuropsychiatric symptoms. Two methodologically similar studies, which assessed the effects of 24 weeks' administration of ginkgo EGB 761 and which included a total of 806 patients with dementia with neuropsychiatric features, demonstrated an improvement (or an attenuation in decline) in cognitive function and neuropsychiatric symptoms, along with improvements in quality of life, physical symptoms, and caregiver distress.[1390,1391] While the heterogeneity of the methods employed in this area means that the jury is still out on the efficacy of ginkgo with regard to the psychological sequelae of dementia, the evidence does look promising.

Ecological Roles of Ginkgolides and Bilobalide and Brain Function

G. biloba is an uncommonly long-lived species, with a potential lifespan of more than 1,500 years. This has often been attributed to its complex chemical defenses. Ginkgo is particularly tolerant of pollution and possesses a high degree of resistance to fungal, viral, and bacterial diseases.[1392,1393] In situ it is also unusually free of herbivorous insects. It has no specialist herbivores; it may be occasionally visited by polyphagous chewing insects, but not generally by sucking insects.[1394] This lack of herbivores may be due to the potent anti-feedant properties of many of its secondary metabolites, most notably the ginkgolides and bilobalide, against a range of insect pests.[1395–1397] Ginkgo extracts also exhibit direct insecticidal properties. For instance, ginkgo proved the most potent of more than 200 plant extracts tested for insecticidal activity against the brown plant-hopper (*Nilaparvata lugens*), with subsequent fractionation showing that this activity was related to the presence of ginkgolides A, B, and C and bilobalide.[1398] Curiously, despite the lethal effects of these components to insects, it is the flavanol glycosides and a range of volatile organic compounds, including a range of monoterpenes and sesquiterpenes, that are preferentially synthesized in response to herbivory by generalist insects or by the application of jasmonic acid.[1392,1397] In contrast, the ginkgolides and bilobalide are induced by bacterial and fungal attack.[1399]

The key mechanism underlying the insecticidal properties of the ginkgolides and bilobalide may well be their potent antagonistic binding properties at insect ionotropic GABA receptors. These receptors also form the target for a range of synthetic insecticides such as dieldrin.[404] Ginkgolides are structurally related to the diterpene picrotoxinin, the classic GABA receptor antagonist, which has deadly insecticidal properties and is used as a GABA receptor probe in neuropharmacological investigations.[1361] However, the ginkgo terpenes are only weak inhibitors of the analogous human $GABA_A$ receptor. This is due to a minor, single-amino-acid difference in the composition of the insect and vertebrate GABA receptor pores, which drastically reduces the potency of the terpenes in this respect in humans.[1013] At the doses encountered by humans the ginkgolides and bilobalide are not likely to exhibit any toxicity related to GABA receptor antagonism.[1013] Indeed, ginkgo terpenes exhibit anticonvulsant properties rather than the convulsant properties expected of reduced GABA activity in mammals. However, GABA antagonism, which could lead to increased activity in cognition-relevant brain areas such as the hippocampus, has been suggested as one mechanism underlying ginkgo's memory-enhancing properties.

The markedly different effects of the ginkgo terpenes in insects and humans, due to a minor morphological difference in a homologous receptor, does illustrate nicely how modulated brain function in humans can be related to a mechanism of action that may prove fatal to insects.

■ *VALERIANA OFFICINALIS* (VALERIAN)

The *Valerianaceae* comprise approximately 350 species of mostly herbaceous plants that originated in the mountainous regions of East Asia and went on to colonize Europe and America, arriving in the latter continent either via the Bering land bridge or across the Atlantic. The majority of species of *Valerianaceae* fall within the *Valeriana* (valerian) genus. The only region of the world that doesn't serve as home to the valerians is New Zealand/Australia. In terms of diversity, the South American Andes is now home to its own extended clade of the family, representing approximately half of all species. Evidence suggests that this relatively youthful group's ancestors arrived in South America from the North approximately 5 million years ago.[1400,1401]

Although they now represent a cosmopolitan group, the vast majority of valerian species have retained the family's trademark chemical signature: synthesis of the valepotriate group of secondary metabolites.[1400] Members of the valerian genus have also been incorporated into the traditional medicine chests around the world; for instance, *V. wallichii* features in the traditional ayurvedic and Tibetan Buddhist pharmacopoeias, *V. fauriei* is used in Chinese and Japanese medicine, *V. capensis* is used in African medicine, *V. officinalis* is a mainstay of European herbalism, and *V. prionophylla* appears in Meso/South American pharmacopoeias. In all of these medical systems the many uses of the valerians include their use as sedatives, anxiolytics, and sleep aids.[1402–1404] Prior to the arrival of synthetic pharmacological sedatives and hypnotics, valerian was a mainstream treatment for anxiety and sleep disturbance in Europe and North America; it was used, for instance, to treat shell-shocked troops in World War I and to calm civilians during air-raids in World War II.[1404]

Valerian's mild sedative and anxiolytic properties are variously attributed to the presence of genus-specific sesquiterpenes, including valerenic acid and valeronone, and/or the valepotriates, a group of iridoids derived from the monoterpene geraniol. The most abundant of these, valtrate, makes up approximately 80% of this class of chemicals in the plant. Root extracts also contain appreciable levels of GABA and a wide range of volatile monoterpenes and sesquiterpenes.[195,1405]

Valerian's terpene constituents bind to a variety of neurotransmitter receptors,[1406] including the 5-HT_{5A} receptor,[1407] which is implicated in circadian rhythms and anxiety. They also act as allosteric modulators of subunit-specific $GABA_A$ channels[1408] and adenosine A_1 receptors[1409] and inhibit the uptake of GABA. Valerenic acid and derivatives have been shown to have potent anxiolytic effects in rodent models,[1069] with these effects underpinned by in vivo binding to a specific subunit (β3) of the $GABA_A$ receptor.[1070] Whole extracts also increase noradrenergic and dopaminergic neurotransmission in vivo, potentially as a downstream consequence of GABAergic activity.[1410,1411] Administration of the whole extract, and the valerenic acid with and without the GABA component, suggested that these two compounds were responsible for the anxiolytic properties demonstrated in the mouse.[1069] However, the valepotriate fraction has also been shown to have independent sedative, anxiolytic, and memory-disturbing properties in mice, again potentially via GABAergic mechanisms.[1068] Similarly, whole extracts of valerian species from around the world, including *V. glechomifolia*, *wallichii*, *officinalis*, and *prionophylla*, have been shown to have antidepressant effects in rodent models,[1403,1410,1411] which may be directly mediated by increased noradrenergic and dopaminergic neurotransmission.[1410,1411] This effect may, in turn, be a downstream consequence of allosteric agonism at GABA receptors.[1410]

The bulk of human research has assessed the effects of valerian extracts on sleep parameters. In one study valerian was shown to be equipotent to benzodiazepines in terms of sleep quality and waking symptoms in insomnia outpatients,[1412] and evidence suggests specific efficacy when coadministered with hops.[1413,1414] However, reviews of valerian's efficacy in sleep disturbance have concluded that the evidence is "promising but not fully conclusive" on the basis of 9 included trials[1415] and that the evidence only "suggested" that valerian improved sleep quality, on the basis of 16 eligible randomized controlled trials.[1416] A recent meta-analysis that included 18 studies[1417] found that valerian extracts significantly improved sleep quality when measured by a simple subjective yes/no question, but that evidence from typical, validated sleep questionnaires was lacking. In general, research in this area is replete with methodological inconsistencies but suggests that valerian is associated with few negative side effects.

The global, traditional use of valerian extracts as anxiolytics and mild sedatives, while confirmed consistently in animal models, has stimulated comparatively little research interest in humans. A single dose of valerian was shown to disrupt vigilance and information-processing task performance after 1 to 2 hours, suggesting sedative effects,[1418] but did not have the same effect the morning after administration.[1418,1419] With regard to anxiolytic properties, reviews of the few studies assessing the efficacy of valerian have been somewhat inconclusive, with only one eligible, methodologically adequate study of the effects of valerian on anxiety identified in a Cochrane review in 2006.[1420] The study in question[1421] failed to find any advantage of valerian over placebo.

Ecological Roles of Valerian Secondary Metabolites and Brain Function

Valerian plant material and extracts have a long history of traditional use as insect deterrents. They also have antimicrobial properties.[1422] However, these effects may be related to their non–genus-specific terpene constituents. No research to date has investigated the specific ecological roles for the iridoid valepotriates and genus-specific sesquiterpenes. Iridoids as a chemical class are found in more than 50 plant families and are thought to mediate the interactions between plants and insects and other invertebrates. In line with this, they are distributed in young and valuable tissue in a pattern suggestive of anti-herbivore properties.[1423] The valepotriates have also been shown to be induced in response to a number of biotic and physical stressors,[1424] again suggesting specific ecological roles. In general, the ecological roles of the valerian secondary metabolites have received little attention.

14 In Conclusion: Comparing and Contrasting the Alkaloids, Phenolics, and Terpenes

The three major groups of secondary metabolites reviewed in the preceding chapters can be differentiated with respect to a number of parameters. The most obvious differences are seen in their ecological roles and their related levels of toxicity. The alkaloids are synthesized by a minority of plants (up to 30%) and appear to play ecological roles almost entirely related to their toxicity, whether this is to herbivores, microbes, or other plants. The phenolics, on the other hand, are ubiquitous, multifunctional molecules that play a diverse range of roles, often of a protective nature, with many exhibiting low toxicity. As an example, they function as antioxidants and antimicrobials and as attractants, providing color and contributing to chemical bouquets, as well as playing a number of roles that blur the boundary between "primary" and "secondary" functions; for instance, partaking in the intercellular and intracellular signaling of the plant. They then also play several purely secondary metabolite roles that are predicated on extending these signaling functions to other organisms, whether these are to symbiotic or pathogenic microbes or to competitor plants. The terpenes as a group fall somewhere between the alkaloids and phenolics and fill some of the middle ground between the two. For example, they can exert potent toxicity in the face of herbivory, the encroachment of competitor plants, or microbial assault, and this is often due to their ability to incorporate themselves into cellular membranes, disturbing their integrity and functioning. However, they can also play benign roles, providing color and representing a significant proportion of the chemicals that make up the many volatile emissions of the plant. This latter role exemplifies their position in the middle ground, as many individual terpenes play dual roles that encompass both toxicity and attraction to insects, with their eventual function dependent on their mode of delivery and quantity delivered.

Naturally, these differing palettes of ecological roles reflect differences in the evolution of the synthetic pathways for each group. In this respect the alkaloids are the comparative newcomers, with phylogenetic analyses and their clade-specific pattern of distribution showing that their synthetic pathways typically developed from primary metabolic pathways following the diversification of the angiosperms a mere 145 million years ago. As an example, the large, monophyletic group of benzylisoquinoline alkaloids are found only in the eudicot clade of angiosperms, placing the evolution of their pathways after the divergence of the monocots, which took place some 90 to 100 million years ago. Alkaloids could therefore be seen as representing something of a recent upgrade to the defensive arsenal of their home plants. Synthesis of the phenolics, on the other hand, is an ancient ability that was enjoyed by plants long before they made their first tentative incursions onto the land 465 million years ago. Indeed, the ability to synthesize antioxidant, ultraviolet light-absorbing flavonoids may well have been one key factor that allowed the early bryophytes to survive in the harsh, dry environment at the water's edge. The subsequent development of the ability to synthesize phenolic lignin polymers, and therefore woody tissue, provided the vascular plants with the structure and physical protection that then allowed them to move away from the water and eventually colonize the vast majority of the earth's terrestrial surface. Naturally, given their extended history, many individual phenolics, or phenolic subgroups, are ubiquitous and are expressed by all plants and are present in all types of plant tissue. The terpenes, again, occupy the middle ground. Having originated solely as primary metabolites, the synthesis of terpenes is an ancient ability. However, the subfamilies of terpene synthase enzymes that produce the terpenes employed in secondary ecological roles arose more recently. Of the seven terpene synthase subfamilies found across plants, the two most ancient produce terpenes solely for use in primary roles, whereas the remaining five subfamilies produce chemicals with ecological roles, with the three most recently diverged subfamilies restricted to the angiosperms.

The answer to the core question posed in this book—"Why do secondary metabolites affect human brain function?"—is inextricably intertwined both with the ecological roles and evolutionary history of the three chemical groups. Indeed, it is also possible to differentiate the alkaloids, phenolics, and terpenes on the basis of how closely their effects on human brain function can be explained by the two hypotheses advanced in Chapter 3. These were that any effects on human brain function are due to the cellular, biochemical, and molecular similarities between plants and humans, or alternatively, they reflect the many similarities between the nervous systems of humans and insects.

In the case of the alkaloids the answer would seem to be fairly straightforward. The evidence summarized in Chapters 4 to 7 suggests that in many cases these compounds owe their defensive properties to interactions with the nervous system of herbivores. These herbivores are primarily insects, a group that constitutes the dominant clade of animals on earth, comprising greater than 1 million species, more than 50% of which survive by eating plant tissue. This group enjoys a close, intimate, and enduring relationship with plants that has been forged in over 400 million years of co-evolution and co-speciation, and it is this relationship that has been the key driver in the development of the plants' many secondary metabolite synthetic pathways. In evolutionary terms, in contrast to the insects, the vertebrates as a family pale into insignificance, and humans have been almost entirely irrelevant to plants. Certainly a plant that didn't have adequate protection against the phytophagous insects and other invertebrates that represent its closest neighbors would have no need to worry about protecting itself against any vertebrate animals. The simple reason, therefore, that many alkaloid secondary metabolites affect human brain function is that, due to a shared genetic heritage, the central nervous system of the human is biochemically, architecturally, and functionally very similar, and in many respects identical, to that of the insect (and other invertebrates). In reality, given that the insects are the dominant clade of animals on earth, it could be said that humans have an elaborated version of the insect nervous system. The similarities include the employment of conserved neurotransmitters, receptors, and signal transduction pathways, all of which form key targets for the alkaloids. More often than not, insects and humans also express broadly similar behavioral modification when the function of their nervous system is modulated by bioactive chemicals, including secondary metabolites. Many of the clearest examples of the concordance in the nervous system or behavioral effects of phytochemicals in insects and humans have been demonstrated with regard to alkaloids—although this may be primarily because the effects of phenolics and terpenes have not been assessed using insect models. Of course, one factor that separates the insects and humans is the dose of alkaloid liable to be encountered. By their very nature alkaloids are toxic, and the comparative dose encountered by an insect feeding on plant tissue will typically be far in excess of the low doses of psychotropic or medicinal alkaloids employed by humans. However, it is worth noting that even seemingly benign alkaloids such as caffeine will readily kill humans at higher doses.

The first thing to note with regard to the phenolics is that, unlike the toxic alkaloids, they have been an integral component of our diet throughout human evolution. Many of their putative beneficial effects in humans would appear to be predicated on the multifarious cellular, biochemical, and molecular similarities between plants and mammals. These similarities also exist courtesy of a mutual genetic inheritance from the last common ancestor of the two taxa. As noted in Chapters 8 and 9, many of the properties of commonly encountered dietary polyphenols, including their effects on inflammatory processes and both cardiovascular and brain function in humans, can be related directly to the multifarious interactions they enjoy with mammalian intracellular signaling pathways. These conserved pathways play crucial roles in cellular responses to stressors in both plants and animals. Well-defined examples include the conserved signaling pathways that regulate the synthesis and activity of the jasmonate hormones in plants and their mammalian counterparts, the prostaglandins, and the TOR kinase signaling pathways that integrate cellular information on current levels of nutrients, energy, and stressors and titrate cellular differentiation, activity, and responses to match the prevailing conditions. While these pathways play analogous roles in the response to stressors in both humans and plants, one major difference in the downstream products of their activity is, for the plant, the synthesis of secondary metabolites, including polyphenols such as the ubiquitous flavonoids. To date, no research has addressed the possibility that flavonoids interact with the same signaling pathways in their home plant as they do in mammalian tissue. However, emerging evidence has confirmed that flavonoids do play a number of other signaling roles in plants, and they have been detected in the nucleus of cells, reinforcing the possibility of wide-ranging signaling roles. It is therefore possible, indeed likely, that any effects in mammalian tissue may simply reflect an attempt by these polyphenols to interact with, and regulate, the conserved stress signaling pathways that contribute to their own synthesis in the plant. In this context the current levels of flavonoids and other polyphenols being produced by the plant would provide a useful source of information to the cell via interactions with the TOR pathways. Conversely, some of the effects of phenolics appear to be predicated on the cellular similarities between humans and the symbiotic bacteria and fungi that live in the plant's rhizosphere. In this instance the flavonoids/polyphenols help manage the relationship between the plant and microbe by binding to microbial receptors that are sensitive to estrogens, thereby regulating the microbes' behaviors. The same phytoestrogen polyphenols also bind to analogous estrogen-sensitive receptors in mammalian cells, unintentionally modulating the various parameters influenced by mammalian estrogenic hormonal signaling, including many aspects of brain function. Notably, TOR signaling is also a conserved feature of fungal cellular functioning, and any modulation by polyphenols of this pathway may also reflect an attempt by the plant to disturb the nutrient/stress signaling within the conserved system in pathogenic fungi. In either event, the effects of polyphenols within mammalian cells can be seen as a direct consequence of unintentional cross-kingdom signaling, predicated on the simple similarities between the various taxa. However, it should be kept in mind that these phytochemicals have been an intrinsic part of our ancestral diet, and their functionality may well have been tailored during human evolution, in much the same way as the many plant-derived "vitamins" that are crucial to our survival.

With regard to the terpenes, as described in Chapters 10 to 13, we can once again return to the insects. As well as deterring insect herbivores, plants also have to attract the insect pollinators that are crucial for the reproduction of 85% of angiosperms. Terpenes make up the lion's share of the volatile attractants employed in this role and appear as a key component in all floral bouquets and other volatile emissions. Again, attraction requires interactions with the insect nervous system, and the selective pressure to use volatile terpenes in ecological roles must have been predicated on the fact that insects endogenously synthesized and utilized the same compounds in their own inter- and intra-species pheromone/allomone chemical communications. Evidence suggests that, for the plant, these terpenes started life as defensive chemicals prior to the evolution of flowers but then assumed the role of attractants along with the advent of insect pollination. This does mean that many of the terpenes that have an effect on human brain function play dual roles with regard to insects. At the low dose associated with many volatile emissions they may act as attractants, with a neutral or possibly beneficial effect on the symbiotic insect's nervous system, while at the higher doses that would be associated with the activation of defensive structures, such as the rupturing of glandular trichomes, they would become toxic deterrents via overactivity within the same neural systems. The comparatively low doses of these volatiles encountered by humans are liable to be safe and well tolerated, and their effects may include modulation of cognitive function or behavior via interactions with the same nervous system mechanisms as those modulated in insects.

Of course, the terpenes as a group also include a number of psychoactive compounds that have no role in attraction but would seem to function only in the deterrence of herbivores. These include many cannabinoids, neoclerodane terpenes such as salvinorin A, and the ginkgolides, all of which interact with aspects of neurotransmission that are equally prevalent in insects and humans. One hint as to why humans tolerate some of these compounds so well is provided by the evidence showing that whereas ginkgolides are deadly toxins to insects due to their antagonistic binding at GABA receptors, they exert 1/10,000th of the same effect at the conserved mammalian GABA receptors due to a tiny morphological difference in the respective receptors. Finally, it is also notable that plants synthesize a wide range of triterpene phytoecdysteroids, many of which appear to have the primary function of disturbing the ecdysteroid system that governs insect growth and development. They do this by mimicking the activity of structurally related, or indeed identical, hormones at ecdysteroid receptors. A number of triterpenes with potential phytoecdysteroid properties then owe their beneficial effects on human brain function to interactions with mammalian receptors for glucocorticoids, which, in turn, are modified triterpenes that are structurally similar to the insect ecdysteroids.

Of course, the two basic hypotheses—"plant/human" and "insect/human" similarities—are not mutually exclusive, and many plant/animal interactions will be driven by the simple chemical similarity of the three taxa. As noted in Chapter 3, the ancestral, primary chemistry of the plant and the animal is largely shared, and functional chemicals are typically derived in both taxa from the same substrates, for instance identical amino acids. This chemical similarity extends to the synthesis by the plant of many of the chemicals we regard as human "neurochemicals," a number of which are employed as plant signaling molecules. Similarly, humans endogenously synthesize a number of chemicals that we think of as plant secondary metabolites (for instance, morphine and the tryptamine hallucinogens) or plant hormones (abscisic acid and possibly salicylic acid) and utilize these as signaling molecules. This chemical similarity may predispose animals to be affected by plant chemicals. However, the very fact that animals can respond to the plant chemical must also have been a necessary condition for the selective evolutionary pressures for the plant to utilize the secondary metabolite in an ecological role in the first place.

Finally, it is also notable that the bioactive properties of plant chemicals are often transferred wholesale to the consuming animal. The obvious examples of this are the "vitamins," which are synthesized by plants and subsequently sequestered by animals from the food chain. Once consumed, these micronutrients play identical, analogous roles as essential antioxidants and cellular enzyme co-factors as those played in the plant. A similar, direct transfer of bioactivity may underlie the benefits of many secondary metabolites with antioxidant properties. Similarly, the antimicrobial properties of many secondary metabolites are also often transferred wholesale to the consuming animal.

The differences between our three groups of phytochemicals extend to whether they owe their bioactivity to a single compound or to a group of compounds. Typically, the effects of an alkaloid-containing extract can be attributed to a single component, and indeed the ability to attribute effects and the ability to isolate an active principal are the key properties that have made the alkaloids the primary source of natural product drug discovery for the past two centuries. In contrast, the bioactivity of terpene extracts cannot typically be traced to a single compound. The components, in this case, are more likely to work most effectively together, through additive or synergistic effects or by polyvalences, whereby either different components contribute a range of different bioactivities that result in the overall effect of the extract or single components exhibit multiple different bioactivities. Examples of synergy include the combined effect of monoterpenes with respect to cholinesterase inhibition, with complex, naturally occurring mixtures being more potent than the sum total of the effects of their constituent parts. The overall psychotropic effect of cannabis, on the other hand, is due to polyvalent effects, with the different bioactivities of the individual cannabinoids contributing to the overall bioactivity of the resin. Tetrahydrocannabinol (THC) may be the principal psychoactive component due to its interactions with cannabinoid receptors, but the overall effects of cannabis are moderated by the independent effects of cannabidiol and the other minor cannabinoids within numerous neurotransmitter systems. Similarly, several herbal medicinal treatments owe their effects to polyvalences that make the effects of the overall extract markedly different from that of any one component. Examples include *Ginkgo biloba, Valeriana officinalis*, *Hypericum perforatum*, and *Salvia miltiorrhiza*.[1425]

Unfortunately, mainstream medicine and the pharmaceutical industry, by and large, do not like complex mixtures of components that owe their effects to poorly understood synergistic and polyvalent activities. What they require are mono-treatments with a well-characterized mode of action. The sad case of Alzheimer's disease illustrates this nicely. Currently the first-line treatments offered to attenuate the cruel cognitive declines seen in dementia are a handful of cholinesterase inhibitors, including the plant-derived alkaloids rivastigmine (a semisynthetic derivative of physostigmine), and galantamine. Alkaloids are toxic by their very nature, and in the case of these treatments they have limited efficacy that has to be carefully balanced with their high level of cholinergic side effects. Indeed, this balancing act is so delicate that in the UK in 2009 the government body responsible for dictating the use of prescription drugs within the National Health Service ruled against the use of these cholinesterase-inhibiting treatments except for patients suffering moderate (as opposed to mild or severe) levels of dementia. This decision was based on evidence of very modest efficacy, high levels of side effects, and the high cost of the treatments.[1426] This left the majority of Alzheimer's sufferers in the UK with no treatment of any sort whatsoever. It was only following a public outcry that the guidance was subsequently amended to allow doctors to prescribe these treatments for mild cases of Alzheimer's disease.[1427] The lack of effective treatments for Alzheimer's disease is a particularly sad case, but it also has to be noted that many plant-derived terpenes have cholinesterase-inhibiting properties, including the complex mixtures of monoterpenes and sesquiterpenes utilized as the secondary metabolites of choice by the wide range of *Lamiaceae* described in Chapter 11. Terpene-containing extracts from *Salvia* species such as *Salvia officinalis* and *Salvia lavandulaefolia* have demonstrable synergistic cholinesterase inhibitory properties both in vitro and in vivo, including, importantly, in the rodent brain after oral consumption. They also exhibit polyvalent effects on a number of other potentially beneficial parameters and have been shown, albeit in comparatively small controlled trials, to consistently improve cognitive function in healthy younger and older adults in a pattern commensurate with enhanced cholinergic function. A single small study involving 30 Alzheimer's patients also demonstrated improved cognitive function. These extracts have the added advantage of being safe for human consumption with no significant side effects. And yet, despite this, there is no interest from the pharmaceutical Industry in carrying out more comprehensive testing of these products. They would prefer to spend the vast amount of money that is required to bring a single new synthetic drug to market (estimated by some, in 2009, to be a staggering $1.3 to $1.7 billion[1428]) rather than spend a tiny fraction of this amount establishing the efficacy of a natural, safe, cholinesterase-inhibiting plant extract that may provide a cheap and effective treatment for the cognitive symptoms of Alzheimer's disease. Of course, the main reason for this is simple straightforward economics; the plant-derived cholinesterase inhibitor would not benefit from the commercial security afforded by the patent protection enjoyed by a synthetic drug for up to 20 years, meaning that the eventual product would have to be sold competitively, at a low price, rather than monopolistically, at a high price.

Beyond pure economics, another rate-limiting factor here is our poor understanding of the many interactions within complex plant-derived extracts. Until we disentangle these the majority of potential treatments will remain relegated to the slightly derided "herbal medicines" aisle. However, this lack of acceptance also reflects, to a great extent, the generally poor-quality evidence base surrounding the efficacy of many herbal extracts. With only a very few exceptions, such as *Ginkgo biloba* and *Hypericum perforatum*, complex herbal extracts have simply failed to attract sufficient research to establish their efficacy either way. Where they have attracted research the literature is typically plagued by inconsistent or poor methodology. Naturally this is largely due to a lack of funding, from both the pharmaceutical industry and governmental bodies, but it also reflects the skepticism of the medical establishment.

It is also difficult, although not impossible, to properly standardize herbal extracts. Often, so-called standardization is achieved simply on the basis of the contents of a single constituent, which is assumed to be the active principal. In the case of many non-alkaloid plant extracts it is difficult to be certain which component this should be. For instance, *Hypericum perforatum* extracts were originally standardized on the basis of their content of hypericin, and it later transpired that this was unlikely to be the principal active component. Standardization on the basis of a single component also tells us nothing about the contents of other, potentially synergistic, constituents. As an example, the cholinesterase-inhibitory properties of *Salvia* extracts are the result of the synergistic and antagonistic contributions of multiple monoterpenes, with high levels of 1,8-cineole and low levels of camphor being the best, but not the only, predictor of the overall potency of the extract. Standardization on the basis of 1,8-cineole alone would be largely meaningless.

Once we have a fuller understanding of the interaction and synergy between active components within herbal extracts, we are faced with the question of how to standardize their levels. However, we have an improving understanding of the effect of abiotic and biotic stressors on the expression of secondary metabolites. The overall and comparative levels of active phytochemicals in plant material depend on a wide range of parameters that are capable of manipulation, including climate, soil composition or growing medium, light levels, and time and season of harvest,[1429] geographic location,[1429,1430] habitat,[1431] salinity,[1432] exposure to herbivores,[1433,1434] pathogens, and ultraviolet light,[1435] hydration status,[1436] and, of course, chemical manipulation of the levels of plant hormones such as the jasmonates and salicylates. Given the sophistication of contemporary agricultural techniques, in particular internal systems such as hydroponics, it would seem feasible to grow well-standardized plants that benefit from augmented and constrained levels of beneficial and detrimental components respectively while retaining the positive synergistic properties of whole extracts. If you want a convincing example of the possibilities look no further than cannabis: between 1993 and 2008 the average content of THC in samples from drug seizures in the United States nearly tripled, from 3.4% to 8.8%.[1207] This was due to selective breeding of "drug-type" plants, along with modified growing techniques that favored the expression of THC over cannabidiol.

Of course, this isn't the only area that requires more research attention. As always when reviewing the literature surrounding the various potentially psychoactive terpene and phenolic phytochemicals, it is disheartening to find how little we actually know about the efficacy of many non-alkaloid compounds and extracts. The previous chapters included summaries of the evidence surrounding only those few compounds, and groups and mixtures of compounds, that have attracted some human research. These represent a tiny proportion of the phytochemicals with psychotropic potential. For instance, Zhang[1303] identified more than 80 traditional herbal extracts that have demonstrated efficacy in animal models of psychiatric disorders. Only a few of these have made the transition to human research. Even for these few interventions there is typically very little research, and what there is can be so heterogeneous and methodologically inconsistent that it is difficult to draw any firm conclusions either way as to efficacy. As an example, despite some good epidemiological evidence and consistent demonstrations of beneficial cardiovascular effects in controlled trials, the entire question of whether the direct administration of flavonoids and other polyphenols can beneficially modulate brain function is really an open question. There are plenty of interesting small-scale studies, many of which show beneficial effects, but there are also plenty of studies that show no effect, and it is possible that there are many more that have fallen afoul of a publication bias that makes it far easier to publish positive results than it is to publish negative results. What is notably lacking are the larger-scale, well-controlled studies that would start to tell us whether these phytochemicals have any efficacy. This lack of evidence in humans is particularly surprising because the literature surrounding the polyphenols is huge. As I write this the single search term "flavonoid" entered into an academic search engine[a] returns over 108,000 papers, mostly detailing basic mechanistic and preclinical research. A good many of these publications theorize that flavonoids are good for the brain, and yet only a few dozen studies, most of which are summarized in Chapter 9, have actually assessed the effects of flavonoids on brain function. Similarly, the search term "resveratrol" returns 20,000-plus papers. A large proportion of these describe laboratory-based in vitro and in/ex vivo studies, many of which have mechanistic relevance to brain function. And yet, literally only a handful of these papers describe studies carried out in humans. This is all the more surprising because both flavonoids and resveratrol are components of foods that we eat as part of our everyday diet, and supplements containing these phytochemicals are considered to be safe and can be found on the shelves of every pharmacy, health shop, and supermarket throughout the land. It seems to me that this huge imbalance in the research effort should be rectified and that some of the enormous resource being directed at establishing the mechanisms by which safe phytochemicals with no firm evidence of efficacy might have their purported effects should be directed toward first finding out if they have any efficacy. Again, it would only take a fraction of the resources expended on bringing a single synthetic drug to market to make gigantic strides in our understanding in this area.

■ FUTURE DIRECTIONS?

We know quite a lot about the mechanisms by which most phytochemicals affect mammalian brain function, but the question of why, rather than how, a plant chemical would have these effects in the first place has been almost entirely ignored. The many strands of research summarized in this book seem to suggest that the reasons are inextricably linked to the evolved ecological roles that secondary metabolites are trying to play for the plant and the simple similarities in biology, predicated on a common evolutionary ancestry, between plants and animals and between the nervous systems of insects and humans.

One thing that becomes apparent when reviewing this entire literature is a number of imbalances in the extant research. These are seen, for instance, in the concentration on basic laboratory research rather than controlled human trials, even for safe interventions; the concentration of the research effort in humans on just one or two of the non-alkaloid herbal extracts; and the general imbalance between the resources and effort poured into "humano-centric" research in comparison to research that furthers our knowledge about terrestrial Earth's dominant taxa, the insects and plants. There are numerous examples highlighted in the preceding chapters of areas of research that have, to date, failed to attract adequate attention, and filling some of these gaps not only will tell us a lot more about why secondary metabolites affect brain function but

a. ISI Web of Science.

may also further our understanding of the utility of secondary metabolites for improving human health. I'll give just two, of many, examples: first, with regard to plants, we now know a great deal about the potential intracellular signaling roles of ubiquitous secondary metabolites such as the flavonoids in mammalian tissue. However, we know next to nothing about their reciprocal signaling roles within the plant itself. A greater understanding of the endogenous roles of these phytochemicals within the conserved signaling pathways in the plant can only increase our understanding of how these phytochemicals might benefit human health and may also throw up entirely new avenues of investigation relevant to the prevention or treatment of cancer, cardiovascular disease, and neuropathologies.

Second, turning to the insects, it is notable that little research, to date, has directly addressed the question of how secondary metabolites affect the insect nervous system and behavior. The little research in this area has tended to aggregate around the use of single phytochemicals as simple neuropharmacological tools that can be used to elucidate the workings of aspects of the insect central nervous system; alternatively, plant-derived drugs have been administered in insect models of drug abuse and addiction. In both cases the research has typically involved alkaloids, and it has been conducted to extrapolate the results back to mammals rather than to understand the effects of the phytochemicals in their intended insect targets. With the exception of a few studies with an agricultural focus investigating the mechanisms underlying the insecticidal properties of secondary metabolites, the effects of most non-alkaloid secondary metabolites have simply never been assessed in insects. However, insect models are simple, economical, time-efficient, and ethically acceptable, and they would provide an excellent starting point for investigations into the molecular mechanisms of action and neuronal and behavioral effects of individual and complex mixtures of phytochemicals from all classes. In particular, they would make ideal tools for the research necessary to bolster our understanding of the complex synergies, antagonisms, and polyvalences that combine to generate the overall physiological effects of many plant extracts, and they could potentially start to open the doorway for the wider use of nontoxic, non-alkaloid phytochemicals in mainstream medicine.

Finally, one notable aspect of most of the research reviewed briefly in this book is that it has typically been conducted within the boundaries of discrete disciplines. Answering the simple question of why plant chemicals modulate brain function requires the integration of thoughts and concepts from a diverse range of disciplines, including molecular biology and biochemistry, plant science, zoology, entomology, pharmacology, medicine, neuroscience, and psychology. The necessary dismantling of discipline "silos" may have some interesting emergent, synergistic properties that go beyond answering this simple question.

REFERENCES

1. Kunzig, R. (2002). La Marmotta. *Discover*, November.
2. Fugazzola Delpino, M. A., D'Eugenio, G., and Pessina, A. (1993). La Marmotta (Anguillara Sabazia, RM). Scavi 1989. Un abitato perilacustre di età neolitica. *Bullettino di Paletnologia Italiana*, ***84***, 342.
3. Bernicchia, A., Fugazzola, M. A., Gemelli, V., Mantovani, B., Lucchetti, A., Cesari, M., et al. (2006). DNA recovered and sequenced from an almost 7000 y-old Neolithic polypore, *Daedaleopsis tricolor*. *Mycological Research*, ***110***, 14–17.
4. Merlin, M. D. (2003). Archaeological evidence for the tradition of psychoactive plant use in the old world. *Economic Botany*, ***57***, 295–323.
5. Akers, B. P., Francisco Ruiz, J., Piper, A., and Ruck, C. A. P. (2011). A prehistoric mural in Spain depicting neurotropic *Psilocybe* mushrooms? *Economic Botany*, ***65***, 121–128.
6. Samorini, G. (1992). The oldest representations of hallucinogenic mushrooms in the world (Sahara Desert, 9000–7000 B.P.). *Integration*, ***2–3***, 69–78.
7. Dillehay, T. D., Rossen, J., Ugent, D., Karathanasis, A., Vasquez, V., and Netherly, P. J. (2010). Early Holocene coca chewing in northern Peru. *Antiquity*, ***84***, 939–953.
8. Krief, S., Hladik, C. M., and Haxaire, C. (2005). Ethnomedicinal and bioactive properties of plants ingested by wild chimpanzees in Uganda. *Journal of Ethnopharmacology*, ***101***, 1–15.
9. Krief, S., Huffman, M. A., Sevenet, T., Hladik, C. M., Grellier, P., Loiseau, P. M., et al. (2006). Bioactive properties of plant species ingested by chimpanzees (*Pan troglodytes schweinfurthii*) in the Kibale National Park, Uganda. *American Journal of Primatology*, ***68***, 51–71.
10. Lacroix, D., Prado, S., Deville, A., Krief, S., Dumontet, V., Kasenene, J., et al. (2009). Hydroperoxy-cycloartane triterpenoids from the leaves of *Markhamia lutea*, a plant ingested by wild chimpanzees. *Phytochemistry*, ***70***, 1239–1245.
11. Krief, S., Martin, M. T., Grellier, P., Kasenene, J., and Sevenet, T. (2004). Novel antimalarial compounds isolated in a survey of self-medicative behavior of wild chimpanzees in Uganda. *Antimicrobial Agents and Chemotherapy*, ***48***, 3196–3199.
12. Gustafsson, E., Krief, S., and Saint Jalme, M. (2011). Neophobia and learning mechanisms: how captive orangutans discover medicinal plants. *Folia Primatologica*, ***82***, 45–55.
13. Forbey, J. S., Harvey, A. L., Huffman, M. A., Provenza, F. D., Sullivan, R., and Tasdemir, D. (2009). Exploitation of secondary metabolites by animals: A response to homeostatic challenges. *Integrative and Comparative Biology*, ***49***, 314–328.
14. Godlaski, T. M. (2011). The god within. *Substance Use & Misuse*, ***46***, 1217–1222.
15. Berlant, S. R. (2005). The entheomycological origin of Egyptian crowns and the esoteric underpinnings of Egyptian religion. *Journal of Ethnopharmacology*, ***102***, 275–288.
16. Schultes, R. E. (1998). Antiquity of the use of New World hallucinogens. *The Heffter Review of Psychedelic Research*, ***1***, 1–7.
17. Pahnke, W. N. (1969). Psychedelic drugs and mystical experience. *International Psychiatry Clinics*, ***5***, 149–162.
18. Hasler, F., Bourquin, D., Brenneisen, R., Bar, T., and Vollenweider, F. X. (1997). Determination of psilocin and 4-hydroxyindole-3-acetic acid in plasma by HPLC-ECD

and pharmacokinetic profiles of oral and intravenous psilocybin in man. *Pharmaceutica Acta Helvetiae*, ***72***, 175–184.

19. Studerus, E., Gamma, A., Kometer, M., and Vollenweider, F. X. (2012). Prediction of psilocybin response in healthy volunteers. *PLoS One*, ***7***.
20. Griffiths, R. R., Richards, W. A., McCann, U., and Jesse, R. (2006). Psilocybin can occasion mystical-type experiences having substantial and sustained personal meaning and spiritual significance. *Psychopharmacology*, ***187***, 268–283.
21. Roberts, T. B., and Hruby, P. J. (2002). Toward an entheogen research agenda. *Journal of Humanistic Psychology*, ***42***, 71–89.
22. Wasson, R. G. (1968). *Soma: Divine Mushroom of Immortality*. New York: Harcourt, Brace, Jovanovich.
23. Wasson, R., Ruck, C., and Hofmann, A. (1978). *The Road to Eleusis: Unveiling the Secret of the Mysteries*. New York: Harcourt Brace Jovanovich.
24. Allegro, J. M. (1970). *The Sacred Mushroom & the Cross*. London: Hodder and Stoughton.
25. La Barre, W. (1972). Hallucinogens and the shamanic origins of religion. In E. T. Furst (Ed.), *Flesh of the Gods: the Ritual Use of Hallucinogens*. New York: Praeger.
26. Escohotado, A. (1999). *A Brief History of Drugs: from the Stone Age to the Stoned Age*. Rochester, Vermont: Park Street Press.
27. Wink, M. (1998). A short history of alkaloids. In M. F. Roberts & M. Wink (Eds.), *Alkaloids: Biochemistry, Ecology, and Medicinal Applications*. New York: Plenum Press.
28. Merlin, M. (1984). *On the Trail of the Ancient Opium Poppy*. Toronto Associated University Presses.
29. Schiff, P. L. (2002). Opium and its alkaloids. *American Journal of Pharmaceutical Education*, ***66***, 186–194.
30. Russo, E. B. (2007). History of cannabis and its preparations in saga, science, and sobriquet. *Chemistry & Biodiversity*, ***4***, 1614–1648.
31. Bertol, E., Fineschi, V., Karch, S. B., Mari, F., and Riezzo, I. (2004). *Nymphaea* cults in ancient Egypt and the New World: a lesson in empirical pharmacology. *Journal of the Royal Society of Medicine*, ***97***, 84–85.
32. Lee, M. R. (1999). The snowdrop (*Galanthus nivalis*): from Odysseus to Alzheimer. *Proceedings of the Royal College of Physicians of Edinburgh*, ***29***, 349–352.
33. Lee, M. R. (2007). *Solanaceae* IV: *Atropa belladonna*, deadly nightshade. *The Journal of the Royal College of Physicians of Edinburgh*, ***37***, 77–84.
34. Lee, M. R. (2006). *Solanaceae* III: henbane, hags and Hawley Harvey Crippen. *The Journal of the Royal College of Physicians of Edinburgh*, ***36***, 366–373.
35. Wasson, R. G., Hofmann, A., and Ruck, C. A. P. (1978). *The Road to Eleusis: Unveiling the Secret of the Mysteries* New York: Harcourt, Brace, Jovanovich.
36. Webster, P. (2000). Mixing the Kykeon. *Eleusis: Journal of Psychoactive Plants and Compounds*.
37. Dioscorides, P., Goodyer, J., and Gunther, R. T. (1934). *The Greek Herbal of Dioscorides*. Oxford: Oxford University Press.
38. Ramoutsaki, I. A., Askitopoulou, H., and Konsolaki, E. (2002). Pain relief and sedation in Roman Byzantine texts: *Mandragoras officinarum, Hyoscyamos niger* and *Atropa belladonna*. In J. C. Diz, A. Franco, D. R. Bacon, J. Rupreht & J. Alvarez (Eds.), *History of Anesthesia* (Vol. 1242, pp. 43–50). Amsterdam: Elsevier Science Bv.
39. Ji, H.-F., Li, X.-J., and Zhang, H.-Y. (2009). Natural products and drug discovery: Can thousands of years of ancient medical knowledge lead us to new and powerful drug combinations in the fight against cancer and dementia? *Embo Reports*, ***10***, 194–200.

40. Davis, A. (1993). Paracelsus: a quincentennial assessment. *Journal of the Royal Society of Medicine*, ***86***, 653–656.
41. Dev, S. (1999). Ancient-modern concordance in Ayurvedic plants: Some examples. *Environmental Health Perspectives*, ***107***, 783–789.
42. Raveenthiran, V. (2011). Knowledge of ancient Hindu surgeons on Hirschsprung disease: evidence from Sushruta Samhita of circa 1200–600 BC. *Journal of Pediatric Surgery*, ***46***, 2204–2208.
43. Torres, C. M. (1995). Archaeological evidence for the antiquity of psychoactive plant use in the Central Andes. *Annuli dei Musei Civici Roverero*, ***11***, 291–326.
44. Lev, E. (2002). Reconstructed materia medica of the Medieval and Ottoman al-Sham. *Journal of Ethnopharmacology*, ***80***, 167–179.
45. Kotrc, R. F., and Walters, K. R. (1979). A bibliography of the Galenic Corpus. A newly researched list and arrangement of the titles of the treatises extant in Greek, Latin, and Arabic. *Transactions & Studies of the College of Physicians of Philadelphia*, ***1***, 256–304.
46. Jacquart, D. (2008). Islamic pharmacology in the Middle Ages: Theories and substances. *European Review*, ***16***, 219–227.
47. Pendergrast, M. (2010). *Uncommon Grounds: The History of Coffee and How It Transformed Our World* New York: Basic Books.
48. Fredholm, B. B. (2011). Notes on the history of caffeine use. *Handbook of Experimental Pharmacology*, 1–9.
49. Dikotter, F., Laamann, L., and Xun, Z. (2002). Narcotic culture: A social history of drug consumption in China. *British Journal of Criminology*, ***42***, 317–336.
50. Tsoucalas, G., Karamanou, M., and Androutsos, G. (2011). The eminent Italian scholar Pietro d'Abano (1250–1315) and his contribution in anatomy. *Italian Journal of Anatomy and Embryology [Archivio Italiano di Anatomia ed Embriologia]*, ***116***, 52–55.
51. Lee, M. R. (2006). The *Solanaceae*: foods and poisons. *The Journal of the Royal College of Physicians of Edinburgh*, ***36***, 162–169.
52. Levack, B. (2006). *The Witch Hunt in Early Modern Europe*. Harlow, UK: Pearson Education Limited.
53. Borzelleca, J. F. (2000). Paracelsus: Herald of modern toxicology. *Toxicological Sciences*, ***53***, 2–4.
54. Galdston, I. (1950). The psychiatry of Paracelsus. *Bulletin of the History of Medicine*, ***24***, 205–218.
55. Webster, C. (1993). Paracelsus, and 500 years of encouraging scientific inquiry. *British Medical Journal*, ***306***, 597–598.
56. Norton, S. (2003). Experimental therapeutics in the renaissance. *Journal of Pharmacology and Experimental Therapeutics*, ***304***, 489–492.
57. Estes, L. L. (1983). The medical origins of the european witch craze: a hypothesis. *Journal of Social History*, ***17***, 271–284.
58. León-Portilla, M. (2002). *Bernardino de Sahagún: The First Anthropologist*. Norman, Oklahoma: University of Oklahoma Press.
59. Diaz, J. L. (2010). Sacred plants and visionary consciousness. *Phenomenology and the Cognitive Sciences*, ***9***, 159–170.
60. Gates, W. (2000). *An Aztec Herbal: The Classic Codex of 1552*. New York: Dover Publications.
61. Goodman, J. (1993). *Tobacco in History: The Cultures of Dependence*. London: Routledge.
62. Guzman, G. (2008). Hallucinogenic mushrooms in Mexico: an overview. *Economic Botany*, ***62***, 404–412.

63. Hofmann, A. (1971). Teonanácatl and Ololiuqui, two ancient magic drugs of Mexico. *Bulletin of Narcotics*.
64. Doll, R. (1999). Tobacco: A medical history. *Journal of Urban Health-Bulletin of the New York Academy of Medicine*, ***76***, 289–313.
65. Crawford, J. (1852). History of coffee. *Journal of the Statistical Society of London*, ***15***, 50–58.
66. Lawson, P. (1993). *The East India Company: A History*. London: Longman.
67. Robins, N. (2006). *The Corporation That Changed the World: How the East India Company Shaped the Modern Multinational*. London: Pluto.
68. Sertürner, F. W. A. F. (1806). Darstellung der reinen Mohnsäure (Opiumsäure) nebst einer chemischen Untersuchung des Opiums mit vorzüglicher Hinsicht auf einen darin neu entdeckten Stoff und die dahin gehörigen Bemerckungen. *J. Pharm. Ärzte Apoth Chem*, ***14***, 47–93.
69. Blakemore, P. R., and White, J. D. (2002). Morphine, the Proteus of organic molecules. *Chemical Communications*, 1159–1168.
70. Zenk, M. H., and Juenger, M. (2007). Evolution and current status of the phytochemistry of nitrogenous compounds. *Phytochemistry*, ***68***, 2757–2772.
71. Cowan, M. W., and Kandel, E. R. (2001). A brief history of synapses and synaptic transmission. In M. W. Cowan, T. C. Südhof & C. F. Stevens (Eds.), *Synapses*. Baltimore: John Hopkins University Press.
72. Newman, D. J., and Cragg, G. M. (2007). Natural products as sources of new drugs over the last 25 years. *Journal of Natural Products*, ***70***, 461–477.
73. Carnwath, T., and Smith, I. (2002). *Heroin Century*. London: Routledge.
74. Way, E. L. (1982). History of opiate use in the Orient and the United-States. *Annals of the New York Academy of Sciences*, ***398***, 12–23.
75. Goldstein, R. A., DesLauriers, C., and Burda, A. M. (2009). Cocaine: history, social implications, and toxicity—A review. *Disease-a-Month*, ***55***, 6–38.
76. Freud, S. (1884). Über coca. *Wien Centralbl Ther*, ***2***, 289–314.
77. Karch, S. B. (1999). Cocaine: history, use, abuse. *Journal of the Royal Society of Medicine*, ***92***, 393–397.
78. Spillane, J. F. (2000). *Cocaine: From Medical Marvel to Modern Menace in the United States, 1884–1920 (Studies in Industry and Society)*. Baltimore: John Hopkins University Press.
79. Gray, J. P. (2011). *Why Our Drug Laws Have Failed and What We Can Do About It: A Judicial Indictment of the War on Drugs*. e-book: Temple University Press.
80. SAMHSA. (2010). National Survey on Drug Use and Health: National Findings *http://www.samhsa.gov*.
81. Palmer, J. D., Soltis, D. E., and Chase, M. W. (2004). The plant tree of life: An overview and some points of view. *American Journal of Botany*, ***91***, 1437–1445.
82. Beerling, D. (2007). *The Emerald Planet*. New York: Oxford University Press.
83. Pimentel, D., and Andow, D. A. (1984). Pest-management and pesticide impacts. *Insect Science and Its Application*, ***5***, 141–149.
84. Harborne, J. R. (1993). *Introduction to Ecological Biochemistry* (4th ed.). London: Elsevier.
85. Tahara, S. (2007). A journey of twenty-five years through the ecological biochemistry of flavonoids. *Bioscience, Biotechnology, and Biochemistry*, ***71***, 1387–1404.
86. Wink, M. (2003). Evolution of secondary metabolites from an ecological and molecular phylogenetic perspective. *Phytochemistry*, ***64***, 3–19.
87. Rattan, R. S. (2010). Mechanism of action of insecticidal secondary metabolites of plant origin. *Crop Protection*, ***29***, 913–920.

88. Chapman, A. D. (2006). *Numbers of Living Species in Australia and the World.* Canberra Australian Biological Resources Study.
89. Odegaard, F. (2000). How many species of arthropods? Erwin's estimate revised. *Biological Journal of the Linnean Society*, ***71***, 583–597.
90. Schoonhoven, L. M., van Loon, J. J. A., and Dicke, M. (2005). *Insect-Plant Biology.* New York: Oxford University Press.
91. Grimaldi, D., and Engel, M. S. (2005). *Evolution of the Insects.* New York: Cambridge University Press.
92. Ehrlich, P. R., and Raven, P. H. (1964). Butterflies and plants: a study in coevolution. *Evolution*, ***18***, 586–608.
93. Conway Morris, S. (2000). The Cambrian "explosion": slow-fuse or megatonnage? *Proceedings of the National Academy of Sciences of the United States of America*, ***97***, 4426–4429.
94. Weng, J. K., and Chapple, C. (2010). The origin and evolution of lignin biosynthesis. *New Phytologist*, ***187***, 273–285.
95. Willis, K. J., and McElwain, J. C. (2002). *The Evolution of Plants.* New York: Oxford University Press.
96. De Bodt, S., Maere, S., and Van de Peer, Y. (2005). Genome duplication and the origin of angiosperms. *Trends in Ecology & Evolution*, ***20***, 591–597.
97. Endress, P. K. (2011). Evolutionary diversification of the flowers in angiosperms. *American Journal of Botany*, ***98***, 370–396.
98. Laurin, M., and Reisz, R. R. (1995). A Re-evaluation of early amniote phylogeny. *Zoological Journal of the Linnean Society*, ***113***, 165–223.
99. Scott, A. C., Stephenson, J., and Chaloner, W. G. (1992). Interaction and coevolution of plants and arthropods during the Paleozoic and Mesozoic. *Philosophical Transactions of the Royal Society of London Series B-Biological Sciences*, ***335***, 129–165.
100. Engel, M. S., and Grimaldi, D. A. (2004). New light shed on the oldest insect. *Nature*, ***427***, 627–630.
101. Clapham, M. E., and Karr, J. A. (2012). Environmental and biotic controls on the evolutionary history of insect body size. *Proceedings of the National Academy of Sciences of the United States of America*, ***109***, 10927–10930.
102. Ren, D., Labandeira, C. C., Santiago-Blay, J. A., Rasnitsyn, A., Shih, C., Bashkuev, A., et al. (2009). A probable pollination mode before angiosperms: Eurasian, long-proboscid scorpionflies. *Science*, ***326***, 840–847.
103. Crepet, W. L. (1996). Timing in the evolution of derived floral characters: Upper Cretaceous (Turonian) taxa with tricolpate and tricolpate-derived pollen. *Review of Palaeobotany and Palynology*, ***90***, 339–359.
104. Ollerton, J., and Coulthard, E. (2009). Evolution of animal pollination. *Science*, ***326***, 808–809.
105. Dotterl, S., and Vereecken, N. J. (2010). The chemical ecology and evolution of bee-flower interactions: a review and perspectives. *Canadian Journal of Zoology-Revue Canadienne De Zoologie*, ***88***, 668–697.
106. Grimaldi, D. (1999). The co-radiations of pollinating insects and angiosperms in the Cretaceous. *Annals of the Missouri Botanical Garden*, ***86***, 373–406.
107. Pichersky, E., and Lewinsohn, E. (2011). Convergent evolution in plant specialized metabolism. In S. S. Merchant, W. R. Briggs & D. Ort (Eds.), *Annual Review of Plant Biology* (Vol. 62, pp. 549–566).
108. Ober, D. (2010). Gene duplications and the time thereafter—examples from plant secondary metabolism. *Plant Biology*, ***12***, 570–577.

109. Jiao, Y., Wickett, N. J., Ayyampalayam, S., Chanderbali, A. S., Landherr, L., Ralph, P. E., et al. (2011). Ancestral polyploidy in seed plants and angiosperms. *Nature*, ***473***, 97–U113.
110. Brenneisen, R. (2007). *Chemistry and Analysis of Phytocannabinoids and Other Cannabis Constituents*.
111. Siebert, D. J. (1994). *Salvia divinorum* and salvinorin A: new pharmacological findings. *Journal of Ethnopharmacology*, ***43***, 53–56.
112. Firn, R. D., and Jones, C. G. (2000). The evolution of secondary metabolism—a unifying model. *Molecular Microbiology*, ***37***, 989–994.
113. Ober, D., and Kaltenegger, E. (2009). Pyrrolizidine alkaloid biosynthesis, evolution of a pathway in plant secondary metabolism. *Phytochemistry*, ***70***, 1687–1695.
114. Biastoff, S., Brandt, W., and Drager, B. (2009). Putrescine N-methyltransferase: The start for alkaloids. *Phytochemistry*, ***70***, 1708–1718.
115. Chen, F., Tholl, D., Bohlmann, J., and Pichersky, E. (2011). The family of terpene synthases in plants: a mid-size family of genes for specialized metabolism that is highly diversified throughout the kingdom. *Plant Journal*, ***66***, 212–229.
116. Theis, N., and Lerdau, M. (2003). The evolution of function in plant secondary metabolites. *International Journal of Plant Sciences*, ***164***, S93–S102.
117. Jenke-Kodama, H., Mueller, R., and Dittmann, E. (2008). Evolutionary mechanisms underlying secondary metabolite diversity. In F. Petersen & R. Amstutz (Eds.), *Progress in Drug Research* (Vol. 65, pp. 119, 121–140).
118. Pollastri, S., and Tattini, M. (2011). Flavonols: old compounds for old roles. *Annals of Botany*, ***108***, 1225–1233.
119. Trigo, J. R. (2011). Effects of pyrrolizidine alkaloids through different trophic levels. *Phytochemistry Reviews*, ***10***, 83–98.
120. Roslin, T., and Salminen, J. P. (2008). Specialization pays off: contrasting effects of two types of tannins on oak specialist and generalist moth species. *Oikos*, ***117***, 1560–1568.
121. Cornell, H. V., and Hawkins, B. A. (2003). Herbivore responses to plant secondary compounds: A test of phytochemical coevolution theory. *American Naturalist*, ***161***, 507–522.
122. Vila, M., Maron, J. L., and Marco, L. (2005). Evidence for the enemy release hypothesis in *Hypericum perforatum*. *Oecologia*, ***142***, 474–479.
123. Walters, D. R. (2011). *Plant Defence: Warding Off Attack by Pathogens, Herbivores, and Parasitic Plants*. Chichester: John Wiley and Sons.
124. Johnson, S. D. (2006). Pollinator-driven speciation in plants. In L. D. Harder & S. C. H. Barrett (Eds.), *Ecology and Evolution of Flowers*. New York: Oxford University Press.
125. Knudsen, J. T., Eriksson, R., Gershenzon, J., and Stahl, B. (2006). Diversity and distribution of floral scent. *Botanical Review*, ***72***, 1–120.
126. Schiestl, F. P. (2010). The evolution of floral scent and insect chemical communication. *Ecology Letters*, ***13***, 643–656.
127. Brodmann, J., Twele, R., Francke, W., Luo, Y.-B., Song, X.-q., and Ayasse, M. (2009). Orchid mimics honey bee alarm pheromone in order to attract hornets for pollination. *Current Biology*, ***19***, 1368–1372.
128. Raguso, R. A. (2008). Wake up and smell the roses: The ecology and evolution of floral scent. In *Annual Review of Ecology Evolution and Systematics* (Vol. 39, pp. 549–569). Palo Alto: Annual Reviews.
129. Schuler, M. A. (2011). P450s in plant–insect interactions. *Biochimica et Biophysica Acta-Proteins and Proteomics*, ***1814***, 36–45.

130. Nelson, D., and Werck-Reichhart, D. (2011). A P450-centric view of plant evolution. *Plant Journal*, ***66***, 194–211.
131. Aniszewski, T. (2007). *Alkaloids—Secrets of Life: Alkaloid Chemistry, Biological Significance, Applications and Ecological Role*. Amsterdam: Elsevier Science.
132. Wink, M. (2010). Introduction. In M. Wink (Ed.), *Annual Plant Reviews, Volume 39: Functions and Biotechnology of Plant Secondary Metabolites, 2nd ed.* Oxford, UK: Wiley-Blackwell.
133. Peiffer, M., Tooker, J. F., Luthe, D. S., and Felton, G. W. (2009). Plants on early alert: glandular trichomes as sensors for insect herbivores. *New Phytologist*, ***184***, 644–656.
134. Schilmiller, A. L., Last, R. L., and Pichersky, E. (2008). Harnessing plant trichome biochemistry for the production of useful compounds. *Plant Journal*, ***54***, 702–711.
135. Biswas, K. K., Foster, A. J., Aung, T., and Mahmoud, S. S. (2009). Essential oil production: relationship with abundance of glandular trichomes in aerial surface of plants. *Acta Physiologiae Plantarum*, ***31***, 13–19.
136. Konno, K. (2011). Plant latex and other exudates as plant defense systems: Roles of various defense chemicals and proteins contained therein. *Phytochemistry*, ***72***, 1510–1530.
137. Wink, M., and Schimmer, O. (2010). Molecular modes of action of defensive secondary metabolites. In M. Wink (Ed.), *Annual Plant Reviews, Volume 39: Functions and Biotechnology of Plant Secondary Metabolites. 2nd ed.* Oxford, UK: Wiley-Blackwell.
138. Opitz, S. E. W., and Muller, C. (2009). Plant chemistry and insect sequestration. *Chemoecology*, ***19***, 117–154.
139. Miller, A. E. M., and Heyland, A. (2010). Endocrine interactions between plants and animals: Implications of exogenous hormone sources for the evolution of hormone signaling. *General and Comparative Endocrinology*, ***166***, 455–461.
140. Volodin, V., Chadin, I., Whiting, P., and Dinan, L. (2002). Screening plants of European North-East Russia for ecdysteroids. *Biochemical Systematics and Ecology*, ***30***, 525–578.
141. Schmelz, E. A., Grebenok, R. J., Ohnmeiss, T. E., and Bowers, W. S. (2000). Phytoecdysteroid turnover in spinach: Long-term stability supports a plant defense hypothesis. *Journal of Chemical Ecology*, ***26***, 2883–2896.
142. Festucci-Buselli, R. A., Contim, L. A. S., Barbosa, L. C. A., Stuart, J., and Otoni, W. C. (2008). Biosynthesis and potential functions of the ecdysteroid 20-hydroxyecdysone—a review. *Botany-Botanique*, ***86***, 978–987.
143. Velarde, R. A., Robinson, G. E., and Fahrbach, S. E. (2006). Nuclear receptors of the honey bee: annotation and expression in the adult brain. *Insect Molecular Biology*, ***15***, 583–595.
144. Poinar, G. O., and Danforth, B. N. (2006). A fossil bee from Early Cretaceous Burmese amber. *Science*, ***314***, 614–614.
145. Leonhardt, S. D., Bluthgen, N., and Schmitt, T. (2011). Chemical profiles of body surfaces and nests from six Bornean stingless bee species. *Journal of Chemical Ecology*, ***37***, 98–104.
146. Agati, G., and Tattini, M. (2010). Multiple functional roles of flavonoids in photoprotection. *New Phytologist*, ***186***, 786–793.
147. Treutter, D. (2006). Significance of flavonoids in plant resistance: a review. *Environmental Chemistry Letters*, ***4***, 147–157.
148. Junker, R. R., and Bluthgen, N. (2010). Floral scents repel facultative flower visitors, but attract obligate ones. *Annals of Botany*, ***105***, 777–782.
149. Gaskett, A. C. (2011). Orchid pollination by sexual deception: pollinator perspectives. *Biological Reviews*, ***86***, 33–75.

150. Jurgens, A., El-Sayed, A. M., and Suckling, D. M. (2009). Do carnivorous plants use volatiles for attracting prey insects? *Functional Ecology*, ***23***, 875–887.
151. Heil, M. (2011). Nectar: generation, regulation, and ecological functions. *Trends in Plant Science*, ***16***, 191–200.
152. Kessler, A., and Baldwin, I. T. (2001). Defensive function of herbivore-induced plant volatile emissions in nature. *Science*, ***291***, 2141–2144.
153. Kessler, A., and Heil, M. (2011). The multiple faces of indirect defences and their agents of natural selection. *Functional Ecology*, ***25***, 348–357.
154. Turlings, T. C. J., Tumlinson, J. H., and Lewis, W. J. (1990). Exploitation of herbivore-induced plant odors by host-seeking parasitic wasps. *Science*, ***250***, 1251–1253.
155. Hilker, M., Kobs, C., Varma, M., and Schrank, K. (2002). Insect egg deposition induces *Pinus sylvestris* to attract egg parasitoids. *Journal of Experimental Biology*, ***205***, 455–461.
156. Heil, M. (2008). Indirect defence via tritrophic interactions. *New Phytologist*, ***178***, 41–61.
157. Bertin, C., Yang, X. H., and Weston, L. A. (2003). The role of root exudates and allelochemicals in the rhizosphere. *Plant and Soil*, ***256***, 67–83.
158. Mandal, S. M., Chakraborty, D., and Dey, S. (2010). Phenolic acids act as signaling molecules in plant–microbe symbioses. *Plant Signaling & Behavior*, ***5***, 359–368.
159. Hassan, S., and Mathesius, U. (2012). The role of flavonoids in root-rhizosphere signalling: opportunities and challenges for improving plant–microbe interactions. *Journal of Experimental Botany*, ***63***, 3429–3444.
160. Kegge, W., and Pierik, R. (2010). Biogenic volatile organic compounds and plant competition. *Trends in Plant Science*, ***15***, 126–132.
161. Frost, C. J., Appel, M., Carlson, J. E., De Moraes, C. M., Mescher, M. C., and Schultz, J. C. (2007). Within-plant signalling via volatiles overcomes vascular constraints on systemic signalling and primes responses against herbivores. *Ecology Letters*, ***10***, 490–498.
162. Heil, M., and Silva Bueno, J. C. (2007). Within-plant signaling by volatiles leads to induction and priming of an indirect plant defense in nature. *Proceedings of the National Academy of Sciences of the United States of America*, ***104***, 5467–5472.
163. Himanen, S. J., Blande, J. D., Klemola, T., Pulkkinen, J., Heijari, J., and Holopainen, J. K. (2010). Birch (*Betula* spp.) leaves adsorb and re-release volatiles specific to neighbouring plants—a mechanism for associational herbivore resistance? *New Phytologist*, ***186***, 722–732.
164. Heil, M., and Karban, R. (2010). Explaining evolution of plant communication by airborne signals. *Trends in Ecology & Evolution*, ***25***, 137–144.
165. Chi, W. C., Fu, S. F., Huang, T. L., Chen, Y. A., Chen, C. C., and Huang, H. J. (2011). Identification of transcriptome profiles and signaling pathways for the allelochemical juglone in rice roots. *Plant Molecular Biology*, ***77***, 591–607.
166. Nishida, N., Tamotsu, S., Nagata, N., Saito, C., and Sakai, A. (2005). Allelopathic effects of volatile monoterpenoids produced by Salvia leucophylla: Inhibition of cell proliferation and DNA synthesis in the root apical meristem of *Brassica campestris* seedlings. *Journal of Chemical Ecology*, ***31***, 1187–1203.
167. Inoue, M., Nishimura, H., Li, H. H., and Mizutani, J. (1992). Allelochemicals from *Polygonum-sachalinense fr schm* (Polygonaceae). *Journal of Chemical Ecology*, ***18***, 1833–1840.
168. Fan, P. H., Hay, A. E., Marston, A., and Hostettmann, K. (2009). Allelopathic potential of phenolic constituents from *Polygonum cuspidatum Sieb. & Zucc* (Polygonaceae). *Planta Medica*, ***75***, 928–928.

169. Murrell, C., Gerber, E., Krebs, C., Parepa, M., Schaffner, U., and Bossdorf, O. (2011). Invasive knotweed affects native plants through allelopathy. *American Journal of Botany*, ***98***, 38–43.
170. Perry, L. G., Weir, T. L., Prithiviraj, B., Paschke, M. W., and Vivanco, J. A. (2006). *Root Exudation and Rhizosphere Biology: Multiple Functions of a Plant Secondary Metabolite.*
171. Li, Z. H., Wang, Q. A., Ruan, X. A., Pan, C. D., and Jiang, D. A. (2010). Phenolics and plant allelopathy. *Molecules*, ***15***, 8933–8952.
172. Stinson, K. A., Campbell, S. A., Powell, J. R., Wolfe, B. E., Callaway, R. M., Thelen, G. C., et al. (2006). Invasive plant suppresses the growth of native tree seedlings by disrupting belowground mutualisms. *PLoS Biology*, ***4***, 727–731.
173. Gonzalez-Lamothe, R., Mitchell, G., Gattuso, M., Diarra, M. S., Malouin, F., and Bouarab, K. (2009). Plant antimicrobial agents and their effects on plant and human pathogens. *International Journal of Molecular Sciences*, ***10***, 3400–3419.
174. Kuhlmann, F., and Mueller, C. (2011). Impacts of ultraviolet radiation on interactions between plants and herbivorous insects: a chemo-ecological perspective. In U. Luttge, W. Beyschlag, B. Budel & D. Francis (Eds.), *Progress in Botany* (Vol. 72, pp. 305–347). Springer-Verlag Berlin.
175. Harborne, J. B. (1999). The comparative biochemistry of phytoalexin induction in plants. *Biochemical Systematics and Ecology*, ***27***, 335–367.
176. Pedras, M. S. C., Yaya, E. E., and Glawischnig, E. (2011). The phytoalexins from cultivated and wild crucifers: Chemistry and biology. *Natural Product Reports*, ***28***, 1381–1405.
177. Iriti, M., and Faoro, F. (2007). Review of innate and specific immunity in plants and animals. *Mycopathologia*, ***164***, 57–64.
178. Zhao, J., Davis, L. C., and Verpoorte, R. (2005). Elicitor signal transduction leading to production of plant secondary metabolites. *Biotechnology Advances*, ***23***, 283–333.
179. Dinan, L. (2001). Phytoecdysteroids: biological aspects. *Phytochemistry*, ***57***, 325–339.
180. Balbi, V., and Devoto, A. (2008). Jasmonate signalling network in *Arabidopsis thaliana*: crucial regulatory nodes and new physiological scenarios. *New Phytologist*, ***177***, 301–318.
181. Avanci, N. C., Luche, D. D., Goldman, G. H., and Goldman, M. H. S. (2010). Jasmonates are phytohormones with multiple functions, including plant defense and reproduction. *Genetics and Molecular Research*, ***9***, 484–505.
182. Pauwels, L., and Goossens, A. (2011). The JAZ proteins: a crucial interface in the jasmonate signaling cascade. *Plant Cell*, ***23***, 3089–3100.
183. Katsir, L., Chung, H. S., Koo, A. J. K., and Howe, G. A. (2008). Jasmonate signaling: a conserved mechanism of hormone sensing. *Current Opinion in Plant Biology*, ***11***, 428–435.
184. Sheard, L. B., Tan, X., Mao, H., Withers, J., Ben-Nissan, G., Hinds, T. R., et al. (2010). Jasmonate perception by inositol-phosphate-potentiated COI1-JAZ co-receptor. *Nature*, ***468***, 400–U301.
185. An, C., and Mou, Z. (2011). Salicylic acid and its function in plant immunity. *Journal of Integrative Plant Biology*, ***53***, 412–428.
186. El Oirdi, M., Abd El Rahman, T., Rigano, L., El Hadrami, A., Rodriguez, M. C., Daayf, F., et al. (2011). *Botrytis cinerea* manipulates the antagonistic effects between immune pathways to promote disease development in tomato. *Plant Cell*, *23*, 2405–2421.
187. Cui, J., Bahrami, A. K., Pringle, E. G., Hernandez-Guzman, G., Bender, C. L., Pierce, N. E., et al. (2005). *Pseudomonas syringae* manipulates systemic plant defenses

against pathogens and herbivores. *Proceedings of the National Academy of Sciences of the United States of America*, ***102***, 1791–1796.

188. Barr, K. L., Hearne, L. B., Briesacher, S., Clark, T. L., and Davis, G. E. (2010). Microbial symbionts in insects influence down-regulation of defense genes in maize. *PLoS One*, ***5***, 10.
189. Chernikova, D., Motamedi, S., Csuros, M., Koonin, E. V., and Rogozin, I. B. (2011). A late origin of the extant eukaryotic diversity: divergence time estimates using rare genomic changes. *Biology Direct*, ***6***, 18.
190. Koonin, E. V. (2010). The incredible expanding ancestor of eukaryotes. *Cell*, ***140***, 606–608.
191. Zhou, X., Lin, Z., and Ma, H. (2010). Phylogenetic Detection of Numerous Gene Duplications Shared by Animals, Fungi and Plants. BioMed Central Ltd.
192. Fritz-Laylin, L. K., Prochnik, S. E., Ginger, M. L., Dacks, J. B., Carpenter, M. L., Field, M. C., et al. (2010). The genome of *Naegleria gruberi* illuminates early eukaryotic versatility. *Cell*, ***140***, 631–642.
193. Koonin, E. V., Fedorova, N. D., Jackson, J. D., Jacobs, A. R., Krylov, D. M., Makarova, K. S., et al. (2004). A comprehensive evolutionary classification of proteins encoded in complete eukaryotic genomes. *Genome Biology*, ***5***, 28.
194. Wilson, J. E. (1984). Some thoughts on the evolutionary basis for the prominent role of ATP and ADP in cellular energy metabolism. *Journal of Theoretical Biology*, ***111***, 615–623.
195. Dewick, P. M. (2009). *Medicinal Natural Products: A Biosynthetic Approach*. Chichester, UK: John Wiley and Sons.
196. Liebeskind, B. J., Hillis, D. M., and Zakon, H. H. (2011). Evolution of sodium channels predates the origin of nervous systems in animals. *Proceedings of the National Academy of Sciences of the United States of America*, ***108***, 9154–9159.
197. Burkhardt, P., Stegmann, C. M., Cooper, B., Kloepper, T. H., Imig, C., Varoqueaux, F., et al. (2011). Primordial neurosecretory apparatus identified in the choanoflagellate *Monosiga brevicollis*. *Proceedings of the National Academy of Sciences of the United States of America*, ***108***, 15264–15269.
198. Brenner, E. D., Stahlberg, R., Mancuso, S., Vivanco, J., Baluska, F., and Van Volkenburgh, E. (2006). Plant neurobiology: an integrated view of plant signaling. *Trends in Plant Science*, ***11***, 413–419.
199. Pelvig, D. P., Pakkenberg, H., Stark, A. K., and Pakkenberg, B. (2008). Neocortical glial cell numbers in human brains. *Neurobiology of Aging*, ***29***, 1754–1762.
200. Hetherington, A. M., and Bardwell, L. (2011). Plant signalling pathways: a comparative evolutionary overview. *Current Biology*, ***21***, R317–319.
201. Suwa, M., Sugihara, M., and Ono, Y. (2011). Functional and structural overview of G-protein-coupled receptors comprehensively obtained from genome sequences. *Pharmaceuticals*, ***4***, 652–664.
202. Kulma, A., and Szopa, J. (2007). Catecholamines are active compounds in plants. *Plant Science*, ***172***, 433–440.
203. Dean, B. (2009). Evolution of the human CNS cholineric system: has this resulted in the emergence of psychiatric disease? *Australian and New Zealand Journal of Psychiatry*, ***43***, 1016–1028.
204. Hoyle, C. H. V. (2011). Evolution of neuronal signalling: Transmitters and receptors. *Autonomic Neuroscience-Basic & Clinical*, ***165***, 28–53.
205. Shiu, S. H., and Bleecker, A. B. (2001). Receptor-like kinases from *Arabidopsis* form a monophyletic gene family related to animal receptor kinases. *Proceedings of the National Academy of Sciences of the United States of America*, ***98***, 10763–10768.

206. Nakagami, H., Pitzschke, A., and Hirt, H. (2005). Emerging MAP kinase pathways in plant stress signalling. *Trends in Plant Science*, ***10***, 339–346.
207. Ausubel, F. M. (2005). Are innate immune signaling pathways in plants and animals conserved? *Nature Immunology*, ***6***, 973–979.
208. Bogre, L., Okresz, L., Henriques, R., and Anthony, R. G. (2003). Growth signalling pathways in *Arabidopsis* and the AGC protein kinases. *Trends in Plant Science*, ***8***, 424–431.
209. Spencer, J. P. E., Vafeiadou, K., Williams, R. J., and Vauzour, D. (2012). Neuroinflammation: modulation by flavonoids and mechanisms of action. *Molecular Aspects of Medicine*, ***33***, 83–97.
210. Williams, R. J., and Spencer, J. P. E. (2012). Flavonoids, cognition, and dementia: Actions, mechanisms, and potential therapeutic utility for Alzheimer disease. *Free Radical Biology and Medicine*, ***52***, 35–45.
211. Ahuja, I., Kissen, R., and Bones, A. M. (2012). Phytoalexins in defense against pathogens. *Trends in Plant Science*, ***17***, 73–90.
212. D'Onofrio, C., Cox, A., Davies, C., and Boss, P. K. (2009). Induction of secondary metabolism in grape cell cultures by jasmonates. *Functional Plant Biology*, ***36***, 323–338.
213. Zhang, H., and Memelink, J. (2009). *Regulation of Secondary Metabolism by Jasmonate Hormones*.
214. Azmitia, E. C. (2007). Serotonin and brain: Evolution, neuroplasticity, and homeostasis. *Pharmacology of Neurogenesis and Neuroenhancement*, ***77***, 31–+.
215. Forde, B. G., and Lea, P. J. (2007). Glutamate in plants: metabolism, regulation, and signalling. *Journal of Experimental Botany*, ***58***, 2339–2358.
216. Kawashima, K., Misawa, H., Moriwaki, Y., Fujii, Y. X., Fujii, T., Horiuchi, Y., et al. (2007). Ubiquitous expression of acetylcholine and its biological functions in life forms without nervous systems. *Life Sciences*, ***80***, 2206–2209.
217. Wessler, I., Kilbinger, H., Bittinger, F., and Kirkpatrick, C. J. (2001). The biological role of non-neuronal acetylcholine in plants and humans. *Japanese Journal of Pharmacology*, ***85***, 2–10.
218. Wang, H. B., Wang, X. C., Zhang, S. Q., and Lou, C. H. (1998). Nicotinic acetylcholine receptor is involved in acetylcholine regulating stomatal movement. *Science in China Series C-Life Sciences*, ***41***, 650–656.
219. Meng, F. X., Miao, L., Zhang, S. Q., and Lou, C. H. (2004). Ca^{2+} is involved in muscarine-acetylcholine-receptor-mediated acetylcholine signal transduction in guard cells of *Vicia faba* L. *Chinese Science Bulletin*, ***49***, 471–475.
220. Meng, F. X., Liu, X., Zhang, S. Q., and Lou, C. H. (2001). Localization of muscarinic acetylcholine receptor in plant guard cells. *Chinese Science Bulletin*, ***46***, 586–588.
221. Bowery, N., Enna, S. J., and Olsen, R. W. (2004). Six decades of GABA. *Biochemical Pharmacology*, ***68***, 1477–1478.
222. Johnston, G. A. R., Hanrahan, J. R., Chebib, M., Duke, R. K., and Mewett, K. N. (2006). Modulation of ionotropic GABA receptors by natural products of plant origin. In S. J. Enna (Ed.), *Advances in Pharmacology* (Vol. 54, pp. 285–316). Academic Press.
223. Dietrich, P., Anschutz, U., Kugler, A., and Becker, D. (2010). Physiology and biophysics of plant ligand-gated ion channels. *Plant Biology*, ***12***, 80–93.
224. Stolarz, M., Krol, E., Dziubinska, H., and Kurenda, A. (2010). Glutamate induces series of action potentials and a decrease in circumnutation rate in *Helianthus annuus*. *Physiologia Plantarum*, ***138***, 329–338.
225. Bouche, N., and Fromm, H. (2004). GABA in plants: just a metabolite? *Trends in Plant Science*, ***9***, 110–115.

226. Fait, A., Fromm, H., Walter, D., Galili, G., and Fernie, A. R. (2008). Highway or byway: the metabolic role of the GABA shunt in plants. *Trends in Plant Science*, ***13***, 14–19.
227. Fait, A., Yellin, A., and Fromm, H. (2006). *GABA and GHB Neurotransmitters in Plants and Animals*.
228. Shelp, B. J., Van Cauwenberghe, O. R., and Bown, A. W. (2003). Gamma aminobutyrate: from intellectual curiosity to practical pest control. *Canadian Journal of Botany-Revue Canadienne De Botanique*, ***81***, 1045–1048.
229. Shelp, B. J., Bown, A. W., and Faure, D. (2006). Extracellular gamma-aminobutyrate mediates communication between plants and other organisms. *Plant Physiology*, ***142***, 1350–1352.
230. Huang, T. F., Jander, G., and de Vos, M. (2011). Non-protein amino acids in plant defense against insect herbivores: Representative cases and opportunities for further functional analysis. *Phytochemistry*, ***72***, 1531–1537.
231. Challet, E. (2007). Minireview: Entrainment of the suprachiasmatic clockwork in diurnal and nocturnal mammals. *Endocrinology*, ***148***, 5648–5655.
232. Pelagio-Flores, R., Munoz-Parra, E., Ortiz-Castro, R., and Lopez-Bucio, J. (2012). Melatonin regulates *Arabidopsis* root system architecture likely acting independently of auxin signaling. *Journal of Pineal Research*, ***53***, 279–288.
233. Cao, J., Cole, I. B., and Murch, S. J. (2006). Neurotransmitters, neuroregulators and neurotoxins in the life of plants. *Canadian Journal of Plant Science*, ***86***, 1183–1188.
234. Pelagio-Flores, R. P.-F. R., Ortiz-Castro, R., Mendez-Bravo, A., Macias-Rodriguez, L., and Lopez-Bucio, J. (2011). Serotonin, a tryptophan-derived signal conserved in plants and animals, regulates root system architecture probably acting as a natural auxin inhibitor in *Arabidopsis thaliana*. *Plant and Cell Physiology*, ***52***, 490–508.
235. Lookadoo, S. E., and Pollard, A. J. (1991). Chemical contents of stinging trichomes of *Cnidoscolus -texanus*. *Journal of Chemical Ecology*, ***17***, 1909–1916.
236. Murch, S. J., Hall, B. A., Le, C. H., and Saxena, P. K. (2010). Changes in the levels of indoleamine phytochemicals during veraison and ripening of wine grapes. *Journal of Pineal Research*, ***49***, 95–100.
237. Kang, K., Kim, Y.-S., Park, S., and Back, K. (2009). Senescence-induced serotonin biosynthesis and its role in delaying senescence in rice leaves. *Plant Physiology*, ***150***, 1380–1393.
238. Ishihara, A., Hashimoto, Y., Tanaka, C., Dubouzet, J. G., Nakao, T., Matsuda, F., et al. (2008). The tryptophan pathway is involved in the defense responses of rice against pathogenic infection via serotonin production. *Plant Journal*, ***54***, 481–495.
239. Park, W. J. (2011). Melatonin as an endogenous plant regulatory signal: debates and perspectives. *Journal of Plant Biology*, ***54***, 143–149.
240. Burnstock, G., and Verkhratsky, A. (2009). Evolutionary origins of the purinergic signalling system. *Acta Physiologica*, ***195***, 415–447.
241. Tanaka, K., Gilroy, S., Jones, A. M., and Stacey, G. (2010). Extracellular ATP signaling in plants. *Trends in Cell Biology*, ***20***, 601–608.
242. Clark, G., and Roux, S. J. (2009). Extracellular nucleotides: Ancient signaling molecules. *Plant Science*, ***177***, 239–244.
243. Iino, M., Nomura, T., Tamaki, Y., Yamada, Y., Yoneyama, K., Takeuchi, Y., et al. (2007). Progesterone: Its occurrence in plants and involvement in plant growth. *Phytochemistry*, ***68***, 1664–1673.
244. Janeczko, A. (2012). The presence and activity of progesterone in the plant kingdom. *Steroids*, ***77***, 169–173.

245. Schumacher, M., Guennoun, R., Robert, F. O., Carelli, C., Gago, N., Ghoumari, A., et al. (2004). Local synthesis and dual actions of progesterone in the nervous system: neuroprotection and myelination. *Growth Hormone & IGF Research*, ***14***, S18–S33.
246. Kaore, S. N., Langade, D. K., Yadav, V. K., Sharma, P., Thawani, V. R., and Sharma, R. (2012). Novel actions of progesterone: what we know today and what will be the scenario in the future? *Journal of Pharmacy and Pharmacology*, ***64***, 1040–1062.
247. Graham, J. D., and Clarke, C. L. (1997). Physiological action of progesterone in target tissues. *Endocrine Reviews*, ***18***, 502–519.
248. Pauli, G. F., Friesen, J. B., Goedecke, T., Farnsworth, N. R., and Glodny, B. (2010). Occurrence of progesterone and related animal steroids in two higher plants. *Journal of Natural Products*, ***73***, 338–345.
249. Simersky, R., Novak, O., Morris, D. A., Pouzar, V., and Strnad, M. (2009). Identification and quantification of several mammalian steroid hormones in plants by UPLC-MS/MS. *Journal of Plant Growth Regulation*, ***28***, 125–136.
250. Janeczko, A., and Skoczowski, A. (2005). Mammalian sex hormones in plants. *Folia Histochemica et Cytobiologica*, ***43***, 71–79.
251. Milanesi, L., and Boland, R. (2004). Presence of estrogen receptor (ER)-like proteins and endogenous ligands for ER in solanaceae. *Plant Science*, ***166***, 397–404.
252. Baudouin, E. (2011). The language of nitric oxide signalling. *Plant Biology*, ***13***, 233–242.
253. Guzik, T. J., Korbut, R., and Adamek-Guzik, T. (2003). Nitric oxide and superoxide in inflammation and immune regulation. *Journal of Physiology and Pharmacology*, ***54***, 469–487.
254. Wink, D. A., Hines, H. B., Cheng, R. Y. S., Switzer, C. H., Flores-Santana, W., Vitek, M. P., et al. (2011). Nitric oxide and redox mechanisms in the immune response. *Journal of Leukocyte Biology*, ***89***, 873–891.
255. Siddiqui, M. H., Al-Whaibi, M. H., and Basalah, M. O. Role of nitric oxide in tolerance of plants to abiotic stress. *Protoplasma*, ***248***, 447–455.
256. Melikian, N., Seddon, M. D., Casadei, B., Chowienczyk, P. J., and Shah, A. M. (2009). Neuronal nitric oxide synthase and human vascular regulation. *Trends in Cardiovascular Medicine*, ***19***, 256–262.
257. Toda, N., Ayajiki, K., and Okamura, T. (2009). Cerebral blood flow regulation by nitric oxide in neurological disorders. *Canadian Journal of Physiology and Pharmacology*, ***87***, 581–594.
258. Szabo, C. (1996). Physiological and pathophysiological roles of nitric oxide in the central nervous system. *Brain Research Bulletin*, ***41***, 131–141.
259. Kitaura, H., Uozumi, N., Tohmi, M., Yamazaki, M., Sakimura, K., Kudoh, M., et al. (2007). Roles of nitric oxide as a vasodilator in neurovascular coupling of mouse somatosensory cortex. *Neuroscience Research*, ***59***, 160–171.
260. Hong, J. K., Yun, B.-W., Kang, J.-G., Raja, M. U., Kwon, E., Sorhagen, K., et al. (2008). Nitric oxide function and signalling in plant disease resistance. *Journal of Experimental Botany*, ***59***, 147–154.
261. Gupta, K. J., Fernie, A. R., Kaiser, W. M., and van Dongen, J. T. (2011). On the origins of nitric oxide. *Trends in Plant Science*, ***16***, 160–168.
262. Lundberg, J. O., and Weitzberg, E. (2010). The biological role of nitrate and nitrite: The times they are a-changin'. Preface. *Nitric Oxide-Biology and Chemistry*, ***22***, 61–63.
263. Foresi, N., Correa-Aragunde, N., Parisi, G., Calo, G., Salerno, G., and Lamattina, L. (2010). Characterization of a nitric oxide synthase from the plant kingdom: NO

generation from the green alga *Ostreococcus tauri* is light irradiance and growth phase dependent. *Plant Cell*, ***22***, 3816–3830.

264. Stefano, G. B., Cadet, P., Kream, R. M., and Zhu, W. (2008). The presence of endogenous morphine signaling in animals. *Neurochemical Research*, ***33***, 1933–1939.
265. Zhu, W., and Stefano, G. B. (2009). Comparative aspects of endogenous morphine synthesis and signaling in animals. In H. Vaudry, E. W. Roubos, G. M. Coast & M. Vallarino (Eds.), *Trends in Comparative Endocrinology and Neurobiology* (Vol. 1163, pp. 330–339). Oxford: Blackwell Publishing.
266. Stefano, G. B., and Kream, R. M. (2010). Dopamine, morphine, and nitric oxide: an evolutionary signaling triad. *CNS Neuroscience & Therapeutics*, ***16***, e124–e137.
267. Wallach, J. V. (2009). Endogenous hallucinogens as ligands of the trace amine receptors: A possible role in sensory perception. *Medical Hypotheses*, ***72***, 91–94.
268. Borowsky, B., Adham, N., Jones, K. A., Raddatz, R., Artymyshyn, R., Ogozalek, K. L., et al. (2001). Trace amines: Identification of a family of mammalian G protein-coupled receptors. *Proceedings of the National Academy of Sciences of the United States of America*, ***98***, 8966–8971.
269. Kleinau, G., Pratzka, J., Nurnberg, D., Gruters, A., Fuhrer-Sakel, D., Krude, H., et al. (2011). Differential modulation of beta-adrenergic receptor signaling by trace amine-associated receptor 1 agonists. *PLoS One*, ***6***, 10.
270. Vladimirov, V., Thiselton, D. L., Kuo, H., McClay, J., Fanous, A., Wormley, B., et al. (2007). A region of 35 kb containing the trace amine associate receptor 6 (TAAR6) gene is associated with schizophrenia in the Irish study of high-density schizophrenia families. *Molecular Psychiatry*, ***12***, 842–853.
271. Danilova, N. (2006). The evolution of immune mechanisms. *Journal of Experimental Zoology Part B: Molecular and Developmental Evolution*, ***306B***, 496–520.
272. Nurnberger, T., Brunner, F., Kemmerling, B., and Piater, L. (2004). Innate immunity in plants and animals: striking similarities and obvious differences. *Immunological Reviews*, ***198***, 249–266.
273. Hayward, A. P., Tsao, J., and Dinesh-Kumar, S. P. (2009). Autophagy and plant innate immunity: Defense through degradation. *Seminars in Cell & Developmental Biology*, ***20***, 1041–1047.
274. Thoma, I., Krischke, M., Loeffler, C., and Mueller, M. J. (2004). The isoprostanoid pathway in plants. *Chemistry and Physics of Lipids*, ***128***, 135–148.
275. Lee, J. Y., and Hwang, D. H. (2008). Dietary fatty acids and eicosanoids. In C. K. Chow (Ed.), *Fatty Acids in Foods and Their Health Implications*. Boca Raton: CRC Press.
276. Tsuboi, K., Sugimoto, Y., and Ichikawa, A. (2002). Prostanoid receptor subtypes. *Prostaglandins & Other Lipid Mediators*, ***68–9***, 535–556.
277. Lee, D.-S., Nioche, P., Hamberg, M., and Raman, C. S. (2008). Structural insights into the evolutionary paths of oxylipin biosynthetic enzymes. *Nature*, ***455***, 363–368.
278. Wasternack, C. (2007). Jasmonates: An update on biosynthesis, signal transduction and action in plant stress response, growth and development. *Annals of Botany*, ***100***, 681–697.
279. Belhadj, A., Saigne, C., Telef, N., Cluzet, S., Bouscaut, J., Corio-Costet, M. F., et al. (2006). Methyl jasmonate induces defense responses in grapevine and triggers protection against *Erysiphe necator*. *Journal of Agricultural and Food Chemistry*, ***54***, 9119–9125.
280. Cohen, S., and Flescher, E. (2009). Methyl jasmonate: a plant stress hormone as an anti-cancer drug. *Phytochemistry*, ***70***, 1600–1609.

281. Li, H. H., Hao, R. L., Wu, S. S., Guo, P. C., Chen, C. J., Pan, L. P., et al. (2011). Occurrence, function and potential medicinal applications of the phytohormone abscisic acid in animals and humans. *Biochemical Pharmacology*, ***82***, 701–712.
282. Bassaganya-Riera, J., Skoneczka, J., Kingston, D. G. J., Krishnan, A., Misyak, S. A., Guri, A. J., et al. (2010). Mechanisms of action and medicinal applications of abscisic acid. *Current Medicinal Chemistry*, ***17***, 467–478.
283. Le Page-Degivry, M. T., Bidard, J. N., Rouvier, E., Bulard, C., and Lazdunski, M. (1986). Presence of abscisic-acid a phytohormone in the mammalian brain. *Proceedings of the National Academy of Sciences of the United States of America*, ***83***, 1155–1158.
284. Scarfi, S., Fresia, C., Ferraris, C., Bruzzone, S., Fruscione, F., Usai, C., et al. (2009). The plant hormone abscisic acid stimulates the proliferation of human hemopoietic progenitors through the second messenger cyclic ADP-ribose. *Stem Cells*, ***27***, 2469–2477.
285. Paterson, J. R., Baxter, G., Dreyer, J. S., Halket, J. M., Flynn, R., and Lawrence, J. R. (2008). Salicylic acid sans aspirin in animals and man: persistence in fasting and biosynthesis from benzoic acid. *Journal of Agricultural and Food Chemistry*, ***56***, 11648–11652.
286. Webb, A. A. R. (2003). The physiology of circadian rhythms in plants. *New Phytologist*, ***160***, 281–303.
287. Dibner, C., Schibler, U., and Albrecht, U. (2010). The mammalian circadian timing system: organization and coordination of central and peripheral clocks. *Annual Review of Physiology*, ***72***, 517–549.
288. Rosbash, M. (2009). The implications of multiple circadian clock origins. *PLoS Biology*, ***7***, 421–425.
289. Jones, M. A., Covington, M. F., DiTacchio, L., Vollmers, C., Panda, S., and Harmer, S. L. (2010). Jumonji domain protein JMJD5 functions in both the plant and human circadian systems. *Proceedings of the National Academy of Sciences of the United States of America*, ***107***, 21623–21628.
290. Lin, J. M., Kilman, V. L., Keegan, K., Paddock, B., Emery-Le, M., Rosbash, M., et al. (2002). A role for casein kinase 2 alpha in the *Drosophila* circadian clock. *Nature*, ***420***, 816–820.
291. Cashmore, A. R. (2003). Cryptochromes: Enabling plants and animals to determine circadian time. *Cell*, ***114***, 537–543.
292. Reiter, L. T., Potocki, L., Chien, S., Gribskov, M., and Bier, E. (2001). A systematic analysis of human disease-associated gene sequences in *Drosophila melanogaster*. *Genome Research*, ***11***, 1114–1125.
293. Bernards, A., and Hariharan, I. K. (2001). Of flies and men—studying human disease in *Drosophila*. *Current Opinion in Genetics & Development*, ***11***, 274–278.
294. Nichols, C. D. (2006). *Drosophila melanogaster* neurobiology, neuropharmacology, and how the fly can inform central nervous system drug discovery. *Pharmacology & Therapeutics*, ***112***, 677–700.
295. Giurfa, M. (2007). Invertebrate cognition: nonelemental learning beyond simple conditioning. In G. North & R. J. Greenspan (Eds.), *Invertebrate Neurobiology*. New York: Cold Spring Harbor Laboratory Press.
296. Menzel, R., and Giurfa, M. (2001). Cognitive architecture of a mini-brain: the honeybee. *Trends in Cognitive Sciences*, ***5***, 62–71.
297. Tomer, R., Denes, A. S., Tessmar-Raible, K., and Arendt, D. (2010). Profiling by image registration reveals common origin of annelid mushroom bodies and vertebrate pallium. *Cell*, ***142***, 800–809.
298. Wolf, F. W., and Heberlein, U. (2003). Invertebrate models of drug abuse. *Journal of Neurobiology*, ***54***, 161–178.

299. Klowden, M. J. (2007). *Physiological Systems in Insects*. London: Academic Press.
300. Menzel, R. (2007). Electrophysiology and optophysiology of complex brain functions in insects. In G. North & R. J. Greenspan (Eds.), *Invertebrate Neurobiology*. New York: Cold Spring Harbor Laboratory Press.
301. Menzel, R., and Manz, G. (2005). Neural plasticity of mushroom body-extrinsic neurons in the honeybee brain. *Journal of Experimental Biology*, ***208***, 4317–4332.
302. Dacher, M., and Gauthier, M. (2008). Involvement of NO-synthase and nicotinic receptors in learning in the honey bee. *Physiology & Behavior*, ***95***, 200–207.
303. Matsumoto, Y., Hatano, A., Unoki, S., and Mizunami, M. (2009). Stimulation of the cAMP system by the nitric oxide-cGMP system underlying the formation of long-term memory in an insect. *Neuroscience Letters*, ***467***, 81–85.
304. Glanzman, D. L. (2007). Simple minds: The neurobiology of invertebrate learning and memory. In G. North & R. J. Greenspan (Eds.), *Invertebrate Neurobiology*. New York: Cold Spring Harbor Laboratory Press.
305. Nassel, D. R., and Winther, A. M. E. *Drosophila* neuropeptides in regulation of physiology and behavior. *Progress in Neurobiology*, ***92***, 42–104.
306. Marder, E. (2007). Searching for insight. In G. North & R. J. Greenspan (Eds.), *Invertebrate Neurobiology*. New York: Cold Spring Harbour Laboratory Press.
307. Watanabe, H., Matsumoto, C. S., Nishino, H., and Mizunami, M. (2011). Critical roles of mecamylamine-sensitive mushroom body neurons in insect olfactory learning. *Neurobiology of Learning and Memory*, ***95***, 1–13.
308. Lee, D., and O'Dowd, D. K. (1999). Fast excitatory synaptic transmission mediated by nicotinic acetylcholine receptors in *Drosophila* neurons. *Journal of Neuroscience*, ***19***, 5311–5321.
309. Gauthier, M. (2010). State of the art on insect nicotinic acetylcholine receptor function in learning and memory. In S. H. Thany (Ed.), *Insect Nicotinic Acetylcholine Receptors* (Vol. 683, pp. 97–115).
310. Guez, D., Zhu, H., Zhang, S. W., and Srinivasan, M. V. (2010). Enhanced cholinergic transmission promotes recall in honeybees. *Journal of Insect Physiology*, ***56***, 1341–1348.
311. Gauthier, M., Cano-Lozano, V., Zaoujal, A., and Richard, D. (1994). Effects of intracranial injections of scopolamine on olfactory conditioning retrieval in the honeybee. *Behavioural Brain Research*, ***63***, 145–149.
312. Zimmerman, J. E., Naidoo, N., Raizen, D. M., and Pack, A. I. (2008). Conservation of sleep: insights from non-mammalian model systems. *Trends in Neurosciences*, ***31***, 371–376.
313. Mustard, J. A., Beggs, K. T., and Mercer, A. R. (2005). Molecular biology of the invertebrate dopamine receptors. *Archives of Insect Biochemistry and Physiology*, ***59***, 103–117.
314. Farooqui, T. (2007). Octopamine-mediated neuromodulation of insect senses. *Neurochemical Research*, ***32***, 1511–1529.
315. Roeder, T. (1999). Octopamine in invertebrates. *Progress in Neurobiology*, ***59***, 533–561.
316. Verlinden, H., Vleugels, R., Marchal, E., Badisco, L., PflÃ¼ger, H.-J., Blenau, W., et al. (2010). The role of octopamine in locusts and other arthropods. *Journal of Insect Physiology*, ***56***, 854–867.
317. Bunzow, J. R., Sonders, M. S., Arttamangkul, S., Harrison, L. M., Zhang, G., Quigley, D. I., et al. (2001). Amphetamine, 3,4-methylenedioxymethamphetamine, lysergic acid diethylamide, and metabolites of the catecholamine neurotransmitters are agonists of a rat trace amine receptor. *Molecular Pharmacology*, ***60***, 1181–1188.

318. Johnson, O., Becnel, J., and Nichols, C. D. (2009). Serotonin 5-HT(2) and 5-HT(1a)-like receptors differentially modulate aggressive behaviors in drosophila melanogaster. *Neuroscience*, ***158***, 1292–1300.
319. Johnson, O., Becnel, J., and Nichols, C. D. (2011). Serotonin receptor activity is necessary for olfactory learning and memory in *Drosophila melanogaster*. *Neuroscience*, ***192***, 372–381.
320. Becnel, J., Johnson, O., Luo, J. N., Nassel, D. R., and Nichols, C. D. (2011). The serotonin 5-HT(7)Dro receptor is expressed in the brain of *Drosophila*, and is essential for normal courtship and mating. *PLoS One*, ***6***, 14.
321. Nilsson, K., Gustafson, L., and Hultberg, B. (2010). Plasma homocysteine and cognition in elderly patients with dementia or other psychogeriatric diseases. *Dementia and Geriatric Cognitive Disorders*, ***30***, 198–204.
322. Devaud, J. M., Couet-Redt, C., Bockaert, J., Grau, Y., and Parmentier, M. L. (2008). Widespread brain distribution of the *Drosophila* metabotropic glutamate receptor. *Neuroreport*, ***19***, 367–371.
323. Daniels, R. W., Gelfand, M. V., Collins, C. A., and Diantonio, A. (2008). Visualizing glutamatergic cell bodies and synapses in *Drosophila* larval and adult CNS. *Journal of Comparative Neurology*, ***508***, 131–152.
324. Manev, H., and Dzitoyeva, S. (2010). GABA-B receptors in *Drosophila*. *Advances in Pharmacology (San Diego, Calif).*, ***58***, 453–464.
325. Gavra, T., and Libersat, F. (2011). Involvement of the opioid system in the hypokinetic state induced in cockroaches by a parasitoid wasp. *Journal of Comparative Physiology A: Neuroethology, Sensory, Neural, and Behavioral Physiology*, ***197***, 279–291.
326. Ford, R., Jackson, D. M., Tetrault, L., Torres, J. C., Assanah, P., Harper, J., et al. (1986). A behavioral role for enkephalins in regulating locomotor activity in the insect *Leucophaea maderae*: Evidence for high affinity kappa-like opioid binding sites. *Comparative Biochemistry and Physiology Part C: Comparative Pharmacology*, ***85***, 61–66.
327. Santoro, C., Hall, L. M., and Zukin, R. S. (1990). Characterization of 2 classes of opioid binding-sites in *Drosophila melanogaster* head membranes. *Journal of Neurochemistry*, ***54***, 164–170.
328. Cooper, P. D., Dennis, S. R., Woodman, J. D., Cowlings, A., and Donnelly, C. (2010). Effect of opioid compounds on feeding and activity of the cockroach, *Periplaneta americana*. *Comparative Biochemistry and Physiology Part C: Toxicology & Pharmacology*, ***151***, 298–302.
329. Stefano, G. B., and Scharrer, B. (1996). The presence of the mu-3 opiate receptor in invertebrate neural tissues. *Comparative Biochemistry and Physiology Part C: Pharmacology, Toxicology and Endocrinology*, ***113***, 369–373.
330. Meldrum, B. S. (2000). Glutamate as a neurotransmitter in the brain: Review of physiology and pathology. *Journal of Nutrition*, ***130***, 1007S–1015S.
331. Nakatani, Y., Matsumoto, Y., Mori, Y., Hirashima, D., Nishino, H., Arikawa, K., et al. (2009). Why the carrot is more effective than the stick: Different dynamics of punishment memory and reward memory and its possible biological basis. *Neurobiology of Learning and Memory*, ***92***, 370–380.
332. Unoki, S., Matsumoto, Y., and Mizunami, M. (2005). Participation of octopaminergic reward system and dopaminergic punishment system in insect olfactory learning revealed by pharmacological study. *European Journal of Neuroscience*, ***22***, 1409–1416.
333. Giannakou, M. E., and Crowther, D. C. (2011). *Drosophila melanogaster* as a model organism for dementia. In P. P. DeDeyn & D. VanDam (Eds.), *Animal Models of Dementia* (pp. 223–240): Humana Press Inc., Totowa, NJ.

334. Rosato, E., Tauber, E., and Kyriacou, C. P. (2006). Molecular genetics of the fruit-fly circadian clock. *European Journal of Human Genetics*, ***14***, 729–738.
335. Sehgal, A., and Mignot, E. (2011). Genetics of sleep and sleep disorders. *Cell*, ***146***, 194–207.
336. Mixson, T. A., Abramson, C. I., and Bozic, J. (2010). The behavior and social communication of honey bees (*Apis mellifera carnica poll.*) under the influence of alcohol. *Psychological Reports*, ***106***, 701–717.
337. Catterson, J. H., Knowles-Barley, S., James, K., Heck, M. M. S., Harmar, A. J., and Hartley, P. S. (2010). Dietary modulation of *Drosophila* sleep-wake behaviour. *PLoS One*, ***5***, 11.
338. Ho, K. S., and Sehgal, A. (2005). *Drosophila melanogaster*: An insect model for fundamental studies of sleep. *Circadian Rhythms*, ***393***, 772–793.
339. Horiuchi, J., and Saitoe, M. (2005). Can flies shed light on our own age-related memory impairment? *Ageing Research Reviews*, ***4***, 83–101.
340. Thamm, M., Balfanz, S., Scheiner, R., Baumann, A., and Blenau, W. (2010). Characterization of the 5-HT1A receptor of the honeybee (*Apis mellifera*) and involvement of serotonin in phototactic behavior. *Cellular and Molecular Life Sciences*, ***67***, 2467–2479.
341. El Hassani, A. K., Dupuis, J. P., Gauthier, M., and Armengaud, C. (2009). Glutamatergic and GABAergic effects of fipronil on olfactory learning and memory in the honeybee. *Invertebrate Neuroscience*, ***9***, 91–100.
342. El Hassani, A. K., Giurfa, M., Gauthier, M., and Armengaud, C. (2008). Inhibitory neurotransmission and olfactory memory in honeybees. *Neurobiology of Learning and Memory*, ***90***, 589–595.
343. Ismail, N., Christine, S., Robinson, G. E., and Fahrbach, S. E. (2008). Pilocarpine improves recognition of nestmates in young honey bees. *Neuroscience Letters*, ***439***, 178–181.
344. Gates, M., and Tschudi, G. (1952). The synthesis of morphine. *Journal of the American Chemical Society*, ***74***, 1109–1110.
345. Stork, G., Niu, D., Fujimoto, A., Koft, E. R., Balkovec, J. M., Tata, J. R., et al. (2001). The first stereoselective total synthesis of quinine. *Journal of the American Chemical Society*, ***123***, 3239–3242.
346. López-Muñoz, F., and Alamo, C. (2012). Contribution of pharmacology to development of monoaminergic hypotheses of depression. In F. López-Muñoz & C. Alamo (Eds.), *Neurobiology of Depression*. Boca Raton: CRC Press.
347. Efferth, T., Fu, Y., Zu, Y., Schwarz, G., Konkimalla, V. S. B., and Wink, M. (2007). Molecular target-guided tumor therapy with natural products derived from traditional Chinese medicine. *Current Medicinal Chemistry*, ***14***, 2024–2032.
348. Cragg, G. M., and Newman, D. J. (2005). Plants as a source of anti-cancer agents. *Journal of Ethnopharmacology*, ***100***, 72–79.
349. Yu, A. M., Idle, J. R., Krausz, K. W., Kupfer, A., and Gonzalez, F. J. (2003). Contribution of individual cytochrome P450 isozymes to the O-demethylation of the psychotropic beta-carboline alkaloids harmaline and harmine. *Journal of Pharmacology and Experimental Therapeutics*, ***305***, 315–322.
350. Haulotte, E., Laurent, P., and Braekman, J.-C. (2012). Biosynthesis of defensive Coccinellidae alkaloids: incorporation of fatty acids in adaline, coccinelline, and harmonine. *European Journal of Organic Chemistry*, 1907–1912.
351. Lin, D. L., Chang, H. C., and Huang, S. H. (2004). Characterization of allegedly musk-containing medicinal products in Taiwan. *Journal of Forensic Sciences*, ***49***, 1187–1193.

352. Svendsen, G. E., and Huntsman, W. D. (1988). A field bioassay of beaver castoreum and some of its components. *American Midland Naturalist*, ***120***, 144–149.
353. Liscombe, D. K., MacLeod, B. P., Loukanina, N., Nandi, O. I., and Facchini, P. J. (2005). Evidence for the monophyletic evolution of benzylisoquinoline alkaloid biosynthesis in angiosperms. *Phytochemistry*, ***66***, 2500–2520.
354. Hanzawa, Y., Imai, A., Michael, A. J., Komeda, Y., and Takahashi, T. (2002). Characterization of the spermidine synthase-related gene family in *Arabidopsis thaliana*. *FEBS Letters*, ***527***, 176–180.
355. Docimo, T., Reichelt, M., Schneider, B., Kai, M., Kunert, G., Gershenzon, J., et al. (2012). The first step in the biosynthesis of cocaine in *Erythroxylum coca*: the characterization of arginine and ornithine decarboxylases. *Plant Molecular Biology*, ***78***, 599–615.
356. Jirschitzka, J., Schmidt, G. W., Reichelt, M., Schneider, B., Gershenzon, J., and D'Auria, J. C. (2012). Plant tropane alkaloid biosynthesis evolved independently in the *Solanaceae* and *Erythroxylaceae*. *Proceedings of the National Academy of Sciences of the United States of America*, ***109***, 10304–10309.
357. Anke, S., Niemüller, D., Moll, S., Hänsch, R., and Ober, D. (2004). Polyphyletic origin of pyrrolizidine alkaloids within the *Asteraceae*. Evidence from differential tissue expression of homospermidine synthase. *Plant Physiology*, ***136***, 4037–4047.
358. Langel, D., Ober, D., and Pelser, P. B. (2011). The evolution of pyrrolizidine alkaloid biosynthesis and diversity in the *Senecioneae*. *Phytochemistry Reviews*, ***10***, 3–74.
359. Wink, M., Schmeller, T., and Latz-Bruning, B. (1998). Modes of action of allelochemical alkaloids: Interaction with neuroreceptors, DNA, and other molecular targets. *Journal of Chemical Ecology*, ***24***, 1881–1937.
360. Ziegler, J., Facchini, P. J., Geißler, R., Schmidt, J., Ammer, C., Kramell, R., et al. (2009). Evolution of morphine biosynthesis in opium poppy. *Phytochemistry*, ***70***, 1696–1707.
361. Erb, M., Lenk, C., Degenhardt, J., and Turlings, T. C. J. (2009). The underestimated role of roots in defense against leaf attackers. *Trends in Plant Science*, ***14***, 653–659.
362. Arab, A., and Trigo, J. R. (2011). Host plant invests in growth rather than chemical defense when attacked by a specialist herbivore. *Journal of Chemical Ecology*, ***37***, 492–495.
363. Steppuhn, A., Gase, K., Krock, B., Halitschke, R., and Baldwin, I. T. (2004). Nicotine's defensive function in nature. *PLoS Biology*, ***2***, 1684–1684.
364. Barbosa, P., Saunders, J., Kemper, J., Trumbule, R., Olechno, J., and Martinat, P. (1986). Plant allelochemicals and insect parasitoids effects of nicotine on *Cotesia congregata* (*Hymenoptera: Braconidae*) and *Hyposoter annulipes* (*Hymenoptera: Ichneumonidae*). *Journal of Chemical Ecology*, ***12***, 1319–1328.
365. Macel, M. (2011). Attract and deter: a dual role for pyrrolizidine alkaloids in plant–insect interactions. *Phytochemistry Reviews*, ***10***, 75–82.
366. Saporito, R. A., Norton, R. A., Andriamaharavo, N. R., Garraffo, H. M., and Spande, T. F. (2011). Alkaloids in the mite *Scheloribates laevigatus*: further alkaloids common to oribatid mites and poison frogs. *Journal of Chemical Ecology*, ***37***, 213–218.
367. Boulogne, I., Petit, P., Ozier-Lafontaine, H., Desfontaines, L., and Loranger-Merciris, G. (2012). Insecticidal and antifungal chemicals produced by plants: a review. *Environmental Chemistry Letters*, 1–23.
368. Joosten, L., and van Veen, J. (2011). Defensive properties of pyrrolizidine alkaloids against microorganisms. *Phytochemistry Reviews*, ***10***, 127–136.
369. Batish, D. R., Singh, H. P., Kaur, M., Kohli, R. K., and Yadav, S. S. (2008). Caffeine affects adventitious rooting and causes biochemical changes in the hypocotyl cuttings of mung bean (*Phaseolus aureus Roxb*). *Acta Physiologiae Plantarum*, ***30***, 401–405.

370. Putnam, A. R. (1988). Allelochemicals from plants as herbicides. *Weed Technology*, 510–518.
371. Nichols, D. E. (2004). Hallucinogens. *Pharmacology & Therapeutics*, ***101***, 131–181.
372. Ray, T. S. (2010). Psychedelics and the human receptorome. *PLoS One*, ***5***.
373. Shorter, D., and Kosten, T. R. (2011). Novel pharmacotherapeutic treatments for cocaine addiction. *BMC Medicine*, ***9***.
374. Kulkarni, S. K., and Dhir, A. (2010). Berberine: a plant alkaloid with therapeutic potential for central nervous system disorders. *Phytotherapy Research*, ***24***, 317–324.
375. Rothman, R. B., Vu, N., Partilla, J. S., Roth, B. L., Hufeisen, S. J., Compton-Toth, B. A., et al. (2003). In vitro characterization of ephedrine-related stereoisomers at biogenic amine transporters and the receptorome reveals selective actions as norepinephrine transporter substrates. *Journal of Pharmacology and Experimental Therapeutics*, ***307***, 138–145.
376. Ma, G. Y., Bavadekar, S. A., Davis, Y. M., Lalchandani, S. G., Nagmani, R., Schaneberg, B. T., et al. (2007). Pharmacological effects of ephedrine alkaloids on human alpha(1)- and alpha(2)-adrenergic receptor subtypes. *Journal of Pharmacology and Experimental Therapeutics*, ***322***, 214–221.
377. Broadley, K. J. (2010). The vascular effects of trace amines and amphetamines. *Pharmacology & Therapeutics*, ***125***, 363–375.
378. Ritz, M. C., Cone, E. J., and Kuhar, M. J. (1990). Cocaine inhibition of ligand binding at dopamine, norepinephrine and serotonin transporters: A structure-activity study. *Life Sciences*, ***46***, 635–645.
379. Pierce, P. A., and Peroutka, S. J. (1989). Hallucinogenic drug-interactions with neurotransmitter receptor-binding sites in human cortex. *Psychopharmacology*, ***97***, 118–122.
380. Bulling, S., Schicker, K., Zhang, Y.-W., Steinkellner, T., Stockner, T., Gruber, C. W., et al. (2012). The mechanistic basis for noncompetitive ibogaine inhibition of serotonin and dopamine transporters. *Journal of Biological Chemistry*, ***287***, 18524–18534.
381. Arias, H. R., Rosenberg, A., Targowska-Duda, K. M., Feuerbach, D., Yuan, X. J., Jozwiak, K., et al. (2010). Interaction of ibogaine with human α3β4-nicotinic acetylcholine receptors in different conformational states. *International Journal of Biochemistry & Cell Biology*, ***42***, 1525–1535.
382. Millan, M. J., Newman-Tancredi, A., Audinot, V., Cussac, D., Lejeune, F., Nicolas, J.-P., et al. (2000). Agonist and antagonist actions of yohimbine as compared to fluparoxan at α2-adrenergic receptors (AR)s, serotonin (5-HT)1A, 5-HT1B, 5-HT1D and dopamine D2 and D3 receptors. Significance for the modulation of frontocortical monoaminergic transmission and depressive states. *Synapse*, ***35***, 79–95.
383. Guillot, T., and Miller, G. (2009). Protective actions of the vesicular monoamine transporter 2 (VMAT2) in monoaminergic neurons. *Molecular Neurobiology*, ***39***, 149–170.
384. Roth, B. L., Lopez, E., Beischel, S., Westkaemper, R. B., and Evans, J. M. (2004). Screening the receptorome to discover the molecular targets for plant-derived psychoactive compounds: a novel approach for CNS drug discovery. *Pharmacology & Therapeutics*, ***102***, 99–110.
385. Larson, B., Harmon, D., Piper, E., Griffis, L., and Bush, L. (1999). Alkaloid binding and activation of D2 dopamine receptors in cell culture. *Journal of Animal Science*, ***77***, 942–947.
386. Ashihara, H., Sano, H., and Crozier, A. (2008). Caffeine and related purine alkaloids: Biosynthesis, catabolism, function and genetic engineering. *Phytochemistry*, ***69***, 841–856.

387. Fredholm, B. B., Battig, K., Holmen, J., Nehlig, A., and Zvartau, E. E. (1999). Actions of caffeine in the brain with special reference to factors that contribute to its widespread use. *Pharmacological Reviews*, ***51***, 83–133.
388. Daly, J. (1993). *Mechanism of Action of Caffeine* (Vol. 97). Raven Press, New York.
389. Kennedy, D. O., and Haskell, C. F. (2011). Cerebral blood flow and behavioural effects of caffeine in habitual and non-habitual consumers of caffeine: A near infrared spectroscopy study. *Biological Psychology*, ***86***, 298–306.
390. Di Matteo, V., Pierucci, M., Di Giovanni, G., Benigno, A., and Esposito, E. (2007). The neurobiological bases for the pharmacotherapy of nicotine addiction. *Current Pharmaceutical Design*, ***13***, 1269–1284.
391. Klinkenberg, I., and Blokland, A. (2010). The validity of scopolamine as a pharmacological model for cognitive impairment: A review of animal behavioral studies. *Neuroscience and Biobehavioral Reviews*, ***34***, 1307–1350.
392. Callaway, J. C., and Grob, C. S. (1998). Ayahuasca preparations and serotonin reuptake inhibitors: A potential combination for severe adverse interactions. *Journal of Psychoactive Drugs*, ***30***, 367–369.
393. Riba, J., Rodriguez-Fornells, A., Urbano, G., Morte, A., Antonijoan, R., Montero, M., et al. (2001). Subjective effects and tolerability of the South American psychoactive beverage Ayahuasca in healthy volunteers. *Psychopharmacology*, ***154***, 85–95.
394. Koenig, X., Kovar, M., Boehm, S., Sandtner, W., and Hilber, K. (2012). Anti-addiction drug ibogaine inhibits hERG channels: a cardiac arrhythmia risk. *Addiction Biology*.
395. Pinna, G., Uzunova, V., Matsumoto, K., Puia, G., Mienville, J. M., Costa, E., et al. (2000). Brain allopregnanolone regulates the potency of the GABA(A) receptor agonist muscimol. *Neuropharmacology*, ***39***, 440–448.
396. Frolund, B., Ebert, B., Kristiansen, U., Liljefors, T., and Krogsgaard-Larsen, P. (2002). GABAA receptor ligands and their therapeutic potentials. *Current Topics in Medicinal Chemistry*, ***2***, 817–832.
397. Johnston, G. A. R., Hanrahan, J. R., Chebib, M., Duke, R. K., and Mewett, K. N. (2006). Modulation of ionotropic GABA receptors by natural products of plant origin. *Advances in Pharmacology (San Diego, Calif.)*, ***54***, 285–316.
398. Gopalakrishnan, A., Sievert, M., and Ruoho, A. E. (2007). Identification of the substrate binding region of vesicular monoamine transporter-2 (VMAT-2) using iodoaminoflisopolol as a novel photoprobe. *Molecular Pharmacology*, ***72***, 1567–1575.
399. Verheij, M. M. M., and Cools, A. R. (2011). Reserpine differentially affects cocaine-induced behavior in low and high responders to novelty. *Pharmacology Biochemistry and Behavior*, ***98***, 43–53.
400. Nelson, D. L., Herbet, A., Pétillot, Y., Pichat, L., Glowinski, J., and Hamon, M. (1979). [3H]harmaline as a specific ligand of MAO A—I. properties of the active site of mao a from rat and bovine brains. *Journal of Neurochemistry*, ***32***, 1817–1827.
401. Lewin, A. H., Miller, G. M., and Gilmour, B. (2011). Trace amine-associated receptor 1 is a stereoselective binding site for compounds in the amphetamine class. *Bioorganic & Medicinal Chemistry*, ***19***, 7044–7048.
402. Eldefrawi, M. E., Eldefrawi, A. T., and O'Brien, R. D. (1970). Mode of action of nicotine in the housefly. *Journal of Agricultural and Food Chemistry*, ***18***, 1113–1116.
403. Lunt, G. G., Robinson, T. N., Miller, T., Knowles, W. P., and Olsen, R. W. (1985). The identification of GABA receptor binding sites in insect ganglia. *Neurochemistry International*, ***7***, 751–754.
404. McGonigle, I., and Lummis, S. C. R. (2010). Molecular characterization of agonists that bind to an insect GABA receptor. *Biochemistry*, ***49***, 2897–2902.

405. Hiripi, L., and Downer, R. G. H. (1993). Characterization of serotonin binding sites in insect (*Locusta migratoria*) brain. *Insect Biochemistry and Molecular Biology*, ***23***, 303–307.
406. Wedemeyer, S., Roeder, T., and Gewecke, M. (1992). Pharmacological characterization of a 5-HT receptor in locust nervous tissue. *European Journal of Pharmacology*, ***223***, 173–178.
407. Rillich, J., Schildberger, K., and Stevenson, P. A. (2011). Octopamine and occupancy: an aminergic mechanism for intruder-resident aggression in crickets. *Proceedings of the Royal Society B-Biological Sciences*, ***278***, 1873–1880.
408. Braun, G., and Bicker, G. (1992). Habituation of an appetitive reflex in the honeybee. *Journal of Neurophysiology*, ***67***, 588–598.
409. Waldrop, B., Christensen, T. A., and Hildebrand, J. G. (1987). GABA-mediated synaptic inhibition of projection neurons in the antennal lobes of the sphinx moth, *Manduca sexta*. *Journal of Comparative Physiology A-Sensory Neural and Behavioral Physiology*, ***161***, 23–32.
410. Gruenewald, B., and Wersing, A. (2008). An ionotropic GABA receptor in cultured mushroom body Kenyon cells of the honeybee and its modulation by intracellular calcium. *Journal of Comparative Physiology A-Neuroethology Sensory Neural and Behavioral Physiology*, ***194***, 329–340.
411. Himmelreich, S., and Gruenewald, B. (2012). Cellular physiology of olfactory learning in the honeybee brain. *Apidologie*, ***43***, 308–321.
412. Demuro, A., Palma, E., Eusebi, F., and Miledi, R. (2001). Inhibition of nicotinic acetylcholine receptors by bicuculline. *Neuropharmacology*, ***41***, 854–861.
413. Benson, J. A. (1988). Bicuculline blocks the response to acetylcholine and nicotine but not to muscarine or GABA in isolated insect neuronal somata. *Brain Research*, ***458***, 65–71.
414. Lozano, V. C., Armengaud, C., and Gauthier, M. (2001). Memory impairment induced by cholinergic antagonists injected into the mushroom bodies of the honeybee. *Journal of Comparative Physiology a-Neuroethology Sensory Neural and Behavioral Physiology*, ***187***, 249–254.
415. Honda, H., Tomizawa, M., and Casida, J. E. (2007). Insect muscarinic acetylcholine receptor: Pharmacological and toxicological profiles of antagonists and agonists. *Journal of Agricultural and Food Chemistry*, ***55***, 2276–2281.
416. Cano Lozano, V., and Gauthier, M. (1998). Effects of the muscarinic antagonists atropine and pirenzepine on olfactory conditioning in the honeybee. *Pharmacol Biochem Behav*, ***59***, 903–907.
417. Kostowski, W. (1968). A note on the effects of some cholinergic and anticholinergic drugs on the aggressive behaviour and spontaneous electrical activity of the central nervous system in the ant, *Formica rufa*. *Journal of Pharmacy and Pharmacology*, ***20***, 381–384.
418. Carnicella, S., Pain, L., and Oberling, P. (2005). Cholinergic effects on fear conditioning II: nicotinic and muscarinic modulations of atropine-induced disruption of the degraded contingency effect. *Psychopharmacology*, ***178***, 533–541.
419. Bartus, R. T., and Dean, R. L. (2009). Pharmaceutical treatment for cognitive deficits in Alzheimer's disease and other neurodegenerative conditions: exploring new territory using traditional tools and established maps. *Psychopharmacology*, ***202***, 15–36.
420. Andretic, R., Kim, Y. C., Jones, F. S., Han, K. A., and Greenspan, R. J. (2008). *Drosophila* D1 dopamine receptor mediates caffeine-induced arousal. *Proceedings of the National Academy of Sciences of the United States of America*, ***105***, 20392–20397.
421. Hendricks, J. C., Finn, S. M., Panckeri, K. A., Chavkin, J., Williams, J. A., Sehgal, A., et al. (2000). Rest in *Drosophila* is a sleep-like state. *Neuron*, ***25***, 129–138.

422. Kucharski, R., and Maleszka, R. (2005). Microarray and real-time PCR analyses of gene expression in the honeybee brain following caffeine treatment. *Journal of Molecular Neuroscience*, ***27***, 269–276.
423. Folkers, E., and Spatz, H. C. (1984). Visual learning performance of *Drosophila melanogaster* is altered by neuropharmaca affecting phosphodiesterase activity and acetylcholine transmission. *Journal of Insect Physiology*, ***30***, 957–965.
424. Shaw, P., Cirelli, C., Greenspan, R., and Tononi, G. (2000). Correlates of sleep and waking in *Drosophila melanogaster*. *Science*, ***287***, 1834.
425. Ho, K., and Sehgal, A. (2005). *Drosophila melanogaster*: an insect model for fundamental studies of sleep. *Methods in Enzymology*, ***393***, 772–793.
426. Nishi, Y., Sasaki, K., and Miyatake, T. (2010). Biogenic amines, caffeine and tonic immobility in *Tribolium castaneum*. *Journal of Insect Physiology*, ***56***, 622–628.
427. Fernandes, F. L., Picanco, M. C., Gontijo, P. C., Fernandes, M. E. D., Pereira, E. J. G., and Semeao, A. A. (2011). Induced responses of *Coffea arabica* to attack of *Coccus viridis* stimulate locomotion of the herbivore. *Entomologia Experimentalis et Applicata*, ***139***, 120–127.
428. Ferré, S. (2008). An update on the mechanisms of the psychostimulant effects of caffeine. *Journal of neurochemistry*, ***105***, 1067–1079.
429. Huang, Z. L., Urade, Y., and Hayaishi, O. (2011). The role of adenosine in the regulation of sleep. *Current Topics in Medicinal Chemistry*, ***11***, 1047–1057.
430. Haskell, C. F., Kennedy, D. O., Wesnes, K. A., and Scholey, A. B. (2005). Cognitive and mood improvements of caffeine in habitual consumers and habitual non-consumers of caffeine. *Psychopharmacology*, ***179***, 813–825.
431. Reissig, C. J., Strain, E. C., and Griffiths, R. R. (2009). Caffeinated energy drinks—A growing problem. *Drug and Alcohol Dependence*, ***99***, 1–10.
432. Nathanson, J. A. (1984). Caffeine and related methylxanthines—possible naturally-occurring pesticides. *Science*, ***226***, 184–187.
433. Makos, M. A., Han, K. A., Heien, M. L., and Ewing, A. G. (2009). Using in vivo electrochemistry to study the physiological effects of cocaine and other stimulants on the *Drosophila melanogaster* dopamine transporter. *ACS Chemical Neuroscience*, ***1***, 74–83.
434. Makos, M. A., Kim, Y. C., Han, K. A., Heien, M. L., and Ewing, A. G. (2009). In vivo electrochemical measurements of exogenously applied dopamine in *Drosophila melanogaster*. *Analytical Chemistry*, ***81***, 1848–1854.
435. Pörzgen, P., Park, S. K., Hirsh, J., Sonders, M. S., and Amara, S. G. (2001). The antidepressant-sensitive dopamine transporter in *Drosophila melanogaster*: a primordial carrier for catecholamines. *Molecular Pharmacology*, ***59***, 83–95.
436. Heberlein, U., Tsai, L. T. Y., Kapfhamer, D., and Lasek, A. W. (2009). *Drosophila*, a genetic model system to study cocaine-related behaviors: A review with focus on LIM-only proteins. *Neuropharmacology*, ***56***, 97–106.
437. Barron, A. B., Maleszka, R., Helliwell, P. G., and Robinson, G. E. (2009). Effects of cocaine on honey bee dance behaviour. *Journal of Experimental Biology*, ***212***, 163–168.
438. McClung, C., and Hirsh, J. (1998). Stereotypic behavioral responses to free-base cocaine and the development of behavioral sensitization in *Drosophila*. *Current Biology*, ***8***, 109–112.
439. Bainton, R. J., Tsai, L. T. Y., Singh, C. M., Moore, M. S., Neckameyer, W. S., and Heberlein, U. (2000). Dopamine modulates acute responses to cocaine, nicotine and ethanol in *Drosophila*. *Current Biology*, ***10***, 187–194.
440. Nathanson, J. A., Hunnicutt, E. J., Kantham, L., and Scavone, C. (1993). Cocaine as a naturally occurring insecticide. *Proceedings of the National Academy of Sciences of the United States of America*, ***90***, 9645–9648.

441. Rothman, R. B., and Baumann, M. H. (2003). Monoamine transporters and psychostimulant drugs. *European Journal of Pharmacology*, ***479***, 23–40.
442. Boghdadi, M. S., and Henning, R. J. (1997). Cocaine: Pathophysiology and clinical toxicology. *Heart & Lung: The Journal of Acute and Critical Care*, ***26***, 466–483.
443. Belzunces, L. P., Vandame, R., and Gu, X. F. (1996). Modulation of honey bee thermoregulation by adrenergic compounds. *Neuroreport*, ***7***, 1601–1604.
444. Rothman, R. B., Baumann, M. H., Dersch, C. M., Romero, D. V., Rice, K. C., Carroll, F. I., et al. (2001). Amphetamine-type central nervous system stimulants release norepinephrine more potently than they release dopamine and serotonin. *Synapse*, ***39***, 32–41.
445. Liles, J. T., Dabisch, P. A., Hude, K. E., Pradhan, L., Varner, K. J., Porter, J. R., et al. (2006). Pressor responses to ephedrine are mediated by a direct mechanism in the rat. *Journal of Pharmacology and Experimental Therapeutics*, ***316***, 95–105.
446. Abourashed, E. A., El-Alfy, A. T., Khan, I. A., and Walker, L. (2003). Ephedra in perspective—a current review. *Phytotherapy Research*, ***17***, 703–712.
447. Brookhart, G. L., Edgecomb, R. S., and Murdock, L. L. (1987). Amphetamine and reserpine deplete brain biogenic-amines and alter blow fly feeding-behavior. *Journal of Neurochemistry*, ***48***, 1307–1315.
448. Andretic, R., van Swinderen, B., and Greenspan, R. J. (2005). Dopaminergic modulation of arousal in *Drosophila*. *Current Biology*, ***15***, 1165–1175.
449. Frischknecht, H. R., and Waser, P. G. (1978). Actions of hallucinogens on ants (*Formica pratensis*)—II. Effects of amphetamine, LSD and delta-9-tetrahydrocannabinol. *General Pharmacology: The Vascular System*, ***9***, 375–380.
450. Griffith, J. D., Cavanaugh, J., Held, J., and Oates, J. A. (1972). Dextroamphetamine: Evaluation of psychomimetic properties in man. *Archives of General Psychiatry*, ***26***, 97.
451. Saudou, F., Boschert, U., Amlaiky, N., Plassat, J. L., and Hen, R. (1992). A family of *Drosophila* serotonin receptors with distinct intracellular signalling properties and expression patterns. *The EMBO journal*, ***11***, 7.
452. Berridge, M. J., and Prince, W. T. (1974). Nature of binding between LSD and a 5-HT receptor—possible explanation for hallucinogenic activity. *British Journal of Pharmacology*, ***51***, 269–278.
453. Blenau, W., May, T., and Erber, J. (1995). Characterization of H-3 LSD binding to a serotonin-sensitive site in honeybee (*Apis mellifera*) brain. *Comparative Biochemistry and Physiology B-Biochemistry & Molecular Biology*, ***112***, 377–384.
454. Blenau, W., Erber, J., and Baumann, A. (1998). Characterization of a dopamine D1 receptor from *Apis mellifera*: Cloning, functional expression, pharmacology, and mRNA localization in the brain. *Journal of Neurochemistry*, ***70***, 15–23.
455. Nichols, C. D., Ronesi, J., Pratt, W., and Sanders-Bush, E. (2002). Hallucinogens and *Drosophila*: linking serotonin receptor activation to behavior. *Neuroscience*, ***115***, 979–984.
456. Waser, P. (1999). *Effects of THC on Brain and Social Organization of Ants*. Totowa, NJ: Humana Press Inc.
457. Kostowski, W., and Tarchalska, B. (1972). The effects of some drugs affecting brain 5-HT on the aggressive behaviour and spontaneous electrical activity of the central nervous system of the ant, *Formica rufa*. *Brain Research*, ***38***, 143–149.
458. Frischknecht, H. R., and Waser, P. G. (1980). Actions of hallucinogens on ants (*Formica pratensis*)—III. Social behavior under the influence of LSD and tetrahydrocannabinol. *General Pharmacology: The Vascular System*, ***11***, 97–106.
459. Passie, T., Halpern, J. H., Stichtenoth, D. O., Emrich, H. M., and Hintzen, A. (2008). The pharmacology of lysergic acid diethylamide: a review. *CNS Neuroscience & Therapeutics*, ***14***, 295–314.

460. Gritsai, O. B., Dubynin, V. A., Pilipenko, V. E., and Petrov, O. P. (2004). Effects of peptide and non-peptide opioids on protective reaction of the cockroach *Periplaneta americana* in the "hot camera". *Journal of Evolutionary Biochemistry and Physiology*, ***40***, 153–160.
461. Zabala, N. A., and Gomez, M. A. (1991). Morphine analgesia, tolerance and addiction in the cricket *Pteronemobius* sp (Orthoptera, Insecta). *Pharmacology Biochemistry and Behavior*, ***40***, 887–891.
462. Zabala, N. A., Miralto, A., Maldonado, H., Nunez, J. A., Jaffe, K., and Calderon, L. D. (1984). Opiate receptor in praying mantis—effect of morphine and naloxone. *Pharmacology Biochemistry and Behavior*, ***20***, 683–687.
463. Dyakonova, V. E., Schurmann, F. W., and Sakharov, D. A. (1999). Effects of serotonergic and opioidergic drugs on escape behaviors and social status of male crickets. *Naturwissenschaften*, ***86***, 435–437.
464. Nunez, J., Maldonado, H., Miralto, A., and Balderrama, N. (1983). The stinging response of the honeybee—effects of morphine, naloxone and some opioid-peptides. *Pharmacology Biochemistry and Behavior*, ***19***, 921–924.
465. Dyakonova, V. E. (2001). Role of opioid peptides in behavior of invertebrates. *Journal of Evolutionary Biochemistry and Physiology*, ***37***, 335–347.
466. Kreek, M. J., Levran, O., Reed, B., Schlussman, S. D., Zhou, Y., and Butelman, E. R. (2012). Opiate addiction and cocaine addiction: underlying molecular neurobiology and genetics. *Journal of Clinical Investigation*, ***122***, 3387–3393.
467. Tan, J., Galligan, J. J., and Hollingworth, R. M. (2007). Agonist actions of neonicotinoids on nicotinic acetylcholine receptors expressed by cockroach neurons. *Neurotoxicology*, ***28***, 829–842.
468. Thany, S. H., and Gauthier, M. (2005). Nicotine injected into the antennal lobes induces a rapid modulation of sucrose threshold and improves short-term memory in the honeybee *Apis mellifera*. *Brain Research*, ***1039***, 216–219.
469. Martin, J. R., Rogers, K. L., Chagneau, C., and Brulet, P. (2007). In vivo bioluminescence imaging of Ca(2+) signalling in the brain of *Drosophila*. *PLoS One*, ***2***, 8.
470. Wolf, F., and Heberlein, U. (2003). Invertebrate models of drug abuse. *Journal of Neurobiology*, ***54***, 161–178.
471. Heishman, S., Kleykamp, B., and Singleton, E. (2010). Meta-analysis of the acute effects of nicotine and smoking on human performance. *Psychopharmacology*, ***210***, 453–469.
472. McBride, J. S., Altman, D. G., Klein, M., and White, W. (1998). Green tobacco sickness. *Tobacco Control*, ***7***, 294–298.
473. van den Beukel, I., van Kleef, R., and Oortgiesen, M. (1998). Differential effects of physostigmine and organophosphates on nicotinic receptors in neuronal cells of different species. *Neurotoxicology*, ***19***, 777–787.
474. Militante, J., Ma, B. W., Akk, G., and Steinbach, J. H. (2008). Activation and block of the adult muscle-type nicotinic receptor by physostigmine: Single-channel studies. *Molecular Pharmacology*, ***74***, 764–776.
475. Prediger, R. D. S., De-Mello, N., and Takahashi, R. N. (2006). Pilocarpine improves olfactory discrimination and social recognition memory deficits in 24-month-old rats. *European Journal of Pharmacology*, ***531***, 176–182.
476. Rupniak, N. M. J., Steventon, M. J., Field, M. J., Jennings, C. A., and Iversen, S. D. (1989). Comparison of the effects of 4 cholinomimetic agents on cognition in primates following disruption by scopolamine or by lists of objects. *Psychopharmacology*, ***99***, 189–195.
477. Duff Sloley, B., and Owen, M. D. (1982). The effects of reserpine on amine concentrations in the nervous system of the cockroach (*Periplaneta americana*). *Insect Biochemistry*, ***12***, 469–476.

478. Banks, C. N., and Adams, M. E. (2012). Biogenic amines in the nervous system of the cockroach, *Periplaneta americana* following envenomation by the jewel wasp, *Ampulex compressa*. *Toxicon*, ***59***, 320–328.
479. Cymborowski, B. (1970). The assumed participation of 5-hydroxytryptamine in regulation of the circadian rhythm of locomotor activity in *Acheta domesticus* L. *Comparative and General Pharmacology*, ***1***, 316–322.
480. Syu, A., Ishiguro, H., Inada, T., Horiuchi, Y., Tanaka, S., Ishikawa, M., et al. (2010). Association of the HSPG2 gene with neuroleptic-induced tardive dyskinesia. *Neuropsychopharmacology*, ***35***, 1155–1164.
481. Liem-Moolenaar, M., de Boer, P., Timmers, M., Schoemaker, R. C., van Hasselt, J. G. C., Schmidt, S., et al. (2011). Pharmacokinetic–pharmacodynamic relationships of central nervous system effects of scopolamine in healthy subjects. *British Journal of Clinical Pharmacology*, ***71***, 886–898.
482. Schultz, W. (2002). Getting formal with dopamine and reward. *Neuron*, ***36***, 241–263.
483. Schultz, W. (2007). Behavioral dopamine signals. *Trends in Neurosciences*, ***30***, 203–210.
484. Koob, G. F., and Volkow, N. D. (2009). Neurocircuitry of addiction. *Neuropsychopharmacology*, ***35***, 217–238.
485. Demos, K. E., Heatherton, T. F., and Kelley, W. M. (2012). Individual differences in nucleus accumbens activity to food and sexual images predict weight gain and sexual behavior. *Journal of Neuroscience*, ***32***, 5549–5552.
486. Knutson, B., Wimmer, G. E., Kuhnen, C. M., and Winkielman, P. (2008). Nucleus accumbens activation mediates the influence of reward cues on financial risk taking. *Neuroreport*, ***19***, 509–513.
487. Marsch, L. A., Bickel, W. K., Badger, G. J., Rathmell, J. P., Swedberg, M. D. B., Jonzon, B., et al. (2001). Effects of infusion rate of intravenously administered morphine on physiological, psychomotor, and self-reported measures in humans. *Journal of Pharmacology and Experimental Therapeutics*, ***299***, 1056–1065.
488. Strang, J., Bearn, J., Farrell, M., Finch, E., Gossop, M., Griffiths, P., et al. (1998). Route of drug use and its implications for drug effect, risk of dependence and health consequences. *Drug and Alcohol Review*, ***17***, 197–211.
489. Askitopoulou, H., Ramoutsaki, I. A., and Konsolaki, E. (2002). Archaeological evidence on the use of opium in the Minoan world. *History of Anesthesia*.Vol. 1242, pp. 23–29.
490. Hagel, J. M., Yeung, E. C., and Facchini, P. J. (2008). Got milk? The secret life of laticifers. *Trends in Plant Science*, ***13***, 631–639.
491. Hamilton, G. R., and Baskett, T. F. (2000). In the arms of Morpheus: the development of morphine for postoperative pain relief. *Canadian Journal of Anaesthesia-Journal Canadien D Anesthesie*, ***47***, 367–374.
492. United Nations Office for Drug Control and Crime Prevention. (2011). *World Drug Report 2011*. Vienna: UNODC.
493. Dubiley, T. A., Rushkevich, Y. E., Koshel, N. M., Voitenko, V. P., and Vaiserman, A. M. (2011). Life span extension in *Drosophila melanogaster* induced by morphine. *Biogerontology*, ***12***, 179–184.
494. Weid, M., Ziegler, J., and Kutchan, T. M. (2004). The roles of latex and the vascular bundle in morphine biosynthesis in the opium poppy, *Papaver somniferum*. *Proceedings of the National Academy of Sciences of the United States of America*, ***101***, 13957–13962.
495. Zabala, N., Miralto, A., Maldonado, H., Nunez, J., Jaffe, K., and de C Calderon, L. (1984). Opiate receptor in praying mantis: effect of morphine and naloxone. *Pharmacology Biochemistry and Behavior*, ***20***, 683–687.

496. Zabala, N. A., and Gómez, M. A. (1991). Morphine analgesia, tolerance and addiction in the cricket *Pteronemobius* sp. (Orthoptera, Insecta). *Pharmacology Biochemistry and Behavior*, ***40***, 887–891.
497. Tekieh, E., Kazemi, M., Dehghani, L., Bahramyian, S., Sadogi, M., Zardooz, H., et al. (2011). Effects of oral morphine on the larvae, pupae and imago development in *Drosophila melanogaster*. *Cell Journal*, ***13***, 149–154.
498. Kavaliers, M., Guglick, M. A., and Hirst, M. (1987). Opioid involvement in the control of feeding in an insect, the American Cockroach. *Life Sciences*, ***40***, 665–672.
499. Pierce, R. C., and Kumaresan, V. (2006). The mesolimbic dopamine system: the final common pathway for the reinforcing effect of drugs of abuse? *Neuroscience & Biobehavioral Reviews*, ***30***, 215–238.
500. Miller, G. M. (2011). The emerging role of trace amine-associated receptor 1 in the functional regulation of monoamine transporters and dopaminergic activity. *Journal of Neurochemistry*, ***116***, 164–176.
501. Wellman, P. J., Miller, D. K., Livermore, C. L., Green, T. A., McMahon, L. R., and Nation, J. R. (1998). Effects of (–)-ephedrine on locomotion, feeding, and nucleus accumbens dopamine in rats. *Psychopharmacology*, ***135***, 133–140.
502. Achat-Mendes, C., Lynch, L. J., Sullivan, K. A., Vallender, E. J., and Miller, G. M. (2012). Augmentation of methamphetamine-induced behaviors in transgenic mice lacking the trace amine-associated receptor 1. *Pharmacology Biochemistry and Behavior*, ***101***, 201–207.
503. Chait, L. (1994). Factors influencing the reinforcing and subjective effects of ephedrine in humans. *Psychopharmacology*, ***113***, 381–387.
504. Johnson, E. L. (1996). Alkaloid content in *Erythroxylum coca* tissue during reproductive development. *Phytochemistry*, ***42***, 35–38.
505. Goldstein, R. A., DesLauriers, C., and Burda, A. M. (2009). Cocaine: history, social implications, and toxicity: A review. *Disease-a-Month*, ***55***, 6–38.
506. Chen, C. Y., and Anthony, J. C. (2004). Epidemiological estimates of risk in the process of becoming dependent upon cocaine: cocaine hydrochloride powder versus crack cocaine. *Psychopharmacology*, ***172***, 78–86.
507. Andraws, R., Chawla, P., and Brown, D. L. (2005). Cardiovascular effects of ephedra alkaloids: A comprehensive review. *Progress in Cardiovascular Diseases*, ***47***, 217–225.
508. Shekelle, P. G., Hardy, M. L., Morton, S. C., Maglione, M., Mojica, W. A., Suttorp, M. J., et al. (2003). Efficacy and safety of ephedra and ephedrine for weight loss and athletic performance. *JAMA: the Journal of the American Medical Association*, ***289***, 1537–1545.
509. Maglione, M., Miotto, K., Iguchi, M., Jungvig, L., Morton, S. C., and Shekelle, P. G. (2005). Psychiatric effects of ephedra use: an analysis of Food and Drug Administration reports of adverse events. *American Journal of Psychiatry*, ***162***, 189–191.
510. Wee, S., Ordway, G. A., and Woolverton, W. L. (2004). Reinforcing effect of pseudoephedrine isomers and the mechanism of action. *European Journal of Pharmacology*, ***493***, 117–125.
511. Martin, W., Sloan, J., Sapira, J., and Jasinski, D. (1971). Physiologic, subjective, and behavioral effects of amphetamine, methamphetamine, ephedrine, phenmetrazine, and methylphenidate in man. *Clinical Pharmacology and Therapeutics*, ***12***, 245.
512. Barker, W. D., and Antia, U. (2007). A study of the use of Ephedra in the manufacture of methamphetamine. *Forensic Science International*, ***166***, 102–109.
513. Marshall, J. F., and O'Dell, S. J. (2012). Methamphetamine influences on brain and behavior: unsafe at any speed? *Trends in Neurosciences*.

514. Sellings, L. H. L., and Clarke, P. B. S. (2003). Segregation of amphetamine reward and locomotor stimulation between nucleus accumbens medial shell and core. *Journal of Neuroscience*, ***23***, 6295–6303.
515. Drevets, W. C., Gautier, C., Price, J. C., Kupfer, D. J., Kinahan, P. E., Grace, A. A., et al. (2001). Amphetamine-induced dopamine release in human ventral striatum correlates with euphoria. *Biological Psychiatry*, ***49***, 81–96.
516. Völlm, B. A., De Araujo, I., Cowen, P. J., Rolls, E. T., Kringelbach, M. L., Smith, K. A., et al. (2004). Methamphetamine activates reward circuitry in drug-naıve human subjects. *Neuropsychopharmacology*, ***29***, 1715–1722.
517. Simmons, M. P., Cappa, J. J., Archer, R. H., Ford, A. J., Eichstedt, D., and Clevinger, C. C. (2008). Phylogeny of the *Celastreae* (Celastraceae) and the relationships of *Catha edulis* (qat) inferred from morphological characters and nuclear and plastid genes. *Molecular Phylogenetics and Evolution*, ***48***, 745–757.
518. Al-Motarreb, A., Al-Habori, M., and Broadley, K. J. (2010). Khat chewing, cardiovascular diseases and other internal medical problems: The current situation and directions for future research. *Journal of Ethnopharmacology*, ***132***, 540–548.
519. Hoffman, R., and Al'Absi, M. (2010). Khat use and neurobehavioral functions: Suggestions for future studies. *Journal of Ethnopharmacology*, ***132***, 554–563.
520. Spivey, A. C., Weston, M., and Woodhead, S. (2001). Celastraceae sesquiterpenoids: biological activity and synthesis. *Chemical Society Reviews*, ***31***, 43–59.
521. Kalix, P., and Braenden, O. (1985). Pharmacological aspects of the chewing of khat leaves. *Pharmacological reviews*, ***37***, 149–164.
522. Patel, N. (2009). Mechanism of action of cathinone: The active ingredient of Khat (*Catha edulis*). *East African Medical Journal*, *77*.
523. Zucchi, R., Chiellini, G., Scanlan, T., and Grandy, D. (2009). Trace amine-associated receptors and their ligands. *British Journal of Pharmacology*, ***149***, 967–978.
524. Motbey, C. P., Hunt, G. E., Bowen, M. T., Artiss, S., and McGregor, I. S. (2012). Mephedrone (4-methylmethcathinone, 'meow'): acute behavioural effects and distribution of Fos expression in adolescent rats. *Addiction Biology*.
525. Kehr, J., Ichinose, F., Yoshitake, S., Goiny, M., Sievertsson, T., Nyberg, F., et al. (2011). Mephedrone, compared with MDMA (ecstasy) and amphetamine, rapidly increases both dopamine and 5-HT levels in nucleus accumbens of awake rats. *British Journal of Pharmacology*, ***164***, 1949–1958.
526. Janzen, D. H., Juster, H. B., and Bell, E. A. (1977). Toxicity of secondary compounds to seed-eating larvae of bruchid beetle *Callosobruchus maculatus*. *Phytochemistry*, ***16***, 223–227.
527. Dimitrijevic, N., Dzitoyeva, S., and Manev, H. (2004). An automated assay of the behavioral effects of cocaine injections in adult *Drosophila*. *Journal of Neuroscience Methods*, ***137***, 181–184.
528. Makos, M. A., Han, K.-A., Heien, M. L., and Ewing, A. G. (2010). Using in vivo electrochemistry to study the physiological effects of cocaine and other stimulants on the *Drosophila melanogaster* dopamine transporter. *ACS Chemical Neuroscience*, ***1***, 74–83.
529. Heberlein, U., Tsai, L. T. Y., Kapfhamer, D., and Lasek, A. W. (2009). *Drosophila*, a genetic model system to study cocaine-related behaviors: A review with focus on LIM-only proteins. *Neuropharmacology*, ***56***, 97–106.
530. Hardie, S. L., Zhang, J. X., and Hirsh, J. (2007). Trace amines differentially regulate adult locomotor activity, cocaine sensitivity, and female fertility in *Drosophila melanogaster*. *Developmental Neurobiology*, ***67***, 1396–1405.

531. Stohs, S. J., Preuss, H. G., and Shara, M. (2011). A review of the receptor-binding properties of p-synephrine as related to its pharmacological effects. *Oxidative Medicine and Cellular Longevity*, ***2011***, 482973.
532. Ashihara, H. (2006). Metabolism of alkaloids in coffee plants. *Brazilian Journal of Plant Physiology*, ***18***, 1–8.
533. Anaya, A. L., Cruz-Ortega, R., and Waller, G. R. (2006). Metabolism and ecology of purine alkaloids. *Frontiers in Bioscience*, ***11***, 2354–2370.
534. Magkos, F., and Kavouras, S. A. (2005). Caffeine use in sports, pharmacokinetics in man, and cellular mechanisms of action. *Critical Reviews in Food Science and Nutrition*, ***45***, 535–562.
535. Latini, S., and Pedata, F. (2008). Adenosine in the central nervous system: release mechanisms and extracellular concentrations. *Journal of Neurochemistry*, ***79***, 463–484.
536. Daly, J. (1993). Mechanism of action of caffeine In Garattini, S (Ed.), *Caffeine, Coffee, and Health* (pp. 97–150). New York: Raven Press.
537. Umemura, T., Higashi, Y., Soga, J., Takemoto, H., Hidaka, T., Nakamura, S., et al. (2006). Acute administration of caffeine on vascular function in humans: As a balance of endothelium-dependent vasodilator and adenosine receptor antagonist. *Journal of Hypertension*, ***24***, 172–172.
538. Quinlan, P. T., Lane, J., Moore, K. L., Aspen, J., Rycroft, J. A., and O'Brien, D. C. (2000). The acute physiological and mood effects of tea and coffee: The role of caffeine level. *Pharmacology Biochemistry and Behavior*, ***66***, 19–28.
539. Smit, H. J., and Rogers, P. J. (2000). Effects of low doses of caffeine on cognitive performance, mood and thirst in low and higher caffeine consumers. *Psychopharmacology*, ***152***, 167–173.
540. Smith, A., Sturgess, W., and Gallagher, J. (1999). Effects of a low dose of caffeine given in different drinks on mood and performance. *Human Psychopharmacology-Clinical and Experimental*, ***14***, 473–482.
541. Durlach, P. J., Edmunds, R., Howard, L., and Tipper, S. P. (2002). A rapid effect of caffeinated beverages on two choice reaction time tasks. *Nutritional Neuroscience*, ***5***, 433–442.
542. Brice, C., and Smith, A. (2001). The effects of caffeine on simulated driving, subjective alertness and sustained attention. *Human Psychopharmacology-Clinical and Experimental*, ***16***, 523–531.
543. Childs, E., and de Wit, H. (2006). Subjective, behavioral, and physiological effects of acute caffeine in light, nondependent caffeine users. *Psychopharmacology*, ***185***, 514–523.
544. James, J. E. (1994). Does caffeine enhance or merely restore degraded psychomotor performance. *Neuropsychobiology*, ***30***, 124–125.
545. Addicott, M. A., and Laurienti, P. J. (2009). A comparison of the effects of caffeine following abstinence and normal caffeine use. *Psychopharmacology*, ***207***, 423–431.
546. Sigmon, S., Herning, R., Better, W., Cadet, J., and Griffiths, R. (2009). Caffeine withdrawal, acute effects, tolerance, and absence of net beneficial effects of chronic administration: cerebral blood flow velocity, quantitative EEG, and subjective effects. *Psychopharmacology*, ***204***, 573–585.
547. Juliano, L. M., and Griffiths, R. R. (2004). A critical review of caffeine withdrawal: empirical validation of symptoms and signs, incidence, severity, and associated features. *Psychopharmacology*, ***176***, 1–29.
548. Hewlett, P., and Smith, A. (2006). Acute effects of caffeine in volunteers with different patterns of regular consumption. *Human Psychopharmacology: Clinical and Experimental*, ***21***, 167–180.

549. Reissig, C., Strain, E., and Griffiths, R. (2009). Caffeinated energy drinks—A growing problem. *Drug and Alcohol Dependence*, ***99***, 1–10.
550. Daly, J. W., and Fredholm, B. B. (1998). Caffeine—an atypical drug of dependence. *Drug and Alcohol Dependence*.
551. Schuh, K. J., and Griffiths, R. R. (1997). Caffeine reinforcement: the role of withdrawal. *Psychopharmacology*, ***130***, 320–326.
552. Sheppard, B. A., Gross, S. C., Pavelka, S. A., Hall, M. J., and Palmatier, M. I. (2012). Caffeine increases the motivation to obtain non-drug reinforcers in rats. *Drug and Alcohol Dependence*.
553. Dack, C., and Reed, P. (2009). Caffeine reinforces flavor preference and behavior in moderate users but not in low caffeine users. *Learning and Motivation*, ***40***, 35–45.
554. Solinas, M., Ferré, S., You, Z. B., Karcz-Kubicha, M., Popoli, P., and Goldberg, S. R. (2002). Caffeine induces dopamine and glutamate release in the shell of the nucleus accumbens. *Journal of Neuroscience*, ***22***, 6321–6324.
555. De Luca, M., Bassareo, V., Bauer, A., and Di Chiara, G. (2007). Caffeine and accumbens shell dopamine. *Journal of Neurochemistry*, ***103***, 157–163.
556. Nehlig, A., and Boyet, S. (2000). Dose–response study of caffeine effects on cerebral functional activity with a specific focus on dependence. *Brain Research*, ***858***, 71–77.
557. Lazarus, M., Shen, H. Y., Cherasse, Y., Qu, W. M., Huang, Z. L., Bass, C. E., et al. (2011). Arousal effect of caffeine depends on adenosine A2A receptors in the shell of the nucleus accumbens. *Journal of Neuroscience*, ***31***, 10067–10075.
558. Haskell, C., Kennedy, D., Wesnes, K., Milne, A., and Scholey, A. (2007). A double-blind, placebo-controlled, multi-dose evaluation of the acute behavioural effects of guaraná in humans. *Journal of Psychopharmacology (Oxford, England)*, ***21***, 65.
559. Rogers, P., Smith, J., Heatherley, S., and Pleydell-Pearce, C. (2008). Time for tea: mood, blood pressure and cognitive performance effects of caffeine and theanine administered alone and together. *Psychopharmacology*, ***195***, 569–577.
560. Haskell, C. F., Kennedy, D. O., Milne, A. L., Wesnes, K. A., and Scholey, A. B. (2008). The effects of L-theanine, caffeine and their combination on cognition and mood. *Biological Psychology*, ***77***, 113–122.
561. Baumann, T. W. (2006). Some thoughts on the physiology of caffeine in coffee: and a glimpse of metabolite profiling. *Brazilian Journal of Plant Physiology*, ***18***, 243–251.
562. Sellier, M. J., Reeb, P., and Marion-Poll, F. (2011). Consumption of bitter alkaloids in *Drosophila melanogaster* in multiple-choice test conditions. *Chemical Senses*, ***36***, 323–334.
563. Matsagas, K., Lim, D. B., Horwitz, M., Rizza, C. L., Mueller, L. D., Villeponteau, B., et al. (2009). Long-term functional side-effects of stimulants and sedatives in *Drosophila melanogaster*. *PLoS One*, ***4***, 11.
564. Guerreiro, O., and Mazzafera, P. (2003). Caffeine and resistance of coffee to the berry borer *Hypothenemus hampei* (Coleoptera: Scolytidae). *Journal of Agricultural and Food Chemistry*, ***51***, 6987–6991.
565. Magalhaes, S. T. V., Guedes, R. N. C., Demuner, A. J., and Lima, E. R. (2008). Effect of coffee alkaloids and phenolics on egg-laying by the coffee leaf miner *Leucoptera coffeella*. *Bulletin of Entomological Research*, ***98***, 483–489.
566. Bustos, A. P., Pohlan, H. A. J., and Schulz, M. (2008). Interaction between coffee (*Coffea arabica* L.) and intercropped herbs under field conditions in the Sierra Norte of Puebla, Mexico. *Journal of Agriculture and Rural Development in the Tropics and Subtropics*, ***109***, 85–93.
567. Hollingsworth, R. G., Armstrong, J. W., and Campbell, E. (2003). Caffeine as a novel toxicant for slugs and snails. *Annals of Applied Biology*, ***142***, 91–97.

568. Dolezelova, E., Nothacker, H. P., Civelli, O., Bryant, P. J., and Zurovec, M. (2007). A *Drosophila* adenosine receptor activates cAMP and calcium signaling. *Insect Biochemistry and Molecular Biology*, ***37***, 318–329.
569. Novakova, M., and Dolezal, T. (2011). Expression of *Drosophila* adenosine deaminase in immune cells during inflammatory response. *PLoS One*, ***6***, 10.
570. Wu, M. N., Ho, K., Crocker, A., Yue, Z., Koh, K., and Sehgal, A. (2009). The effects of caffeine on sleep in *Drosophila* require PKA activity, but not the adenosine receptor. *Journal of Neuroscience*, ***29***, 11029–11037.
571. Hagen, E. H., Sullivan, R. J., Schmidt, R., Morris, G., Kempter, R., and Hammerstein, P. (2009). Ecology and neurobiology of toxin avoidance and the paradox of drug reward. *Neuroscience*, ***160***, 69–84.
572. Sullivan, R. J., Hagen, E. H., and Hammerstein, P. (2008). Revealing the paradox of drug reward in human evolution. *Proceedings of the Royal Society B-Biological Sciences*, ***275***, 1231–1241.
573. Appenzeller, T. (2012). Human migrations: Eastern odyssey. *Nature*, ***485***, 24.
574. Goebel, T., Waters, M. R., and O'Rourke, D. H. (2008). The late Pleistocene dispersal of modern humans in the Americas. *Science*, ***319***, 1497–1502.
575. Singaravelan, N., Nee'man, G., Inbar, M., and Izhaki, I. (2005). Feeding responses of free-flying honeybees to secondary compounds mimicking floral nectars. *Journal of Chemical Ecology*, ***31***, 2791–2804.
576. Perry, C. J., and Barron, A. B. (2013). Neural mechanisms of reward in insects. *Annual Review of Entomology*, ***58***, 543–562
577. Studerus, E., Kometer, M., Hasler, F., and Vollenweider, F. X. (2011). Acute, subacute and long-term subjective effects of psilocybin in healthy humans: a pooled analysis of experimental studies. *Journal of Psychopharmacology*, ***25***, 1434–1452.
578. Lewin, L. (1927). *Phantastica.* Stilke.
579. Hollister, L. E. (1968). *Chemical Psychoses: LSD and Related Drugs.* Thomas.
580. Jaffe, J. H. (1990). Drug addiction and drug abuse. *The Pharmacological Basis of Therapeutics*, ***8***, 522–573.
581. Fantegrossi, W. E., Murnane, K. S., and Reissig, C. J. (2008). The behavioral pharmacology of hallucinogens. *Biochemical Pharmacology*, ***75***, 17–33.
582. Moreno, J. L., Holloway, T., Albizu, L., Sealfon, S. C., and González-Maeso, J. (2011). Metabotropic glutamate mGlu2 receptor is necessary for the pharmacological and behavioral effects induced by hallucinogenic 5-HT2A receptor agonists. *Neuroscience Letters*, ***493***, 76–79.
583. Behrendt, R. P. (2003). Hallucinations: Synchronisation of thalamocortical γ oscillations underconstrained by sensory input. *Consciousness and Cognition*, ***12***, 413–451.
584. Gonzalez-Maeso, J., Ang, R. L., Yuen, T., Chan, P., Weisstaub, N. V., Lopez-Gimenez, J. F., et al. (2008). Identification of a serotonin/glutamate receptor complex implicated in psychosis. *Nature*, ***452***, 93–U99.
585. Ballanger, B., Strafella, A. P., van Eimeren, T., Zurowski, M., Rusjan, P. M., Houle, S., et al. (2010). Serotonin 2A receptors and visual hallucinations in Parkinson disease. *Archives of Neurology*, ***67***, 416–421.
586. Meltzer, H. Y., Mills, R., Revell, S., Williams, H., Johnson, A., Bahr, D., et al. (2010). Pimavanserin, a serotonin(2A) receptor inverse agonist, for the treatment of Parkinson's disease psychosis. *Neuropsychopharmacology*, ***35***, 881–892.
587. Carhart-Harris, R. L., Erritzoe, D., Williams, T., Stone, J. M., Reed, L. J., Colasanti, A., et al. (2012). Neural correlates of the psychedelic state as determined by fMRI studies with psilocybin. *Proceedings of the National Academy of Sciences of the United States of America*, ***109***, 2138–2143.

588. Halberstadt, A. L., and Geyer, M. A. (2011). Multiple receptors contribute to the behavioral effects of indoleamine hallucinogens. *Neuropharmacology*, ***61***, 364–381.
589. Schindler, E. A. D., Dave, K. D., Smolock, E. M., Aloyo, V. J., and Harvey, J. A. (2011). Serotonergic and dopaminergic distinctions in the behavioral pharmacology of (±)-1-(2,5-dimethoxy-4-iodophenyl)-2-aminopropane (DOI) and lysergic acid diethylamide (LSD). *Pharmacology Biochemistry and Behavior*.
590. Halpern, J. H., and Pope Jr, H.G. (1999). Do hallucinogens cause residual neuropsychological toxicity? *Drug and Alcohol Dependence*, ***53***, 247–256.
591. Amsterdam, J., Opperhuizen, A., and Brink, W. (2011). Harm potential of magic mushroom use: A review. *Regulatory Toxicology and Pharmacology*, ***59***, 423–429.
592. Giacomelli, S., Palmery, M., Romanelli, L., Cheng, C. Y., and Silvestrini, B. (1998). Lysergic acid diethylamide (LSD) is a partial agonist of D-2 dopaminergic receptors and it potentiates dopamine-mediated prolactin secretion in lactotrophs in vitro. *Life Sciences*, ***63***, 215–222.
593. Hoffmeister, F. (1975). Negative reinforcing properties of some psychotropic-drugs in drug-naive rhesus-monkeys. *Journal of Pharmacology and Experimental Therapeutics*, ***192***, 468–477.
594. Nichols, C. D. (2007). 5-HT2 receptors in *Drosophila* are expressed in the brain and modulate aspects of circadian behaviors. *Developmental Neurobiology*, ***67***, 752–763.
595. Terry, M., Steelman, K. L., Guilderson, T., Dering, P., and Rowe, M. W. (2006). Lower Pecos and Coahuila peyote: new radiocarbon dates. *Journal of Archaeological Science*, ***33***, 1017–1021.
596. Halpern, J. H. (2004). Hallucinogens and dissociative agents naturally growing in the United States. *Pharmacology & Therapeutics*, ***102***, 131–138.
597. Osmond, H., and Smythies, J. (1952). Schizophrenia: a new approach. *Journal of Mental Science*, ***98***, 309–315.
598. Palenicek, T., Balikova, M., Bubenikov-Valesova, V., and Horacek, J. (2008). Mescaline effects on rat behavior and its time profile in serum and brain tissue after a single subcutaneous dose. *Psychopharmacology*, ***196***, 51–62.
599. Monte, A. P., Waldman, S. R., MaronaLewicka, D., Wainscott, D. B., Nelson, D. L., SandersBush, E., et al. (1997). Dihydrobenzofuran analogues of hallucinogens.4. Mescaline derivatives. *Journal of Medicinal Chemistry*, ***40***, 2997–3008.
600. Kovacic, P., and Somanathan, R. (2009). Novel, unifying mechanism for mescaline in the central nervous system: electrochemistry, catechol redox metabolite, receptor, cell signaling and structure activity relationships. *Oxidative Medicine and Cellular Longevity*, ***2***, 181–190.
601. Oepen, G., Fuenfgeld, M., Harrington, A., Hermle, L., and Botsch, H. (1989). Right-hemisphere involvement in mescaline-induced psychosis. *Psychiatry Research*, ***29***, 335–336.
602. Hermle, L., Funfgeld, M., Oepen, G., Botsch, H., Borchardt, D., Gouzoulis, E., et al. (1992). Mescaline-induced psychopathological, neuropsychological, and neurometabolic effects in normal subjects—experimental psychosis as a tool for psychiatric research. *Biological Psychiatry*, ***32***, 976–991.
603. Hermle, L., Gouzoulis-Mayfrank, E., and Spitzer, M. (1998). Blood flow and cerebral laterality in the mescaline model of psychosis. *Pharmacopsychiatry*, ***31***, 85–91.
604. van Dongen, P. W. J., and de Groot, A. N. J. A. (1995). History of ergot alkaloids from ergotism to ergometrine. *European Journal of Obstetrics & Gynecology and Reproductive Biology*, ***60***, 109–116.
605. Schardl, C. L., Grossman, R. B., Nagabhyru, P., Faulkner, J. R., and Mallik, U. P. (2007). Loline alkaloids: Currencies of mutualism. *Phytochemistry*, ***68***, 980–996.

606. Caporael, L. R. (1976). Ergotism: the satan loosed in Salem? *Science*, ***192***, 21–26.
607. Hofmann, A. (1980). *LSD, My Problem Child.* McGraw-Hill, New York.
608. Schultes, R. E. (1941). *A Contribution to Our Knowledge of Rivea Corymbosa: The Narcotic Ololiuqui of the Aztecs.*Botanical Museum of Harvard University.
609. Osmond, H. (1955). Ololiuqui: The Ancient Aztec narcotic remarks on the effects of *Rivea corymbosa* (Ololiuqui). *British Journal of Psychiatry*, ***101***, 526–537.
610. Steiner, U., and Leistner, E. (2011). Ergoline alkaloids in convolvulaceous host plants originate from epibiotic clavicipitaceous fungi of the genus *Periglandula. Fungal Ecology.*
611. Markert, A., Steffan, N., Ploss, K., Hellwig, S., Steiner, U., Drewke, C., et al. (2008). Biosynthesis and accumulation of ergoline alkaloids in a mutualistic association between *Ipomoea asarifolia* (*Convolvulaceae*) and a clavicipitalean fungus. *Plant Physiology*, ***147***, 296–305.
612. Heim, E., Heimann, H., and Lukacs, G. (1968). [Psychotomimetic effects of the Mexican drug "Ololiuqui"]. *Psychopharmacologia*, ***13***, 35–48.
613. Klinke, H. B., Muller, I. B., Steffenrud, S., and Dahl-Sorensen, R. (2010). Two cases of lysergamide intoxication by ingestion of seeds from Hawaiian Baby Woodrose. *Forensic Science International*, ***197***, 5.
614. White, J. F., and Torres, M. S. (2010). Is plant endophyte-mediated defensive mutualism the result of oxidative stress protection? *Physiologia Plantarum*, ***138***, 440–446.
615. Pennell, C., Popay, A., Ball, O. J. P., Hume, D., and Baird, D. (2005). Occurrence and impact of pasture mealybug (*Balanococcus poae*) and root aphid (*Aploneura lentisci*) on ryegrass (*Lolium* spp.) with and without infection by *Neotyphodium* fungal endophytes. *New Zealand Journal of Agricultural Research*, ***48***, 329–337.
616. Potter, D. A., Stokes, J. T., Redmond, C. T., Schardl, C. L., and Panaccione, D. G. (2008). Contribution of ergot alkaloids to suppression of a grass-feeding caterpillar assessed with gene knockout endophytes in perennial ryegrass. *Entomologia Experimentalis et Applicata*, ***126***, 138–147.
617. Clement, S. L., Hu, J. G., Stewart, A. V., Wang, B. R., and Elberson, L. R. (2011). Detrimental and neutral effects of a wild grass-fungal endophyte symbiotum on insect preference and performance. *Journal of Insect Science*, ***11***, 13.
618. Spatafora, J. W., Sung, G. H., Sung, J. M., Hywel-Jones, N. L., and White, J. F. (2007). Phylogenetic evidence for an animal pathogen origin of ergot and the grass endophytes. *Molecular Ecology*, ***16***, 1701–1711.
619. Quednow, B. B., Geyer, M. A., and Halberstadt, A. L. (2010). 6-serotonin and schizophrenia. *Handbook of Behavioral Neuroscience*, ***21***, 585–620.
620. Vollenweider, F. X., Vontobel, P., Hell, D., and Leenders, K. L. (1999). 5-HT modulation of dopamine release in basal ganglia in psilocybin-induced psychosis in man—A PET study with [^{11}C] raclopride. *Neuropsychopharmacology*, ***20***, 424–433.
621. Leary, T., Litwin, G. H., and Metzner, R. (1963). Reactions to psilocybin administered in a supportive environment. *Journal of Nervous and Mental Disease*, ***137***, 561–573.
622. Pahnke, W. N. (1963). *Drugs and Mysticism: An Analysis of the Relationship Between Psychedelic Drugs and the Mystical Consciousness.* Harvard University.
623. Doblin, R. (1991). Pahnke Good-Friday experiment—a long-term follow-up and methodological critique. *Journal of Transpersonal Psychology*, ***23***, 1–28.
624. Griffiths, R. R., Richards, W. A., Johnson, M. W., McCann, U. D., and Jesse, R. (2008). Mystical-type experiences occasioned by psilocybin mediate the attribution of personal meaning and spiritual significance 14 months later. *Journal of Psychopharmacology*, ***22***, 621–632.

625. Griffiths, R. R., Johnson, M. W., Richards, W. A., Richards, B. D., McCann, U., and Jesse, R. (2011). Psilocybin occasioned mystical-type experiences: immediate and persisting dose-related effects. *Psychopharmacology*, ***218***, 649–665.
626. Gouzoulis-Mayfrank, E., Schreckenberger, M., Sabri, O., Arning, C., Thelen, B., Spitzer, M., et al. (1999). Neurometabolic effects of psilocybin, 3,4-methylenedioxye thylamphetamine (MDE) and d-methamphetamine in healthy volunteers - A double-blind, placebo-controlled PET study with [F-18]FDG. *Neuropsychopharmacology*, ***20***, 565–581.
627. Torres, C. M., Repke, D. B., Chan, K., McKenna, D., Llagostera, A., and Schultes, R. E. (1991). Snuff powders from pre-Hispanic San Pedro de Atacama—chemical and contextual analysis. *Current Anthropology*, ***32***, 640–649.
628. Pochettino, M. L., Cortella, A. R., and Ruiz, M. (1999). Hallucinogenic snuff from northwestern Argentina: Microscopical identification of *Anadenanthera colubrina var. cebil* (Fabaceae) in powdered archaeological material. *Economic Botany*, ***53***, 127–132.
629. McKenna, D. J., and Towers, G. H. N. (1984). Biochemistry and pharmacology of tryptamines and beta-carbolines—a minireview. *Journal of Psychoactive Drugs*, ***16***, 347–358.
630. Schultes, R. E., and Hofmann, A. (1979). *Plants of the Gods: Origins of Hallucinogenic Use*. New York: McGraw-Hill Book Company).
631. Tupper, K. W. (2008). The globalization of ayahuasca: Harm reduction or benefit maximization? *International Journal of Drug Policy*, ***19***, 297–303.
632. Dos Santos, R. G., Valle, M., Bouso, J. C., Nomdedéu, J. F., Rodríguez-Espinosa, J., McIlhenny, E. H., et al. (2011). Autonomic, neuroendocrine, and immunological effects of ayahuasca: A comparative study with D-amphetamine. *Journal of Clinical Psychopharmacology*, ***31***, 717.
633. Smith, R. L., Canton, H., Barrett, R. J., and Sanders-Bush, E. (1998). Agonist properties of N,N-dimethyltryptamine at serotonin 5-HT2A and 5-HT2C receptors. *Pharmacology Biochemistry and Behavior*, ***61***, 323–330.
634. Cozzi, N. V., Gopalakrishnan, A., Anderson, L. L., Feih, J. T., Shulgin, A. T., Daley, P. F., et al. (2009). Dimethyltryptamine and other hallucinogenic tryptamines exhibit substrate behavior at the serotonin uptake transporter and the vesicle monoamine transporter. *Journal of Neural Transmission*, ***116***, 1591–1599.
635. Shen, H. W., Wu, C., Jiang, X. L., and Yu, A. M. (2010). Effects of monoamine oxidase inhibitor and cytochrome P450 2D6 status on 5-methoxy-N,N-dimethyltryptamine metabolism and pharmacokinetics. *Biochemical Pharmacology*, ***80***, 122–128.
636. Brammer, M. K., Gilmore, D. L., and Matsumoto, R. R. (2006). Interactions between 3,4-methylenedioxymethamphetamine and sigma(1) receptors. *European Journal of Pharmacology*, ***553***, 141–145.
637. Fontanilla, D., Johannessen, M., Hajipour, A. R., Cozzi, N. V., Jackson, M. B., and Ruoho, A. E. (2009). The hallucinogen N, N-dimethyltryptamine (DMT) is an endogenous sigma-1 receptor regulator. *Science Signalling*, ***323***, 934.
638. Strassman, R. J., Qualls, C. R., and Berg, L. M. (1996). Differential tolerance to biological and subjective effects of four closely spaced doses of N,N-dimethyltryptamine in humans. *Biological Psychiatry*, ***39***, 784–795.
639. Gouzoulis-Mayfrank, E., Heekeren, K., Neukirch, A., Stoll, M., Stock, C., Obradovic, M., et al. (2005). Psychological effects of (S)-ketamine and N,N-dimethyltryptamine (DMT): A double-blind, cross-over study in healthy volunteers. *Pharmacopsychiatry*, ***38***, 301–311.

640. Ott, J. (2001). Pharmanopo-psychonautics: Human intranasal, sublingual, intrarectal, pulmonary and oral pharmacology of bufotenine. *Journal of Psychoactive Drugs*, ***33***, 273–281.
641. Riba, J., Valle, M., Urbano, G., Yritia, M., Morte, A., and Barbanoj, M. J. (2003). Human pharmacology of ayahuasca: Subjective and cardiovascular effects, monoamine metabolite excretion, and pharmacokinetics. *Journal of Pharmacology and Experimental Therapeutics*, ***306***, 73–83.
642. Barbanoj, M. J., Riba, J., Clos, S., Gimenez, S., Grasa, E., and Romero, S. (2008). Daytime Ayahuasca administration modulates REM and slow-wave sleep in healthy volunteers. *Psychopharmacology*, ***196***, 315–326.
643. Riba, J., Romero, S., Grasa, E., Mena, E., Carrio, I., and Barbanoj, M. J. (2006). Increased frontal and paralimbic activation following ayahuasca, the pan-amazonian inebriant. *Psychopharmacology*, ***186***, 93–98.
644. Vollenweider, F. X., Leenders, K. L., Scharfetter, C., Maguire, P., Stadelmann, O., and Angst, J. (1997). Positron emission tomography and fluorodeoxyglucose studies of metabolic hyperfrontality and psychopathology in the psilocybin model of psychosis. *Neuropsychopharmacology*, ***16***, 357–372.
645. Bouso, J. C., González, D., Fondevila, S., Cutchet, M., Fernández, X., Barbosa, P. C. R., et al. (2012). Personality, psychopathology, life attitudes and neuropsychological performance among ritual users of ayahuasca: A longitudinal study. *PLoS One*, ***7***, e42421.
646. Fábregas, J. M., González, D., Fondevila, S., Cutchet, M., Fernández, X., Barbosa, P. C. R., et al. (2010). Assessment of addiction severity among ritual users of ayahuasca. *Drug and Alcohol Dependence*, ***111***, 257–261.
647. Alper, K. R. (2001). Ibogaine: a review. *The Alkaloids: Chemistry and Biology*, ***56***, 1–38.
648. Alper, K. R., Lotsof, H. S., Frenken, G., Luciano, D. J., and Bastiaans, J. (1999). Treatment of acute opioid withdrawal with ibogaine. *American Journal on Addictions*, ***8***, 234–242.
649. Robson, MJ., Noorbakhsh, B., Seminerio, MJ., and Matsumoto, RR. (2012). Sigma-1 receptors: potential targets for the treatment of substance abuse. *Current Pharmaceutical Design*, ***18***, 902–919.
650. Popik, P. (1996). Facilitation of memory retrieval by the "anti-addictive" alkaloid, ibogaine. *Life Sciences*, ***59***, PL379–PL385.
651. Helsley, S., Fiorella, D., Rabin, R. A., and Winter, J. C. (1997). Effects of ibogaine on performance in the 8-arm radial maze. *Pharmacology Biochemistry and Behavior*, ***58***, 37–41.
652. Gill, R. I. S., Ellis, B. E., and Isman, M. B. (2003). Tryptamine-induced resistance in tryptophan decarboxylase transgenic poplar and tobacco plants against their specific herbivores. *Journal of Chemical Ecology*, ***29***, 779–793.
653. Thomas, J. C., Saleh, E. F., Alammar, N., and Akroush, A. M. (1998). The indole alkaloid tryptamine impairs reproduction in *Drosophila melanogaster*. *Journal of Economic Entomology*, ***91***, 841–846.
654. Miles, D. H., Ly, A. M., Randle, S. A., Hedin, P. A., and Burks, M. L. (1987). Alkaloidal insect antifeedants from *Virola-calophylla warb*. *Journal of Agricultural and Food Chemistry*, ***35***, 794–797.
655. Corcuera, L. J. (1984). Effects of indole alkaloids from *Gramineae* on aphids. *Phytochemistry*, ***23***, 539–541.
656. Daly, J. W., Garraffo, H. M., Spande, T. F., Yeh, H. J. C., Peltzer, P. M., Cacivio, P. M., et al. (2008). Indolizidine 239Q and quinolizidine 275I. Major alkaloids in two Argentinian bufonid toads (*Melanophryniscus*). *Toxicon*, ***52***, 858–870.

657. Clarke, B. T. (2007). The natural history of amphibian skin secretions, their normal functioning and potential medical applications. *Biological Reviews*, ***72***, 365–379.
658. Weil, A. T., and Davis, W. (1994). Bufo alvarius: a potent hallucinogen of animal origin. *Journal of Ethnopharmacology*, ***41***, 1–8.
659. Hunziker, A. T. (2001). The Genera of Solanaceae Illustrated, Arranged According to a New System. Gantner. Ruggell, Liechtenstein.
660. Llamazares, A. M., Sarasola, C. M., and Funes, F. (2004). Principales plantas sagradas de Sudamérica. *El Lenguaje de los Dioses: Arte, Chamanismo y Cosmovisión en Sudamérica*, 259–285.
661. Luna-Cavazos, M., Bye, R., and Jiao, M. (2009). The origin of *Datura metel* (Solanaceae): genetic and phylogenetic evidence. *Genetic Resources and Crop Evolution*, ***56***, 263–275.
662. Mummenhoff, K., and Franzke, A. (2007). Gone with the bird: late Tertiary and Quaternary intercontinental long-distance dispersal and allopolyploidization in plants. *Systematics and Biodiversity*, ***5***, 255–260.
663. Clarkson, J. J., Knapp, S., Garcia, V. F., Olmstead, R. G., Leitch, A. R., and Chase, M. W. (2004). Phylogenetic relationships in *Nicotiana* (Solanaceae) inferred from multiple plastid DNA regions. *Molecular Phylogenetics and Evolution*, ***33***, 75–90.
664. Tu, T. Y., Volis, S., Dillon, M. O., Sun, H., and Wen, J. (2010). Dispersals of *Hyoscyameae* and *Mandragoreae* (Solanaceae) from the New World to Eurasia in the early Miocene and their biogeographic diversification within Eurasia. *Molecular Phylogenetics and Evolution*, ***57***, 1226–1237.
665. Geeta, R., and Gharaibeh, W. (2007). Historical evidence for a pre-Columbian presence of *Datura* in the Old World and implications for a first millennium transfer from the New World. *Journal of Biosciences*, ***32***, 1227–1244.
666. Diaz, J., Collings, P., and Siegel, R. (1977). On the use of *Tagetes lucida* and *Nicotiana rustica* as a Huichol smoking mixture: the Aztec "Yahutli" with suggestive hallucinogenic effects. *Economic Botany*, ***31***, 16–23.
667. VanPool, C. S. (2009). The signs of the sacred: identifying shamans using archaeological evidence. *Journal of Anthropological Archaeology*, ***28***, 177–190.
668. Dobkin de Rios, M., and du Toit, B. (1976). Suggested hallucinogenic motifs in New World massive earthworks. *Drugs, Rituals and Altered States of Consciousness*.
669. Janiger, O., and de Rios, M. D. (1973). Suggestive hallucinogenic properties of tobacco. *Medical Anthropology Newsletter*, 6–11.
670. Tushingham, S., Ardura, D., Eerkens, J. W., Palazoglu, M., Shahbaz, S., and Fiehn, O. (2013). Hunter-gatherer tobacco smoking: earliest evidence from the Pacific Northwest Coast of North America. *Journal of Archaeological Science*, ***40***, 1397–1407.
671. Ratsch, A., Steadman, K. J., and Bogossian, F. (2010). The pituri story: a review of the historical literature surrounding traditional Australian Aboriginal use of nicotine in Central Australia. *Journal of Ethnobiology and Ethnomedicine*, ***6***, 26.
672. Guindon, G. E., and Boisclair, D. (2003). Past, current and future trends in tobacco use. WHO Tobacco Control Papers.
673. Benowitz, N. (2008). Clinical pharmacology of nicotine: implications for understanding, preventing, and treating tobacco addiction. *Clinical Pharmacology & Therapeutics*, ***83***, 531–541.
674. Benowitz, N. L. (2010). Nicotine addiction. *New England Journal of Medicine*, ***362***, 2295.
675. Heishman, S. J., Kleykamp, B. A., and Singleton, E. G. (2010). Meta-analysis of the acute effects of nicotine and smoking on human performance. *Psychopharmacology*, ***210***, 453–469.

676. Poorthuis, R. B., Goriounova, N. A., Couey, J. J., and Mansvelder, H. D. (2009). Nicotinic actions on neuronal networks for cognition: General principles and long-term consequences. *Biochemical Pharmacology*, ***78***, 668–676.
677. Mansvelder, H. D., van Aerde, K. I., Couey, J. J., and Brussaard, A. B. (2006). Nicotinic modulation of neuronal networks: from receptors to cognition. *Psychopharmacology*, ***184***, 292–305.
678. Levin, E. D., Conners, C. K., Silva, D., Hinton, S. C., Meck, W. H., March, J., et al. (1998). Transdermal nicotine effects on attention. *Psychopharmacology*, ***140***, 135–141.
679. Fisher, D. J., Grant, B., Smith, D. M., Borracci, G., Labelle, A., and Knott, V. J. (2012). Nicotine and the hallucinating brain: Effects on mismatch negativity (MMN) in schizophrenia. *Psychiatry Research*, ***196***, 181–187.
680. Zhang, X., Chen, D., Xiu, M., Haile, C., He, S., Luo, X., et al. (2012). Cigarette smoking, psychopathology and cognitive function in first-episode drug-naive patients with schizophrenia: a case-control study. *Psychological Medicine*, ***1***, 1–10.
681. Corvin, A., O'Mahony, E., O'Regan, M., Comerford, C., O'Connell, R., Craddock, N., et al. (2001). Cigarette smoking and psychotic symptoms in bipolar affective disorder. *British Journal of Psychiatry*, ***179***, 35–38.
682. Wiles, N. J., Zammit, S., Bebbington, P., Singleton, N., Meltzer, H., and Lewis, G. (2006). Self-reported psychotic symptoms in the general population Results from the longitudinal study of the British National Psychiatric Morbidity Survey. *British Journal of Psychiatry*, ***188***, 519–526.
683. Zador, E., and Jones, D. (1986). The biosynthesis of a novel nicotine alkaloid in the trichomes of *Nicotiana stocktonii*. *Plant Physiology*, ***82***, 479–484.
684. Zulak, K., Liscombe, D., Ashihara, H., and Facchini, P. (2006). Alkaloids. *Plant Secondary Metabolism in Diet and Human Health*. Blackwell Publishing, Oxford, pp. 102–136.
685. Walter, A., and Hummel, G. M. (2008). Root growth of *Nicotiana attenuata* is decreased immediately after simulated leaf herbivore attack. *Plant Signaling & Behavior*, ***3***, 236–237.
686. Voelckel, C., and Baldwin, I. T. (2004). Generalist and specialist lepidopteran larvae elicit different transcriptional responses in *Nicotiana attenuata*, which correlate with larval FAC profiles. *Ecology Letters*, ***7***, 770–775.
687. Musser, R. O., Hum-Musser, S. M., Eichenseer, H., Peiffer, M., Ervin, G., Murphy, J. B., et al. (2002). Herbivory: Caterpillar saliva beats plant defences—A new weapon emerges in the evolutionary arms race between plants and herbivores. *Nature*, ***416***, 599–600.
688. Jones, A. K., and Sattelle, D. B. (2010). Diversity of insect nicotinic acetylcholine receptor subunits. In S. H. Thany (Ed.), *Insect Nicotinic Acetylcholine Receptors* (Vol. 683, pp. 25–43). Berlin: Springer-Verlag Berlin.
689. Hou, J., Kuromi, H., Fukasawa, Y., Ueno, K., Sakai, T., and Kidokoro, Y. (2004). Repetitive exposures to nicotine induce a hyper-responsiveness via the cAMP/PKA/CREB signal pathway in *Drosophila*. *Journal of Neurobiology*, ***60***, 249–261.
690. Köhler, A., Pirk, C. W. W., and Nicolson, S. W. (2012). Honeybees and nectar nicotine: Deterrence and reduced survival versus potential health benefits. *Journal of Insect Physiology*, ***58***, 286–292.
691. Singaravelan, N., Inbar, M., Ne'eman, G., Distl, M., Wink, M., and Izhaki, I. (2006). The effects of nectar-nicotine on colony fitness of caged honeybees. *Journal of Chemical Ecology*, ***32***, 49–58.
692. De Feo, V. (2004). The ritual use of *Brugmansia* species in traditional Andean medicine in northern Peru. *Economic Botany*, ***58***, 221–229.

693. Lee, M. (2006). The mandrake. *Journal of the Royal College of Physicians of Edinburgh*, ***36***, 278–285.
694. Carter, A. J. (1996). Narcosis and nightshade. *British Medical Journal*, ***313***, 1630–1632.
695. Zhang, W. W., Song, M. K., Cui, Y. Y., Wang, H., Zhu, L., Niu, Y. Y., et al. (2008). Differential neuropsychopharmacological influences of naturally occurring tropane alkaloids anisodamine versus scopolamine. *Neuroscience Letters*, ***443***, 241–245.
696. Brown, J. H., and Taylor, P. (1996). Muscarinic receptor agonists and antagonists. *The Pharmacological Basis of Therapeutics*.New York: McGraw-Hill, pp. 141–160.
697. Gaudreau, J. D., and Gagnon, P. (2005). Psychotogenic drugs and delirium pathogenesis: the central role of the thalamus. *Medical Hypotheses*, ***64***, 471–475.
698. Perry, E., and Perry, R. (1995). Acetylcholine and hallucinations—disease-related compared to drug-induced alterations in human consciousness. *Brain and Cognition*, ***28***, 240–258.
699. Wiebe, T. H., Sigurdson, E. S., and Katz, L. Y. (2008). Angel's Trumpet (*Datura stramonium*) poisoning and delirium in adolescents in Winnipeg, Manitoba: Summer 2006. *Paediatrics & Child Health*, ***13***, 193.
700. Gauss, C. J. (1906). Geburten in künstlichem Dämmerschlaf. *Archives of Gynecology and Obstetrics*, ***78***, 579–631.
701. Geis, G. (1961). The status of interrogation drugs in the United States. *Journal of Forensic Medicine*, ***8***, 29.
702. Nachum, Z., Shupak, A., and Gordon, C. R. (2006). Transdermal scopolamine for prevention of motion sickness—Clinical pharmacokinetics and therapeutic applications. *Clinical Pharmacokinetics*, ***45***, 543–566.
703. Hamborg-Petersen, B., Nielsen, M., and Thordal, C. (2009). Toxic effect of scopolamine eye drops in children. *Acta Ophthalmologica*, ***62***, 485–488.
704. Wilkinson, J. A. (1987). Side effects of transdermal scopolamine. *Journal of Emergency Medicine*, ***5***, 389–392.
705. Fredrickson, A., Snyder, P. J., Cromer, J., Thomas, E., Lewis, M., and Maruff, P. (2008). The use of effect sizes to characterize the nature of cognitive change in psychopharmacological studies: an example with scopolamine. *Human Psychopharmacology-Clinical and Experimental*, ***23***, 425–436.
706. Parrott, A. C. (1989). Transdermal scopolamine—a review of its effects upon motion sickness, psychological performance, and physiological functioning. *Aviation Space and Environmental Medicine*, ***60***, 1–9.
707. Thienel, R., Kellermann, T., Schall, U., Voss, B., Reske, M., Halfter, S., et al. (2009). Muscarinic antagonist effects on executive control of attention. *International Journal of Neuropsychopharmacology*, ***12***, 1307.
708. Dumas, J. A., Saykin, A. J., McAllister, T. W., McDonald, B. C., Hynes, M. L., and Newhouse, P. A. (2008). Nicotinic versus muscarinic blockade alters verbal working memory-related brain activity in older women. *American Journal of Geriatric Psychiatry*, ***16***, 272.
709. Schon, K., Atri, A., Hasselmo, M. E., Tricarico, M. D., LoPresti, M. L., and Stern, C. E. (2005). Scopolamine reduces persistent activity related to long-term encoding in the parahippocampal gyrus during delayed matching in humans. *Journal of Neuroscience*, ***25***, 9112–9123.
710. Sperling, R., Greve, D., Dale, A., Killiany, R., Holmes, J., Rosas, H. D., et al. (2002). Functional MRI detection of pharmacologically induced memory impairment. *Proceedings of the National Academy of Sciences of the United States of America*, ***99***, 455–460.

711. Rosier, A. M., Cornette, L., Dupont, P., Bormans, G., Mortelmans, L., and Orban, G. A. (1999). Regional brain activity during shape recognition impaired by a scopolamine challenge to encoding. *European Journal of Neuroscience*, **11**, 3701–3714.
712. Drachman, D. A., and Leavitt, J. (1974). Human memory and the cholinergic system: A relationship to aging? *Archives of Neurology*, **30**, 113.
713. Wink, M. (1993). Allelochemical properties or the raison d'etre of alkaloids. *The Alkaloids (G. Cordell, Hrg.)*.
714. Alves, M. N., Sartoratto, A., and Trigo, J. R. (2007). Scopolamine in *Brugmansia suaveolens* (Solanaceae): Defense, allocation, costs, and induced response. *Journal of Chemical Ecology*, **33**, 297–309.
715. Krug, E., and Proksch, P. (1993). Influence of dietary alkaloids on survival and growth of *Spodoptera littoralis*. *Biochemical Systematics and Ecology*, **21**, 749–756.
716. Khan, M. B., and Harborne, J. B. (1990). Induced alkaloid defence in *Atropa acuminata* in response to mechanical and herbivore leaf damage. *Chemoecology*, **1**, 77–80.
717. Shonle, I., and Bergelson, J. (2000). Evolutionary ecology of the tropane alkaloids of *Datura stramonium* L.(Solanaceae). *Evolution*, **54**, 778–788.
718. Kitamura, Y., Tominaga, Y., and Ikenaga, T. (2004). Winter cherry bugs feed on plant tropane alkaloids and de-epoxidize scopolamine to atropine. *Journal of Chemical Ecology*, **30**, 2085–2090.
719. Grant, V. (1983). Behavior of hawkmoths on flowers of *Datura meteloides*. *Botanical Gazette*, 280–284.
720. Buhl, E., Schildberger, K., and Stevenson, P. A. (2008). A muscarinic cholinergic mechanism underlies activation of the central pattern generator for locust flight. *Journal of Experimental Biology*, **211**, 2346–2357.
721. Hoffmann, K., Wirmer, A., Kunst, M., Gocht, D., and Heinrich, R. (2007). Muscarinic excitation in grasshopper song control circuits is limited by acetylcholinesterase activity. *Zoological Science*, **24**, 1028–1035.
722. Terazima, E., and Yoshino, M. (2010). Modulatory action of acetylcholine on the Na(+)-dependent action potentials in Kenyon cells isolated from the mushroom body of the cricket brain. *Journal of Insect Physiology*, **56**, 1746–1754.
723. Ismail, N., Robinson, G. E., and Fahrbach, S. E. (2006). Stimulation of muscarinic receptors mimics experience-dependent plasticity in the honey bee brain. *Proceedings of the National Academy of Sciences of the United States of America*, **103**, 207–211.
724. Dobrin, S. E., Herlihy, J. D., Robinson, G. E., and Fahrbach, S. E. (2011). Muscarinic regulation of Kenyon cell dendritic arborizations in adult worker honey bees. *Arthropod Structure & Development*, **40**, 409–419.
725. Wustenberg, D. G., and Grunewald, B. (2004). Pharmacology of the neuronal nicotinic acetylcholine receptor of cultured Kenyon cells of the honeybee, *Apis mellifera*. *Journal of Comparative Physiology A-Neuroethology Sensory Neural and Behavioral Physiology*, **190**, 807–821.
726. Lozano, V. r. C. and Gauthier, M. (1998). Effects of the muscarinic antagonists atropine and pirenzepine on olfactory conditioning in the honeybee. *Pharmacology Biochemistry and Behavior*, **59**, 903–907.
727. Dworacek, B., and Rupreht, J. (2002). Physostigmine: short history and its impact on anaesthesiology of present days. *International Congress Series*, **1242**, 87–93.
728. Coelho Filho, J., and Birks, J.. (2001). Physostigmine for dementia due to Alzheimer's disease. *Cohrane Database of Systematic Reviews*.
729. Heinrich, M., and Lee Teoh, H. (2004). Galanthamine from snowdrop—the development of a modern drug against Alzheimer's disease from local Caucasian knowledge. *Journal of Ethnopharmacology*, **92**, 147–162.

730. Loy, C., and Schneider, L. (2006). Galantamine for Alzheimer's disease and mild cognitive impairment. *Cochrane Database of Systematic Reviews*, 84.
731. Li, J., Wu, H. M., Zhou, R. L., Liu, G. J., and Dong, B. R. (2008). Huperzine A for Alzheimer's disease. *Cochrane Database of Systematic Reviews*, CD005592.
732. Birks, J., Grimley Evans, J., Iakovidou, V., Tsolaki, M., and Holt, F. (2009). Rivastigmine for Alzheimer's disease. *Cochrane Database of Systematic Reviews.*
733. Bond, M., Rogers, G., Peters, J., Anderson, R., Hoyle, M., Miners, A., et al. (2012). The effectiveness and cost-effectiveness of donepezil, galantamine, rivastigmine and memantine for the treatment of Alzheimer's disease (review of Technology Appraisal No. 111): a systematic review and economic model.
734. Rafii, M., Walsh, S., Little, J., Behan, K., Reynolds, B., Ward, C., et al. (2011). A phase II trial of huperzine A in mild to moderate Alzheimer disease. *Neurology*, ***76***, 1389–1394.
735. Bowsher, C. S., M. Tobin, A. (2008). *Plant Biochemistry*. New York: Garland Science.
736. Crozier, A., Jaganath, I. B., and Clifford, M. N. (2007). Phenols, polyphenols and tannins: an overview. In *Plant Secondary Metabolites* (pp. 1–24). Blackwell Publishing Ltd.
737. Maffei, M. E. (2010). Sites of synthesis, biochemistry and functional role of plant volatiles. *South African Journal of Botany*, ***76***, 612–631.
738. Glinwood, R., Ninkovic, V., and Pettersson, J. (2011). Chemical interaction between undamaged plants—Effects on herbivores and natural enemies. *Phytochemistry*, ***72***, 1683–1689.
739. Liberato, M., de Morais, S. M., Siqueira, S. M. C., de Menezes, J., Ramos, D. N., Machado, L. K. A., et al. (2011). Phenolic content and antioxidant and antiacetylcholinesterase properties of honeys from different floral origins. *Journal of Medicinal Food*, ***14***, 658–663.
740. Adler, L. S. (2000). The ecological significance of toxic nectar. *Oikos*, ***91***, 409–420.
741. Oberdörster, E., Clay, M. A., Cottam, D. M., Wilmot, F. A., McLachlan, J. A., and Milner, M. J. (2001). Common phytochemicals are ecdysteroid agonists and antagonists: a possible evolutionary link between vertebrate and invertebrate steroid hormones. *Journal of Steroid Biochemistry and Molecular Biology*, ***77***, 229–238.
742. Musser, R. O., Cipollini, D. F., Hum-Musser, S. M., Williams, S. A., Brown, J. K., and Felton, G. W. (2005). Evidence that the caterpillar salivary enzyme glucose oxidase provides herbivore offense in solanaceous plants. *Archives of Insect Biochemistry and Physiology*, ***58***, 128–137.
743. Smith, C. M., and Boyko, E. V. (2007). The molecular bases of plant resistance and defense responses to aphid feeding: current status. *Entomologia Experimentalis et Applicata*, ***122***, 1–16.
744. Bezemer, T. M., and van Dam, N. M. (2005). Linking aboveground and belowground interactions via induced plant defenses. *Trends in Ecology & Evolution*, ***20***, 617–624.
745. Cesco, S., Neumann, G., Tomasi, N., Pinton, R., and Weisskopf, L. (2010). Release of plant-borne flavonoids into the rhizosphere and their role in plant nutrition. *Plant and Soil*, ***329***, 1–25.
746. Weston, L. A., and Mathesius, U. (2013). Flavonoids: their structure, biosynthesis and role in the rhizosphere, including allelopathy. *Journal of Chemical Ecology*, 1–15.
747. Martens, S., and Mithofer, A. (2005). Flavones and flavone synthases. *Phytochemistry*, ***66***, 2399–2407.
748. Perez-Jimenez, J., Neveu, V., Vos, F., and Scalbert, A. (2010). Identification of the 100 richest dietary sources of polyphenols: an application of the Phenol-Explorer database. *European Journal of Clinical Nutrition*, ***64***, S112–S120.

749. Perez-Jimenez, J., Neveu, V., Vos, F., and Scalbert, A. (2010). Systematic analysis of the content of 502 polyphenols in 452 foods and beverages: an application of the Phenol-Explorer database. *Journal of Agricultural and Food Chemistry*, ***58***, 4959–4969.
750. Chun, O. K., Chung, S. J., and Song, W. O. (2007). Estimated dietary flavonoid intake and major food sources of US adults. *Journal of Nutrition*, ***137***, 1244–1252.
751. Zamora-Ros, R., Andres-Lacueva, C., Lamuela-Raventós, R. M., Berenguer, T., Jakszyn, P., Barricarte, A., et al. (2010). Estimation of dietary sources and flavonoid intake in a Spanish adult population (EPIC-Spain). *Journal of the American Dietetic Association*, ***110***, 390–398.
752. Johannot, L., and Somerset, S. M. (2006). Age-related variations in flavonoid intake and sources in the Australian population. *Public Health Nutrition*, ***9***, 1045–1054.
753. Ovaskainen, M.-L., Torronen, R., Koponen, J. M., Sinkko, H., Hellstrom, J., Reinivuo, H., et al. (2008). Dietary intake and major food sources of polyphenols in Finnish adults. *Journal of Nutrition*, ***138***, 562–566.
754. Perez-Jimenez, J., Fezeu, L., Touvier, M., Arnault, N., Manach, C., Hercberg, S., et al. (2011). Dietary intake of 337 polyphenols in French adults. *American Journal of Clinical Nutrition*, ***93***, 1220–1228.
755. Djousse, L., Hopkins, P. N., North, K. E., Pankow, J. S., Arnett, D. K., and Ellison, R. C. (2011). Chocolate consumption is inversely associated with prevalent coronary heart disease: The National Heart, Lung, and Blood Institute Family Heart Study. *Clinical Nutrition*, ***30***, 182–187.
756. Deka, A., and Vita, J. A. (2011). Tea and cardiovascular disease. *Pharmacological Research*, ***64***, 136–145.
757. Bauer, S. R., Ding, E. L., and Smit, L. A. (2011). Cocoa consumption, cocoa flavonoids, and effects on cardiovascular risk factors: an evidence-based review. *Current Cardiovascular Risk Reports*, ***5***, 120–127.
758. McCullough, M. L., Peterson, J. J., Patel, R., Jacques, P. F., Shah, R., and Dwyer, J. T. (2012). Flavonoid intake and cardiovascular disease mortality in a prospective cohort of US adults. *American Journal of Clinical Nutrition*, ***95***, 454–464.
759. Medina-Remon, A., Zamora-Ros, R., Rotches-Ribalta, M., Andres-Lacueva, C., Martinez-Gonzalez, M. A., Covas, M. I., et al. (2011). Total polyphenol excretion and blood pressure in subjects at high cardiovascular risk. *Nutrition Metabolism and Cardiovascular Diseases*, ***21***, 323–331.
760. Kay, C. D., Hooper, L., Kroon, P. A., Rimm, E. B., and Cassidy, A. (2012). Relative impact of flavonoid composition, dose and structure on vascular function: A systematic review of randomised controlled trials of flavonoid-rich food products. *Molecular Nutrition & Food Research*.
761. Monagas, M., Khan, N., Andres-Lacueva, C., Casas, R., Urpi-Sarda, M., Llorach, R., et al. (2009). Effect of cocoa powder on the modulation of inflammatory biomarkers in patients at high risk of cardiovascular disease. *American Journal of Clinical Nutrition*, ***90***, 1144–1150.
762. Shrime, M. G., Bauer, S. R., McDonald, A. C., Chowdhury, N. H., Coltart, C. E. M., and Ding, E. L. (2011). Flavonoid-rich cocoa consumption affects multiple cardiovascular risk factors in a meta-analysis of short-term studies. *Journal of Nutrition*, ***141***, 1982–1988.
763. Kuriyama, S., Hozawa, A., Ohmori, K., Shimazu, T., Matsui, T., Ebihara, S., et al. (2006). Green tea consumption and cognitive function: a cross-sectional study from the Tsurugaya Project. *American Journal of Clinical Nutrition*, ***83***, 355–361.
764. Arab, L., Liu, W., and Elashoff, D. (2009). Green and black tea consumption and risk of stroke: A meta-analysis. *Stroke*, ***40***, 1786–1792.

765. Hollman, P. C. H., Geelen, A., and Kromhout, D. (2010). Dietary flavonol intake may lower stroke risk in men and women. *Journal of Nutrition*, ***140***, 600–604.
766. Commenges, D., Scotet, V., Renaud, S., Jacqmin-Gadda, H., Barberger-Gateau, P., and Dartigues, J. F. (2000). Intake of flavonoids and risk of dementia. *European Journal of Epidemiology*, ***16***, 357–363.
767. Ng, T. P., Feng, L., Niti, M., Kua, E. H., and Yap, K. B. (2008). Tea consumption and cognitive impairment and decline in older Chinese adults. *American Journal of Clinical Nutrition*, ***88***, 224–231.
768. Barberger-Gateau, P., Raffaitin, C., Letenneur, L., Berr, C., Tzourio, C., Dartigues, J. F., et al. (2007). Dietary patterns and risk of dementia: The Three-City Cohort Study. *Neurology*, ***69***, 1921–1930.
769. Dai, Q., Borenstein, A. R., Wu, Y. G., Jackson, J. C., and Larson, E. B. (2006). Fruit and vegetable juices and Alzheimer's disease: The Kame Project. *American Journal of Medicine*, ***119***, 751–759.
770. Letenneur, L., Proust-Lima, C., Le Gouge, A., Dartigues, J. F., and Barberger-Gateau, P. (2007). Flavonoid intake and cognitive decline over a 10-year period. *American Journal of Epidemiology*, ***165***, 1364–1371.
771. Nurk, E., Refsum, H., Drevon, C. A., Tell, G. S., Nygaard, H. A., Engedal, K., et al. (2009). Intake of flavonoid-rich wine, tea, and chocolate by elderly men and women is associated with better cognitive test performance. *Journal of Nutrition*, ***139***, 120–127.
772. Devore, E. E., Kang, J. H., Breteler, M., and Grodstein, F. (2012). Dietary intakes of berries and flavonoids in relation to cognitive decline. *Annals of Neurology*.
773. Kesse-Guyot, E., Fezeu, L., Andreeva, V. A., Touvier, M., Scalbert, A., Hercberg, S., et al. (2012). Total and specific polyphenol intakes in midlife are associated with cognitive function measured 13 years later. *Journal of Nutrition*, ***142***, 76–83.
774. Sofi, F., Abbate, R., Gensini, G. F., and Casini, A. (2010). Accruing evidence on benefits of adherence to the Mediterranean diet on health: an updated systematic review and meta-analysis. *American Journal of Clinical Nutrition*, ***92***, 1189–1196.
775. Singh, M., Arseneault, M., Sanderson, T., Murthy, V., and Ramassamy, C. (2008). Challenges for research on polyphenols from foods in Alzheimer's disease: Bioavailability, metabolism, and cellular and molecular mechanisms. *Journal of Agricultural and Food Chemistry*, ***56***, 4855–4873.
776. Steffen, L. M. (2006). Eat your fruit and vegetables. *The Lancet*, ***367***, 278–279.
777. Smith, A. G., Croft, M. T., Moulin, M., and Webb, M. E. (2007). Plants need their vitamins too. *Current Opinion in Plant Biology*, ***10***, 266–275.
778. Kennedy, D. O., and Haskell, C. F. (2011). Vitamins and cognition: what is the evidence? *Drugs*, ***71***, 1957–1971.
779. Nishikimi, M., Kawai, T., and Yagi, K. (1992). Guinea pigs possess a highly mutated gene for L-gulono-gamma-lactone oxidase, the key enzyme for L-ascorbic acid biosynthesis missing in this species. *Journal of Biological Chemistry*, ***267***, 21967–21972.
780. Tanaka, T., Tateno, Y., and Gojobori, T. (2005). Evolution of vitamin B-6 (pyridoxine) metabolism by gain and loss of genes. *Molecular Biology and Evolution*, ***22***, 243–250.
781. Grune, T., Lietz, G., Palou, A., Ross, A. C., Stahl, W., Tang, G. W., et al. (2010). Beta-carotene is an important vitamin A source for human. *Journal of Nutrition*, ***140***, 2268S–2285S.
782. Schweigert, F. J., Raila, J., Wichert, B., and Kienzle, E. (2002). Cats absorb beta-carotene, but it is not converted to vitamin A. *Journal of Nutrition*, ***132***, 1610S–1612S.

783. Banhegyi, G., Braun, L., Csala, M., Puskas, F., and Mandl, J. (1997). Ascorbate metabolism and its regulation in animals. *Free Radical Biology and Medicine*, ***23***, 793–803.
784. Pauling, L. (1970). Evolution and the need for ascorbic acid. *Proceedings of the National Academy of Sciences of the United States of America*, ***67***, 1643–1648.
785. Brunet, M., Guy, F., Pilbeam, D., Lieberman, D. E., Likius, A., Mackaye, H. T., et al. (2005). New material of the earliest hominid from the Upper Miocene of Chad. *Nature*, ***434***, 752–755.
786. Milton, K. (2000). Back to basics: why foods of wild primates have relevance for modern human health. *Nutrition*, ***16***, 480–483.
787. Cordain, L., Eaton, S. B., Sebastian, A., Mann, N., Lindeberg, S., Watkins, B. A., et al. (2005). Origins and evolution of the Western diet: health implications for the 21st century. *American Journal of Clinical Nutrition*, ***81***, 341–354.
788. Benzie, I. F. F. (2003). Evolution of dietary antioxidants. *Comparative Biochemistry and Physiology A-Molecular & Integrative Physiology*, ***136***, 113–126.
789. Eaton, S. B., and Konner, M. J. (1997). Paleolithic nutrition revisited: A twelve-year retrospective on its nature and implications. *European Journal of Clinical Nutrition*, ***51***, 207–216.
790. Pérez-López, F. R., Chedraui, P., Haya, J., and Cuadros, J. L. (2009). Effects of the Mediterranean diet on longevity and age-related morbid conditions. *Maturitas*, ***64***, 67–79.
791. Lindeberg, S., Jonsson, T., Granfeldt, Y., Borgstrand, E., Soffman, J., Sjostrom, K., et al. (2007). A Palaeolithic diet improves glucose tolerance more than a Mediterranean-like diet in individuals with ischaemic heart disease. *Diabetologia*, ***50***, 1795–1807.
792. Rusznyak, S., and Szent-Györgyi, A. (1936). Vitamin P: flavonols as vitamins. *Nature*, ***138***, 27.
793. Spencer, J. P. E. (2009). Flavonoids and brain health: multiple effects underpinned by common mechanisms. *Genes and Nutrition*, ***4***, 243–250.
794. De Vos, M., Van Oosten, V. R., Van Poecke, R. M. P., Van Pelt, J. A., Pozo, M. J., Mueller, M. J., et al. (2005). Signal signature and transcriptome changes of *Arabidopsis* during pathogen and insect attack. *Molecular Plant-Microbe Interactions*, ***18***, 923–937.
795. Vlot, A. C., Dempsey, D. A., and Klessig, D. F. (2009). Salicylic acid, a multifaceted hormone to combat disease. In *Annual Review of Phytopathology* (Vol. 47, pp. 177–206). Palo Alto: Annual Reviews.
796. Traw, M. B., and Bergelson, J. (2003). Interactive effects of jasmonic acid, salicylic acid, and gibberellin on induction of trichomes in Arabidopsis. *Plant Physiology*, ***133***, 1367–1375.
797. Moore, J. W., Loake, G. J., and Spoel, S. H. (2011). Transcription dynamics in plant immunity. *Plant Cell*, ***23***, 2809–2820.
798. Schultz, J. C. (2002). Shared signals and the potential for phylogenetic espionage between plants and animals. *Integrative and Comparative Biology*, ***42***, 454–462.
799. Lin, L., and Tan, R. X. (2011). Cross-kingdom actions of phytohormones: a functional scaffold exploration. *Chemical Reviews*, ***111***, 2734–2760.
800. Pierpoint, W. S. (1997). The natural history of salicylic acid: Plant product and mammalian medicine. *Interdisciplinary Science Reviews*, ***22***, 45–52.
801. Pourcel, L., and Grotewold, E. (2009). *Participation of Phytochemicals in Plant Development and Growth*. New York: Springer.
802. Taylor, L. P., and Grotewold, E. (2005). Flavonoids as developmental regulators. *Current Opinion in Plant Biology*, ***8***, 317–323.

803. Peer, W. A., and Murphy, A. S. (2008). *Flavonoids as signal molecules.* New York: Springer.
804. Venkateshwaran, M., Volkening, J. D., Sussman, M. R., and Ané, J.-M. (2013). Symbiosis and the social network of higher plants. *Current Opinion in Plant Biology, **16***, 118–127.
805. Howitz, K. T., and Sinclair, D. A. (2008). Xenohormesis: Sensing the chemical cues of other species. *Cell, **133***, 387–391.
806. Greiss, S., and Gartner, A. (2009). Sirtuin/Sir2 phylogeny, evolutionary considerations and structural conservation. *Molecules and cells, **28***, 407–415.
807. Howitz, K. T., Bitterman, K. J., Cohen, H. Y., Lamming, D. W., Lavu, S., Wood, J. G., et al. (2003). Small molecule activators of sirtuins extend *Saccharomyces cerevisiae* lifespan. *Nature, **425***, 191–196.
808. Wood, J. G., Rogina, B., Lavu, S., Howitz, K., Helfand, S. L., Tatar, M., et al. (2004). Sirtuin activators mimic caloric restriction and delay ageing in metazoans. *Nature, **430***, 686–689.
809. Fox, J. E. (2004). Chemical communication threatened by endocrine-disrupting chemicals. *Environmental Health Perspectives, **112***, 648.
810. Poulin, M.-J., Simard, J., Catford, J.-G., Librie, F., and Piché, Y. (1997). Response of symbiotic endomycorrhizal fungi to estrogens and antiestrogens. *Molecular Plant-Microbe Interactions, **10***, 481–487.
811. Gyorgypal, Z., and Kondorosi, A. (1991). Homology of the ligand-binding regions of Rhizobium symbiotic regulatory protein NodD and vertebrate nuclear receptors. *Molecular & General Genetics, **226***, 337–340.
812. Baker, M. E. (1995). *Endocrine activity of plant-derived compounds: an evolutionary perspective.* Paper presented at the Proceedings of the Society for Experimental Biology and Medicine. Society for Experimental Biology and Medicine, New York.
813. Buer, C. S., Imin, N., and Djordjevic, M. A. (2010). Flavonoids: New roles for old molecules. *Journal of Integrative Plant Biology, **52***, 98–111.
814. Zhang, T., Liu, Y., Yang, T., Zhang, L., Xu, S., Xue, L., et al. (2006). Diverse signals converge at MAPK cascades in plant. *Plant Physiology and Biochemistry, **44***, 274–283.
815. Brown, J., and Auger, K. (2011). Phylogenomics of phosphoinositide lipid kinases: perspectives on the evolution of second messenger signaling and drug discovery. *BMC Evolutionary Biology, **11***, 4.
816. Forzani, C., Carreri, A., van Bentem, S. D., Lecourieux, D., Lecourieux, F., and Hirt, H. (2011). The *Arabidopsis* protein kinase Pto-interacting 1–4 is a common target of the oxidative signal-inducible 1 and mitogen-activated protein kinases. *FEBS Journal, **278***, 1126–1136.
817. Robaglia, C., Thomas, M., and Meyer, C. (2012). Sensing nutrient and energy status by SnRK1 and TOR kinases. *Current Opinion in Plant Biology, **15***, 301–307.
818. Zoncu, R., Efeyan, A., and Sabatini, D. M. (2011). mTOR: from growth signal integration to cancer, diabetes and ageing. *Nature Reviews Molecular Cell Biology, **12***, 21–35.
819. van Dam, T. J., Zwartkruis, F. J., Bos, J. L., and Snel, B. (2011). Evolution of the TOR pathway. *Journal of Molecular Evolution, **73***, 209–220.
820. Ren, D., Liu, Y., Yang, K.-Y., Han, L., Mao, G., Glazebrook, J., et al. (2008). A fungal-responsive MAPK cascade regulates phytoalexin biosynthesis in *Arabidopsis. Proceedings of the National Academy of Sciences of the United States of America, **105***, 5638–5643.
821. Nadarajah, K., and Sidek, H. M. (2010). The green MAPKs. *Asian Journal of Plant Sciences, **9***, 1–10.

822. Ritsema, T., van Zanten, M., Leon-Reyes, A., Voesenek, L., Millenaar, F. F., Pieterse, C. M. J., et al. (2010). Kinome profiling reveals an interaction between jasmonate, salicylate and light control of hyponastic petiole growth in *Arabidopsis thaliana. PLoS One*, ***5***, 10.
823. Spencer, J. P. (2009). Flavonoids and brain health: multiple effects underpinned by common mechanisms. *Genes & Nutrition*, ***4***, 243–250.
824. Gamet-Payrastre, L., Manenti, S., Gratacap, M.-P., Tulliez, J., Chap, H., and Payrastre, B. (1999). Flavonoids and the inhibition of PKC and PI 3-kinase. *General Pharmacology: The Vascular System*, ***32***, 279–286.
825. Delledonne, M. (2005). NO news is good news for plants. *Current Opinion in Plant Biology*, ***8***, 390–396.
826. Huang, X., Stettmaier, K., Michel, C., Hutzler, P., Mueller, M. J., and Durner, J. (2004). Nitric oxide is induced by wounding and influences jasmonic acid signaling in *Arabidopsis thaliana. Planta*, ***218***, 938–946.
827. Hollender, C., and Liu, Z. C. (2008). Histone deacetylase genes in *Arabidopsis* development. *Journal of Integrative Plant Biology*, ***50***, 875–885.
828. Wang, C. Z., Gao, F., Wu, J. G., Dai, J. L., Wei, C. H., and Li, Y. (2010). *Arabidopsis* putative deacetylase AtSRT2 regulates basal defense by suppressing PAD4, EDS5 and SID2 expression. *Plant and Cell Physiology*, ***51***, 1820–1820.
829. Conseil, G., Baubichon-Cortay, H., Dayan, G., Jault, J. M., Barron, D., and Di Pietro, A. (1998). Flavonoids: A class of modulators with bifunctional interactions at vicinal ATP- and steroid-binding sites on mouse P-glycoprotein. *Proceedings of the National Academy of Sciences of the United States of America*, ***95***, 9831–9836.
830. Reid, G., Wielinga, P., Zelcer, N., van der Heijden, I., Kuil, A., de Haas, M., et al. (2003). The human multidrug resistance protein MRP4 functions as a prostaglandin efflux transporter and is inhibited by nonsteroidal antiinflammatory drugs. *Proceedings of the National Academy of Sciences of the United States of America*, ***100***, 9244–9249.
831. Wu, C. P., Calcagno, A. M., Hladky, S. B., Ambudkar, S. V., and Barrand, M. A. (2005). Modulatory effects of plant phenols on human multidrug-resistance proteins 1, 4 and 5 (ABCC1, 4 and 5). *FEBS Journal*, ***272***, 4725–4740.
832. Hooper, L., Kay, C., Abdelhamid, A., Kroon, P. A., Cohn, J. S., Rimm, E. B., et al. (2012). Effects of chocolate, cocoa, and flavan-3-ols on cardiovascular health: a systematic review and meta-analysis of randomized trials. *American Journal of Clinical Nutrition*, ***95***, 740–751.
833. Youdim, K. A., Qaiser, M. Z., Begley, D. J., Rice-Evans, C. A., and Abbott, N. J. (2004). Flavonoid permeability across an in situ model of the blood–brain barrier. *Free Radical Biology and Medicine*, ***36***, 592–604.
834. Spencer, J. P. (2010). The impact of fruit flavonoids on memory and cognition. *British Journal of Nutrition*, ***104***, S40–S47.
835. Miller, M. G., and Shukitt-Hale, B. (2012). Berry fruit enhances beneficial signaling in the brain. *Journal of Agricultural and Food Chemistry*, ***60***, 5709–5715.
836. Smoliga, J. M., Baur, J. A., and Hausenblas, H. A. (2011). Resveratrol and health—A comprehensive review of human clinical trials. *Molecular Nutrition & Food Research*, ***55***, 1129–1141.
837. Ishrat, T., Hoda, M. N., Khan, M. B., Yousuf, S., Ahmad, M., Khan, M. M., et al. (2009). Amelioration of cognitive deficits and neurodegeneration by curcumin in rat model of sporadic dementia of Alzheimer's type (SDAT). *European Neuropsychopharmacology*, ***19***, 636–647.
838. Stangl, D., and Thuret, S. (2009). Impact of diet on adult hippocampal neurogenesis. *Genes & Nutrition*, ***4***, 271–282.

839. Wasowski, C., and Marder, M. (2012). Flavonoids as GABAA receptor ligands: the whole story? *Journal of Experimental Pharmacology*, ***2012***, 9–24.
840. Karton, Y., Jiang, J.-l., Ji, X.-d., Melman, N., Olah, M. E., Stiles, G. L., et al. (1996). Synthesis and biological activities of flavonoid derivatives as A3 adenosine receptor antagonists. *Journal of Medicinal Chemistry*, ***39***, 2293–2301.
841. Katavic, P. L., Lamb, K., Navarro, H., and Prisinzano, T. E. (2007). Flavonoids as opioid receptor ligands: identification and preliminary structure-activity relationships. *Journal of Natural Products*, ***70***, 1278–1282.
842. Lee, B.-H., Choi, S.-H., Shin, T.-J., Pyo, M. K., Hwang, S.-H., Lee, S.-M., et al. (2011). Effects of quercetin on α9α10 nicotinic acetylcholine receptor-mediated ion currents. *European Journal of Pharmacology*, ***650***, 79–85.
843. Andero, R., Daviu, N., Escorihuela, R. M., Nadal, R., and Armario, A. (2012). 7, 8-dihydroxyflavone, a TrkB receptor agonist, blocks long-term spatial memory impairment caused by immobilization stress in rats. *Hippocampus*, ***22***, 399–408.
844. Han, Y.-S., Bastianetto, S., Dumont, Y., and Quirion, R. (2006). Specific plasma membrane binding sites for polyphenols, including resveratrol, in the rat brain. *Journal of Pharmacology and Experimental Therapeutics*, ***318***, 238–245.
845. Hou, D.-X., and Kumamoto, T. (2010). Flavonoids as protein kinase inhibitors for cancer chemoprevention: direct binding and molecular modeling. *Antioxidants & Redox Signaling*, ***13***, 691–719.
846. Frank-Cannon, T. C., Alto, L. T., McAlpine, F. E., and Tansey, M. G. (2009). Does neuroinflammation fan the flame in neurodegenerative diseases. *Molecular Neurodegeneration*, ***4***, 47.
847. García-Lafuente, A., Guillamón, E., Villares, A., Rostagno, M. A., and Martínez, J. A. (2009). Flavonoids as anti-inflammatory agents: implications in cancer and cardiovascular disease. *Inflammation Research*, ***58***, 537–552.
848. Peluso, I., Raguzzini, A., and Serafini, M. (2013). Effect of flavonoids on circulating levels of TNF-α and IL-6 in humans: A systematic review and meta-analysis. *Molecular Nutrition & Food Research*.
849. Manach, C., Scalbert, A., Morand, C., Rémésy, C., and Jiménez, L. (2004). Polyphenols: food sources and bioavailability. *American Journal of Clinical Nutrition*, ***79***, 727–747.
850. Crozier, A., Del Rio, D., and Clifford, M. N. (2010). Bioavailability of dietary flavonoids and phenolic compounds. *Molecular Aspects of Medicine*, ***31***, 446–467.
851. Serra, A., Macià, A., Romero, M.-P., Valls, J., Bladé, C., Arola, L., et al. (2010). Bioavailability of procyanidin dimers and trimers and matrix food effects in in vitro and in vivo models. *British Journal of Nutrition*, ***103***, 944.
852. Manach, C., Williamson, G., Morand, C., Scalbert, A., and Rémésy, C. (2005). Bioavailability and bioefficacy of polyphenols in humans. I. Review of 97 bioavailability studies. *American Journal of Clinical Nutrition*, ***81***, 230S–242S.
853. Grimm, T., Skrabala, R., Chovanová, Z., Muchová, J., Sumegová, K., Liptáková, A., et al. (2006). Single and multiple dose pharmacokinetics of maritime pine bark extract (pycnogenol) after oral administration to healthy volunteers. *BMC clinical pharmacology*, ***6***, 4.
854. Andres-Lacueva, C., Monagas, M., Khan, N., Izquierdo-Pulido, M., Urpi-Sarda, M., Permanyer, J., et al. (2008). Flavanol and flavonol contents of cocoa powder products: influence of the manufacturing process. *Journal of Agricultural and Food Chemistry*, ***56***, 3111–3117.
855. Hollenberg, N. K. (2006). Vascular action of cocoa flavanols in humans: the roots of the story. *Journal of Cardiovascular Pharmacology*, ***47***, S99–S102.

856. Johnson, R., Bryant, S., and Huntley, A. L. (2012). Green tea and green tea catechin extracts: An overview of the clinical evidence. *Maturitas*, ***73***, 280–287.
857. Schoonees, A., Visser, J., Musekiwa, A., and Volmink, J. (2012). Pycnogenol (R) for the treatment of chronic disorders. *Cochrane Database of Systematic Reviews*.
858. Knopman, D., Boland, L., Mosley, T., Howard, G., Liao, D., Szklo, M., et al. (2001). Cardiovascular risk factors and cognitive decline in middle-aged adults. *Neurology*, ***56***, 42–48.
859. Barnes, D. E., Alexopoulos, G. S., Lopez, O. L., Williamson, J. D., and Yaffe, K. (2006). Depressive symptoms, vascular disease, and mild cognitive impairment: findings from the Cardiovascular Health Study. *Archives of General Psychiatry*, ***63***, 273.
860. Cukierman-Yaffe, T., Gerstein, H. C., Williamson, J. D., Lazar, R. M., Lovato, L., Miller, M. E., et al. (2009). Relationship between baseline glycemic control and cognitive function in individuals with type 2 diabetes and other cardiovascular rIsk factors: the Action to Control Cardiovascular Risk in Diabetes-Memory in Diabetes (ACCORD-MIND) trial. *Diabetes Care*, ***32***, 221–226.
861. Bruehl, H., Sweat, V., Hassenstab, J., Polyakov, V., and Convit, A. (2010). Cognitive impairment in nondiabetic middle-aged and older adults is associated with insulin resistance. *Journal of Clinical and Experimental Neuropsychology*, ***32***, 487–493.
862. Pearson, S., Schmidt, M., Patton, G., Dwyer, T., Blizzard, L., Otahal, P., et al. (2010). Depression and insulin resistance cross-sectional associations in young adults. *Diabetes Care*, ***33***, 1128–1133.
863. Sorond, F. A., Lipsitz, L. A., Hollenberg, N. K., and Fisher, N. D. (2008). Cerebral blood flow response to flavanol-rich cocoa in healthy elderly humans. *Neuropsychiatric Disease and Treatment*, ***4***, 433.
864. Francis, S., Head, K., Morris, P., and Macdonald, I. (2006). The effect of flavanol-rich cocoa on the fMRI response to a cognitive task in healthy young people. *Journal of Cardiovascular Pharmacology*, ***47***, S215–S220.
865. Wightman, E. L., Haskell, C. F., Forster, J. S., Veasey, R. C., and Kennedy, D. O. (2012). Epigallocatechin gallate, cerebral blood flow parameters, cognitive performance and mood in healthy humans: a double-blind, placebo-controlled, crossover investigation. *Human Psychopharmacology: Clinical and Experimental*, ***27***, 177–186.
866. Field, D. T., Williams, C. M., and Butler, L. T. (2011). Consumption of cocoa flavanols results in an acute improvement in visual and cognitive functions. *Physiology & Behavior*, ***103***, 255–260.
867. Scholey, A., Downey, L. A., Ciorciari, J., Pipingas, A., Nolidin, K., Finn, M., et al. (2012). Acute neurocognitive effects of epigallocatechin gallate (EGCG). *Appetite*, ***58***, 767–770.
868. Desideri, G., Kwik-Uribe, C., Grassi, D., Necozione, S., Ghiadoni, L., Mastroiacovo, D., et al. (2012). Benefits in cognitive function, blood pressure, and insulin resistance through cocoa flavanol consumption in elderly subjects with mild cognitive impairment. Novelty and significance. The Cocoa, Cognition, and Aging (CoCoA) Study. *Hypertension*, ***60***, 794–801.
869. Crews, W. D., Harrison, D. W., and Wright, J. W. (2008). A double-blind, placebo-controlled, randomized trial of the effects of dark chocolate and cocoa on variables associated with neuropsychological functioning and cardiovascular health: clinical findings from a sample of healthy, cognitively intact older adults. *American Journal of Clinical Nutrition*, ***87***, 872–880.
870. Pase, M. P., Scholey, A. B., Pipingas, A., Kras, M., Nolidin, K., Gibbs, A., et al. (2013). Cocoa polyphenols enhance positive mood states but not cognitive performance: A randomized, placebo-controlled trial. *Journal of Psychopharmacology*.

871. Borgwardt, S., Hammann, F., Scheffler, K., Kreuter, M., Drewe, J., and Beglinger, C. (2012). Neural effects of green tea extract on dorsolateral prefrontal cortex. *European Journal of Clinical Nutrition.*
872. Park, S. K., Jung, I. C., Lee, W. K., Lee, Y. S., Park, H. K., Go, H. J., et al. (2011). A combination of green tea extract and L-theanine improves memory and attention in subjects with mild cognitive impairment: A double-blind placebo-controlled study. *Journal of Medicinal Food*, ***14***, 334–343.
873. Kohama, T., and Negami, M. (2013). Effect of low-dose French maritime pine bark extract on climacteric syndrome in 170 perimenopausal women: a randomized, double-blind, placebo-controlled trial. *Journal of Reproductive Medicine*, ***58***, 39–46.
874. Yang, H.-M., liao, M.-F., Zhu, S.-Y., liao, M.-N., and Rohdewald, P. (2007). A randomised, double-blind, placebo-controlled trial on the effect of Pycnogenol® on the climacteric syndrome in peri-menopausal women. *Acta Obstetricia et Gynecologica Scandinavica*, ***86***, 978–985.
875. Ryan, J., Croft, K., Mori, T., Wesnes, K., Spong, J., Downey, L., et al. (2008). An examination of the effects of the antioxidant Pycnogenol® on cognitive performance, serum lipid profile, endocrinological and oxidative stress biomarkers in an elderly population. *Journal of Psychopharmacology*, ***22***, 553–562.
876. Pipingas, A., Silberstein, R. B., Vitetta, L., Rooy, C. V., Harris, E. V., Young, J. M., et al. (2008). Improved cognitive performance after dietary supplementation with a *Pinus radiata* bark extract formulation. *Phytotherapy Research*, ***22***, 1168–1174.
877. Theadom, A., Mahon, S., Barker-Collo, S., McPherson, K., Rush, E., Vandal, A. C., et al. (2013). Enzogenol for cognitive functioning in traumatic brain injury: a pilot placebo-controlled RCT. *European Journal of Neurology.*
878. Kalt, W., Blumberg, J. B., McDonald, J. E., Vinqvist-Tymchuk, M. R., Fillmore, S. A., Graf, B. A., et al. (2008). Identification of anthocyanins in the liver, eye, and brain of blueberry-fed pigs. *Journal of Agricultural and Food Chemistry*, ***56***, 705–712.
879. Andres-Lacueva, C., Shukitt-Hale, B., Galli, R. L., Jauregui, O., Lamuela-Raventos, R. M., and Joseph, J. A. (2005). Anthocyanins in aged blueberry-fed rats are found centrally and may enhance memory. *Nutritional Neuroscience*, ***8***, 111–120.
880. Shukitt-Hale, B., Lau, F. C., and Joseph, J. A. (2008). Berry fruit supplementation and the aging brain. *Journal of Agricultural and Food Chemistry*, ***56***, 636–641.
881. Krikorian, R., Shidler, M. D., Nash, T. A., Kalt, W., Vinqvist-Tymchuk, M. R., Shukitt-Hale, B., et al. (2010). Blueberry supplementation improves memory in older adults. *Journal of Agricultural and Food Chemistry*, ***58***, 3996–4000.
882. Stalmach, A., Edwards, C. A., Wightman, J. D., and Crozier, A. (2011). Identification of (poly) phenolic compounds in Concord grape juice and their metabolites in human plasma and urine after juice consumption. *Journal of Agricultural and Food Chemistry*, ***59***, 9512–9522.
883. Krikorian, R., Nash, T. A., Shidler, M. D., Shukitt-Hale, B., and Joseph, J. A. (2010). Concord grape juice supplementation improves memory function in older adults with mild cognitive impairment. *British Journal of Nutrition*, ***103***, 730.
884. Krikorian, R., Boespflug, E. L., Fleck, D. E., Stein, A. L., Wightman, J. D., Shidler, M. D., et al. (2012). Concord grape juice supplementation and neurocognitive function in human aging. *Journal of Agricultural and Food Chemistry*, ***60***, 5736–5742.
885. Hendrickson, S. J., and Mattes, R. D. (2008). No acute effects of grape juice on appetite, implicit memory and mood. *Food & Nutrition Research*, ***52***.
886. Fang, H., Tong, W., Shi, L. M., Blair, R., Perkins, R., Branham, W., et al. (2001). Structure–activity relationships for a large diverse set of natural, synthetic, and environmental estrogens. *Chemical Research in Toxicology*, ***14***, 280–294.

887. Rochester, J. R., and Millam, J. R. (2009). Phytoestrogens and avian reproduction: Exploring the evolution and function of phytoestrogens and possible role of plant compounds in the breeding ecology of wild birds. *Comparative Biochemistry and Physiology - Part A: Molecular & Integrative Physiology*, ***154***, 279–288.
888. le Maire, A., Bourguet, W., and Balaguer, P. (2010). A structural view of nuclear hormone receptor: endocrine disruptor interactions. *Cellular and Molecular Life Sciences*, ***67***, 1219–1237.
889. Collins-Burow, B. M., Burow, M. E., Duong, B. N., and McLachlan, J. A. (2000). Estrogenic and antiestrogenic activities of flavonoid phytochemicals through estrogen receptor binding-dependent and -independent mechanisms. *Nutrition and Cancer*, ***38***, 229–244.
890. Collins-Burow, B. M., Antoon, J. W., Frigo, D. E., Elliott, S., Weldon, C. B., Boue, S. M., et al. (2012). Antiestrogenic activity of flavonoid phytochemicals mediated via the c-Jun N-terminal protein kinase pathway. Cell-type specific regulation of estrogen receptor alpha. *Journal of Steroid Biochemistry and Molecular Biology*, ***132***, 186–193.
891. Shor, D., Sathyapalan, T., Atkin, S. L., and Thatcher, N. J. (2012). Does equol production determine soy endocrine effects? *European Journal of Nutrition*, ***51***, 389–398.
892. Barnes, S. (2010). The biochemistry, chemistry and physiology of the isoflavones in soybeans and their food products. *Lymphatic Research and Biology*, ***8***, 89–98.
893. Guha, N., Kwan, M. L., Quesenberry, C. P., Jr., Weltzien, E. K., Castillo, A. L., and Caan, B. J. (2009). Soy isoflavones and risk of cancer recurrence in a cohort of breast cancer survivors: the Life After Cancer Epidemiology study. *Breast Cancer Res Treat*, ***118***, 395–405.
894. Anderson, L. N., Cotterchio, M., Boucher, B. A., and Kreiger, N. (2013). Phytoestrogen intake from foods, during adolescence and adulthood, and risk of breast cancer by estrogen and progesterone receptor tumor subgroup among Ontario women. *International Journal of Cancer*, ***132***, 1683–1692.
895. Boucher, B. A., Cotterchio, M., Anderson, L. N., Kreiger, N., Kirsh, V. A., and Thompson, L. U. (2013). Use of isoflavone supplements is associated with reduced postmenopausal breast cancer risk. *International Journal of Cancer*, ***132***, 1439–1450.
896. Dong, J. Y., and Qin, L. Q. (2011). Soy isoflavones consumption and risk of breast cancer incidence or recurrence: a meta-analysis of prospective studies. *Breast Cancer Research and Treatment*, ***125***, 315–323.
897. Zhang, Y. B., Chen, W. H., Guo, J. J., Fu, Z. H., Yi, C., Zhang, M., et al. (2013). Soy isoflavone supplementation could reduce body weight and improve glucose metabolism in non-Asian postmenopausal women: A meta-analysis. *Nutrition*, ***29***, 8–14.
898. Taku, K., Melby, M. K., Kronenberg, F., Kurzer, M. S., and Messina, M. (2012). Extracted or synthesized soybean isoflavones reduce menopausal hot flash frequency and severity: systematic review and meta-analysis of randomized controlled trials. *Menopause*, ***19***, 776–790.
899. Beavers, D. P., Beavers, K. M., Miller, M., Stamey, J., and Messina, M. J. (2012). Exposure to isoflavone-containing soy products and endothelial function: A Bayesian meta-analysis of randomized controlled trials. *Nutrition Metabolism and Cardiovascular Diseases*, ***22***, 182–191.
900. Taku, K., Lin, N., Cai, D. L., Hu, J. W., Zhao, X. H., Zhang, Y. M., et al. (2010). Effects of soy isoflavone extract supplements on blood pressure in adult humans: systematic review and meta-analysis of randomized placebo-controlled trials. *Journal of Hypertension*, ***28***, 1971–1982.
901. Daniel, J. M. (2013). Estrogens, estrogen receptors, and female cognitive aging: The impact of timing. *Hormones and Behavior*, ***63***, 231–237.

902. Sumien, N., Chaudhari, K., Sidhu, A., and Forster, M. J. (2013). Does phytoestrogen supplementation affect cognition differentially in males and females? *Brain Research*.
903. Sarkaki, A., Badavi, M., Aligholi, H., and Moghaddam, A. Z. (2009). Preventive effects of soy meal (+/- isoflavone) on spatial cognitive deficiency and body weight in an ovariectomized animal model of Parkinson's disease. *Pak J Biol Sci*, ***12***, 1338–1345.
904. Neese, S., Wang, V., Doerge, D., Woodling, K., Andrade, J., Helferich, W., et al. (2010). Impact of dietary genistein and aging on executive function in rats. *Neurotoxicology and teratology*, ***32***, 200–211.
905. Lee, Y. B., Lee, H. J., Won, M. H., Hwang, I. K., Kang, T. C., Lee, J. Y., et al. (2004). Soy isoflavones improve spatial delayed matching-to-place performance and reduce cholinergic neuron loss in elderly male rats. *Journal of Nutrition*, ***134***, 1827–1831.
906. Bryant, M., Cassidy, A., Hill, C., Powell, J., Talbot, D., and Dye, L. (2005). Effect of consumption of soy isoflavones on behavioural, somatic and affective symptoms in women with premenstrual syndrome. *British Journal of Nutrition*, ***93***, 731–739.
907. Basaria, S., Wisniewski, A., Dupree, K., Bruno, T., Song, M. Y., Yao, F., et al. (2009). Effect of high-dose isoflavones on cognition, quality of life, androgens, and lipoprotein in post-menopausal women. *Journal of Endocrinological Investigation*, ***32***, 150–155.
908. Hooper, L., Ryder, J. J., Kurzer, M. S., Lampe, J. W., Messina, M. J., Phipps, W. R., et al. (2009). Effects of soy protein and isoflavones on circulating hormone concentrations in pre- and post-menopausal women: a systematic review and meta-analysis. *Human Reproduction Update*, ***15***, 423–440.
909. Lamport, D. J., Dye, L., Wightman, J. D., and Lawton, C. L. (2012). The effects of flavonoid and other polyphenol consumption on cognitive performance: A systematic research review of human experimental and epidemiological studies. *Nutrition and Aging*, ***1***, 5–25.
910. Greendale, G. A., Huang, M.-H., Leung, K., Crawford, S. L., Gold, E. B., Wight, R., et al. (2012). Dietary phytoestrogen intakes and cognitive function during the menopausal transition: results from the Study of Women's Health Across the Nation Phytoestrogen Study. *Menopause*, ***19***, 894–903.
911. Henderson, V., John, J. S., Hodis, H., Kono, N., McCleary, C., Franke, A., et al. (2012). Long-term soy isoflavone supplementation and cognition in women A randomized, controlled trial. *Neurology*, ***78***, 1841–1848.
912. Hamilton, R. T., Rettberg, J. R., Mao, Z., To, J., Zhao, L., Appt, S. E., et al. (2011). Hippocampal responsiveness to 17β-estradiol and equol after long-term ovariectomy: Implication for a therapeutic window of opportunity. *Brain Research*, ***1379***, 11–22.
913. Thorp, A., Sinn, N., Buckley, J., Coates, A., and Howe, P. (2009). Soya isoflavone supplementation enhances spatial working memory in men. *British Journal of Nutrition*, ***102***, 1348–1354.
914. Chung, I. M., Park, M. R., Chun, J. C., and Yun, S. J. (2003). Resveratrol accumulation and resveratrol synthase gene expression in response to abiotic stresses and hormones in peanut plants. *Plant Science*, ***164***, 103–109.
915. Gottstein, D., and Gross, D. (1992). Phytoalexins of woody plants. *Trees-Structure and Function*, ***6***, 55–68.
916. Baur, J. A., Pearson, K. J., Price, N. L., Jamieson, H. A., Lerin, C., Kalra, A., et al. (2006). Resveratrol improves health and survival of mice on a high-calorie diet. *Nature*, ***444***, 337–342.
917. Udenigwe, C. C., Ramprasath, V. R., Aluko, R. E., and Jones, P. J. H. (2008). Potential of resveratrol in anticancer and anti-inflammatory therapy. *Nutrition Reviews*, ***66***, 445–454.

918. Nijveldt, R. J., van Nood, E., van Hoorn, D. E. C., Boelens, P. G., van Norren, K., and van Leeuwen, P. A. M. (2001). Flavonoids: a review of probable mechanisms of action and potential applications. *American Journal of Clinical Nutrition*, ***74***, 418–425.
919. Soleas, G. J., Diamandis, E. P., and Goldberg, D. M. (1997). Resveratrol: A molecule whose time has come? And gone? *Clinical Biochemistry*, ***30***, 91–113.
920. Saiko, P., Szakmary, A., Jaeger, W., and Szekeres, T. (2008). Resveratrol and its analogs: Defense against cancer, coronary disease and neurodegenerative maladies or just a fad? *Mutation Research-Reviews in Mutation Research*, ***658***, 68–94.
921. Fan, E. G., Zhang, L. J., Jiang, S., and Bai, Y. H. (2008). Beneficial effects of resveratrol on atherosclerosis. *Journal of Medicinal Food*, ***11***, 610–614.
922. Joseph, J. A., Fisher, D. R., Cheng, V., Rimando, A. M., and Shukitt-Hale, B. (2008). Cellular and behavioral effects of stilbene resveratrol analogues: Implications for reducing the deleterious effects of aging. *Journal of Agricultural and Food Chemistry*, ***56***, 10544–10551.
923. Ates, O., Cayli, S., Altinoz, E., Gurses, I., Yucel, N., Sener, M., et al. (2005). *Neuroprotection by resveratrol against traumatic brain injury in rats*. Paper presented at the 19th Annual Congress of Turkish Neurosurgery Society, Antalya, Turkey.
924. Jin, F., Wu, Q., Lu, Y. F., Gong, Q. H., and Shi, J. S. (2009). Neuroprotective effect of resveratrol on 6-OHDA-induced Parkinson's disease in rats. *European Journal of Pharmacology*, ***600***, 78–82.
925. Sharma, M., and Gupta, Y. K. (2002). Chronic treatment with trans rosveratrol prevents intracerebroventricular streptozotocin induced cognitive impairment and oxidative stress in rats. *Life Sciences*, ***71***, 2489–2498.
926. Sonmez, U., Sonmez, A., Erbil, G., Tekmen, I., and Baykara, B. (2007). Neuroprotective effects of resveratrol against traumatic brain injury in immature rats. *Neuroscience Letters*, ***420***, 133–137.
927. Yu, W., Fu, Y.-C., and Wang, W. (2012). Cellular and molecular effects of resveratrol in health and disease. *Journal of Cellular Biochemistry*, ***113***, 752–759.
928. Ghanim, H., Sia, C. L., Abuaysheh, S., Korzeniewski, K., Patnaik, P., Marumganti, A., et al. (2010). An antiinflammatory and reactive oxygen species suppressive effects of an extract of *Polygonum cuspidatum* containing resveratrol. *Journal of Clinical Endocrinology & Metabolism*, ***95***, E1–E8.
929. Ghanim, H., Sia, C. L., Korzeniewski, K., Lohano, T., Abuaysheh, S., Marumganti, A., et al. (2011). A resveratrol and polyphenol preparation suppresses oxidative and inflammatory stress response to a high-fat, high-carbohydrate meal. *Journal of Clinical Endocrinology & Metabolism*, ***96***, 1409–1414.
930. Yoshino, J., Conte, C., Fontana, L., Mittendorfer, B., Imai, S. I., Schechtman, K. B., et al. (2012). Resveratrol supplementation does not improve metabolic function in nonobese women with normal glucose tolerance. *Cell Metabolism*, ***16***, 658–664.
931. Wong, R. H. X., Howe, P. R. C., Buckley, J. D., Coates, A. M., Kunz, I., and Berry, N. M. (2011). Acute resveratrol supplementation improves flow-mediated dilatation in overweight/obese individuals with mildly elevated blood pressure. *Nutrition, Metabolism and Cardiovascular Diseases*, ***21***, 851–856.
932. Kennedy, D. O., Wightman, E. L., Reay, J. L., Lietz, G., Okello, E. J., Wilde, A., et al. (2010). Effects of resveratrol on cerebral blood flow variables and cognitive performance in humans: a double-blind, placebo-controlled, crossover investigation. *American Journal of Clinical Nutrition*, ***91***, 1590–1597.
933. Ammon, H., and Wahl, M. (1991). Pharmacology of *Curcuma longa*. *Planta Medica*, ***57***, 1–7.

934. Aggarwal, B., and Sung, B. (2009). Pharmacological basis for the role of curcumin in chronic diseases: an age-old spice with modern targets. *Trends in pharmacological sciences*, ***30***, 85–94.
935. Gupta, S. C., Patchva, S., and Aggarwal, B. B. (2013). Therapeutic roles of curcumin: Lessons learned from clinical trials. *AAPS Journal*, ***15***, 195–218.
936. Lim, G. P., Chu, T., Yang, F., Beech, W., Frautschy, S. A., and Cole, G. M. (2001). The curry spice curcumin reduces oxidative damage and amyloid pathology in an Alzheimer transgenic mouse. *Journal of Neuroscience*, ***21***, 8370.
937. Natarajan, C., and Bright, J. (2002). Curcumin inhibits experimental allergic encephalomyelitis by blocking IL-12 signaling through Janus kinase-STAT pathway in T lymphocytes. *Journal of Immunology*, ***168***, 6506.
938. Calabrese, V., Scapagnini, G., Colombrita, C., Ravagna, A., Pennisi, G., Giuffrida Stella, A., et al. (2003). Redox regulation of heat shock protein expression in aging and neurodegenerative disorders associated with oxidative stress: a nutritional approach. *Amino Acids*, ***25***, 437–444.
939. Zhang, L., Xu, T., Wang, S., Yu, L., Liu, D., Zhan, R., et al. (2012). Curcumin produces antidepressant effects via activating MAPK/ERK-dependent brain-derived neurotrophic factor expression in the amygdala of mice. *Behavioural Brain Research*, ***235***, 67–72.
940. Ishrat, T., Hoda, M. N., Khan, M. B., Yousuf, S., Ahmad, M., Khan, M. M., et al. (2009). Amelioration of cognitive deficits and neurodegeneration by curcumin in rat model of sporadic dementia of Alzheimer's type (SDAT). *European Neuropsychopharmacology*.
941. Ahmed, T., and Gilani, A. H. (2009). Inhibitory effect of curcuminoids on acetylcholinesterase activity and attenuation of scopolamine-induced amnesia may explain medicinal use of turmeric in Alzheimer's disease. *Pharmacology Biochemistry and Behavior*, ***91***, 554–559.
942. Ng, T. P., Chiam, P. C., Lee, T., Chua, H. C., Lim, L., and Kua, E. H. (2006). Curry consumption and cognitive function in the elderly. *American Journal of Epidemiology*, ***164***, 898–906.
943. Ringman, J. M., Frautschy, S. A., Teng, E., Begum, A. N., Bardens, J., Beigi, M., et al. (2012). Oral curcumin for Alzheimer's disease: tolerability and efficacy in a 24-week randomized, double blind, placebo-controlled study. *Alzheimer's Research & Therapy*, ***4***, 43.
944. Baum, L., Lam, C. W. K., Cheung, S. K.-K., Kwok, T., Lui, V., Tsoh, J., et al. (2008). Six-month randomized, placebo-controlled, double-blind, pilot clinical trial of curcumin in patients with Alzheimer disease. *Journal of Clinical Psychopharmacology*, ***28***, 110–113.
945. Cropley, V., Croft, R., Silber, B., Neale, C., Scholey, A., Stough, C., et al. (2012). Does coffee enriched with chlorogenic acids improve mood and cognition after acute administration in healthy elderly? A pilot study. *Psychopharmacology*, ***219***, 737–749.
946. Crews, W. D., Harrison, D. W., Griffin, M. L., Addison, K., Yount, A. M., Giovenco, M. A., et al. (2005). A double-blinded, placebo-controlled, randomized trial of the neuropsychologic efficacy of cranberry juice in a sample of cognitively intact older adults: Pilot study findings. *Journal of Alternative and Complementary Medicine*, ***11***, 305–309.
947. Zhang, K., and Zuo, Y. (2004). GC-MS determination of flavonoids and phenolic and benzoic acids in human plasma after consumption of cranberry juice. *Journal of Agricultural and Food Chemistry*, ***52***, 222–227.

948. Jaramillo, M. A., and Manos, P. S. (2001). Phylogeny and patterns of floral diversity in the genus *Piper* (Piperaceae). *American Journal of Botany*, ***88***, 706–716.
949. Cairney, S., Maruff, P., and Clough, A. R. (2002). The neurobehavioural effects of kava. *Australian and New Zealand Journal of Psychiatry*, ***36***, 657–662.
950. Sarris, J., LaPorte, E., and Schweitzer, I. (2011). Kava: a comprehensive review of efficacy, safety, and psychopharmacology. *Australian and New Zealand Journal of Psychiatry*, ***45***, 27–35.
951. Seitz, U., Schüle, A., and Gleitz, J. (1997). 3[H]-monoamine uptake inhibition properties of kava pyrones. *Planta Medica*, ***63***, 548.
952. Singh, Y. N., and Singh, N. N. (2002). Therapeutic potential of kava in the treatment of anxiety disorders. *CNS Drugs*, ***16***, 731–743.
953. Wu, D., Yu, L., Nair, M., De Witt, D., and Ramsewak, R. (2002). Cyclooxygenase enzyme inhibitory compounds with antioxidant activities from *Piper methysticum* (kava kava) roots. *Phytomedicine*, ***9***, 41–47.
954. Ligresti, A., Villano, R., Allarà, M., Ujváry, I., and Di Marzo, V. (2012). Kavalactones and the endocannabinoid system: The plant-derived yangonin is a novel CB1 receptor ligand. *Pharmacological Research*, ***66***, 163–169.
955. Pittler, M. H., and Ernst, E. (2003). Kava extract versus placebo for treating anxiety. *Cochrane Database of Systematic Reviews*, ***1***.
956. Witte, S., Loew, D., and Gaus, W. (2005). Meta-analysis of the efficacy of the acetonic kava-kava extract WS® 1490 in patients with non-psychotic anxiety disorders. *Phytotherapy Research*, ***19***, 183–188.
957. Sarris, J., Kavanagh, D., Byrne, G., Bone, K., Adams, J., and Deed, G. (2009). The Kava Anxiety Depression Spectrum Study (KADSS): a randomized, placebo-controlled crossover trial using an aqueous extract of *Piper methysticum*. *Psychopharmacology*, ***205***, 399–407.
958. LaPorte, E., Sarris, J., Stough, C., and Scholey, A. (2011). Neurocognitive effects of kava (*Piper methysticum*): a systematic review. *Human Psychopharmacology: Clinical and Experimental*, ***26***, 102–111.
959. Sarris, J., Scholey, A., Schweitzer, I., Bousman, C., LaPorte, E., Ng, C., et al. (2012). The acute effects of kava and oxazepam on anxiety, mood, neurocognition; and genetic correlates: a randomized, placebo-controlled, double-blind study. *Human Psychopharmacology: Clinical and Experimental*, ***27***, 262–269.
960. Lebot, V., McKenna, D. J., Johnston, E., Zheng, Q. Y., and McKern, D. (1999). Morphological, phytochemical, and genetic variation in hawaiian cultivars of 'Awa (Kava, *Piper methysticum*, piperaceae). *Economic Botany*, ***53***, 407–418.
961. Singh, S. K., and Khurma, U. R. (2008). Assessing the potential of kava (*Piper methysticum* Forst) and wild kava (*Piper aduncum* L.) as organic amendments for managing root-knot nematodes. *South Pacific Journal of Natural and Applied Sciences*, ***26***, 33–38.
962. Xuan, T., Elzaawely, A., Fukuta, M., and Tawata, S. (2006). Herbicidal and fungicidal activities of lactones in kava (*Piper methysticum*). *Journal of Agricultural and Food Chemistry*, ***54***, 720–725.
963. Khanh, T., Chung, I., Tawata, S., and Xuan, T. (2006). Weed suppression by *Passiflora edulis* and its potential allelochemicals. *Weed Research*, ***46***, 296–303.
964. Di Carlo, G., Borrelli, F., Ernst, E., and Izzo, A. (2001). St John's wort: Prozac from the plant kingdom. *Trends in Pharmacological Sciences*, ***22***, 292–297.
965. Beerhues, L. (2006). Hyperforin. *Phytochemistry*, ***67***, 2201–2207.
966. Meruelo, D., Lavie, G., and Lavie, D. (1988). Therapeutic agents with dramatic antiretroviral activity and little toxicity at effective doses: aromatic polycyclic diones

hypericin and pseudohypericin. *Proceedings of the National Academy of Sciences of the United States of America*, ***85***, 5230.

967. Zanoli, P. (2004). Role of hyperforin in the pharmacological activities of St. John's Wort. *CNS Drug Reviews*, ***10***, 203–218.
968. Medina, M. A., Martínez-Poveda, B., Amores-Sánchez, M. I., and Quesada, A. R. (2006). Hyperforin: More than an antidepressant bioactive compound? *Life Sciences*, ***79***, 105–111.
969. Butterweck, V., Jurgenliemk, G., Nahrstedt, A., and Winterhoff, H. (2000). Flavonoids from *Hypericum perforatum* show antidepressant activity in the forced swimming test. *Planta Medica*, ***66***, 3–6.
970. Butterweck, V., and Schmidt, M. (2007). St. John's wort: role of active compounds for its mechanism of action and efficacy. *WMW Wiener Medizinische Wochenschrift*, ***157***, 356–361.
971. Filippini, R., Piovan, A., Borsarini, A., and Caniato, R. (2010). Study of dynamic accumulation of secondary metabolites in three subspecies of *Hypericum perforatum*. *Fitoterapia*, ***81***, 115–119.
972. Butterweck, V. (2003). Mechanism of action of St John's wort in depression: what is known? *CNS Drugs*, ***17***, 539–562.
973. Kumar, V., Mdzinarishvili, A., Kiewert, C., Abbruscato, T., Bickel, U., Schyf, C., et al. (2006). NMDA receptor-antagonistic properties of hyperforin, a constituent of St. John's Wort. *Journal of Pharmacological Sciences*.
974. Pilkington, K., Rampes, H., and Richardson, J. (2006). Complementary medicine for depression. *Expert Review of Neurotherapeutics*, ***6***, 1741–1751.
975. Linde, K., Berner, M. M., and Kriston, L. (2008). St John's wort for major depression. *Cochrane Database of Systematic Reviews*, CD000448.
976. Guillet, G., Podeszfinski, C., Regnault-Roger, C., Arnason, J., and Philogène, B. J. R. (2000). Behavioral and biochemical adaptations of generalist and specialist herbivorous insects feeding on *Hypericum perforatum* (Guttiferae). *Environmental Entomology*, ***29***, 135–139.
977. Ballou, L. M., and Lin, R. Z. (2008). Rapamycin and mTOR kinase inhibitors. *Journal of Chemical Biology*, ***1***, 27–36.
978. Van Aller, G. S., Carson, J. D., Tang, W., Peng, H., Zhao, L., Copeland, R. A., et al. (2011). Epigallocatechin gallate (EGCG), a major component of green tea, is a dual phosphoinositide-3-kinase/mTOR inhibitor. *Biochemical and Biophysical Research Communications*, ***406***, 194–199.
979. Brüning, A. (2012). Inhibition of mTOR signaling by quercetin in cancer treatment and prevention. *Anti-cancer Agents in Medicinal Chemistry*.
980. Zhang, X., Chen, L. X., Ouyang, L., Cheng, Y., and Liu, B. (2012). Plant natural compounds: targeting pathways of autophagy as anti-cancer therapeutic agents. *Cell Proliferation*, ***45***, 466–476.
981. Ghosh, H. S., McBurney, M., and Robbins, P. D. (2010). SIRT1 negatively regulates the mammalian target of rapamycin. *PLoS One*, ***5***, e9199.
982. Liu, M., Wilk, S. A., Wang, A., Zhou, L., Wang, R.-H., Ogawa, W., et al. (2010). Resveratrol inhibits mTOR signaling by promoting the interaction between mTOR and DEPTOR. *Journal of Biological Chemistry*, ***285***, 36387–36394.
983. Poulose, S. M., Bielinski, D. F., Gomes, S. M., Carrihill-Knoll, K., Rabin, B. M., and Shukitt-Hale, B. (2010). Protective effects of berries and walnuts against the accelerated aging and age-associated stress caused by irradiation in critical regions of rat brain. *FASEB Journal*, ***24***.

984. Bassham, D. C. (2009). Function and regulation of macroautophagy in plants. *Biochimica et Biophysica Acta (BBA) - Molecular Cell Research*, **1793**, 1397–1403.
985. Karowe, D. N., and Radi, J. K. (2011). Are the phytoestrogens genistein and daidzein anti-herbivore defenses? A test using the gypsy moth (*Lymantria dispar*). *Journal of Chemical Ecology*, **37**, 830–837.
986. Wasserman, M. D., Chapman, C. A., Milton, K., Gogarten, J. F., Wittwer, D. J., and Ziegler, T. E. (2012). Estrogenic plant consumption predicts red colobus monkey (*Procolobus rufomitratus*) hormonal state and behavior. *Hormones and Behavior*.
987. Fox, J. E., Starcevic, M., Jones, P. E., Burow, M. E., and McLachlan, J. A. (2004). Phytoestrogen signaling and symbiotic gene activation are disrupted by endocrine-disrupting chemicals. *Environmental Health Perspectives*, **112**, 672.
988. Makoi, J. H., and Ndakidemi, P. A. (2010). Biological, ecological and agronomic significance of plant phenolic compounds in rhizosphere of the symbiotic legumes. *African Journal of Biotechnology*, **6**.
989. Paris, M., Pettersson, K., Schubert, M., Bertrand, S., Pongratz, I., Escriva, H., et al. (2008). An amphioxus orthologue of the estrogen receptor that does not bind estradiol: insights into estrogen receptor evolution. *Bmc Evolutionary Biology*, **8**, 219.
990. Hirvonen, J., Rajalin, A.-M., Wohlfahrt, G., Adlercreutz, H., Wähälä, K., and Aarnisalo, P. (2011). Transcriptional activity of estrogen-related receptor γ (ERRγ) is stimulated by the phytoestrogen equol. *Journal of Steroid Biochemistry and Molecular Biology*, **123**, 46–57.
991. Markov, G. V., and Laudet, V. (2011). Origin and evolution of the ligand-binding ability of nuclear receptors. *Molecular and Cellular Endocrinology*, **334**, 21–30.
992. Boué, S. M., Burow, M. E., Wiese, T. E., Shih, B. Y., Elliott, S., Carter-Wientjes, C. H., et al. (2010). Estrogenic and antiestrogenic activities of phytoalexins from red kidney bean (*Phaseolus vulgaris* L.). *Journal of Agricultural and Food Chemistry*, **59**, 112–120.
993. Heitman, J., Movva, N., and Hall, M. N. (1991). Targets for cell cycle arrest by the immunosuppressant rapamycin in yeast. *Science (New York, NY)*, **253**, 905.
994. Cruz, M. C., Cavallo, L. M., Görlach, J. M., Cox, G., Perfect, J. R., Cardenas, M. E., et al. (1999). Rapamycin antifungal action is mediated via conserved complexes with FKBP12 and TOR kinase homologs in *Cryptococcus neoformans*. *Molecular and Cellular Biology*, **19**, 4101–4112.
995. Harrison, D. E., Strong, R., Sharp, Z. D., Nelson, J. F., Astle, C. M., Flurkey, K., et al. (2009). Rapamycin fed late in life extends lifespan in genetically heterogeneous mice. *Nature*, **460**, 392–395.
996. Jung, H., Hwang, I., Sung, W., Kang, H., Kang, B., Seu, Y., et al. (2005). Fungicidal effect of resveratrol on human infectious fungi. *Archives of Pharmacal Research*, **28**, 557–560.
997. Araujo, C., and Leon, L. (2001). Biological activities of *Curcuma longa* L. *Memórias do Instituto Oswaldo Cruz*, **96**, 723–728.
998. Augustin, J. M., Kuzina, V., Andersen, S. B., and Bak, S. (2011). Molecular activities, biosynthesis and evolution of triterpenoid saponins. *Phytochemistry*, **72**, 435–457.
999. Lee, S., Peterson, C. J., and Coats, J. R. (2003). Fumigation toxicity of monoterpenoids to several stored product insects. *Journal of Stored Products Research*, **39**, 77–85.
1000. Regnault-Roger, C. (1997). The potential of botanical essential oils for insect pest control. *Integrated Pest Management Reviews*, **2**, 25–34.
1001. Grodnitzky, J. A., and Coats, J. R. (2002). QSAR evaluation of monoterpenoids' insecticidal activity. *Journal of Agricultural and Food Chemistry*, **50**, 4576–4580.

1002. Abdelgaleil, S. A. M. (2010). Molluscicidal and insecticidal potential of monoterpenes on the white garden snail, *Theba pisana* (Muller) and the cotton leafworm, *Spodoptera littoralis* (Boisduval). *Applied Entomology and Zoology*, ***45***, 425–433.
1003. Echeverrigaray, S., Zacaria, J., and Beltrao, R. (2010). Nematicidal activity of monoterpenoids against the root-knot nematode *Meloidogyne incognita*. *Phytopathology*, ***100***, 199–203.
1004. Sparg, S. G., Light, M. E., and van Staden, J. (2004). Biological activities and distribution of plant saponins. *Journal of Ethnopharmacology*, ***94***, 219–243.
1005. Davies, T. G. E., O'Reilly, A. O., Field, L. M., Wallace, B. A., and Williamson, M. S. (2008). Knockdown resistance to DDT and pyrethroids: from target-site mutations to molecular modelling. *Pest Management Science*, ***64***, 1126–1130.
1006. Tong, F., and Coats, J. R. (2012). Quantitative structure–activity relationships of monoterpenoid binding activities to the housefly GABA receptor. *Pest Management Science*.
1007. Bloomquist, J. R. (2003). Chloride channels as tools for developing selective insecticides. *Archives of Insect Biochemistry and Physiology*, ***54***, 145–156.
1008. Priestley, C. M., Williamson, E. M., Wafford, K. A., and Sattelle, D. B. (2003). Thymol, a constituent of thyme essential oil, is a positive allosteric modulator of human GABA(A) receptors and a homo-oligomeric GABA receptor from *Drosophila melanogaster*. *British Journal of Pharmacology*, ***140***, 1363–1372.
1009. Waliwitiya, R., Belton, P., Nicholson, R. A., and Lowenberger, C. A. (2010). Effects of the essential oil constituent thymol and other neuroactive chemicals on flight motor activity and wing beat frequency in the blowfly *Phaenicia sericata*. *Pest Management Science*, ***66***, 277–289.
1010. Waliwitiya, R., Belton, P., Nicholson, R. A., and Lowenberger, C. A. (2012). Plant terpenoids: acute toxicities and effects on flight motor activity and wing beat frequency in the blow fly *Phaenicia sericata*. *Journal of Economic Entomology*, ***105***, 72–84.
1011. Ozoe, Y., and Akamatsu, M. (2001). Non-competitive GABA antagonists: probing the mechanisms of their selectivity for insect versus mammalian receptors. *Pest Management Science*, ***57***, 923–931.
1012. Ozoe, Y., Akamatsu, M., Higata, T., Ikeda, I., Mochida, K., Koike, K., et al. (1998). Picrodendrin and related terpenoid antagonists reveal structural differences between ionotropic GABA receptors of mammals and insects. *Bioorganic & Medicinal Chemistry*, ***6***, 481–492.
1013. Thompson, A. J., McGonigle, I., Duke, R., Johnston, G. A. R., and Lummis, S. C. R. (2012). A single amino acid determines the toxicity of *Ginkgo biloba* extracts. *FASEB Journal*, ***26***, 1884–1891.
1014. Blenau, W., and Baumann, A. (2001). Molecular and pharmacological properties of insect biogenic amine receptors: Lessons from *Drosophila melanogaster* and *Apis mellifera*. *Archives of Insect Biochemistry and Physiology*, ***48***, 13–38.
1015. Kostyukovsky, M., Rafaeli, A., Gileadi, C., Demchenko, N., and Shaaya, E. (2002). Activation of octopaminergic receptors by essential oil constituents isolated from aromatic plants: possible mode of action against insect pests. *Pest Management Science*, ***58***, 1101–1106.
1016. Enan, E. (2001). Insecticidal activity of essential oils: octopaminergic sites of action. *Comparative Biochemistry and Physiology C-Toxicology & Pharmacology*, ***130***, 325–337.
1017. Price, D. N., and Berry, M. S. (2006). Comparison of effects of octopamine and insecticidal essential oils on activity in the nerve cord, foregut, and dorsal unpaired median neurons of cockroaches. *Journal of Insect Physiology*, ***52***, 309–319.

1018. Enan, E. E. (2005). Molecular response of *Drosophila melanogaster* tyramine receptor cascade to plant essential oils. *Insect Biochemistry and Molecular Biology*, ***35***, 309–321.
1019. Lei, J., Leser, M., and Enan, E. (2010). Nematicidal activity of two monoterpenoids and SER-2 tyramine receptor of *Caenorhabditis elegans*. *Biochemical Pharmacology*, ***79***, 1062–1071.
1020. Anderson, J. A., and Coats, J. R. (2012). Acetylcholinesterase inhibition by nootkatone and carvacrol in arthropods. *Pesticide Biochemistry and Physiology*, ***102***, 124–128.
1021. Picollo, M. I., Toloza, A. C., Mougabure Cueto, G., Zygadlo, J., and Zerba, E. (2008). Anticholinesterase and pediculicidal activities of monoterpenoids. *Fitoterapia*, ***79***, 271–278.
1022. Ryan, M. F., and Byrne, O. (1988). Plant-insect coevolution and inhibition of acetylcholinesterase. *Journal of Chemical Ecology*, ***14***, 1965–1975.
1023. Abdelgaleil, S. A. M., Mohamed, M. I. E., Badawy, M. E. I., and El-arami, S. A. A. (2009). Fumigant and contact toxicities of monoterpenes to *Sitophilus oryzae* (L.) and *Tribolium castaneum* (Herbst) and their inhibitory effects on acetylcholinesterase activity. *Journal of Chemical Ecology*, ***35***, 518–525.
1024. Keane, S., and Ryan, M. F. (1999). Purification, characterisation, and inhibition by monoterpenes of acetylcholinesterase from the waxmoth, *Galleria mellonella* (L.). *Insect Biochemistry and Molecular Biology*, ***29***, 1097–1104.
1025. Bathori, M., Toth, N., Hunyadi, A., Marki, A., and Zador, E. (2008). Phytoecdysteroids and anabolic-androgenic steroids: Structure and effects on humans. *Current Medicinal Chemistry*, ***15***, 75–91.
1026. Dinan, L. (2009). The Karlson Lecture. Phytoecdysteroids: What use are they? *Archives of Insect Biochemistry and Physiology*, ***72***, 126–141.
1027. Dinan, L., Savchenko, T., Whiting, P., and Sarker, S. D. (1999). Plant natural products as insect steroid receptor agonists and antagonists. *Pesticide Science*, ***55***, 331–335.
1028. Chaieb, I. (2010). Saponins as insecticides: a review. *Tunisian Journal of Plant Protection*, ***5***, 39–50.
1029. Leung, K. W., and Wong, A. S. T. (2010). Pharmacology of ginsenosides: a literature review. *Chinese Medicine*, ***5***, 20.
1030. Savchenko, T., Whiting, P., Germade, A., and Dinan, L. (2000). Ecdysteroid agonist and antagonist activities in species of the Solanaceae. *Biochemical Systematics and Ecology*, ***28***, 403–419.
1031. Dinan, L., Savchenko, T., and Whiting, P. (2001). On the distribution of phytoecdysteroids in plants. *Cellular and Molecular Life Sciences*, ***58***, 1121–1132.
1032. Zhang, A.-h., Lei, F.-j., Fang, S.-w., Jia, M.-h., and Zhang, L.-x. (2011). Effects of ginsenosides on the growth and activity of antioxidant enzymes in American ginseng seedlings. *Journal of Medicinal Plants Research*, 5, 3217–3223.
1033. Griffin, S. G., Wyllie, S. G., Markham, J. L., and Leach, D. N. (1999). The role of structure and molecular properties of terpenoids in determining their antimicrobial activity. *Flavour and Fragrance Journal*, ***14***, 322–332.
1034. Schmiderer, C., Grassi, P., Novak, J., and Franz, C. (2008). Diversity of essential oil glands of Spanish sage (*Salvia lavandulifolia* Vahl, Lamiaceae). *Natural Product Communications*, ***3***, 1155–1160.
1035. Schmiderer, C., Grassi, P., Novak, J., Weber, M., and Franz, C. (2008). Diversity of essential oil glands of clary sage (*Salvia sclarea* L., Lamiaceae). *Plant Biology*, ***10***, 433–440.
1036. Granero, A. M., Sanz, J. M. G., Gonzalez, F. J. E., Vidal, J. L. M., Dornhaus, A., Ghani, J., et al. (2005). Chemical compounds of the foraging recruitment pheromone in bumblebees. *Naturwissenschaften*, ***92***, 371–374.

1037. Kaufman, P. E., Mann, R. S., and Butler, J. F. (2010). Evaluation of semiochemical toxicity to *Aedes aegypti*, *Ae. albopictus* and *Anopheles quadrimaculatus* (Diptera: Culicidae). *Pest Management Science*, ***66***, 497–504.

1038. Ryan, M. F., Awde, J., and Moran, S. (1992). Insect pheromones as reversible competitive inhibitors of acetylcholinesterase. *Invertebrate Reproduction & Development*, ***22***, 31–38.

1039. Tarelli, G., Zerba, E., and Alzogaray, R. A. (2009). Toxicity to vapor exposure and topical application of essential oils and monoterpenes on *Musca domestica* (Diptera: Muscidae). *Journal of Economic Entomology*, ***102***, 1383–1388.

1040. Willmer, P. G., Nuttman, C. V., Raine, N. E., Stone, G. N., Pattrick, J. G., Henson, K., et al. (2009). Floral volatiles controlling ant behaviour. *Functional Ecology*, ***23***, 888–900.

1041. Offenberg, J., Nielsen, M. G., MacIntosh, D. J., Havanon, S., and Aksornkoae, S. (2004). Evidence that insect herbivores are deterred by ant pheromones. *Proceedings of the Royal Society of London. Series B: Biological Sciences*, ***271***, S433–S435.

1042. Tripathi, A. K., Upadhyay, S., Bhuiyan, M., and Bhattacharya, P. (2009). A review on prospects of essential oils as biopesticide in insect-pest management. *Journal of Pharmacognosy and Phytotherapy*, ***1***, 052–063.

1043. Pohlit, A. M., Lopes, N. P., Gama, R. A., Tadei, W. P., and Neto, V. F. D. (2011). Patent literature on mosquito repellent inventions which contain plant essential oils - A review. *Planta Medica*, ***77***, 598–617.

1044. Pohlit, A. M., Rezende, A. R., Baldin, E. L. L., Lopes, N. P., and Neto, V. F. D. (2011). Plant extracts, isolated phytochemicals, and plant-derived agents which are lethal to arthropod vectors of human tropical diseases - A review. *Planta Medica*, ***77***, 618–630.

1045. Orhan, I., Kartal, M., Naz, Q., Ejaz, A., Yilmaz, G., Kan, Y., et al. (2007). Antioxidant and anticholinesterase evaluation of selected Turkish Salvia species. *Food Chemistry*, ***103***, 1247–1254.

1046. Orhan, I., and Aslan, M. (2009). Appraisal of scopolamine-induced antiamnesic effect in mice and in vitro antiacetylcholinesterase and antioxidant activities of some traditionally used Lamiaceae plants. *Journal of Ethnopharmacology*, ***122***, 327–332.

1047. Orhan, I., Kartal, M., Kan, Y., and Sener, B. (2008). Activity of essential oils and individual components against acetyl and butyrylcholinesterase. *Zeitschrift Fur Naturforschung Section C-A Journal of Biosciences*, ***63***, 547–553.

1048. Senol, F. S., Orhan, I., Celep, F., Kahraman, A., Dogan, M., Yilmaz, G., et al. (2010). Survey of 55 Turkish Salvia taxa for their acetylcholinesterase inhibitory and antioxidant activities. *Food Chemistry*, ***120***, 34–43.

1049. Kennedy, D. O., Pace, S., Haskell, C., Okello, E. J., Milne, A., and Scholey, A. B. (2006). Effects of cholinesterase inhibiting sage (*Salvia officinalis*) on mood, anxiety and performance on a psychological stressor battery. *Neuropsychopharmacology*, ***31***, 845–852.

1050. Scholey, A. B., Tildesley, N. T. J., Ballard, C. G., Wesnes, K. A., Tasker, A., Perry, E. K., et al. (2008). An extract of *Salvia* (sage) with anticholinesterase properties improves memory and attention in healthy older volunteers. *Psychopharmacology*, ***198***, 127–139.

1051. Tildesley, N. T. J., Kennedy, D. O., Perry, E. K., Ballard, C. G., Savelev, S., Wesnes, K. A., et al. (2003). *Salvia lavandulaefolia* (Spanish Sage) enhances memory in healthy young volunteers. *Pharmacology Biochemistry and Behavior*, ***75***, 669–674.

1052. Tildesley, N. T. J., Kennedy, D. O., Perry, E. K., Ballard, C. G., Wesnes, K. A., and Scholey, A. B. (2005). Positive modulation of mood and cognitive performance following administration of acute doses of *Salvia lavandulaefolia* essential oil to healthy young volunteers. *Physiology & Behavior*, ***83***, 699–709.

1053. Kennedy, D. O., Dodd, F. L., Robertson, B. C., Okello, E. J., Reay, J. L., Scholey, A. B., et al. (2011). Monoterpenoid extract of sage (*Salvia lavandulaefolia*) with cholinesterase-inhibiting properties improves cognitive performance and mood in healthy adults. *Journal of Psychopharmacology*, ***25***, 1088–1100.
1054. Savelev, S., Okello, E., Perry, N. S. L., Wilkins, R. M., and Perry, E. K. (2003). Synergistic and antagonistic interactions of anticholinesterase terpenoids in *Salvia lavandulaefolia* essential oil. *Pharmacology Biochemistry and Behavior*, ***75***, 661–668.
1055. Savelev, S. U., Okello, E. J., and Perry, E. K. (2004). Butyryl- and acetyl-cholinesterase inhibitory activities in essential oils of *Salvia* species and their constituents. *Phytotherapy Research*, ***18***, 315–324.
1056. Park, T. J., Seo, H. K., Kang, B. J., and Kim, K. T. (2001). Noncompetitive inhibition by camphor of nicotinic acetylcholine receptors. *Biochemical Pharmacology*, ***61***, 787–793.
1057. Ren, Y. H., Houghton, P. J., Hider, R. C., and Howes, M. J. R. (2004). Novel diterpenoid acetylcholinesterase inhibitors from *Salvia miltiorrhiza*. *Planta Medica*, ***70***, 201–204.
1058. Zhou, Y. Q., Li, W. X., Xu, L., and Chen, L. Y. (2011). In *Salvia miltiorrhiza*, phenolic acids possess protective properties against amyloid beta-induced cytotoxicity, and tanshinones act as acetylcholinesterase inhibitors. *Environmental Toxicology and Pharmacology*, ***31***, 443–452.
1059. Wong, K. K. K., Ho, M. T. W., Lin, H. Q., Lau, K. F., Rudd, J. A., Chung, R. C. K., et al. (2010). Cryptotanshinone, an acetylcholinesterase inhibitor from *Salvia miltiorrhiza*, ameliorates scopolamine-induced amnesia in Morris Water Maze Task. *Planta Medica*, ***76***, 228–234.
1060. Kim, D. H., Jeon, S. J., Jung, J. W., Lee, S., Yoon, B. H., Shin, B. Y., et al. (2007). Tanshinone congeners improve memory impairments induced by scopolamine on passive avoidance tasks in mice. *European Journal of Pharmacology*, ***574***, 140–147.
1061. Choudhary, M. I., Nawaz, S. A., Zaheer ul, H., Lodhi, M. A., Ghayur, M. N., Jalil, S., et al. (2005). Withanolides, a new class of natural cholinesterase inhibitors with calcium antagonistic properties. *Biochemical and Biophysical Research Communications, 334*, 276–287.
1062. Das, A., Shanker, G., Nath, C., Pal, R., Singh, S., and Singh, H. K. (2002). A comparative study in rodents of standardized extracts of *Bacopa monniera* and *Ginkgo biloba*—Anticholinesterase and cognitive enhancing activities. *Pharmacology Biochemistry and Behavior*, ***73***, 893–900.
1063. Perry, N. S. L., Houghton, P. J., Jenner, P., Keith, A., and Perry, E. K. (2002). *Salvia lavandulaefolia* essential oil inhibits cholinesterase in vivo. *Phytomedicine*, ***9***, 48–51.
1064. Bhatnagar, M., Suhalka, P., Sukhwal, P., Jain, A., and Sharma, D. (2012). Inhibition of acetylcholinesterase and NO synthase activity in the mice brain: Effect of a *Withania somnifera* leaf juice. *Neurophysiology*, ***44***, 301–308.
1065. Tong, F., and Coats, J. R. (2010). Effects of monoterpenoid insecticides on (3)H -TBOB binding in house fly GABA receptor and (36)Cl(-) uptake in American cockroach ventral nerve cord. *Pesticide Biochemistry and Physiology*, ***98***, 317–324.
1066. Sharma, A., Cardoso-Taketa, A., García, G., and Villarreal, M. (2012). A systematic updated review of scientifically tested selected plants used for anxiety disorders. *Botanics: Targets and Therapy*, ***2***, 21–39.
1067. Hall, A. C., Turcotte, C. M., Betts, B. A., Yeung, W.-Y., Agyeman, A. S., and Burk, L. A. (2004). Modulation of human GABAA and glycine receptor currents by menthol and related monoterpenoids. *European Journal of Pharmacology*, ***506***, 9–16.

1068. Maurmann, N., Reolon, G. K., Rech, S. B., Fett-Neto, A. G., and Roesler, R. (2011). A valepotriate fraction of *Valeriana glechomifolia* shows sedative and anxiolytic properties and impairs recognition but not aversive memory in mice. *Evidence-Based Complementary and Alternative Medicine*, 1–7.

1069. Murphy, K., Kubin, Z. J., Shepherd, J. N., and Ettinger, R. H. (2010). *Valeriana officinalis* root extracts have potent anxiolytic effects in laboratory rats. *Phytomedicine*, ***17***, 674–678.

1070. Benke, D., Barberis, A., Kopp, S., Altmann, K. H., Schubiger, M., Vogt, K. E., et al. (2009). GABA(A) receptors as in vivo substrate for the anxiolytic action of valerenic acid, a major constituent of valerian root extracts. *Neuropharmacology*, ***56***, 174–181.

1071. Kavvadias, D., Monschein, V., Sand, P., Riederer, P., and Schreier, P. (2003). Constituents of sage (*Salvia officinalis*) with in vitro affinity to human brain benzodiazepine receptor. *Planta Medica*, ***69***, 113–117.

1072. Johnston, G. A. (1996). $GABA_A$ receptor pharmacology. *Pharmacology & Therapeutics*, ***69***, 173–198.

1073. Kazmi, I., Afzal, M., Gupta, G., and Anwar, F. (2012). Antiepileptic potential of ursolic acid stearoyl glucoside by GABA receptor stimulation. *CNS Neuroscience & Therapeutics*, ***18***, 799–800.

1074. Pemminati, S., Gopalakrishna, H., Rai, A., Shetty, S., Vinod, A., Pai, P. G., et al. (2011). Anxiolytic effect of chronic administration of ursolic acid in rats. *Journal of Applied Pharmaceutical Sciences*, ***1***, 68–71.

1075. Taviano, M., Miceli, N., Monforte, M., Tzakou, O., and Galati, E. (2007). Ursolic acid plays a role in *Nepeta sibthorpii* Bentham CNS depressing effects. *Phytotherapy Research*, ***21***, 382–385.

1076. Awad, R., Muhammad, A., Durst, T., Trudeau, V. L., and Arnason, J. T. (2009). Bioassay-guided fractionation of lemon balm (*Melissa officinalis* L.) using an in vitro measure of GABA transaminase activity. *Phytotherapy Research*, ***23***, 1075–1081.

1077. Johnston, G., Chebib, M., Duke, R., Fernandez, S., Hanrahan, J., Hinton, T., et al. (2009). Herbal products and GABA receptors. *Encyclopedia of Neuroscience*, 1095–1101.

1078. Chebib, M., and Johnston, G. (2000). GABA-activated ligand gated ion channels: medicinal chemistry and molecular biology. *Journal of Medicinal Chemistry*, ***43***, 1427–1447.

1079. Pertwee, R. (2006). The pharmacology of cannabinoid receptors and their ligands: an overview. *International Journal of Obesity*, ***30***, S13–S18.

1080. Roth, B. L., Baner, K., Westkaemper, R., Siebert, D., Rice, K. C., Steinberg, S., et al. (2002). Salvinorin A: A potent naturally occurring nonnitrogenous kappa opioid selective agonist. *Proceedings of the National Academy of Sciences of the United States of America*, ***99***, 11934–11939.

1081. Galeotti, N., Di Cesare Mannelli, L., Mazzanti, G., Bartolini, A., and Ghelardini, C. (2002). Menthol: a natural analgesic compound. *Neuroscience Letters*, *322*, 145–148.

1082. Cinar, O. G., Kirmizibekmez, H., Akaydin, G., and Yesilada, E. (2011). Investigation of in vitro opioid receptor binding activities of some Turkish *Salvia* species. *Records of Natural Products*, ***5***, 281–289.

1083. Park, T.-J., Park, Y.-S., Lee, T.-G., Ha, H., and Kim, K.-T. (2003). Inhibition of acetylcholine-mediated effects by borneol. *Biochemical Pharmacology*, ***65***, 83–90.

1084. Kennedy, D. O., Wake, G., Savelev, S., Tildesley, N. T. J., Perry, E. K., Wesnes, K. A., et al. (2003). Modulation of mood and cognitive performance following acute administration of single doses of *Melissa officinalis* (lemon balm) with human CNS nicotinic and muscarinic receptor-binding properties. *Neuropsychopharmacology*, ***28***, 1871–1881.

1085. Perry, N., Court, G., Bidet, N., Court, J., and Perry, E. (1996). European herbs with cholinergic activities: Potential in dementia therapy. *International Journal of Geriatric Psychiatry*, ***11***, 1063–1069.
1086. Wake, G., Court, J., Pickering, A., Lewis, R., Wilkins, R., and Perry, E. (2000). CNS acetylcholine receptor activity in European medicinal plants traditionally used to improve failing memory. *Journal of Ethnopharmacology*, ***69***, 105–114.
1087. Schliebs, R., Liebmann, A., Bhattacharya, S. K., Kumar, A., Ghosal, S., and Bigl, V. (1997). Systemic administration of defined extracts from *Withania somnifera* (Indian ginseng) and Shilajit differentially affects cholinergic but not glutamatergic and GABAergic markers in rat brain. *Neurochemistry International*, ***30***, 181.
1088. Schmelz, E. A., Grebenok, R. J., Galbraith, D. W., and Bowers, W. S. (1999). Insect-induced synthesis of phytoecdysteroids in spinach, *Spinacia oleracea*. *Journal of Chemical Ecology*, ***25***, 1739–1757.
1089. Gorelick-Feldman, J. I. (2009). *Phytoecdysteroids: Understanding Their Anabolic Activity*. Rutgers University, Graduate School, New Brunswick, NJ.
1090. Burris, T. P., Busby, S. A., and Griffin, P. R. (2012). Targeting orphan nuclear receptors for treatment of metabolic diseases and autoimmunity. *Chemistry & Biology*, ***19***, 51–59.
1091. Steigerová, J., Rárová, L., Oklešt'ková, J., Křížová, K., Levková, M., Šváchová, M., et al. (2012). Mechanisms of natural brassinosteroid-induced apoptosis of prostate cancer cells. *Food and Chemical Toxicology*, ***50***, 4068–4076.
1092. Fahrbach, S. E., Smagghe, G., and Velarde, R. A. (2012). Insect nuclear receptors. *Annual Review of Entomology*, ***57***, 83–106.
1093. Drew, B. T., and Sytsma, K. J. (2012). Phylogenetics, biogeography, and staminal evolution in the tribe *Mentheae* (Lamiaceae). *American Journal of Botany*, ***99***, 933–953.
1094. Jenks, A. A., Walker, J. B., and Kim, S.-C. (2011). Evolution and origins of the Mazatec hallucinogenic sage, *Salvia divinorum* (Lamiaceae): a molecular phylogenetic approach. *Journal of Plant Research*, ***124***, 593–600.
1095. Bohlmann, J., Meyer-Gauen, G., and Croteau, R. (1998). Plant terpenoid synthases: Molecular biology and phylogenetic analysis. *Proceedings of the National Academy of Sciences of the United States of America*, ***95***, 4126–4133.
1096. Tissier, A. (2012). Glandular trichomes: what comes after expressed sequence tags? *The Plant Journal*, ***70***, 51–68.
1097. Bottega, S., and Corsi, G. (2000). Structure, secretion and possible functions of calyx glandular hairs of *Rosmarinus officinalis* L. (Labiatae). *Botanical Journal of the Linnean Society*, ***132***, 325–335.
1098. Zandi-Sohani, N. (2011). Efficiency of Labiateae plants essential oils against adults of cotton whitefly (*Bemisia tabaci*). *Indian Journal of Agricultural Sciences*, ***81***, 1164–1167.
1099. Cetin, H., Cinbilgel, I., Yanikoglu, A., and Gokceoglu, M. (2006). Larvicidal activity of some Labiatae (Lamiaceae) plant extracts from Turkey. *Phytotherapy Research*, ***20***, 1088–1090.
1100. Pavela, R. (2004). Insecticidal activity of certain medicinal plants. *Fitoterapia*, ***75***, 745–749.
1101. Walker, J. B., Sytsma, K. J., Treutlein, J., and Wink, M. (2004). *Salvia* (Lamiaceae) is not monophyletic: implications for the systematics, radiation, and ecological specializations of Salvia and tribe Mentheae. *American Journal of Botany*, ***91***, 1115–1125.
1102. Valnet, J. (1990). *Aromatherapie*. Paris: Maloine S.A. Publishers.
1103. Crellin, J. K., and Philpott, J. (1997). *A Reference Guide to Medicinal Plants: Herbal Medicine Past and Present* (Vol. 2). Duke University Press Books.

1104. Perry, E. K., Pickering, A. T., Wang, W. W., Houghton, P. J., and Perry, N. S. (1999). Medicinal plants and Alzheimer's disease: from ethnobotany to phytotherapy. *Journal of Pharmacy and Pharmacology*, ***51***, 527–534.
1105. Grieve, M. (1971). *A Modern Herbal* (Vol. 2). Dover Publications.
1106. Kennedy, D. O., and Scholey, A. B. (2006). The psychopharmacology of European herbs with cognition-enhancing properties. *Current Pharmaceutical Design*, ***12***, 4613–4623.
1107. Leung, A. Y., and Foster, S. (1996). *Encyclopedia of Common Natural Ingredients Used in Food, Drugs,and Cosmetics*. John Wiley & Sons, Inc.
1108. Grausgruber-Groger, S., Schmiderer, C., Steinborn, R., and Novak, J. (2012). Seasonal influence on gene expression of monoterpene synthases in *Salvia officinalis* (Lamiaceae). *Journal of Plant Physiology*, ***169***, 353–359.
1109. Mantle, D., Eddeb, F., and Pickering, A. T. (2000). Comparison of relative antioxidant activities of British medicinal plant species in vitro. *Journal of Ethnopharmacology*, ***72***, 47–51.
1110. Hohmann, J., Zupko, I., Redei, D., Csanyi, M., Falkay, G., Mathe, I., et al. (1999). Protective effects of the aerial parts of *Salvia officinalis*, *Melissa officinalis* and *Lavandula angustifolia* and their constituents against enzyme-dependent and enzyme-independent lipid peroxidation. *Planta Medica*, ***65***, 576–578.
1111. Lu, Y., and Foo, L. Y. (1999). Rosmarinic acid derivatives from *Salvia officinalis*. *Phytochemistry*, ***51***, 91–94.
1112. Lu, Y., and Foo, L. Y. (2001). Salvianolic acid L, a potent phenolic antioxidant from *Salvia officinalis*. *Tetrahedron Letters*, ***42***, 8223–8225.
1113. Rutherford, D. M., Nielsen, M. P. C., Hansen, S. K., Witt, M. R., Bergendorff, O., and Sterner, O. (1992). Isolation and identification from *Salvia officinalis* of 2 diterpenes which inhibit tert-butylbicyclophosphoro s-35 thionate binding to chloride channel of rat cerebrocortical membranes in vitro. *Neuroscience Letters*, ***135***, 224–226.
1114. Hold, K. M., Sirisoma, N. S., Ikeda, T., Narahashi, T., and Casida, J. E. (2000). Alpha-thujone (the active component of absinthe): gamma-aminobutyric acid type A receptor modulation and metabolic detoxification. *Proceedings of the National Academy of Sciences of the United States of America*, ***97***, 3826–3831.
1115. Perry, N. S. L., Houghton, P. J., Theobald, A., Jenner, P., and Perry, E. K. (2000). In vitro inhibition of human erythrocyte acetylcholinesterase by *Salvia lavandulaefolia* essential oil and constituent terpenes. *Journal of Pharmacy and Pharmacology*, ***52***, 895–902.
1116. Perry, N. S. L., Houghton, P. J., Sampson, J., Theobald, A. E., Hart, S., Lis-Balchin, M., et al. (2001). In vitro activity of *S. lavandulaefolia* (Spanish sage) relevant to treatment of Alzheimer's disease. *Journal of Pharmacy and Pharmacology*, ***53***, 1347–1356.
1117. Eidi, M., Eidi, A., and Bahar, M. (2006). Effects of *Salvia officinalis* L. (sage) leaves on memory retention and its interaction with the cholinergic system in rats. *Nutrition*, ***22***, 321–326.
1118. Moss, L., Rouse, M., Wesnes, K. A., and Moss, M. (2010). Differential effects of the aromas of *Salvia* species on memory and mood. *Human Psychopharmacology-Clinical and Experimental*, ***25***, 388–396.
1119. Perry, N. S. L., Bollen, C., Perry, E. K., and Ballard, C. (2003). *Salvia* for dementia therapy: review of pharmacological activity and pilot tolerability clinical trial. *Pharmacology Biochemistry and Behavior*, ***75***, 651–659.
1120. Akhondzadeh, S., Noroozian, M., Mohammadi, M., Ohadinia, S., Jamshidi, A. H., and Khani, M. (2003). *Salvia officinalis* extract in the treatment of patients with mild to moderate Alzheimer's disease: a double-blind, randomized and placebo-controlled trial. *Journal of Clinical Pharmacy and Therapeutics*, ***28***, 53–59.

1121. Valdés, L. J., Diaz, J., and Paul, A. G. (1983). Ethnopharmacology of *ska María Pastora* (*Salvia divinorum*, Epling AND Játiva-M.). *Journal of Ethnopharmacology*, *7*, 287–312.
1122. Fontana, G., Savona, G., Rodriguez, B., Dersch, C. M., Rothman, R. B., and Prisinzano, T. E. (2008). Synthetic studies of neoclerodane diterpenoids from *Salvia splendens* and evaluation of opioid receptor affinity. *Tetrahedron*, ***64***, 10041–10048.
1123. Wasson, R. G. (1962). A new Mexican psychotropic drug from the mint family. *Botanical Museum Leaflets Harvard University*, ***20***, 77–84.
1124. Ortega, A., Blount, J. F., and Manchand, P. S. (1982). Salvinorin, a new trans-neoclerodane diterpene from *Salvia divinorum* (Labiatae). *Journal of the Chemical Society-Perkin Transactions1*, 2505–2508.
1125. Valdes, L. J., Butler, W. M., Hatfield, G. M., Paul, A. G., and Koreeda, M. (1984). Divinorin-A, a psychotropic terpenoid, and divinorin-B from the hallucinogenic Mexican mint *Salvia divinorum*. *Journal of Organic Chemistry*, ***49***, 4716–4720.
1126. Cunningham, C. W., Rothman, R. B., and Prisinzano, T. E. (2011). Neuropharmacology of the naturally occurring κ-opioid hallucinogen salvinorin A. *Pharmacological Reviews*, ***63***, 316–347.
1127. Baker, L. E., Panos, J. J., Killinger, B. A., Peet, M. M., Bell, L. M., Haliw, L. A., et al. (2009). Comparison of the discriminative stimulus effects of salvinorin A and its derivatives to U69,593 and U50,488 in rats. *Psychopharmacology*, ***203***, 203–211.
1128. Butelman, E. R., Rus, S., Prisinzano, T. E., and Kreek, M. J. (2010). The discriminative effects of the kappa-opioid hallucinogen salvinorin A in nonhuman primates: dissociation from classic hallucinogen effects. *Psychopharmacology*, ***210***, 253–262.
1129. Munro, T. A., Rizzacasa, M. A., Roth, B. L., Toth, B. A., and Yan, F. (2005). Studies toward the pharmacophore of salvinorin A, a potent kappa opioid receptor agonist. *Journal of Medicinal Chemistry*, ***48***, 345–348.
1130. Lee, D. Y. W., Ma, Z. Z., Liu-Chen, L. Y., Wang, Y. L., Chen, Y., Carlezon, W. A., et al. (2005). New neoclerodane diterpenoids isolated from the leaves of *Salvia divinorum* and their binding affinities for human kappa opioid receptors. *Bioorganic & Medicinal Chemistry*, ***13***, 5635–5639.
1131. Harding, W. W., Tidgewell, K., Schmidt, M., Shah, K., Dersch, C. M., Snyder, J., et al. (2005). Salvinicins A and B, new neoclerodane diterpenes from *Salvia divinorum*. *Organic Letters*, *7*, 3017–3020.
1132. Listos, J., Merska, A., and Fidecka, S. (2011). Pharmacological activity of salvinorin A, the major component of *Salvia divinorum*. *Pharmacological Reports*, ***63***, 1305–1309.
1133. Ford, J. A., Watkins, W. C., and Blumenstein, L. (2011). Correlates of *Salvia divinorum* use in a national sample: Findings from the 2009 National Survey on Drug Use and Health. *Addictive Behaviors*, ***36***, 1032–1037.
1134. Nyi, P. P., Lai, E. P., Lee, D. Y., Biglete, S. A., Torrecer, G. I., and Anderson, I. B. (2010). Influence of age on *Salvia divinorum* use: Results of an Internet survey. *Journal of Psychoactive Drugs*, ***42***, 385–392.
1135. Lange, J. E., Daniel, J., Homer, K., Reed, M. B., and Clapp, J. D. (2010). *Salvia divinorum*: Effects and use among YouTube users. *Drug and Alcohol Dependence*, ***108***, 138–140.
1136. Sumnall, H., Measham, F., Brandt, S., and Cole, J. (2011). *Salvia divinorum* use and phenomenology: results from an online survey. *Journal of Psychopharmacology*, ***25***, 1496–1507.
1137. Gonzalez, D., Riba, J., Bouso, J. C., Gomez-Jarabo, G., and Barbanoj, M. J. (2006). Pattern of use and subjective effects of *Salvia divinorum* among recreational users. *Drug and Alcohol Dependence*, ***85***, 157–162.

1138. Hooker, J. M., Xu, Y., Schiffer, W., Shea, C., Carter, P., and Fowler, J. S. (2008). Pharmacokinetics of the potent hallucinogen, salvinorin A in primates parallels the rapid onset and short duration of effects in humans. *NeuroImage*, ***41***, 1044–1050.
1139. Johnson, M. W., MacLean, K. A., Reissig, C. J., Prisinzano, T. E., and Griffiths, R. R. (2011). Human psychopharmacology and dose-effects of salvinorin A, a kappa opioid agonist hallucinogen present in the plant *Salvia divinorum*. *Drug and Alcohol Dependence*, ***115***, 150–155.
1140. Addy, P. H. (2012). Acute and post-acute behavioral and psychological effects of salvinorin A in humans. *Psychopharmacology*, ***220***, 195–204.
1141. Wang, B. Q. (2010). *Salvia miltiorrhiza*: Chemical and pharmacological review of a medicinal plant. *Journal of Medicinal Plants Research*, ***4***, 2813–2820.
1142. Cheng, T. O. (2007). Cardiovascular effects of Danshen. *International Journal of Cardiology*, ***121***, 9–22.
1143. Xiao, Y., Gao, S. H., Di, P., Chen, J. F., Chen, W. S., and Zhang, L. (2009). Methyl jasmonate dramatically enhances the accumulation of phenolic acids in *Salvia miltiorrhiza* hairy root cultures. *Physiologia Plantarum*, ***137***, 1–9.
1144. Wang, X.-Y., Cui, G.-H., Huang, L.-Q., and Qiu, D.-Y. (2007). Effects of methyl jasmonate on accumulation and release of tanshinones in suspension cultures of *Salvia miltiorrhiza* hairy root. *Journal of Chinese Materia Medica*, ***32***, 300–302.
1145. Wang, J. W., and Wu, J. Y. (2010). Tanshinone biosynthesis in *Salvia miltiorrhiza* and production in plant tissue cultures. *Applied Microbiology and Biotechnology*, ***88***, 437–449.
1146. Wu, J. Y., Ng, J., Shi, M., and Wu, S. J. (2007). Enhanced secondary metabolite (tanshinone) production of *Salvia miltiorrhiza* hairy roots in a novel root-bacteria coculture process. *Applied Microbiology and Biotechnology*, ***77***, 543–550.
1147. Lee, J., Ji, Y., Lee, S., and Lee, I. (2007). Effect of *Saliva miltiorrhiza* bunge on antimicrobial activity and resistant gene regulation against methicillin-resistant *Staphylococcus aureus* (MRSA). *Journal of Microbiology - Seoul*, ***45***, 350.
1148. Wang, T., Fu, F., Han, B., Zhang, L., and Zhang, X. (2012). Danshensu ameliorates the cognitive decline in streptozotocin-induced diabetic mice by attenuating advanced glycation end product-mediated neuroinflammation. *Journal of Neuroimmunology*, ***245***, 79–86.
1149. Zhang, Q. L., Sun, Y. B., Bai, B., and Huang, H. (2010). Effects of *Salvia miltiorrhiza* Bge.f.alba on neuronal regeneration following cerebral ischemia/reperfusion. *Neural Regeneration Research*, ***5***, 1066–1070.
1150. Yoo, K. Y., and Park, S. Y. (2012). Terpenoids as potential anti-Alzheimer's disease therapeutics. *Molecules*, ***17***, 3524–3538.
1151. Loizzo, M. R., Tundis, R., Conforti, F., Menichini, F., Bonesi, M., Nadjafi, F., et al. (2010). *Salvia leriifolia* Benth (Lamiaceae) extract demonstrates in vitro antioxidant properties and cholinesterase inhibitory activity. *Nutrition Research*, ***30***, 823–830.
1152. Dastmalchi, K., Damien Dorman, H. J., Oinonen, P. P., Darwis, Y., Laakso, I., and Hiltunen, R. (2008). Chemical composition and in vitro antioxidative activity of a lemon balm (*Melissa officinalis* L.) extract. *LWT - Food Science and Technology*, ***41***, 391–400.
1153. Tittel, G., Wagner, H., and Bos, R. (1982). Chemical-composition of the essential oil from *Melissa*. *Planta Medica*, ***46***, 91–98.
1154. Pereira, R., Fachinetto, R., de Souza Prestes, A., Puntel, R., Santos da Silva, G., Heinzmann, B., et al. (2009). Antioxidant effects of different extracts from *Melissa officinalis*, *Matricaria recutita* and *Cymbopogon citratus*. *Neurochemical Research*, ***34***, 973–983.

1155. Dastmalchi, K., Ollilainen, V., Lackman, P., Gennäs, G. B.a., Dorman, H. J. D., Järvinen, P. P., et al. (2009). Acetylcholinesterase inhibitory guided fractionation of *Melissa officinalis* L. *Bioorganic & Medicinal Chemistry*, ***17***, 867–871.
1156. Salah, S. M., and Jäger, A. K. (2005). Screening of traditionally used Lebanese herbs for neurological activities. *Journal of Ethnopharmacology*, ***97***, 145–149.
1157. Abuhamdah, S., Huang, L., Elliott, M., Howes, M., Ballard, C., Holmes, C., et al. (2008). Pharmacological profile of an essential oil derived from *Melissa officinalis* with anti-agitation properties: focus on ligand-gated channels. *Journal of Pharmacy and Pharmacology*, ***60***, 377–384.
1158. Ibarra, A., Feuillere, N., Roller, M., Lesburgere, E., and Beracochea, D. (2010). Effects of chronic administration of *Melissa officinalis* L. extract on anxiety-like reactivity and on circadian and exploratory activities in mice. *Phytomedicine*, ***17***, 397–403.
1159. Lopez, V., Martin, S., Gomez-Serranillos, M. P., Carretero, M. E., Jager, A. K., and Calvo, M. I. (2009). Neuroprotective and neurological properties of *Melissa officinalis*. *Neurochemical Research*, ***34***, 1955–1961.
1160. Gertsch, J., Leonti, M., Raduner, S., Racz, I., Chen, J. Z., Xie, X. Q., et al. (2008). Beta-caryophyllene is a dietary cannabinoid. *Proceedings of the National Academy of Sciences of the United States of America*, ***105***, 9099–9104.
1161. Zeraatpishe, A., Oryan, S., Bagheri, M. H., Pilevarian, A. A., Malekirad, A. A., Baeeri, M., et al. (2011). Effects of *Melissa officinalis* L. on oxidative status and DNA damage in subjects exposed to long-term low-dose ionizing radiation. *Toxicology and Industrial Health*, ***27***, 205–212.
1162. Kennedy, D. O., Scholey, A. B., Tildesley, N. T. J., Perry, E. K., and Wesnes, K. A. (2002). Modulation of mood and cognitive performance following acute administration of *Melissa officinalis* (lemon balm). *Pharmacology Biochemistry and Behavior*, ***72***, 953–964.
1163. Kennedy, D., Little, W., and Scholey, A. (2004). Attenuation of laboratory-induced stress in humans after acute administration of *Melissa officinalis* (lemon balm). *Psychosomatic Medicine*, ***66***, 607.
1164. Ballard, C. G., O'Brien, J. T., Reichelt, K., and Perry, E. K. (2002). Aromatherapy as a safe and effective treatment for the management of agitation in severe dementia: The results of a double-blind, placebo-controlled trial with *Melissa*. *Journal of Clinical Psychiatry*, ***63***, 553–558.
1165. Burns, A., Perry, E., Holmes, C., Francis, P., Morris, J., Howes, M.-J., et al. (2011). A double-blind placebo-controlled randomized trial of *Melissa officinalis* oil and donepezil for the treatment of agitation in Alzheimer's disease. *Dementia and Geriatric Cognitive Disorders*, ***31***, 158–164.
1166. Akhondzadeh, S., Noroozian, M., Mohammadi, M., Ohadinia, S., Jamshidi, A. H., and Khani, M. (2003). *Melissa officinalis* extract in the treatment of patients with mild to moderate Alzheimer's disease: a double-blind, randomised, placebo-controlled trial. *Journal of Neurology Neurosurgery and Psychiatry*, ***74***, 863–866.
1167. Martínez, A. L., González-Trujano, M. E., Chávez, M., and Pellicer, F. (2012). Anti-nociceptive effectiveness of triterpenes from rosemary in visceral nociception. *Journal of Ethnopharmacology*, ***142***, 28–34.
1168. Begum, A., Sandhya, S., Ali, S. S., Vinod, K. R., Reddy, S., and Banji, D. (2013). An in-depth review on the medicinal flora *Rosmarinus officinalis* (Lamiaceae). *Acta Scientiarum Polonorum Technologia Alimentaria*, ***12***, 61–73.
1169. Miraldi, E., Giachetti, D., Mazzoni, G., and Biagi, M. (2012). Quali-quantitative analysis of eight *Rosmarinus officinalis* essential oils of different origin. First report. *Journal of the Siena Academy of Sciences*, ***2***, 42–43.

1170. Jiang, Y., Wu, N., Fu, Y.-J., Wang, W., Luo, M., Zhao, C.-J., et al. (2011). Chemical composition and antimicrobial activity of the essential oil of rosemary. *Environmental Toxicology and Pharmacology*, ***32***, 63–68.
1171. Pengelly, A., Snow, J., Mills, S. Y., Scholey, A., Wesnes, K., and Butler, L. R. (2012). Short-term study on the effects of rosemary on cognitive function in an elderly population. *Journal of Medicinal Food*, ***15***, 10–17.
1172. Okamura, N., Haraguchi, H., Hashimoto, K., and Yagi, A. (1994). Flavonoids in *Rosmarinus officinalis* leaves. *Phytochemistry*, ***37***, 1463–1466.
1173. Juhás, Š., Bukovská, A., Čikoš, Š., Czikková, S., Fabian, D., and Koppel, J. (2009). Anti-inflammatory effects of *Rosmarinus officinalis* essential oil in mice. *Acta Veterinaria Brno*, ***78***, 121–127.
1174. Zendehdel, M., Beizaee, S., Taati, M., and Bashiri, A. (2011). Antinociceptive mechanisms of *Rosmarinus officinalis* extract in mice using writhing test. *International Journal of Veterinary Research*, **5**, 240–246.
1175. Martínez, A. L., González-Trujano, M. E., Pellicer, F., López-Muñoz, F. J., and Navarrete, A. (2009). Antinociceptive effect and GC/MS analysis of *Rosmarinus officinalis* L. essential oil from its aerial parts. *Planta Medica*, ***75***, 508–511.
1176. Hosseinzadeh, H., and Nourbakhsh, M. (2003). Effect of *Rosmarinus officinalis* L. aerial parts extract on morphine withdrawal syndrome in mice. *Phytotherapy Research*, ***17***, 938–941.
1177. Ventura-Martínez, R., Rivero-Osorno, O., Gómez, C., and González-Trujano, M. E. (2011). Spasmolytic activity of *Rosmarinus officinalis* L. involves calcium channels in the guinea pig ileum. *Journal of Ethnopharmacology*, ***137***, 1528–1532.
1178. Orhan, I., Aslan, S., Kartal, M., Şener, B., and Hüsnü Can Başer, K. (2008). Inhibitory effect of Turkish *Rosmarinus officinalis* L. on acetylcholinesterase and butyrylcholinesterase enzymes. *Food Chemistry*, ***108***, 663–668.
1179. Machado, D. G., Cunha, M. P., Neis, V. B., Balen, G. O., Colla, A. R., Grando, J., et al. (2012). *Rosmarinus officinalis* L. hydroalcoholic extract, similar to fluoxetine, reverses depressive-like behavior without altering learning deficit in olfactory bulbectomized mice. *Journal of Ethnopharmacology*, ***143***, 158–169.
1180. Hosseinzadeh, H., Karimi, G., and Noubakht, M. (2004). Effects of *Rosmarinus officinalis* L. aerial parts essential oil on intact memory and scopolamine-induced learning deficits in rats performing the morris water maze task. *Journal of Medicinal Plants*, **3**, 68–68.
1181. Zanella, C. A., Treichel, H., Cansian, R. L., and Roman, S. S. (2012). The effects of acute administration of the hydroalcoholic extract of rosemary (*Rosmarinus officinalis* L.) (Lamiaceae) in animal models of memory. *Brazilian Journal of Pharmaceutical Sciences*, ***48***, 389–397.
1182. Diego, M. A., Jones, N. A., Field, T., Hernandez-Reif, M., Schanberg, S., Kuhn, C., et al. (1998). Aromatherapy positively affects mood, EEG patterns of alertness and math computations. *International Journal of Neuroscience*, ***96***, 217–224.
1183. Moss, M., Cook, J., Wesnes, K., and Duckett, P. (2003). Aromas of rosemary and lavender essential oils differentially affect cognition and mood in healthy adults. *International Journal of Neuroscience*, ***113***, 15–38.
1184. Moss, M., and Oliver, L. (2012). Plasma 1, 8-cineole correlates with cognitive performance following exposure to rosemary essential oil aroma. *Therapeutic Advances in Psychopharmacology*, **2**, 103–113.
1185. Rozman, V., Kalinovic, I., and Korunic, Z. (2007). Toxicity of naturally occurring compounds of Lamiaceae and Lauraceae to three stored-product insects. *Journal of Stored Products Research*, ***43***, 349–355.

1186. Ntalli, N., Ferrari, F., Giannakou, I., and Menkissoglu-Spiroudi, U. (2010). Phytochemistry and nematicidal activity of the essential oils from 8 Greek Lamiaceae aromatic plants and 13 terpene components. *Journal of Agricultural and Food Chemistry*, ***58***, 7856–7863.
1187. Koliopoulos, G., Pitarokili, D., Kioulos, E., Michaelakis, A., and Tzakou, O. (2010). Chemical composition and larvicidal evaluation of *Mentha*, *Salvia*, and *Melissa* essential oils against the West Nile virus mosquito *Culex pipiens*. *Parasitology Research*, ***107***, 327–335.
1188. Yildirim, E., Kordali, S., and Yazici, G. (2011). Insecticidal effects of essential oils of eleven plant species from Lamiaceae on *Sitophilus granarius* (L.) (Coleoptera: Curculionidae). *Romanian Biotechnological Letters*, ***16***, 6702–6709.
1189. Oshaghi, M., Ghalandari, R., Vatandoost, H., Shayeghi, M., Kamali-Nejad, M., Tourabi-Khaledi, H., et al. (2003). Repellent effect of extracts and essential oil of *Citrus limon* (Rutaceae) and *Melissa officinalis* (Labiatae) against main malaria vector, *Anopheles stephensi* (Diptera: Culicidae) in Iran. *Iran J Public Health*, ***32***, 47–52.
1190. de Almeida, L. F. R., Frei, F., Mancini, E., De Martino, L., and De Feo, V. (2010). Phytotoxic activities of Mediterranean essential oils. *Molecules*, ***15***, 4309–4323.
1191. Kordali, S., Kesdek, M., and Cakir, A. (2007). Toxicity of monoterpenes against larvae and adults of Colorado potato beetle, *Leptinotarsa decemlineata* Say (Coleoptera: Chrysomelidae). *Industrial Crops and Products*, ***26***, 278–297.
1192. Stamopoulos, D. C., Damos, P., and Karagianidou, G. (2007). Bioactivity of five monoterpenoid vapours to *Tribolium confusum* (du Val) (Coleoptera: Tenebrionidae). *Journal of Stored Products Research*, ***43***, 571–577.
1193. Schmidt, S., and Walter, G. H. (2011). Adapting to cope with eucalypt oils: mandibular extensions in pergid sawfly larvae and potential preadaptations in its sister family *Argidae* (Insecta, Hymenoptera, Symphyta). *Journal of Morphology*, ***272***, 1314–1324.
1194. Damiani, N., Gende, L. B., Bailac, P., Marcangeli, J. A., and Eguaras, M. J. (2009). Acaricidal and insecticidal activity of essential oils on *Varroa destructor* (Acari: Varroidae) and *Apis mellifera* (Hymenoptera: Apidae). *Parasitology Research*, ***106***, 145–152.
1195. Rohloff, J., and Bones, A. (2005). Volatile profiling of *Arabidopsis thaliana* - Putative olfactory compounds in plant communication. *Phytochemistry*, ***66***, 1941–1955.
1196. Aazza, S., Lyoussi, B., and Miguel, M. G. (2011). Antioxidant and antiacetylcholinesterase activities of some commercial essential oils and their major compounds. *Molecules*, ***16***, 7672–7690.
1197. Mallavadhani, U. V., Mahapatra, A., Raja, S. S., and Manjula, C. (2003). Antifeedant activity of some pentacyclic triterpene acids and their fatty acid ester analogues. *Journal of Agricultural and Food Chemistry*, ***51***, 1952–1955.
1198. Chandramu, C., Manohar, R. D., Krupadanam, D. G. L., and Dashavantha, R. V. (2003). Isolation, characterization and biological activity of betulinic acid and ursolic acid from *Vitex negundo* L. *Phytotherapy Research*, ***17***, 129–134.
1199. Varanda, E. M., Zúñiga, G. E., Salatino, A., Roque, N. F., and Corcuera, L. J. (1992). Effect of ursolic acid from epicuticular waxes of *Jacaranda decurrens* on *Schizaphis graminum*. *Journal of Natural Products*, ***55***, 800–803.
1200. Muñoz, E., Escalona, D., Salazar, J. R., Alarcon, J., and Céspedes, C. L. (2013). Insect growth regulatory effects by diterpenes from *Calceolaria talcana* Grau & Ehrhart (Calceolariaceae: Scrophulariaceae) against *Spodoptera frugiperda* and *Drosophila melanogaster*. *Industrial Crops and Products*, ***45***, 283–292.
1201. Machado, D. G., Neis, V. B., Balen, G. O., Colla, A., Cunha, M. P., Dalmarco, J. B., et al. (2012). Antidepressant-like effect of ursolic acid isolated from *Rosmarinus officinalis* L. in mice: Evidence for the involvement of the dopaminergic system. *Pharmacology Biochemistry and Behavior*, ***103***, 204–211.

1202. Siebert, D. J. (2004). Localization of salvinorin A and related compounds in glandular trichomes of the psychoactive sage, *Salvia divinorum*. *Annals of Botany*, ***93***, 763–771.
1203. Klein Gebbinck, E. A., Jansen, B. J., and de Groot, A. (2002). Insect antifeedant activity of clerodane diterpenes and related model compounds. *Phytochemistry*, ***61***, 737–770.
1204. Valdes, L. J. (1986). Loliolide from *Salvia divinorum*. *Journal of Natural Products*, ***49***, 171–171.
1205. Clarke, R. C., and Watson, D. P. (2007). Cannabis and natural cannabis medicines. In M. A. ElSohly (Ed.), *Marijuana and the Cannabinoids* (pp. 1–15). Towtowa, New Jersey: Humana Press Inc.
1206. Chandra, S., Lata, H., Galal, A., Khan, I., and ElSohly, M. (2011). Botany of *Cannabis sativa* L.: identification, cultivation and processing. *Planta Medica*, ***77***, P_4.
1207. Mehmedic, Z., Chandra, S., Slade, D., Denham, H., Foster, S., Patel, A. S., et al. (2010). Potency trends of Δ9-THC and other cannabinoids in confiscated cannabis preparations from 1993 to 2008. *Journal of Forensic Sciences*, ***55***, 1209–1217.
1208. Sirikantaramas, S., Taura, F., Tanaka, Y., Ishikawa, Y., Morimoto, S., and Shoyama, Y. (2005). Tetrahydrocannabinolic acid synthase, the enzyme controlling marijuana psychoactivity, is secreted into the storage cavity of the glandular trichomes. *Plant and Cell Physiology*, ***46***, 1578–1582.
1209. Taura, F., Sirikantaramas, S., Shoyama, Y., Yoshikai, K., Shoyama, Y., and Morimoto, S. (2007). Cannabidiolic-acid synthase, the chemotype-determining enzyme in the fiber-type *Cannabis sativa*. *FEBS Letters*, ***581***, 2929–2934.
1210. Brenneisen, R. (2007). Chemistry and analysis of phytocannabinoids and other Cannabis constituents. In M. A. ElSohly (Ed.), *Marijuana and the Cannabinoids* (pp. 17–49). Totowa, New Jersey: Humana Press.
1211. Hillig, K. W., and Mahlberg, P. G. (2004). A chemotaxonomic analysis of cannabinoid variation in *Cannabis* (Cannabaceae). *American Journal of Botany*, ***91***, 966–975.
1212. Fellermeier, M., Eisenreich, W., Bacher, A., and Zenk, M. H. (2001). Biosynthesis of cannabinoids. *European Journal of Biochemistry*, ***268***, 1596–1604.
1213. Brown, A. (2009). Novel cannabinoid receptors. *British Journal of Pharmacology*, ***152***, 567–575.
1214. Gaoni, Y., and Mechoulam, R. (1964). Isolation, structure, and partial synthesis of an active constituent of hashish. *Journal of the American Chemical Society*, ***86***, 1646–1647.
1215. Devane, W. A., Dysarz, F. A., Johnson, M. R., Melvin, L. S., and Howlett, A. C. (1988). Determination and characterization of a cannabinoid receptor in rat brain. *Molecular Pharmacology*, ***34***, 605–613.
1216. Devane, W. A., Hanus, L., Breuer, A., Pertwee, R. G., Stevenson, L. A., Griffin, G., et al. (1992). Isolation and structure of a brain constituent that binds to the cannabinoid receptor. *Science (New York, NY)*, ***258***, 1946.
1217. McPartland, J. M., Agraval, J., Gleeson, D., Heasman, K., and Glass, M. (2006). Cannabinoid receptors in invertebrates. *Journal of Evolutionary Biology*, ***19***, 366–373.
1218. Gyombolai, P., Pap, D., Turu, G., Catt, K. J., Bagdy, G., and Hunyady, L. (2012). Regulation of endocannabinoid release by G proteins: A paracrine mechanism of G-protein-coupled receptor action. *Molecular and Cellular Endocrinology*, ***353***, 29–36.
1219. Tsou, K., Brown, S., Sanudo-Pena, M. C., Mackie, K., and Walker, J. M. (1998). Immunohistochemical distribution of cannabinoid CB1 receptors in the rat central nervous system. *Neuroscience*, ***83***, 393–411.
1220. Rivers, J. R.-J., and Ashton, J. C. (2010). The development of cannabinoid CBII receptor agonists for the treatment of central neuropathies. *Central Nervous System Agents in Medicinal Chemistry*, ***10***, 47–64.

1221. Fride, E. (2002). Endocannabinoids in the central nervous system—an overview. *Prostaglandins Leukotrienes and Essential Fatty Acids*, ***66***, 221–234.
1222. D'Souza, D. C., Sewell, R. A., and Ranganathan, M. (2009). Cannabis and psychosis/schizophrenia: human studies. *European Archives of Psychiatry and Clinical Neuroscience*, ***259***, 413–431.
1223. Green, B. O. B., Kavanagh, D., and Young, R. (2003). Being stoned: a review of self-reported cannabis effects. *Drug and Alcohol Review*, ***22***, 453–460.
1224. Oleson, E. B., and Cheer, J. F. (2012). A brain on cannabinoids: the role of dopamine release in reward seeking. *Cold Spring Harbor Perspectives in Medicine*, ***2***.
1225. Maldonado, R., Berrendero, F., Ozaita, A., and Robledo, P. (2011). Neurochemical basis of cannabis addiction. *Neuroscience*, ***181***, 1–17.
1226. Hill, A. J., Williams, C. M., Whalley, B. J., and Stephens, G. J. (2012). Phytocannabinoids as novel therapeutic agents in CNS disorders. *Pharmacology & Therapeutics*, ***133***, 79–97.
1227. Russo, E. B. (2011). Taming THC: potential cannabis synergy and phytocannabinoid-terpenoid entourage effects. *British Journal of Pharmacology*, ***163***, 1344–1364.
1228. Russo, E., and Guy, G. W. (2006). A tale of two cannabinoids: The therapeutic rationale for combining tetrahydrocannabinol and cannabidiol. *Medical Hypotheses*, ***66***, 234–246.
1229. Fusar-Poli, P., Crippa, J. A., Bhattacharyya, S., Borgwardt, S. J., Allen, P., Martin-Santos, R., et al. (2009). Distinct effects of delta-9-tetrahydrocannabinol and cannabidiol on neural activation during emotional processing. *Archives of General Psychiatry*, ***66***, 95.
1230. Bhattacharyya, S., Morrison, P. D., Fusar-Poli, P., Martin-Santos, R., Borgwardt, S., Winton-Brown, T., et al. (2009). Opposite effects of δ-9-tetrahydrocannabinol and cannabidiol on human brain function and psychopathology. *Neuropsychopharmacology*, ***35***, 764–774.
1231. Bisogno, T., Hanuš, L., De Petrocellis, L., Tchilibon, S., Ponde, D. E., Brandi, I., et al. (2009). Molecular targets for cannabidiol and its synthetic analogues: effect on vanilloid VR1 receptors and on the cellular uptake and enzymatic hydrolysis of anandamide. *British Journal of Pharmacology*, ***134***, 845–852.
1232. Carrier, E. J., Auchampach, J. A., and Hillard, C. J. (2006). Inhibition of an equilibrative nucleoside transporter by cannabidiol: A mechanism of cannabinoid immunosuppression. *Proceedings of the National Academy of Sciences of the United States of America*, ***103***, 7895–7900.
1233. Russo, E. B., Burnett, A., Hall, B., and Parker, K. K. (2005). Agonistic properties of cannabidiol at 5-HT1a receptors. *Neurochemical Research*, ***30***, 1037–1043.
1234. Zanelati, T., Biojone, C., Moreira, F., Guimaraes, F., and Joca, S. (2009). Antidepressant-like effects of cannabidiol in mice: possible involvement of 5-HT1A receptors. *British Journal of Pharmacology*, ***159***, 122–128.
1235. Johnson, J. R., Burnell-Nugent, M., Lossignol, D., Ganae-Motan, E. D., Potts, R., and Fallon, M. T. (2010). Multicenter, double-blind, randomized, placebo-controlled, parallel-group study of the efficacy, safety, and tolerability of THC:CBD extract and THC extract in patients with intractable cancer-related pain. *Journal of Pain and Symptom Management*, ***39***, 167–179.
1236. Mechoulam, R. (2009). Plant cannabinoids: a neglected pharmacological treasure trove. *British Journal of Pharmacology*, ***146***, 913–915.
1237. Zhou, W., Yoshioka, M., and Yokogoshi, H. (2009). Sub-chronic effects of s-limonene on brain neurotransmitter levels and behavior of rats. *Journal of nutritional science and vitaminology*, ***55***, 367–373.

1238. de Almeida, A. A. C., Costa, J. P., de Carvalho, R. B. F., de Sousa, D. P., and de Freitas, R. M. (2012). Evaluation of acute toxicity of a natural compound (+)-limonene epoxide and its anxiolytic-like action. *Brain Research*, ***1448***, 56–62.
1239. McPartland, J. M., Blanchon, D. J., and Musty, R. E. (2008). Cannabimimetic effects modulated by cholinergic compounds. *Addiction Biology*, ***13***, 411–415.
1240. SAMHSA. (2005). Summary of Findings from the 2005 National Household Survey on Drug Abuse. *http://www.samhsa.gov*.
1241. Anthony, J. C., Warner, L. A., and Kessler, R. C. (1994). Comparative epidemiology of dependence on tobacco, alcohol, controlled substances, and inhalants: basic findings from the National Comorbidity Survey. *Experimental and Clinical Psychopharmacology*, ***2***, 244.
1242. Trigo, J. M., Martin-García, E., Berrendero, F., Robledo, P., and Maldonado, R. (2010). The endogenous opioid system: a common substrate in drug addiction. *Drug and Alcohol Dependence*, ***108***, 183–194.
1243. Malouf, R., and Grimley Evans, J. (2003). Vitamin B6 for cognition. *Cochrane Database of Systematic Reviews* ***CD004393***.
1244. Tanda, G., and Goldberg, S. R. (2003). Cannabinoids: reward, dependence, and underlying neurochemical mechanisms-a review of recent preclinical data. *Psychopharmacology*, ***169***, 115–134.
1245. van Hell, H. H., Jager, G., Bossong, M. G., Brouwer, A., Jansma, J. M., Zuurman, L., et al. (2012). Involvement of the endocannabinoid system in reward processing in the human brain. *Psychopharmacology*, ***219***, 981–990.
1246. Bossong, M. G., van Berckel, B. N. M., Boellaard, R., Zuurman, L., Schuit, R. C., Windhorst, A. D., et al. (2008). Δ9-Tetrahydrocannabinol induces dopamine release in the human striatum. *Neuropsychopharmacology*, ***34***, 759–766.
1247. Serrano, A., and Parsons, L. H. (2011). Endocannabinoid influence in drug reinforcement, dependence and addiction-related behaviors. *Pharmacology & Therapeutics*, ***132***, 215–241.
1248. Katsidoni, V., Anagnostou, I., and Panagis, G. (2012). Cannabidiol inhibits the reward-facilitating effect of morphine: involvement of 5-HT1A receptors in the dorsal raphe nucleus. *Addiction Biology*, n/a-n/a.
1249. van Hell, H. H., Vink, M., Ossewaarde, L., Jager, G., Kahn, R. S., and Ramsey, N. F. (2010). Chronic effects of cannabis use on the human reward system: An fMRI study. *European Neuropsychopharmacology*, ***20***, 153–163.
1250. Pate, D. W. (1994). Chemical ecology of Cannabis. *Journal of the International Hemp Association*, ***1***, 29–32.
1251. Eisohly, H. N., Turner, C. E., Clark, A. M., and Eisohly, M. A. (1982). Synthesis and antimicrobial activities of certain cannabichromene and cannabigerol related compounds. *Journal of Pharmaceutical Sciences*, ***71***, 1319–1323.
1252. McPartland, J. (1984). Pathogenicity of *Phomopsis ganjae* on *Cannabis sativa* and the fungistatic effect of cannabinoids produced by the host. *Mycopathologia*, ***87***, 149–153.
1253. McPartland, J. M. (1997). Cannabis as repellent and pesticide. *Journal of the International Hemp Association*, ***4***, 87–92.
1254. Gottwald, R. (2002). Entomological studies on hemp (*Cannabis sativa* L.). *Gesunde Pflanzen*, ***54***, 146–152.
1255. Jalees, S., Sharma, S. K., Rahman, S. J., and Verghese, T. (1993). Evaluation of insecticidal properties of an indigenous plant, *Cannabis sativa* Linn., against mosquito larvae under laboratory conditions. *Journal of Entomological Research (New Delhi)* ***17***, 117–120.

1256. Badshah, H., Khan, A. S., Farid, A., Zeb, A., and Khan, A. (2005). Toxic effects of palpoluck *Polygonum hydropepper* L. and Bhang *Cannabis sativa* L. plants extracts against termites *Heterotermes indicola* (Wasmann) and *Coptotermes heimi* (Wasmann) (Isoptera: Rhinotermitidae). *Songklanakarin Journal of Science and Technology*, ***27*** 705–710.

1257. Mukhtar, T., Kayani, M. Z., and Hussain, M. A. (2013). Nematicidal activities of *Cannabis sativa* L. and *Zanthoxylum alatum* Roxb. against *Meloidogyne incognita*. *Industrial Crops and Products*, ***42***, 447–453.

1258. Rothschild, M., Rowan, M., and Fairbairn, J. (1977). Storage of cannabinoids by *Arctia caja* and *Zonocerus elegans* fed on chemically distinct strains of *Cannabis sativa.Nature*, ***266***, 650–651.

1259. Rothschild, M., and Fairbairn, J. (1980). Ovipositing butterfly (*Pieris brassicae* L.) distinguishes between aqueous extracts of two strains of *Cannabis sativa* L. and THC and CBD. *Nature, UK*, ***286***, 56–59.

1260. Bolognini, D., Rock, E. M., Cluny, N. L., Cascio, M. G., Limebeer, C. L., Duncan, M., et al. (2013). Cannabidiolic acid prevents vomiting in *Suncus murinus* and nausea-induced behaviour in rats by enhancing 5-HT1A receptor activation. *British Journal of Pharmacology*, ***168***, 1456–1470.

1261. Eichler, M., Spinedi, L., Unfer-Grauwiler, S., Bodmer, M., Surber, C., Luedi, M., et al. (2012). Heat exposure of *Cannabis sativa* extracts affects the pharmacokinetic and metabolic profile in healthy male subjects. *Planta Medica*, ***78***, 686–691.

1262. Panossian, A., and Wikman, G. (2009). Evidence-based efficacy of adaptogens in fatigue, and molecular mechanisms related to their stress-protective activity. *Current Clinical Pharmacology*, ***4***, 198.

1263. Pawar, V. S., and Shivakumar, H. (2012). A current status of adaptogens: natural remedy to stress. *Asian Pacific Journal of Tropical Disease*, ***2, Supplement 1***, S480–S490.

1264. Buckingham, J. C. (2006). Glucocorticoids: exemplars of multi-tasking. *British Journal of Pharmacology*, ***147***, S258–S268.

1265. Francis, G., Kerem, Z., Makkar, H. P. S., and Becker, K. (2002). The biological action of saponins in animal systems: a review. *British Journal of Nutrition*, ***88***, 587–605.

1266. Wen, J., and Zimmer, E. A. (1996). Phylogeny and biogeography of *Panax* L. (the Ginseng Genus, Araliaceae): Inferences from ITS sequences of nuclear ribosomal DNA. *Molecular Phylogenetics and Evolution*, ***6***, 167–177.

1267. Choi, H. I., Kim, N. H., Lee, J., Choi, B. S., Kim, K. D., Park, J. Y., et al. (2012). Evolutionary relationship of *Panax ginseng* and *P. quinquefolius* inferred from sequencing and comparative analysis of expressed sequence tags. *Genetic Resources and Crop Evolution*, 1–11.

1268. Nair, R., Sellaturay, S., and Sriprasad, S. (2012). The history of ginseng in the management of erectile dysfunction in ancient China (3500–2600 BCE). *Indian Journal of Urology*, ***28***, 15.

1269. Yun, T. (2001). Brief introduction of *Panax ginseng* CA Meyer. *Journal of Korean Medical Science*, ***16***, 3–5.

1270. Cho, I. H. (2012). Effects of *Panax ginseng* in neurodegenerative diseases. *Journal of Ginseng Research*, ***36***, 342–353.

1271. Lu, J. M., Yao, Q. Z., and Chen, C. Y. (2009). Ginseng compounds: an update on their molecular mechanisms and medical applications. *Current Vascular Pharmacology*, ***7***, 293–302.

1272. Chen, C. F., Chiou, W. F., and Zhang, J. T. (2008). Comparison of the pharmacological effects of *Panax ginseng* and *Panax quinquefolium*. *Acta Pharmacologica Sinica*, ***29***, 1103–1108.

1273. Kennedy, D. O., and Scholey, A. B. (2003). Ginseng: potential for the enhancement of cognitive performance and mood. *Pharmacology Biochemistry and Behavior*, ***75***, 687–700.
1274. Jia, Y. L., Zhang, S. K., Huang, F. Y., and Leung, S. W. (2012). Could ginseng-based medicines be better than nitrates in treating ischemic heart disease? A systematic review and meta-analysis of randomized controlled trials. *Complementary Therapies in Medicine*, ***20***, 155–166.
1275. Jang, D. J., Lee, M. S., Shin, B. C., Lee, Y. C., and Ernst, E. (2008). Red ginseng for treating erectile dysfunction: a systematic review. *British Journal of Clinical Pharmacology*, ***66***, 444–450.
1276. Kim, H. G., Yoo, S. R., Park, H. J., Lee, N. H., Shin, J. W., Sathyanath, R., et al. (2011). Antioxidant effects of *Panax ginseng* C.A. Meyer in healthy subjects: A randomized, placebo-controlled clinical trial. *Food and Chemical Toxicology*, ***49***, 2229–2235.
1277. Coleman, C., Hebert, J., and Reddy, P. (2003). The effects of *Panax ginseng* on quality of life. *Journal of Clinical Pharmacy and Therapeutics*, ***28***, 5–15.
1278. Dang, H. X., Chen, Y., Liu, X. M., Wang, Q., Wang, L. W., Jia, W., et al. (2009). Antidepressant effects of ginseng total saponins in the forced swimming test and chronic mild stress models of depression. *Progress in Neuro-Psychopharmacology & Biological Psychiatry*, ***33***, 1417–1424.
1279. Jesky, R., and Hailong, C. (2011). Are herbal compounds the next frontier for alleviating learning and memory impairments? An integrative look at memory, dementia and the promising therapeutics of traditional chinese medicines. *Phytotherapy Research*, ***25***, 1105–1118.
1280. Kennedy, D. O., Scholey, A. B., and Wesnes, K. A. (2001). Differential, dose-dependent changes in cognitive performance following acute administration of a *Ginkgo biloba/Panax ginseng* combination to healthy young volunteers. *Nutritional Neuroscience*, ***4***, 399–412.
1281. Kennedy, D. O., Scholey, A. B., and Wesnes, K. A. (2002). Modulation of cognition and mood following administration of single doses of *Ginkgo biloba*, ginseng, and a ginkgo/ginseng combination to healthy young adults. *Physiology & Behavior*, ***75***, 739–751.
1282. Kennedy, D. O., Haskell, C. F., Wesnes, K. A., and Scholey, A. B. (2004). Improved cognitive performance in human volunteers following administration of guarana (*Paullinia cupana*) extract: comparison and interaction with *Panax ginseng*. *Pharmacology Biochemistry and Behavior*, ***79***, 401–411.
1283. Sünram-Lea, S., Birchall, R., Wesnes, K., and Petrini, O. (2005). The effect of acute administration of 400 mg of *Panax ginseng* on cognitive performance and mood in healthy young volunteers. *Current Topics in Nutraceutical Research*, ***3***, 65–74.
1284. Kennedy, D. O., Scholey, A. B., Drewery, L., Marsh, V. R., Moore, B., and Ashton, H. (2003). Electroencephalograph effects of single doses of *Ginkgo biloba* and *Panax ginseng* in healthy young volunteers. *Pharmacology Biochemistry and Behavior*, ***75***, 701–709.
1285. Reay, J. L., Kennedy, D. O., and Scholey, A. B. (2005). Single doses of *Panax ginseng* (G115) reduce blood glucose levels and improve cognitive performance during sustained mental activity. *Journal of Psychopharmacology*, ***19***, 357–365.
1286. Reay, J. L., Kennedy, D. O., and Scholey, A. B. (2006). Effects of *Panax ginseng*, consumed with and without glucose, on blood glucose levels and cognitive performance during sustained "mentally demanding" tasks. *Journal of Psychopharmacology*, ***20***, 771.
1287. Reay, J. L., Scholey, A. B., and Kennedy, D. O. (2010). *Panax ginseng* (G115) improves aspects of working memory performance and subjective ratings of calmness in

healthy young adults. *Human Psychopharmacology-Clinical and Experimental*, ***25***, 462–471.

1288. Kennedy, D., Reay, J., and Scholey, A. (2007). Effects of 8 weeks administration of Korean *Panax ginseng* extract on the mood and cognitive performance of healthy individuals. *Journal of Ginseng Research*, ***31***, 34–43.
1289. Scholey, A., Ossoukhova, A., Owen, L., Ibarra, A., Pipingas, A., He, K., et al. (2010). Effects of American ginseng (*Panax quinquefolius*) on neurocognitive function: an acute, randomised, double-blind, placebo-controlled, crossover study. *Psychopharmacology*, ***212***, 345–356.
1290. Lee, M. S., Yang, E. J., Kim, J. I., and Ernst, E. (2009). Ginseng for cognitive function in Alzheimer's disease: A systematic review. *Journal of Alzheimers Disease*, ***18***, 339–344.
1291. Iqbal, M., and Datta, A. K. (2007). Ashwagandha (*Withania somnifera*)—The elixir of life. *Plant Archives*, ***7***, 449–455.
1292. Mirjalili, M. H., Moyano, E., Bonfill, M., Cusido, R. M., and Palazon, J. (2009). Steroidal lactones from *Withania somnifera*, an ancient plant for novel medicine. *Molecules*, ***14***, 2373–2393.
1293. Chaurasiya, N. D., Uniyal, G. C., Lal, P., Misra, L., Sangwan, N. S., Tuli, R., et al. (2007). Analysis of withanolides in root and leaf of *Withania somnifera* by HPLC with photodiode array and evaporative light scattering detection. *Phytochemical Analysis*, ***19***, 148–154.
1294. Singh, N., Bhalla, M., de Jager, P., and Gilca, M. (2011). An overview on Ashwagandha: a rasayana (rejuvenator) of ayurveda. *African Journal of Traditional, Complementary and Alternative Medicines*, ***8***.
1295. Alam, N., Hossain, M., Khalil, M. I., Moniruzzaman, M., Sulaiman, S. A., and Gan, S. H. (2012). Recent advances in elucidating the biological properties of *Withania somnifera* and its potential role in health benefits. *Phytochemistry Reviews*, 1–16.
1296. Berghe, W. V., Sabbe, L., Kaileh, M., Haegeman, G., and Heyninck, K. (2012). Molecular insight in the multifunctional activities of withaferin A. *Biochemical Pharmacology*.
1297. Baitharu, I., Jain, V., Deep, S. N., Hota, K. B., Hota, S. K., Prasad, D., et al. (2013). *Withania somnifera* root extract ameliorates hypobaric hypoxia-induced memory impairment in rats. *Journal of Ethnopharmacology*, ***145***, 431–441.
1298. Kiasalari, Z., Khalili, M., and Aghaei, M. (2009). Effect of withania somnifera on levels of sex hormones in the diabetic male rats. *Iranian Journal of Reproductive Medicine*, ***7***, 163–168.
1299. Ahmad, M. K., Mahdi, A. A., Shukla, K. K., Islam, N., Rajender, S., Madhukar, D., et al. (2010). *Withania somnifera* improves semen quality by regulating reproductive hormone levels and oxidative stress in seminal plasma of infertile males. *Fertility and Sterility*, ***94***, 989–996.
1300. Mehta, A., Binkley, P., Gandhi, S., and Ticku, M. (1991). Pharmacological effects of *Withania somnifera* root extract on GABAA receptor complex. *The Indian Journal of Medical Research*, ***94***, 312.
1301. Bhattarai, J. P., Ah Park, S., and Han, S. K. (2010). The methanolic extract of *Withania somnifera* ACTS on GABAA receptors in gonadotropin releasing hormone (GnRH) neurons in mice. *Phytotherapy Research*, ***24***, 1147–1150.
1302. Kulkarni, S., Akula, K. K., and Dhir, A. (2008). Effects of *Withania sonmifera* Dunal root extract against pentylenetetrazol seizure threshold in mice: Possible involvement of GABAergic system. *Indian Journal of Experimental Biology*, ***46***, 465.
1303. Zhang, Z.-J. (2004). Therapeutic effects of herbal extracts and constituents in animal models of psychiatric disorders. *Life Sciences*, ***75***, 1659–1700.

1304. Khan, Z., and Ghosh, A. (2011). Withaferin-A displays enhanced anxiolytic efficacy without tolerance in rats following sub chronic administration. *African Journal of Biotechnology*, ***10***, 12973–12978.

1305. Baitharu, I., Deep, S. N., Jain, V., Prasad, D., and Ilavazhagan, G. (2013). Inhibition of glucocorticoid receptors ameliorates hypobaric hypoxia-induced memory impairment in rat. *Behavioural Brain Research*, ***240***, 76–86.

1306. Soman, S., Korah, P., Jayanarayanan, S., Mathew, J., and Paulose, C. (2012). Oxidative stress-induced NMDA receptor alteration leads to spatial memory deficits in temporal lobe epilepsy: Ameliorative effects of *Withania somnifera* and withanolide A. *Neurochemical Research*, ***37***, 1915–1927.

1307. Andrade, C., Aswath, A., Chaturvedi, S., Srinivasa, M., and Raguram, R. (2000). A double-blind, placebo-controlled evaluation of the anxiolytic efficacy of an ethanolic extract of *Withania somnifera*. *Indian Journal of Psychiatry*, ***42***, 295.

1308. Chandrasekhar, K., Kapoor, J., and Anishetty, S. (2012). A prospective, randomized double-blind, placebo-controlled study of safety and efficacy of a high-concentration full-spectrum extract of Ashwagandha root in reducing stress and anxiety in adults. *Indian Journal of Psychological Medicine*, ***34***, 255.

1309. Cooley, K., Szczurko, O., Perri, D., Mills, E. J., Bernhardt, B., Zhou, Q., et al. (2009). Naturopathic care for anxiety: a randomized controlled trial ISRCTN78958974. *PLoS One*, ***4***, e6628.

1310. Russo, A., and Borrelli, F. (2005). *Bacopa monniera*, a reputed nootropic plant: an overview. *Phytomedicine*, ***12***, 305–317.

1311. Deepak, M., Sangli, G., Arun, P., and Amit, A. (2005). Quantitative determination of the major saponin mixture bacoside A in *Bacopa monnieri* by HPLC. *Phytochemical Analysis*, ***16***, 24–29.

1312. Viji, V., and Helen, A. (2011). Inhibition of pro-inflammatory mediators: role of *Bacopa monniera* (L.) Wettst. *Inflammopharmacology*, ***19***, 283–291.

1313. Saraf, M. K., Prabhakar, S., Khanduja, K. L., and Anand, A. (2011). *Bacopa monniera* attenuates scopolamine-induced impairment of spatial memory in mice. *Evidence-Based Complementary and Alternative Medicine*, ***2011***.

1314. Abbas, M., Subhan, F., Mohani, N., Rauf, K., Ali, G., and Khan, M. (2011). The involvement of opioidergic mechanisms in the activity of *Bacopa monnieri* extract and its toxicological studies. *African Journal of Pharmacy and Pharmacology*, ***5***, 1120–1124.

1315. Prabhakar, S., Saraf, M. K., Pandhi, P., and Anand, A. (2008). *Bacopa monniera* exerts antiamnesic effect on diazepam-induced anterograde amnesia in mice. *Psychopharmacology*, ***200***, 27–37.

1316. Mathew, J., Balakrishnan, S., Antony, S., Abraham, P. M., and Paulose, C. S. (2012). Decreased GABA receptor in the cerebral cortex of epileptic rats: effect of *Bacopa monnieri* and Bacoside-A. *Journal of Biomedical Science*, ***19***.

1317. Sheikh, N., Ahmad, A., Siripurapu, K. B., Kuchibhotla, V. K., Singh, S., and Palit, G. (2007). Effect of *Bacopa monniera* on stress-induced changes in plasma corticosterone and brain monoamines in rats. *Journal of Ethnopharmacology*, ***111***, 671–676.

1318. Kumar, T., Srivastav, M., and Singh, R. (2011). Randomized control double blind crossover study to clinically asses the effect of Brahmi (*Bacopa monniera*) on quality of life in human volunteers. *Journal of Pharmacy Research*, ***4***, 1213–1215.

1319. Nathan, P. J., Clarke, J., Lloyd, J., Hutchison, C., Downey, L., and Stough, C. (2001). The acute effects of an extract of *Bacopa monniera* (Brahmi) on cognitive function in healthy normal subjects. *Human Psychopharmacology: Clinical and Experimental*, ***16***, 345–351.

1320. Downey, L. A., Kean, J., Nemeh, F., Lau, A., Poll, A., Gregory, R., et al. (2012). An acute, double-blind, placebo-controlled crossover study of 320-mg and 640-mg doses of a special extract of *Bacopa monnieri* (CDRI 08) on sustained cognitive performance. *Phytotherapy Research* [in press].

1321. Stough, C., Lloyd, J., Clarke, J., Downey, L. A., Hutchison, C. W., Rodgers, T., et al. (2001). The chronic effects of an extract of *Bacopa monniera* (Brahmi) on cognitive function in healthy human subjects. *Psychopharmacology*, ***156***, 481–484.

1322. Stough, C., Downey, L. A., Lloyd, J., Silber, B., Redman, S., Hutchison, C., et al. (2008). Examining the nootropic effects of a special extract of *Bacopa monniera* on human cognitive functioning: 90-day double-blind placebo-controlled randomized trial. *Phytotherapy Research*, ***22***, 1629–1634.

1323. Barbhaiya, H. C., Desai, R. P., Saxena, V. S., Pravina, K., Wasim, P., Geetharani, P., et al. (2008). Efficacy and tolerability of BacoMind® on memory improvement in elderly participants—a double-blind placebo-controlled study. *Pharmacology and Toxicology*, ***3***, 425–434.

1324. Raghav, S., Singh, H., Dalal, P., Srivastava, J., and Asthana, O. (2006). Randomized controlled trial of standardized *Bacopa monniera* extract in age-associated memory impairment. *Indian Journal of Psychiatry*, ***48***, 238.

1325. Calabrese, C., Gregory, W. L., Leo, M., Kraemer, D., Bone, K., and Oken, B. (2008). Effects of a standardized *Bacopa monnieri* extract on cognitive performance, anxiety, and depression in the elderly: A randomized, double-blind, placebo-controlled trial. *Journal of Alternative and Complementary Medicine*, ***14***, 707–713.

1326. Morgan, A., and Stevens, J. (2010). Does *Bacopa monnieri* improve memory performance in older persons? Results of a randomized, placebo-controlled, double-blind trial. *Journal of Alternative and Complementary Medicine*, ***16***, 753–759.

1327. Roodenrys, S., Booth, D., Bulzomi, S., Phipps, A., Micallef, C., and Smoker, J. (2002). Chronic effects of Brahmi (*Bacopa monnieri*) on human memory. *Neuropsychopharmacology*, ***27***, 279–281.

1328. Pase, M. P., Kean, J., Sarris, J., Neale, C., Scholey, A. B., and Stough, C. (2012). The cognitive-enhancing effects of *Bacopa monnieri*: a systematic review of randomized, controlled human clinical trials. *Journal of Alternative and Complementary Medicine*, ***18***, 647–652.

1329. Peth-Nui, T., Wattanathorn, J., Muchimapura, S., Tong-Un, T., Piyavhatkul, N., Rangseekajee, P., et al. (2012). Effects of 12-week *Bacopa monnieri* consumption on attention, cognitive processing, working memory, and functions of both cholinergic and monoaminergic systems in healthy elderly volunteers. *Evidence-Based Complementary and Alternative Medicine*.

1330. Macchioni, F., Siciliano, T., Magi, M., Cecchi, F., Cioni, P. L., and Braca, A. (2008). Activity of aqueous extract of *Panax notoginseng* flower buds against *Aedes albopictus* larvae and pupae. *Bulletin of Insectology*, ***61***, 31.

1331. Enriz, R. D., Baldoni, H. A., Zamora, M. A., Jáuregui, E. A., Sosa, M. E., Tonn, C. E., et al. (2000). Structure-antifeedant activity relationship of clerodane diterpenoids. Comparative study with withanolides and azadirachtin. *Journal of Agricultural and Food Chemistry*, ***48***, 1384–1392.

1332. Mareggiani, G., Picollo, M., Zerba, E., Burton, G., Tettamanzi, M., Benedetti-Doctorovich, M., et al. (2000). Antifeedant activity of withanolides from *Salpichroa origanifolia* on *Musca domestica*. *Journal of Natural Products*, ***63***, 1113–1116.

1333. Ascher, K. R. S., Eliyahu, M., Glotter, E., Goldman, A., Kirson, I., Abraham, A., et al. (1987). The antifeedant effect of some new withanolides on three insect species, *Spodoptera littoralis, Epilachna varivestis* and *Tribolium castaneum*. *Phytoparasitica*, ***15***, 15–29.

1334. Bansal, S. K., Singh, K. V., Sharma, S., and Sherwani, M. R. K. (2011). Comparative larvicidal potential of different plant parts of *Withania somnifera* against vector mosquitoes in the semi-arid region of Rajasthan. *Journal of Environmental Biology*, ***32***, 71–75.
1335. Baumann, T. W., and Meier, C. M. (1993). Chemical defence by withanolides during fruit development in *Physalis peruviana*. *Phytochemistry*, ***33***, 317–321.
1336. Osbourn, A. (1996). Saponins and plant defence: A soap story. *Trends in Plant Science*, ***1***, 4–9.
1337. Tennessen, J. M., Baker, K. D., Lam, G., Evans, J., and Thummel, C. S. (2011). The *Drosophila* estrogen-related receptor directs a metabolic switch that supports developmental growth. *Cell Metabolism*, ***13***, 139–148.
1338. Hahm, E. R., Lee, J., Huang, Y., and Singh, S. V. (2011). Withaferin a suppresses estrogen receptor-α expression in human breast cancer cells. *Molecular Carcinogenesis*, ***50***, 614–624.
1339. Bajguz, A. (2007). Metabolism of brassinosteroids in plants. *Plant Physiology and Biochemistry*, ***45***, 95–107.
1340. Zullo, M. A. T., and Adam, G. (2002). Brassinosteroid phytohormones: structure, bioactivity and applications. *Brazilian Journal of Plant Physiology*, ***14***, 143–181.
1341. Bhardwaj, R., Arora, N., Uppal, P., Sharma, I., and Kanwar, M. (2011). Prospects of brassinosteroids in medicinal applications. *Brassinosteroids: A Class of Plant Hormone*, 439–458.
1342. Esposito, D., Komarnytsky, S., Shapses, S., and Raskin, I. (2011). Anabolic effect of plant brassinosteroid. *FASEB Journal*, ***25***, 3708–3719.
1343. Janeczko, A., Budziszewska, B., Skoczowski, A., and Dybała, M. (2008). Specific binding sites for progesterone and 17β-estradiol in cells of *Triticum aestivum L. Acta Biochimica Polonica*, ***55***, 707–711.
1344. Park, Y. H., Kim, Y. C., Park, S. U., Lim, H. S., Kim, J. B., Cho, B. K., et al. (2012). Age-dependent distribution of fungal endophytes in *Panax ginseng* roots cultivated in Korea. *Journal of Ginseng Research*, ***36***, 327–333.
1345. Nicol, R. W., Yousef, L., Traquair, J. A., and Bernards, M. A. (2003). Ginsenosides stimulate the growth of soilborne pathogens of American ginseng. *Phytochemistry*, ***64***, 257–264.
1346. Bernards, M. A., Yousef, L. F., and Nicol, R. W. (2006). The allelopathic potential of ginsenosides. *Allelochemicals: Biological Control of Plant Pathogens and Diseases*, 157–175.
1347. Kochan, E., Wasiela, M., and Sienkiewicz, M. (2012). The production of ginsenosides in hairy root cultures of American ginseng, *Panax quinquefolium* L. and their antimicrobial activity. *In Vitro Cellular & Developmental Biology-Plant*, 1–6.
1348. Ying, Y. X., Ding, W. L., and Li, Y. (2012). Characterization of soil bacterial communities in rhizospheric and nonrhizospheric soil of *Panax ginseng*. *Biochemical Genetics*, ***20***, 848–859.
1349. Vaccarini, C. E., and Bonetto, G. M. (2000). Selective phytotoxic activity of withanolides from *Iochroma australe* to crop and weed species. *Journal of Chemical Ecology*, ***26***, 2187–2196.
1350. Chandra, S., Chatterjee, P., Dey, P., and Bhattacharya, S. (2012). Allelopathic effect of Ashwagandha against the germination and radicle growth of *Cicer arietinum* and *Triticum aestivum*. *Pharmacognosy Research*, ***4***, 166.
1351. Javaid, A., Shafique, S., and Shafique, S. (2008). Herbicidal activity of *Datura metel* L. against *Phalaris minor* Retz. *Pakistan Journal of Weed Science Research*, ***14***, 209–220.
1352. Gong, W., Chen, C., Dobes, C., Fu, C. X., and Koch, M. A. (2008). Phylogeography of a living fossil: Pleistocene glaciations forced *Ginkgo biloba* L. (Ginkgoaceae) into

two refuge areas in China with limited subsequent postglacial expansion. *Molecular Phylogenetics and Evolution*, ***48***, 1094–1105.

1353. Smith, J., and Luo, Y. (2004). Studies on molecular mechanisms of *Ginkgo biloba* extract. *Applied Microbiology and Biotechnology*, ***64***, 465–472.
1354. Berger, P. (2001). Ginkgo leaf extract. *Medical Herbalism*, ***2***, 5–6.
1355. DeFeudis, F. V., and Drieu, K. (2004). "Stress-alleviating" and "vigilance-enhancing" actions of *Ginkgo biloba* extract (EGb 761). *Drug Development Research*, ***62***, 1–25.
1356. Chan, P. C., Xia, Q. S., and Fu, P. P. (2007). *Ginkgo biloba* leaf extract: Biological, medicinal, and toxicological effects. *Journal of Environmental Science and Health Part C-Environmental Carcinogenesis & Ecotoxicology Reviews*, ***25***, 211–244.
1357. Fehske, C., Leuner, K., and Müller, W. (2009). *Ginkgo biloba* extract (EGb761®) influences monoaminergic neurotransmission via inhibition of NE uptake, but not MAO activity after chronic treatment. *Pharmacological Research*, ***60***, 68–73.
1358. Eckert, A. (2012). Mitochondrial effects of *Ginkgo biloba* extract. *International Psychogeriatrics*, ***24***, S18–S20.
1359. Thompson, A. J., Jarvis, G. E., Duke, R. K., Johnston, G. A. R., and Lummis, S. C. R. (2011). Ginkgolide B and bilobalide block the pore of the 5-HT3 receptor at a location that overlaps the picrotoxin binding site. *Neuropharmacology*, ***60***, 488–495.
1360. Huang, S. H., Lewis, T. G., Lummis, S. C. R., Thompson, A. J., Chebib, M., Johnston, G. A. R., et al. (2012). Mixed antagonistic effects of the ginkgolides at recombinant human ρ_1 $GABA_C$ receptors. *Neuropharmacology*.
1361. Ivic, L., Sands, T. T., Fishkin, N., Nakanishi, K., Kriegstein, A. R., and Strømgaard, K. (2003). Terpene trilactones from *Ginkgo biloba* are antagonists of cortical glycine and GABAA receptors. *Journal of Biological Chemistry*, ***278***, 49279–49285.
1362. Yoshitake, T., Yoshitake, S., and Kehr, J. (2010). The *Ginkgo biloba* extract EGb 761® and its main constituent flavonoids and ginkgolides increase extracellular dopamine levels in the rat prefrontal cortex. *British Journal of Pharmacology*, ***159***, 659–668.
1363. Kehr, J., Yoshitake, S., Ijiri, S., Koch, E., Nöldner, M., and Yoshitake, T. (2012). *Ginkgo biloba* leaf extract (EGb 761®) and its specific acylated flavonol constituents increase dopamine and acetylcholine levels in the rat medial prefrontal cortex: possible implications for the cognitive enhancing properties of EGb 761®. *International Psychogeriatrics*, ***24***, S25–S34.
1364. Silberstein, R., Pipingas, A., Song, J., Camfield, D., Nathan, P., and Stough, C. (2011). Examining brain-cognition effects of *Ginkgo biloba* extract: Brain activation in the left temporal and left prefrontal cortex in an object working memory task. *Evidence-Based Complementary and Alternative Medicine*.
1365. Mashayekh, A., Pham, D. L., Yousem, D. M., Dizon, M., Barker, P. B., and Lin, D. D. M. (2011). Effects of *Ginkgo biloba* on cerebral blood flow assessed by quantitative MR perfusion imaging: a pilot study. *Neuroradiology*, ***53***, 185–191.
1366. Santos, R., Galduroz, J., Barbieri, A., Castiglioni, M., Ytaya, L., and Bueno, O. (2003). Cognitive performance, SPECT, and blood viscosity in elderly non-demented people using *Ginkgo biloba*. *Pharmacopsychiatry*, ***36***, 127–133.
1367. Hindmarch, I. (1986). Activity of *Ginkgo biloba* extract on short-term-memory. *Presse Medicale*, ***15***, 1592–1594.
1368. Rigney, U., Kimber, S., and Hindmarch, I. (1999). The effects of acute doses of standardized *Ginkgo biloba* extract on memory and psychomotor performance in volunteers. *Phytotherapy Research*, ***13***, 408–415.
1369. Warot, D., Lacomblez, L., Danjou, P., Weiller, E., Payan, C., and Puech, A. J. (1991). Comparative effects of *Ginkgo biloba* extracts on psychomotor performances and memory in healthy volunteers. *Therapie*, ***46***, 33–36.

1370. Kennedy, D. O., Scholey, A. B., and Wesnes, K. A. (2000). The dose-dependent cognitive effects of acute administration of *Ginkgo biloba* to healthy young volunteers. *Psychopharmacology*, ***151***, 416–423.
1371. Kennedy, D. O., Jackson, P. A., Haskell, C. F., and Scholey, A. B. (2007). Modulation of cognitive performance following single doses of 120 mg *Ginkgo biloba* extract administered to healthy young volunteers. *Human Psychopharmacology-Clinical and Experimental*, **22**, 559–566.
1372. Stough, C., Clarke, J., Lloyd, J., and Nathan, P. J. (2001). Neuropsychological changes after 30-day *Ginkgo biloba* administration in healthy participants. *International Journal of Neuropsychopharmacology*, **4**, 131–134.
1373. Hartley, D. E., Heinze, L., Elsabagh, S., and File, S. E. (2003). Effects on cognition and mood in postmenopausal women of 1-week treatment with *Ginkgo biloba*. *Pharmacology Biochemistry and Behavior*, **75**, 711–720.
1374. Mix, J. A., and Crews, W. D. (2000). An examination of the efficacy of *Ginkgo biloba* extract EGb 761 on the neuropsychologic functioning of cognitively intact older adults. *Journal of Alternative and Complementary Medicine*, **6**, 219–229.
1375. Mix, J. A., and Crews, W. D. (2002). A double-blind, placebo-controlled, randomized trial of *Ginkgo biloba* extract EGb 761 (R) in a sample of cognitively intact older adults: neuropsychological findings. *Human Psychopharmacology-Clinical and Experimental*, **17**, 267–277.
1376. Solomon, P. R., Adams, F., Silver, A., Zimmer, J., and DeVeaux, R. (2002). Ginkgo for memory enhancement: A randomized controlled trial. *JAMA, Journal of the American Medical Association*, ***288***, 835–840.
1377. Moulton, P. L., Boyko, L. N., Fitzpatrick, J. L., and Petros, T. V. (2001). The effect of *Ginkgo biloba* on memory in healthy male volunteers. *Physiology & Behavior*, ***73***, 659–665.
1378. Crews, W., Harrison, D. W., Griggin, M., Falwell, K. D., Crist, T., Longest, L., et al. (2005). The neuropsychological efficacy of Ginkgo preparations in healthy and cognitively intact adults: a comprehensive review. *HerbalGram*, ***67***, 43–62.
1379. Laws, K. R., Sweetnam, H., and Kondel, T. K. (2012). Is *Ginkgo biloba* a cognitive enhancer in healthy individuals? A meta-analysis. *Human Psychopharmacology: Clinical and Experimental.*
1380. Canter, P. H., and Ernst, E. (2007). *Ginkgo biloba* is not a smart drug: an updated systematic review of randomised clinical trials testing the nootropic effects of *G. biloba* extracts in healthy people. *Human Psychopharmacology: Clinical and Experimental*, **22**, 265–278.
1381. Kaschel, R. (2011). Specific memory effects of *Ginkgo biloba* extract EGb 761 in middle-aged healthy volunteers. *Phytomedicine*, ***18***, 1202–1207.
1382. Grass-Kapanke, B., Busmane, A., Lasmanis, A., Hoerr, R., and Kaschel, R. (2011). Effects of *Ginkgo biloba* special extract EGb 761® in very mild cognitive impairment (vMCI). *Neuroscience & Medicine*, **2**, 48–56.
1383. Birks, J.,., Grimley Evans, J., and Van Dongen, M. (2002). *Ginkgo biloba* for cognitive impairment and dementia. *Cochrane Database of Systematic Reviews* CD003120.
1384. Birks, J., and Grimley Evans, J. (2009). *Ginkgo biloba* for cognitive impairment and dementia. *Cochrane Database of Systematic Reviews*. CD003120.
1385. Snitz, B. E., O'Meara, E. S., Carlson, M. C., Arnold, A. M., Ives, D. G., Rapp, S. R., et al. (2009). *Ginkgo biloba* for preventing cognitive decline in older adults. *JAMA, The Journal of the American Medical Association*, **302**, 2663–2670.
1386. Vellas, B., Coley, N., Ousset, P. J., Berrut, G., Dartigues, J. F., Dubois, B., et al. (2012). Long-term use of standardised *Ginkgo biloba* extract for the prevention of

Alzheimer's disease (GuidAge): a randomised placebo-controlled trial. *Lancet Neurology*, ***11***, 851–859.

1387. Wang, B. S., Wang, H., Song, Y. Y., Qi, H., Rong, Z. X., Wang, B. S., et al. (2010). Effectiveness of standardized *Ginkgo biloba* extract on cognitive symptoms of dementia with a six-month treatment: a bivariate random effect meta-analysis. *Pharmacopsychiatry*, ***43***, 86–91.
1388. Weinmann, S., Roll, S., Schwarzbach, C., Vauth, C., and Willich, S. N. (2010). Effects of *Ginkgo biloba* in dementia: systematic review and meta-analysis. *BMC Geriatrics*, ***10***, 14.
1389. Janssen, I. M., Sturtz, S., Skipka, G., Zentner, A., Velasco Garrido, M., Garrido, M. V., et al. (2010). *Ginkgo biloba* in Alzheimer's disease: a systematic review. *Wiener Medizinische Wochenschrift (1946)*, *160*, 539–546.
1390. Ihl, R., Bachinskaya, N., Korczyn, A. D., Vakhapova, V., Tribanek, M., Hoerr, R., et al. (2011). Efficacy and safety of a once-daily formulation of *Ginkgo biloba* extract EGb 761 in dementia with neuropsychiatric features: a randomized controlled trial. *International Journal of Geriatric Psychiatry*, ***26***, 1186–1194.
1391. Herrschaft, H., Nacu, A., Likhachev, S., Sholomov, I., Hoerr, R., and Schlaefke, S. (2012). *Ginkgo biloba* extract EGb 761 in dementia with neuropsychiatric features: A randomised, placebo-controlled trial to confirm the efficacy and safety of a daily dose of 240 mg. *Journal of Psychiatric Research*, ***46*** , 716–723.
1392. Mohanta, T. K., Occhipinti, A., Zebelo, S. A., Foti, M., Fliegmann, J., Bossi, S., et al. (2012). *Ginkgo biloba* responds to herbivory by activating early signaling and direct defenses. *PLoS One*, *7*.
1393. Mazzanti, G., Mascellino, M. T., Battinelli, L., Coluccia, D., Manganaro, M., and Saso, L. (2000). Antimicrobial investigation of semipurified fractions of *Ginkgo biloba* leaves. *Journal of Ethnopharmacology*, ***71***, 83–88.
1394. Wheeler, A. (1975). Insect associates of *Ginkgo biloba*. *Entomological News*, ***86***, 37–44.
1395. Matsumoto, T., and Sei, T. (1987). Antifeedant activities of *Ginkgo biloba* L. components against the larva of *Pieris rapae crucivora*. *Agricultural and Biological Chemistry*, ***51***, 249–250.
1396. Fu-shun, Y., Evans, K., Stevens, L., Beek, T. A., and Schoonhoven, L. (1990). Deterrents extracted from the leaves of *Ginkgo biloba*: effects on feeding and contact chemoreceptors. *Entomologia Experimentalis et Applicata*, ***54***, 57–64.
1397. Van Den Boom, C. E. M., Beek, T. A., and Dicke, M. (2003). Differences among plant species in acceptance by the spider mite *Tetranychus urticae* Koch. *Journal of Applied Entomology*, ***127***, 177–183.
1398. Ahn, Y. J., Kwon, M., Park, H. M., and Han, C. K. (1997). *Potent insecticidal activity of* Ginkgo biloba*-derived trilactone terpenes against* Nilaparvata lugens. Paper presented at the ACS Symposium Series.
1399. Kang, S.-M., Min, J.-Y., Kim, Y.-D., Karigar, C. S., Kim, S.-W., Goo, G.-H., et al. (2009). Effect of biotic elicitors on the accumulation of bilobalide and ginkgolides in *Ginkgo biloba* cell cultures. *Journal of Biotechnology*, ***139***, 84–88.
1400. Bell, C. D. (2004). Preliminary phylogeny of Valerianaceae (Dipsacales) inferred from nuclear and chloroplast DNA sequence data. *Molecular Phylogenetics and Evolution*, ***31***, 340–350.
1401. Bell, C. D., and Donoghue, M. J. (2005). Phylogeny and biogeography of Valerianaceae (Dipsacales) with special reference to the South American valerians. *Organisms Diversity & Evolution*, ***5***, 147–159.
1402. Herrera-Arellano, A., Luna-Villegas, G., Cuevas-Uriostegui, M. L., Alvarez, L., Vargas-Pineda, G., Zamilpa-Alvarez, A., et al. (2001). Polysomnographic evaluation

of the hypnotic effect of *Valeriana edulis* standardized extract in patients suffering from insomnia. *Planta Medica*, ***67***, 695–699.

1403. Holzmann, I., Cechinel Filho, V., Mora, T. C., Cáceres, A., Martínez, J. V., Cruz, S. M., et al. (2011). Evaluation of behavioral and pharmacological effects of hydroalcoholic extract of *Valeriana prionophylla* Standl. from Guatemala. *Evidence-Based Complementary and Alternative Medicine.*
1404. Sharma, M., Jain, U., Patel, A., and Gupta, N. (2010). A comprehensive pharmacognostic report on valerian. *International Journal of Pharmaceutical Sciences and Research*, ***1***, 6–40.
1405. Houghton, P. (1999). The scientific basis for the reputed activity of valerian. *Journal of Pharmacy and Pharmacology*, ***51***, 505–512.
1406. Marder, M., Viola, H., Wasowski, C., Fernández, S., Medina, J., and Paladini, A. (2003). 6-Methylapigenin and hesperidin: new valeriana flavonoids with activity on the CNS. *Pharmacology Biochemistry and Behavior*, ***75***, 537–545.
1407. Dietz, B., Mahady, G., Pauli, G., and Farnsworth, N. (2005). Valerian extract and valerenic acid are partial agonists of the 5-HT5a receptor in vitro. *Molecular Brain Research*, ***138***, 191–197.
1408. Khom, S., Baburin, I., Timin, E., Hohaus, A., Trauner, G., Kopp, B., et al. (2007). Valerenic acid potentiates and inhibits GABAA receptors: Molecular mechanism and subunit specificity. *Neuropharmacology*, ***53***, 178–187.
1409. Lacher, S., Mayer, R., Sichardt, K., Nieber, K., and Müller, C. (2007). Interaction of valerian extracts of different polarity with adenosine receptors: Identification of isovaltrate as an inverse agonist at A1 receptors. *Biochemical Pharmacology*, ***73***, 248–258.
1410. Sah, S. P., Mathela, C. S., and Chopra, K. (2011). Antidepressant effect of *Valeriana wallichii* patchouli alcohol chemotype in mice: Behavioural and biochemical evidence. *Journal of Ethnopharmacology*, ***135***, 197–200.
1411. Müller, L. G., Salles, L. A., Stein, A. C., Betti, A. H., Sakamoto, S., Cassel, E., et al. (2012). Antidepressant-like effect of *Valeriana glechomifolia* Meyer (Valerianaceae) in mice. *Progress in Neuro-Psychopharmacology and Biological Psychiatry*, ***36***, 101–109.
1412. Ziegler, G., Ploch, M., Miettinen-Baumann, A., and Collet, W. (2002). Efficacy and tolerability of valerian extract LI 156 compared with oxazepam in the treatment of non-organic insomnia: a randomized, double-blind, comparative clinical study. *European Journal of Medical Research*, ***7***.
1413. Schmitz, M., and Jackel, M. (1998). Comparative study for assessing quality of life of patients with exogenous sleep disorders (temporary sleep onset and sleep interruption disorders) treated with a hops-valerian preparation and a benzodiazepine drug. *Wiener Medizinische Wochenschrift*, ***148***, 291–298.
1414. Fussel, A., Wolf, A., and Brattstrom, A. (2000). Effect of a fixed valerian-hop extract combination (Ze 91019) on sleep polygraphy in patients with non-organic insomnia: a pilot study. *European Journal of Medical Research*, ***5***, 385–390.
1415. Stevinson, C., and Ernst, E. (2000). Valerian for insomnia: a systematic review of randomized clinical trials. *Sleep Medicine*, ***1***, 91–99.
1416. Bent, S., Padula, A., Moore, D., Patterson, M., and Mehling, W. (2006). Valerian for sleep: a systematic review and meta-analysis. *American Journal of Medicine*, ***119***, 1005–1012.
1417. Fernández-San-Martín, M. I., Masa-Font, R., Palacios-Soler, L., Sancho-Gómez, P., Calbó-Caldentey, C., and Flores-Mateo, G. (2010). Effectiveness of valerian on

insomnia: A meta-analysis of randomized placebo-controlled trials. *Sleep Medicine*, ***11***, 505–511.
1418. Gerhard, U., Linnenbrink, N., Georghiadou, C., and Hobi, V. (1996). Vigilanzminderende Effekte zweier pflanzlicher Schlafmittel [Vigilance-decreasing effects of 2 plant-derived sedatives]. *Revue Suisse Medecine Praxis*, ***85***.
1419. Kuhlmann, J., Berger, W., Podzuweit, H., and Schmidt, U. (1999). The influence of valerian treatment on "reaction time, alertness and concentration" in volunteers. *Pharmacopsychiatry* ***32***.
1420. Miyasaka, L., Atallah, A., and Soares, B. (2006). Valerian for anxiety disorders. *Cochrane Database of Systematic Reviews (Online)*.
1421. Andreatini, R., Sartori, V. A., Seabra, M. L. V., and Leite, J. R. (2002). Effect of valepotriates (valerian extract) in generalized anxiety disorder: a randomized placebo-controlled pilot study. *Phytotherapy Research*, ***16***, 650–654.
1422. Letchamo, W., Ward, W., Heard, B., and Heard, D. (2004). Essential oil of *Valeriana officinalis* L. cultivars and their antimicrobial activity as influenced by harvesting time under commercial organic cultivation. *Journal of Agricultural and Food Chemistry*, ***52***, 3915–3919.
1423. Alba, C., Prioreschi, R., and Quintero, C. (2012). Population and leaf-level variation of iridoid glycosides in the invasive weed *Verbascum thapsus* L.(common mullein): implications for herbivory by generalist insects. *Chemoecology*, 1–10.
1424. Kittipongpatana, N., Davis, D. L., and Porter, J. R. (2002). Methyl jasmonate increases the production of valepotriates by transformed root cultures of *Valerianella locusta*. *Plant Cell Tissue and Organ Culture*, ***71***, 65–75.
1425. Houghton, P. (2009). Synergy and polyvalence: paradigms to explain the activity of herbal products. *Evaluation of Herbal Medicinal Products*, 85–94.
1426. NICE technology appraisal guidance 111. (2009). Donepezil, galantamine, rivastigmine (review) and memantine for the treatment of Alzheimer's disease (amended).
1427. NICE technology appraisal guidance 111. (2011). Donepezil, galantamine, rivastigmine (review) and memantine for the treatment of Alzheimer's disease (amended).
1428. Collier, R. (2009). Rapidly rising clinical trial costs worry researchers. *Canadian Medical Association Journal*, ***180***, 277–278.
1429. Santos-Gomes, P. C., Seabra, R. M., Andrade, P. B., and Fernandes-Ferreira, M. (2002). Phenolic antioxidant compounds produced by in vitro shoots of sage (*Salvia officinalis* L.). *Plant Science*, ***162***, 981–987.
1430. Putievsky, E., Ravid, U., and Dudai, N. (1986). The influence of season and harvest frequency on essential oil and herbal yields from a pure clone of sage (*Salvia officinalis*) grown under cultivated conditions. *Journal of Natural Products*, ***49***, 326–329.
1431. Ben Farhat, M., Jordan, M. J., Chaouech-Hamada, R., Landoulsi, A., and Sotomayor, J. A. (2009). Variations in essential oil, phenolic compounds, and antioxidant activity of Tunisian cultivated *Salvia officinalis* L. *Journal of Agricultural and Food Chemistry*, ***57***, 10349–10356.
1432. Ben Taarit, M., Msaada, K., Hosni, K., Hammami, M., Kchouk, M. E., and Marzouk, B. (2009). Plant growth, essential oil yield and composition of sage (*Salvia officinalis* L.) fruits cultivated under salt stress conditions. *Industrial Crops and Products*, ***30***, 333–337.
1433. Vuorinen, T., Reddy, G. V. P., Nerg, A.-M., and Holopainen, J. K. (2004). Monoterpene and herbivore-induced emissions from cabbage plants grown at elevated atmospheric CO_2 concentration. *Atmospheric Environment*, ***38***, 675–682.
1434. Dicke, M., van Loon, J. J. A., and Soler, R. (2009). Chemical complexity of volatiles from plants induced by multiple attack. *Nature Chemical Biology*, ***5***, 317–324.

1435. Langcake, P., and Pryce, R. J. (1976). The production of resveratrol by *Vitis vinifera* and other members of the Vitaceae as a response to infection or injury. *Physiological Plant Pathology*, ***9***, 77–86.

1436. Bettaieb, I., Zakhama, N., Wannes, W. A., Kchouk, M. E., and Marzouk, B. (2009). Water deficit effects on *Salvia officinalis* fatty acids and essential oils composition. *Scientia Horticulturae*, ***120***, 271–275.

INDEX